D1109299

Identification and Quantification of Drugs, Metabolites and Metabolizing Enzymes by LC–MS

PROGRESS IN PHARMACEUTICAL AND BIOMEDICAL ANALYSIS

Volume 1 **Pharmaceutical and Biomedical Applications of Liquid Chromatography**
edited by **C.M. Riley, W.J. Lough and I.W. Wainer**

Volume 2 **Pharmaceutical and Biomedical Applications of Capillary Electrophoresis**
edited by **S.M. Lunte and D.M. Radzik**

Volume 3 **Development and Validation of Analytical Methods**
edited by **C.M. Riley and T.W. Rosanske**

Volume 4 **Identification and Determination of Impurities in Drugs**
edited by **S. Görög**

Volume 5 **High Throughput Bioanalytical Sample Preparation. Methods and Automation Strategies**
edited by **D.A. Wells**

Volume 6 **Identification and Quantification of Drugs, Metabolites and Metabolizing Enzymes by LC–MS**
edited by **S. Chowdhury**

PROGRESS IN PHARMACEUTICAL AND BIOMEDICAL ANALYSIS 6

Identification and Quantification of Drugs, Metabolites and Metabolizing Enzymes by LC–MS

Edited by

Swapan K. Chowdhury
Drug Metabolism Department
Schering Plough Research Institute
Kenilworth, NJ,
USA

ELSEVIER

2005

Amsterdam – Boston – Heidelberg – London – New York – Oxford – Paris – San Diego
San Francisco – Singapore – Sydney – Tokyo

ELSEVIER B.V.
Radarweg 29
P.O. Box 211, 1000 AE
Amsterdam, The Netherlands

ELSEVIER Inc.
525 B Street, Suite 1900
San Diego, CA 92101-4495
USA

ELSEVIER Ltd
The Boulevard, Langford Lane
Kidlington, Oxford OX5 1GB
UK

ELSEVIER Ltd
84 Theobalds Road
London WC1X 8RR
UK

First edition 2005

Library of Congress Cataloging in Publication Data
A catalog record is available from the Library of Congress.

British Library Cataloguing in Publication Data
A catalogue record is available from the British Library.

ISBN:0-444-51710-3

♾ The paper used in this publication meets the requirements of ANSI/NISO Z39.48-1992 (Permanence of Paper).
Printed in The Netherlands

Preface

Impressed by a series of presentations on current issues in drug metabolism at the 2003 Eastern Analytical Symposium that I had the privilege to coordinate, a colleague suggested that I capture these talks in a book. Although, initially skeptical by the immense task of editing a book, the lack of discussion in the literature in a consolidated format critical to the current use of liquid chromatography–mass spectrometry (LC–MS) in drug discovery and development motivated me to plunge into arranging and editing this manuscript. Fortunately, many of the speakers from the two sessions and several other expert drug metabolism practitioners agreed to contribute to this text.

When outlining prospective chapters for this book, I asked myself what could be helpful to a drug metabolism scientist? Clearly, a detailed discussion on the issues and problems that are encountered in today's drug metabolism laboratories utilizing LC–MS instrumentation is a great starting point. LC–MS is becoming more powerful with respect to sensitivity, selectivity, ion mass-to-charge ratio resolution and mass accuracy which are aided by the ruggedness and ease of use of these instruments. As new techniques of transferring from liquid to gas phase and measuring masses of drug molecules and metabolites become more prevalent, so do the technical challenges of putting these techniques into proper use as well as the task of consolidating emerging applications. Thus, a series of chapters addressing how to utilize LC–MS instrumentation for current drug metabolism problems that may be encountered in today's practice are part of this collection.

This book is intended for those beginning to use LC–MS for drug metabolism studies as well as for those considered to be advanced practitioners. Emphasis is placed on techniques and applications that are of current importance as well as future applications. In a naïve sense, most LC–MS applications for drug metabolism studies can be summarized into providing two pieces of information: what drug-derived analytes are in a sample and how much is there? With this in mind, Korfmacher, Majumdar, Jemal, Chu and Nomeir elegantly describe various aspects of high-throughput quantitative determination of drugs and metabolites in biological matrices and address many practical issues faced in today's laboratory. Hop and Prakash, Grotz *et al.*, Anari *et al.*, Zhu *et al.*, Ramanathan *et al.* and Chowdhury *et al.*, thoroughly discuss many facets of metabolite identification using novel technologies and data-acquisition software. A discussion of the role of LC–MS is further expanded by Ghosal *et al.* into characterization of cytochrome P450 isozymes and UGT-glucuronosyltransferases (phenotyping) responsible for metabolism of xenobiotics. Reaction phenotyping is rapidly becoming an integral part of drug discovery and development efforts since it provides a basis for predicting drug–drug interactions in the clinic and assists in designing better and safer medicines. This chapter endeavors to provide a detailed overview of drug metabolizing enzymes together with procedures for identification of their role in the metabolism of xenobiotics. The introduction of this book is composed by K. Alton who reviews how modern LC–MS tools are transforming the pharmaceutical industry, compares experiences in the pre-LC–MS-era to those from today's laboratories, and provides insights into future uses of LC–MS.

Drug metabolism scientists with a broad range of training and technical experience in LC–MS should benefit from the information presented in this book.

Swapan K. Chowdhury

Contents

List of contributors

Kevin B. Alton, Department of Drug Metabolism and Pharmacokinetics, Schering Plough Research Institute, 2015 Galloping Hills Road, Kenilworth, NJ 07033, USA

M. Reza Anari, Department of Drug Metabolism, Merck Research Laboratories, WP75A-203, Sumneytown Pike, West Point, PA 19486, USA

Thomas A. Baillie, Department of Drug Metabolism, Merck Research Laboratories, WP75A-203, Sumneytown Pike, West Point, PA 19486, USA

N. Blumenkrantz, Department of Drug Metabolism and Pharmacokinetics, Schering Plough Research Institute, 2015 Galloping Hills Road, Kenilworth, NJ 07033, USA

Chandra Prakash, Pfizer Global Research and Development, Eastern Point Road, Groton, CT 06340, USA

Swapan K. Chowdhury, Department of Drug Metabolism and Pharmacokinetics, Schering Plough Research Institute, 2015 Galloping Hills Road, Kenilworth, NJ 07033, USA

Inhou Chu, Department of Drug Metabolism and Pharmacokinetics, Schering Plough Research Institute, 2015 Galloping Hills Road, Kenilworth, NJ 07033, USA

Nigel A. Clarke, Quest Diagnostics Nichols Research Institute, 33608 Ortega Highway San Juan Capistrano, CA 92675, USA

Kathleen A. Cox, Department of Drug Metabolism and Pharmacokinetics, Schering-Plough Research Institute 2015 Galloping Hill Road, Kenilworth, NJ 07033, USA

Anima Ghosal, Department of Drug Metabolism and Pharmacokinetics, Schering Plough Research Institute, 2015 Galloping Hill Road, Kenilworth, NJ 07033, USA

V.S. Gopaul, J&J PRD, Preclinical PK and Metabolism OCD Bldg, 1001 Route 202, Raritan, NJ, 08869, USA

Diane E. Grotz, Department of Drug Metabolism and Pharmacokinetics, Schering-Plough Research Institute, 2015 Galloping Hill Road, Kenilworth, NJ 07033, USA

Cornelis E.C.A. Hop, Department of Pfizer Global Research and Development, Eastern Point Road, Groton, CT 06340, USA

Mohammed Jemal, Department of Drug Disposition and Bioanalytical Sciences, Bristol–Myers Squibb Pharmaceutical Research Institute, 1 Squibb Drive, New Brunswick, NJ 08903, USA

Narendra S. Kishnani, Department of Drug Metabolism and Pharmacokinetics, Schering Plough Research Institute, 2015 Galloping Hill Road, Kenilworth, NJ 07033, USA

Walter A. Korfmacher, Department of Drug Metabolism and Pharmacokinetics, Schering Plough Research Institute, 2015 Galloping Hills Road, Kenilworth, NJ 07033, USA

K.M. Kulmatycki, Department of Drug Metabolism and Pharmacokinetics, Schering Plough Research Institute, 2015 Galloping Hills Road, Kenilworth, NJ 07033, USA

Tapan K. Majumdar, Department of Preclinical Safety, Novartis Pharmaceuticals Corporation, East Hanover, NJ 07936, USA

Amin A. Nomeir, Department of Drug Metabolism and Pharmacokinetics, Schering Plough Research Institute, 2015 Galloping Hills Road, Kenilworth, NJ 07033, USA

Ragulan Ramanathan, Department of Drug Metabolism and Pharmacokinetics, Schering Plough Research Institute, 2015 Galloping Hills Road, Kenilworth, NJ 07033, USA

Gary L. Skiles, Amgen, Inc., PKDM M/S 1-1-B, One Amgen Center Dr, Thousand Oaks, CA 91320

Philip R. Tiller, Department of Drug Metabolism, Merck Research Laboratories, WP75A-203, Sumneytown Pike, West Point, PA 19486, USA

R. Zhong, Department of Drug Metabolism and Pharmacokinetics, Schering Plough Research Institute, 2015 Galloping Hills Road, Kenilworth, NJ 07033, USA

Donglu Zhang, Biotransformation Department, Bristol–Myers Pharmaceutical Research Institute, Princeton, NJ 08543, USA

Mingshe Zhu, Biotransformation Department, Bristol–Myers Pharmaceutical Research Institute, Princeton, NJ 08543, USA

Identification and Quantification of Drugs, Metabolites and Metabolizing
Enzymes by LC–MS
Swapan K. Chowdhury, editor.

Chapter 1

LC–MS AND DRUG METABOLISM: A JOURNEY BACK IN TIME, PRESENT TRENDS AND FUTURE DIRECTIONS

Kevin B. Alton

1.1. Introduction

While preparing the introduction to this insightful collection of contemporary thinking about the role of modern liquid chromatograph–mass spectrometer (LC–MS) in Drug Metabolism, I came across an undergraduate paper that I had written nearly 40 years ago in fulfillment of requirements for a University Physics course. Intrigued by the possibility that one could influence the trajectory of an accelerated ion in a predictable manner by the mere presence of an electromagnetic field, I pored over the literature available from the early 1900s up to 1967 and prepared a very minor masterpiece (in my own mind) entitled "The History, Development, and Utilization of Mass Spectrometry as an Analytical Tool." Little did I know in 1967 that much of my professional career in the Pharmaceutical Industry would be dominated by this "analytical tool" for the very specific purpose of detecting, identifying or quantifying drug metabolites. There is a certain irony in that I have now been asked by the Editor to offer my own thoughts regarding the present state-of-affairs and the future landscape for LC–MS in the study of drug metabolism.

Digressing for a moment, I would encourage anyone reading this monograph and not familiar with the ground-breaking efforts of Sir J.J. Thomson, Francis Aston and A.J. Dempster to locate their seminal papers on "positive ray analysis" in Philosophical Magazine [1–3] and Physical Review [4]. The extraordinary insight of these pioneers in mass spectrometry never fails to amaze me. For those prone to allergies, before you jump headlong into a pile of these dusty old references, you may want to consider taking another marvel of modern pharmaceutical science, that being a non-sedating antihistamine.

My training in Drug Metabolism really began with triflubazam, a 1,5-benzodiazepine anxiolytic agent, which in the end like so many others, never made it to market. As a freshman in this field, this was the first molecule for which mass spectrometry played a large role in the detection and structural characterization of its many interesting metabolites. Using a combination of direct-inlet MS, GC–MS, IR and NMR, we were able to identify seven urinary metabolites in humans [5]. I offer this example not in self-promotion but rather to give some context to the readership among which there may be many new to the field of LC–MS or Drug Metabolism. What I've failed to mention is that this research alone took nearly 4 years and involved the extraction of over 60ℓ of urine in order to isolate enough material to achieve unequivocal identification of these metabolites. Furthermore, quantitation of drug in any biomatrix involved manual injection of extracted and derivatized sample onto a packed GC column that typically yielded less than 15 values (standards and unknowns) per day. Clearly, advances in mass spectrometry, most specifically LC–MS, have greatly accelerated our ability to quantify, profile and characterize drug metabolites. This has been largely due to the order-of-magnitude increases in MS sensitivity and selectivity as well as vast improvements in instrument control and automation attributed to the remarkable processing speed and data storage capacity of personal computers.

Ever since the introduction of atmospheric pressure ionization (API) techniques in the mid-1970s [6] that enabled the simple interface of a liquid chromatograph to a mass spectrometer, the use of LC–MS has increasingly dominated the analysis of drug and metabolites in various matrices. Quadrupole instruments of varying design led the way for LC–MS systems capable of quantifying drug in thousands of biological samples per week. In the first four chapters of this treatise, Drs. Korfmacher, Majumdar, Jemal, Chu and Nomeir will walk you through various aspects of high-throughput quantitative LC–MS determination of drug in biofluids as employed by the Pharmaceutical Industry in contemporary drug discovery and development settings.

The next six chapters [5–10] focus on metabolite characterization as practiced in four different pharmaceutical laboratories. Specific strategies for structural identification are developed for the reader using a representative sampling of modern instruments as well as novel technical approaches to solve interesting structural or analytical problems. Last, but certainly not least, Chapter 12 takes aim at summarizing an enormous wealth of information that is available regarding the major role that CYP P450 isoforms and uridine diphosphate (UDP)-glucuronosyltransferases play in metabolizing xenobiotics. In the remainder of this preface, I endeavor to lay the foundation for present trends and future directions for LC–MS in Drug Metabolism.

Varying configurations of quadrupole mass spectrometers have been the mainstay in the analysis of drug metabolites for well over a decade. More recently, the linear ion trap (LIT) along with hybrid and pure bred time-of-flight (TOF) mass spectrometers possessing even higher sensitivity and selectivity, have enabled the structural characterization of high potency drugs with increasingly smaller sample size. Since TOF mass spectrometers also operate at high resolution (5000–20 000), these instruments achieve high enough mass accuracy (<5 ppm) to compute elemental composition of mass ions and their fragment ions. With improvements in the user-friendliness of software to control LC–TOFMS systems, this technology has secured a strong position in the Drug Metabolism laboratory and in the future is expected to play a much wider role in metabolite and biomarker identification. LITs and hybrid LITs with their high sensitivity and fast scan speed, will contend for domination in automated metabolite characterization through data-dependent scans or multiple scans simultaneously performed in MS and MS/MS modes. These features are of particular advantage in drug discovery settings, where rapid metabolite profiling is required to speed up the lead optimization process and candidate selection for development. Another technical advance that has moved to the forefront in a modern Drug Metabolism laboratory is that offered by Fourier transform mass spectrometry (FTMS). Recent improvements in FTMS instrument ruggedness and software control, coupled with its ultra-high resolution and high sensitivity, makes this instrument invaluable to biomarker identification. Unlike drug metabolites, where an investigator iteratively searches for variations on the starting molecule, undiscovered biomarkers are true unknowns buried in an endogenous sea of background noise. FTMS systems can routinely provide sub-ppm accuracy in *m/z* measurements leading to a very reliable determination of elemental composition to significantly narrow down putative structures. Hopefully, the bioanalytical LC–MS folks will be up to the task of routinely developing methods to measure the temporal changes of an endogenous substance detectable in placebo as well as in treatment arms of a study. Such is the likely future and final validation step for metabonomics.

Drug Metabolism as practiced in the Pharmaceutical Industry is a multidisciplinary research effort involving preclinical, nonclinical and clinical absorption, distribution, metabolism and excretion (ADME) studies on developmental candidates in support of worldwide registration of new therapeutic agents. A typical ADME program involves radio-tracer ($^{3}H/^{14}C$) studies in mice, rats, dogs, monkeys or humans to characterize metabolites and the routes of excretion. Advances in mass spectral technology have fundamentally changed the overall landscape in Drug Metabolism laboratories. Of particular note in comparison to the "old days", improved instrument sensitivity and selectivity have reduced our dependency on the administration of radiolabeled drug to detect and characterize metabolites. This has important implications from a standpoint of radioisotope waste management as well as obviating conduct of a radiolabeled human absorption, metabolism and excretion (AME) study prior to proof-of-concept clinical trials. In my group, we recently initiated a strategy which provides early feedback critical to each Drug Safety & Metabolism development program by routinely profiling plasma and urinary metabolites from unlabeled first-in-human single- and multiple-dose clinical studies using LC–MS techniques. Identification of human metabolites early in a clinical program ensures that (a) all major human metabolites are adequately covered in non-clinical safety assessments, (b) the enzymes responsible for major in vivo biotransformation pathways in humans are identified, thus leading to the design of an appropriate clinical drug–drug interaction program, and (c) any metabolite-related issues (i.e., structural alerts, human-specific metabolites) that can impact further development of the drug are addressed early in the program. In practical terms, this accelerated view of human metabolism in a clinical setting becomes the driving force to shape the direction of each Drug Metabolism program thereby greatly improving upon how we invest time and resources.

In the absence of radioactivity or reference standards for each human metabolite, the aforementioned exercise is largely qualitative in nature. With the possibility of widely different MS responses for metabolites of divergent structure, the potential exists for missing metabolites or failing to correctly define a major metabolite. However, judicious selection of LC–MS and LC–MS/MS experiments (such as, an accurate LC–MS with mass defect filter) together with a direct comparison of matrix profiles from drug versus placebo-administered subjects can often distinguish drug-derived from endogenous interfering peaks. If the quantitative determination of these human metabolites is required for exposure evaluation in preclinical species, then validated high-throughput bioanalytical LC–MS/MS method(s) can be developed for a more robust assessment.

Although plasma and urine samples from a rising single- or multiple-dose safety and tolerability clinical study can provide a robust qualitative assessment of putative human metabolites, radiolabeled studies will still continue to be widely used to determine excretion patterns in urine and feces (dose mass balance). Liquid chromatography with online inductively coupled plasma mass spectrometry (LC–ICP/MS) has clearly demonstrated the potential for quantitative determination of metabolites of bromine containing compounds [7,8]. For example, since Br is normally not present as an endogenous analyte, each Br atom can be selectively detected by ICP/MS and displayed as a "bromatogram." Any drug-derived molecule containing the same number of Br atoms would behave similarly, thereby normalizing detector response and facilitating its quantitation. Qualitative characterization of drug and metabolites in the same sample can be performed simultaneously by splitting a small portion of the LC effluent to a conventional API/MS system [8]. The limitation of ICP/MS for selective detection of pharmaceutically relevant compounds not containing elements like S, Cl, Br, F, or I may be overcome by administering drugs synthetically enriched with ^{13}C or ^{15}N. With additional improvements in sample introduction, ICP/MS coupled with an API/MS system could potentially provide an analytical platform for simultaneously producing mass balance data from urine and feces as well as metabolite identification without measuring a single alpha, beta or gamma decay event. This would not only facilitate but also enhance the study of human metabolism early in a development program and obviate administering a radionuclide to humans. It should be noted that the rapid throughput realized by modern "mega" performance LC systems is largely incompatible with most commercial in-line radiometric detectors due to poor counting statistics associated with very short detector cell residence time. Whether this technical problem can be overcome without resorting to stop-flow techniques or fraction collection remains to be seen, however, LC–ICP/MS presently offers an attractive alternative which shouldn't suffer from this limitation.

One last thought regarding the changing landscape in the LC–MS analysis of human therapeutic agents directly involves the physico-chemical nature of many new chemical entities (NCEs) entering drug development. Aside from the mass spectral characterization of therapeutic macromolecules (proteins, antibodies, etc.), which would engender the publication of another monograph, a more recent challenge has been to follow the "other half" of large (400–1500 Da) molecules that end up in our Drug Metabolism laboratories. In many cases these compounds possess a linear or branched backbone comprised of numerous hetero-cyclic systems, which can undergo cleavage into one or more large metabolic products. Traditionally, only a single side or functionality within a drug would be targeted for isotope incorporation. Recently, regulatory agencies have been quick to point this out as a deficiency and expect that adequate exposure to all major human metabolites be established in preclinical drug safety. This has fundamentally changed the synthetic strategy for producing isotopically enriched drugs, which meet the requirements of tracking each metabolically relevant chunk of drug while at the same time not compromising the simultaneous use of stable isotope internal standards prepared for quantitative mass spectrometry.

I would like to thank the Editor for giving me this opportunity to wax poetic about times long past and hopefully wax prophetic about the future of LC–MS in Drug Metabolism. I'll now leave it up to each of the outstanding collection of scientists contributing to this book to share their own research and experiences.

References

1. Thomson, J.J., 1911. Rays of positive electricity. Philos. Mag. 20, 752.
2. Thomson, J.J., 1911. Rays of positive electricity. Philos. Mag. 21, 225.
3. Aston, F.W., 1919. A positive ray spectrograph. Philos. Mag. 38, 707.
4. Dempster, A.J., 1918. A new method of positive ray analysis. Phys. Rev. 2, 316.
5. Alton, K.B., Grimes, R.M., Shaw, C., Patrick, J.E., McGuire, J.L., 1975. Biotransformation of a 1,5-benzodiazepine, triflubazam, by man. Drug Metab. Disp. 3, 352–360.
6. Horning, E.C., Carroll, D.I., Dzidic, I., Haegele, K.D., Horning, M.G., Stillwell, R.N., **1974.** Atmospheric pressure ionization (API) mass spectrometry. Solvent-mediated ionization of samples introduced in solution and in a liquid chromatograph effluent stream. J. Chromatogr. Sci. 12(11), 725–729.
7. Nicholson, J.K., Lindon, J.C., Scarfe, G., Wilson, I.D., Abou-Shakra, F., Sage, A.B., Castro-Perez, J., **2001.** High-performance liquid chromatography coupled to inductively coupled plasma mass spectrometry (HPLC-ICP-MS) and orthogonal acceleration time-of-flight mass spectrometry (oa-TOF-MS) for the simultaneous analysis and identification of the metabolites of 2-bromo-4-trifluoromethylacetaniline in rat urine. Anal. Chem. 73, 1491–1494.
8. Olivia, C., Nicholson, J.K., Lenz, E.M., Abou-Shakra, F., Castro-Perez, J., Sage, A.B., Wilson, I.D., 2000. Directly coupled liquid chromatography with inductively coupled plasma mass spectrometry and orthogonal acceleration time-of-flight mass spectrometry for the identification of drug metabolites in urine: application to diclofenac using chlorine and sulfur detection. Rapid Commun. Mass Spectrom. 14(24), 2377–2384.

Identification and Quantification of Drugs, Metabolites and Metabolizing
Enzymes by LC–MS
Swapan K. Chowdhury, editor.

Chapter 2

STRATEGIES FOR INCREASING THROUGHPUT FOR PK SAMPLES IN A DRUG DISCOVERY ENVIRONMENT

Walter A. Korfmacher

2.1. Introduction

The need for increasing throughput when analyzing samples in a drug discovery environment is driven by the need to investigate large numbers of new chemical entities (NCEs) that are provided by medicinal chemists in their search for new drugs. The successful development of new drugs is an increasingly difficult and expensive undertaking. The time from "test tube" to market for an NCE is typically 12–15 years with an overall cost that is approaching $1 billion/new drug that gets FDA approval to go on the market (Kaitin, 2003; Sinko, 1999). Therefore, pharmaceutical companies are continually looking for ways to be more efficient in terms of their research efforts so that the time from the discovery of NCEs to the approval of new drugs can be reduced.

The time from "test tube" to development is a critical step in the process of new drug discovery. It is during this time that thousands of compounds that have "in vitro activity" are tested in various in vitro and in vivo screens in order to select the few that will proceed into development for more rigorous safety testing followed by clinical studies. This critical new drug discovery phase can be envisioned as a series of stages as shown in Fig. 1. This figure shows this phase of new drug discovery as a series of screens that new compounds must pass through in order to select the few that then become development candidates. If done correctly, the initial screens will be higher throughput and will be followed by lower throughput screens in a manner that filters out the problem compounds, but allows the good compounds to pass on to the next stage.

Another view of this phase of the new drug discovery process can be seen in Fig. 2 (Korfmacher, 2003). In this figure, the new drug discovery process is viewed as an iterative process where the drug metabolism representative takes information from the various steps and feeds it back to the medicinal chemists so that improved versions of the NCE can be made. This stage in the process is often referred to as the "lead optimization" stage. If this process is performed correctly, the NCE that goes into development has a greater likelihood of success than one that bypassed

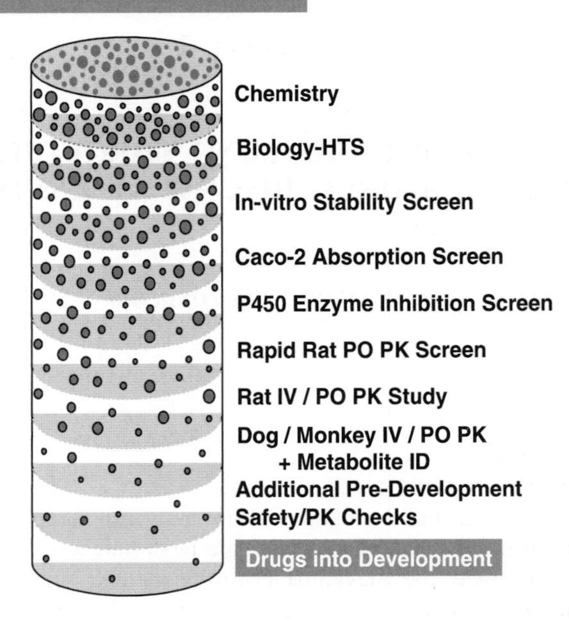

Figure 1.
Schematic showing a series of screens that can be used for "sifting out" the problematic compounds (using DMPK criteria) using the higher throughput screens in order to find the smaller number of compounds that are suitable for more extensive testing in the lower throughput screens. (Adapted from W.A. Korfmacher, "Bioanalytical Assays in a Drug Discovery Environment" in "Using Mass Spectrometry for Drug Metabolism Studies", 2004. W. Korfmacher. Ed., CRC Press. Copyright © 2004, with permission from Taylor and Francis Group, LLC.)

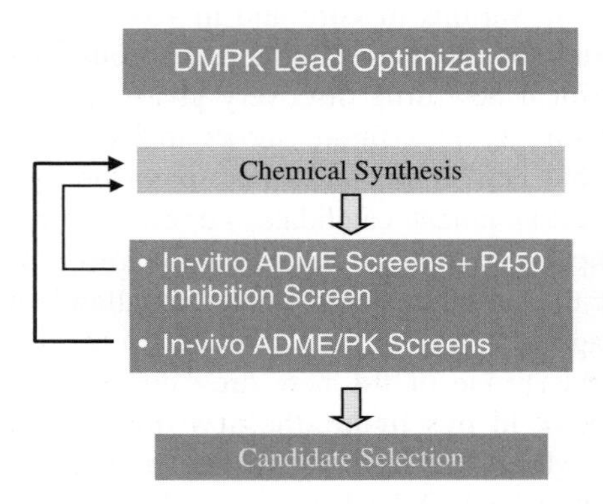

Figure 2.
Scheme showing the iterative nature of lead optimization as a process leading to the selection of compounds for development. Reprinted from Korfmacher (2003) copyright 2003, with permission from Thompson Scientific

these steps and had a minimal evaluation of its drug metabolism and pharmacokinetic (DMPK) properties evaluated before going into development.

The importance of early pharmacokinetic (PK) screening was discussed recently in a review by Newton and Lockey (2003). Additional monographs that discuss the utility of drug metabolism involvement in the drug discovery process have been authored by Roberts (2001), Riley *et al.* (2002), Eddershaw *et al.* (2000), Yan and Caldwell (2001), Sinko (1999), Prokai and Prokai-Tatrai (1999), and Thompson (2000, 2001). These reports and others stress the importance of early PK screening as an integral part of the new drug discovery process. The importance of HPLC–MS/MS as the analytical tool of choice for drug discovery PK studies was described by Korfmacher *et al.* (1997) and has been the subject of multiple review articles in more recent years, including current reviews by Ackermann *et al.* (2002a) and Kostiainen *et al.* (2003). This chapter describes some analytical strategies that can be used for early PK screening and also discusses some pitfalls that should be avoided.

2.2. Strategies for Early PK Assays

One important concept that should be considered in developing a strategy for early PK assays is cycle time. Cycle time in this context can be defined as the time between the initiation of a study and the delivery of the PK report to the discovery team or the corporate database. For early PK studies, there are typically two phases: the in-life phase and the assay/report phase. The in-life phase includes compound formulation, dosing, and sample collection. The assay/report phase includes the sample analysis and the PK report preparation. As shown in Fig. 3, the assay/report phase can be further broken down into smaller steps. What is important to realize is that any one of these steps can become the bottleneck in the process. Therefore, it is important to ensure that each of these steps is performed as efficiently as possible. This concept is important because a common error is to focus on one or two steps while ignoring the others. In this chapter, we will discuss various strategies for the efficient use of HPLC–MS/MS technology as well as describe an integrated system for producing higher throughput PK data in a drug discovery environment.

Sample tracking can be viewed in at least two ways. One may want to know where a given set of plasma samples is being stored while awaiting sample preparation. This can be accomplished by using fairly simple login procedures or in a more sophisticated manner by utilizing barcoding technology and sample tracking systems such as the WATSON® LIMS software that is common to many DMPK laboratories. A second type of sample tracking is to track the samples that come into the lab and then record their number and type in order to provide lab productivity information; this type of tracking can be performed using Excel® and is not difficult to implement for a small lab or group.

Standard curve preparation is an important step for any analytical laboratory. It has become easier to perform this step as simple liquid-handling robots have become available for the laboratory. For example, these robots can be programmed to dilute

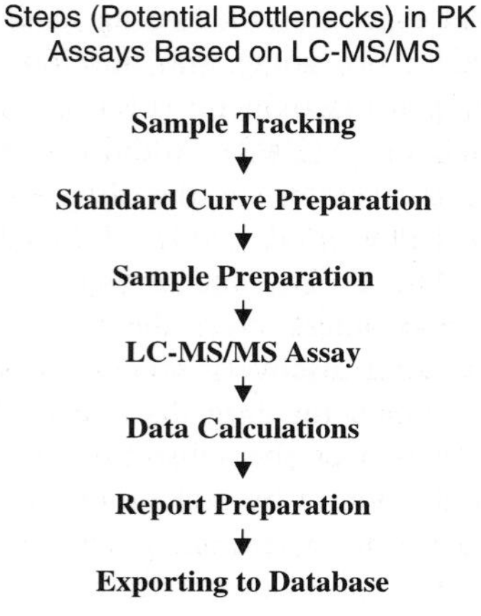

Figure 3.
Schematic showing the various steps involved with discovery PK studies from sample receipt to database entry

stock solutions and to make working solutions and then to add appropriate aliquots of the working solutions to the matrix that is to be assayed. In this way, one can semi-automate a task that would otherwise become tedious.

The question of whether or not to use quality control (QC) samples in an exploratory assay is often answered differently by various laboratories. In our laboratory, QC samples are not used for various PK screening assays (e.g. the cassette-accelerated rapid rat screen (vide infra)), but are employed for discovery assays of compounds that are being tested in more definitive PK studies, such as a single rising dose study or a radiolabeled-NCE PK study.

Sample preparation is a topic that has received a lot of attention in the last several years. In a drug discovery environment, it is important that the sample preparation procedure be as generic as possible so that it can be used for the thousands of compounds/year that need to be assayed in an efficient manner. Ackermann *et al.* (2002a) provided a good summary of various procedures in a recent overview of HPLC–MS/MS and how it is used in a drug discovery environment for bioanalytical support efforts. Currently, the most popular sample preparation procedure in the drug discovery arena is protein precipitation.

The advantage of protein precipitation is that it is a very generic process that works for most compounds and is one that can be semi-automated using 96-well plates and robots such as the Tomtec Quadra 96® system. In a typical procedure, 40–50 μl of plasma would be added to each well of the 96-well plate and the robot would add 150 μl of acetonitrile or methanol that contained the internal standard for the assay. This would be followed by vortexing and centrifugation, then the robot would be used to transfer an aliquot of the supernatant into a clean 96-well plate, which would then be covered and placed into an autosampler that would be used to inject the samples into the HPLC–MS/MS system.

The technology for mass spectrometers has continued to change. As shown in Fig. 4, there are multiple types of mass spectrometers that are now available for both qualitative and quantitative assays. The most commonly used MS tool for quantitative assays is LC–MS/MS based on the triple quadrupole mass spectrometer. Protein precipitation as a sample preparation technique has been aided by the continuing improvement in the sensitivity of triple quadrupole mass spectrometers. The earlier models allowed one to get down to perhaps 10 ng/ml as the limit of quantitation (LOQ) for many discovery compounds, the next generation allowed us to reach 1 ng/ml in most cases; the newest models appear to be able to reach 0.1 ng/ml for many compounds. This increase in sensitivity has followed the need for lower LOQs as more potent compounds are discovered for various project areas. Furthermore, this continued improvement in the triple quadrupole technology has allowed this type of mass spectrometer to remain as the technique of choice for routine quantitation of drug discovery PK samples.

One of the concerns that has emerged regarding HPLC–MS/MS PK assays in both drug discovery and development is the issue of matrix ion suppression. Several recent reports in the literature have discussed various aspects of this potential problem (Bonfiglio *et al.*, 1999; Hsieh *et al.*, 2001b; Jemal *et al.*, 2003; King *et al.*, 2000; Matuszewski *et al.*, 1998; Mei *et al.*, 2003; Miller-Stein *et al.*, 2000; Muller *et al.*, 2002; Schuhmacher *et al.*, 2003; Tiller and Romanyshyn, 2002; Tong *et al.*, 2002; Zheng *et al.*, 2002). In two of the early reports by King *et al.* (Bonfiglio *et al.*, 1999; King *et al.*, 2000), the authors proposed the use of a post-column infusion system (Fig. 5) to test whether or not matrix ion suppression was an issue for a particular assay. In the reports by King *et al.* (Bonfiglio *et al.*, 1999; King *et al.*, 2000), the authors concluded that protein precipitation could cause significant matrix ion suppression problems for some assays. The post-column infusion system, shown in Fig. 5, is a useful technique for testing an assay to see if it will have a problem with matrix ion suppression. The procedure is straightforward; one infuses the analyte of interest and the internal standard into the eluant from the HPLC column before it enters the MS ion source. By comparing the injection of mobile phase or a solvent blank to a sample extract made from drug-free plasma (plasma from control animals), one can easily observe what part of the chromatogram is free of matrix ion suppression effects. This concept was demonstrated in a recent publication by Muller *et al.* (2002) who stated that matrix ion suppression is an issue that needs to be considered when performing high throughput assays using "fast LC methods."

Recently, Hsieh *et al.* (2003a) described the utility of fast HPLC–MS/MS for the analysis of drug discovery compounds. The goal of this work was to show that fast gradients could be used in a drug discovery environment and that the data that were obtained were equivalent to the results from using longer HPLC assays. The methodology described in this work provides an excellent example of the proper way to demonstrate that faster HPLC techniques can be employed and one can still get the correct results. In this report, the authors assayed the levels of the test compound plus a known hydroxyl (M+16) metabolite. As shown in Fig. 6, the authors first demonstrated that baseline separation of the test compound and the known metabolite could still be obtained while changing from the standard HPLC conditions that

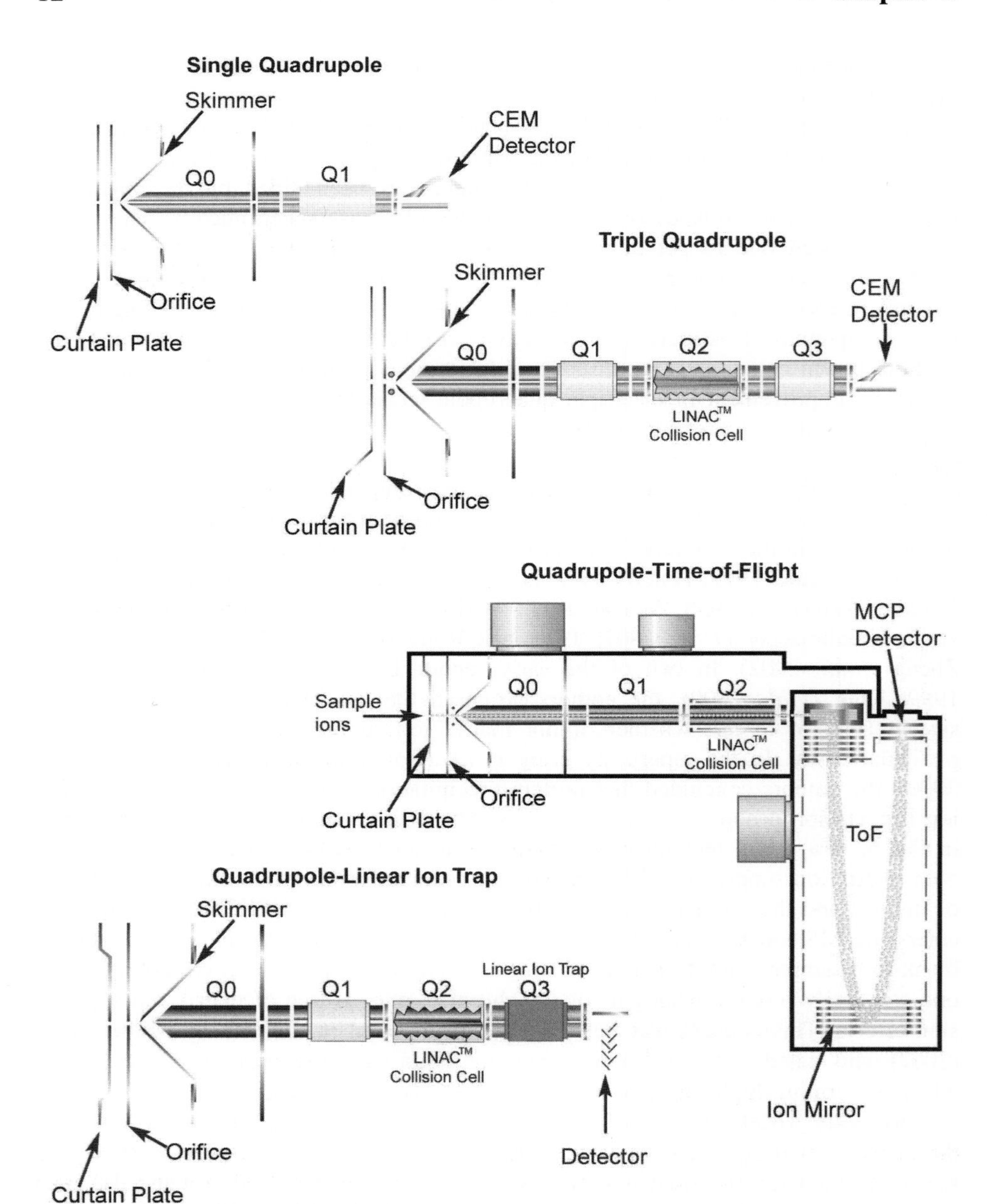

Figure 4.
Different types of mass spectrometers that are used for various drug metabolism assays. Figure provided by Sciex

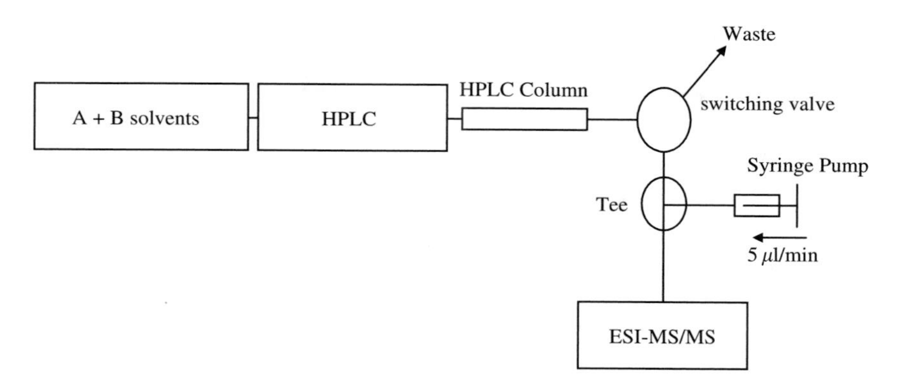

Figure 5.
Scheme showing the post-column infusion system for assessing matrix effects. Adapted from Bonfiglio *et al.* (1999) copyright 2003, with permission from John Wiley and Sons Ltd

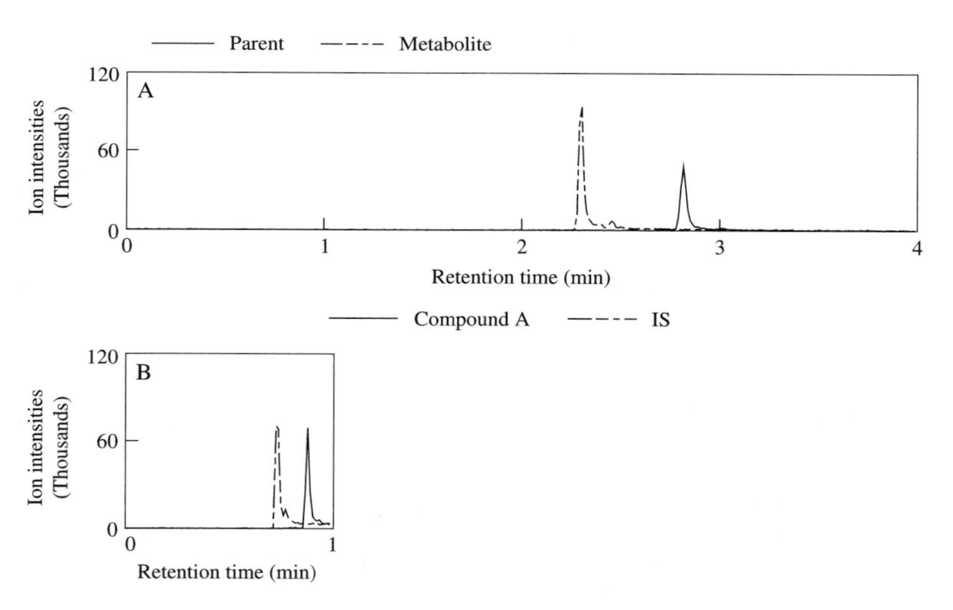

Figure 6.
LC–MS/MS chromatogram of a test compound (parent) and its M+16 metabolite using standard column chromatography (A) or mini-bore column chromatography (B). Reprinted from Hsieh *et al.* (2003a) copyright 2003, with permission from Elsevier

needed 4 min/sample to the new assay based on a shorter mini-bore column (2 × 30 mm) and a ballistic gradient that was complete in 1 min. The authors then demonstrated, as shown in Fig. 7, that under electrospray ionization (ESI) conditions, while there was a significant matrix ion suppression effect on the analyte signal

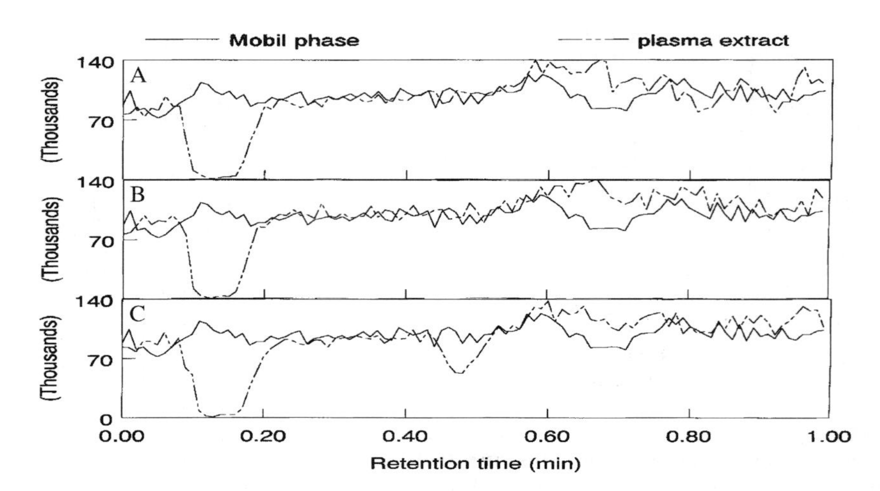

Figure 7.
Post-column infusion chromatograms under positive ESI–MS conditions using the mini-bore column and testing extracts from (A) rat, (B) dog, and (C) monkey blank plasma samples and comparing the mobile phase injection to the plasma matrices. Reprinted from Hsieh *et al.* (2003a) copyright 2003, with permission from Elsevier

from ca. 0.1 to 0.25 min, for the remainder of the chromatogram window, the matrix effect was minor. This test was performed for blank sample extracts obtained from rat, dog, and monkey plasma. This test is a valuable way to ensure that matrix ion suppression will not be a problem for the assay that is being developed. The authors also demonstrated that similar infusion chromatograms were obtained for the M+16 metabolite as well as the internal standard; thereby demonstrating that all three compounds eluted in the part of the chromatographic time window that did not have a matrix ion suppression problem.

Some reports have stated that atmospheric pressure chemical ionization (APCI) is less prone to matrix ionization suppression effects than is the ESI source (Bonfiglio *et al.*, 1999; King *et al.*, 2000). Hsieh *et al.* (2003a) also investigated the extent of matrix ion suppression when using the fast gradient with protein precipitation as the sample preparation procedure and the APCI source. As shown in Fig. 8, the amount of matrix ion suppression of the analyte signal was minimal when using HPLC–APCI–MS/MS conditions, although a signal reduction could be seen in the 0–0.2 min portion of the infusion chromatograms based on rat, dog, or monkey plasma extracts. As a final test of the faster HPLC conditions, the authors showed that in a four-way comparison, as shown in Fig. 9, the PK data were essentially the same for both the analyte and the metabolite from an individual set of monkey samples, regardless of whether or not the assay was based on APCI or ESI, using either standard or fast chromatography. Therefore, one can certainly use protein precipitation and fast gradient chromatography as long as one can demonstrate that the analytes and internal standard elute during chromatographic time windows are free of matrix ion suppression problems.

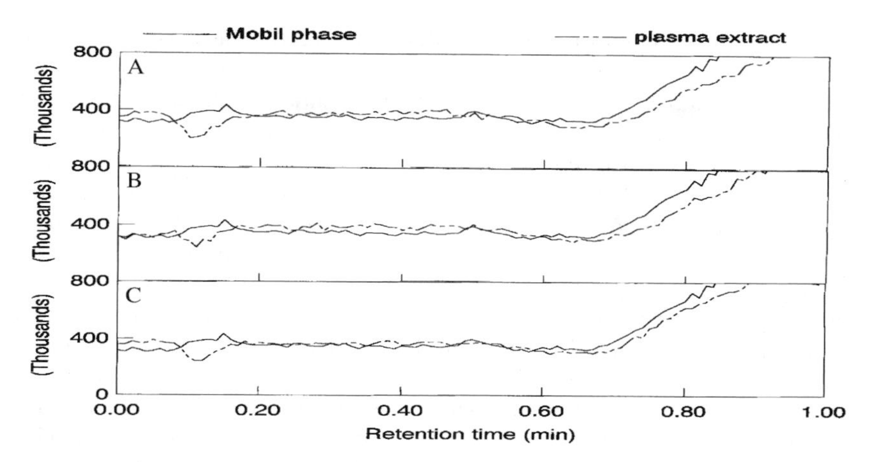

Figure 8.
Post-column infusion chromatograms under positive APCI–MS conditions using the mini-bore column and testing extracts from (A) rat, (B) dog, and (C) monkey blank plasma samples and comparing the mobile phase injection to the plasma matrices. Reprinted from Hsieh *et al.* (2003a) copyright 2003, with permission from Elsevier

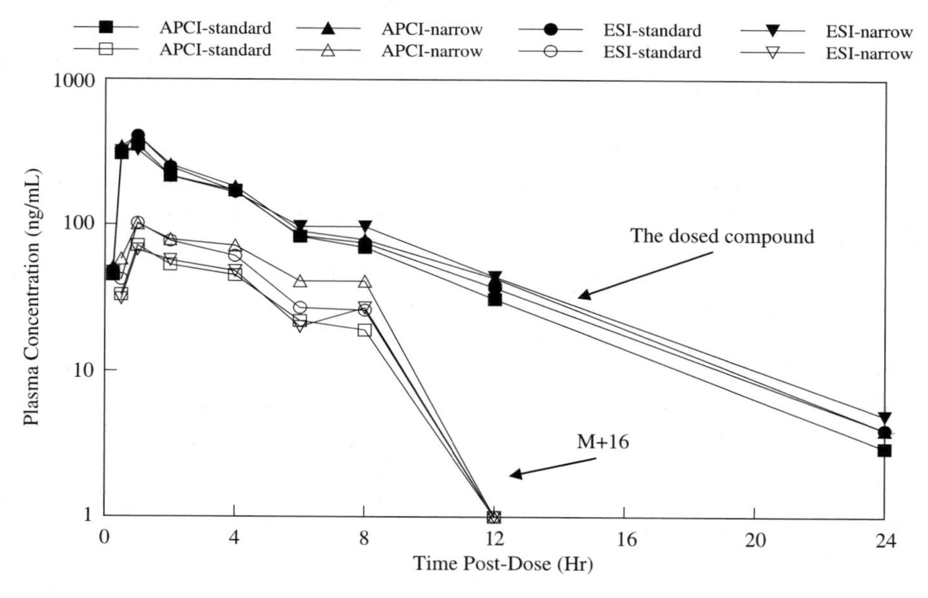

Figure 9.
Four-way comparison of the analytical results obtained from assaying one set of samples from a PK study using either ESI–MS or APCI–MS and either standard or mini-bore (narrow) chromatography. Reprinted from Hsieh *et al.* (2003a) copyright 2003, with permission from Elsevier

In another report on matrix effects, Mei *et al.* (2003) describe a very interesting study that looked at matrix effects that can be caused by the tube that is used to store the plasma sample. In this report, the authors studied multiple brands of sample tubes and also compared ionization sources from three major MS vendors in order to compare sensitivity to matrix effects based on ionization source type (ESI vs. APCI) and design. The authors studied the matrix effects on a series of compounds that varied in terms of chromatographic retention time. The authors found at least one brand of sample tube that contained a leachable polymeric compound that caused significant ion suppression problems for compounds that eluted at the latter part of a chromatogram on a typical reversed-phase system. The authors concluded that the extent of the matrix ion suppression varied depending on the ionization mode and the vendor source design. Interestingly, the authors also noted that the anticoagulant, Li-heparin could cause ionization enhancement for compounds that eluted early in the chromatogram. In a recent report by Shou and Naidong (2003), it was concluded that a commonly used dosing vehicle, PEG 400, could result in matrix ion suppression for some compounds. Therefore, as these reports demonstrate, it is clear that additional effort will need to be applied to the understanding of the extent of the matrix ion suppression issue and how best to avoid it when using LC–MS/MS assays.

Another area that has received a lot of attention in the literature is the variety of sample preparation techniques (Ackermann *et al.*, 2002a, b; Cox *et al.*, 2002; Mallet *et al.*, 2001; Zeng *et al.*, 2002). Ackermann (Ackermann *et al.*, 2002a, b) discussed various procedures for on-line sample preparation techniques including column-switching systems and "turbulent flow chromatography." Most of these on-column systems are based on a short extraction column followed by a longer analytical column. One drawback of this setup is the sequential nature, which can lead to a longer than desired total assay time for each sample. One solution is to employ two extraction columns in a staggered fashion so as to decrease the overall/sample cycle time. For example, Fig. 10 shows a 12-port valve configuration setup in an "alternate/regenerate backflush configuration" (Ackermann *et al.*, 2002a); this system allows one to load the next sample while the previous sample is going through the analytical column and the other extraction column is being regenerated.

In a series of articles, Hsieh (Hsieh *et al.*, 2000, 2001a, 2002a–d, 2003b) described single-column direct plasma injection systems that can be utilized to provide simplified sample preparation procedures. As shown in Fig. 11, the potential advantage of the single-column system is its simplicity. In one application of this system, Hsieh (Hsieh *et al.*, 2002b, c) demonstrated that six compounds could be assayed in a single analysis as shown in Fig. 12; while this was applied to pooled plasma samples, it shows that the assay could have been applied to a cassette-dosing sample.

Another chromatographic development that shows promise for speeding up HPLC–MS/MS assays is the monolithic column. Several reports have discussed the monolithic column and how it might be used for various pharmaceutical applications (Barbarin *et al.*, 2003; Dams *et al.*, 2002; Hsieh *et al.*, 2002d, 2003b; Lutz *et al.*, 2002; McCalley, 2002; Peng *et al.*, 2003; Smith and McNair, 2003; Tanaka *et al.*, 2001; van Nederkassel *et al.*, 2003; Wu *et al.*, 2001). The potential advantage of monolithic columns is that they can be used in high-flowrate applications; increasing the flowrate in an HPLC system

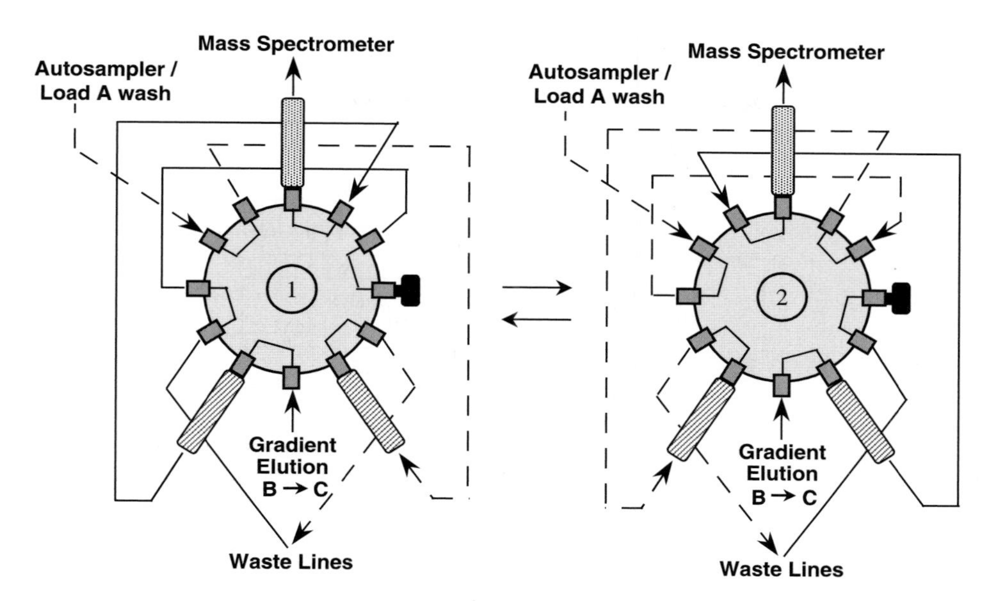

Figure 10.
Diagram of a 12-port valve configured for column switching using an alternate/regenerate back-flush configuration. In this arrangement, dual desalting columns are connected to a single analytical column. Reprinted from Ackermann *et al.* (2002a) copyright 2002, with permission from Bentham Science Publishers

can dramatically reduce the assay time. Hsieh (Hsieh *et al.*, 2002d) described the use of a monolithic column in an HPLC–MS/MS assay for a discovery compound and a metabolite. In their assay, a double (containing the compound and the metabolite) standard curve was used to assay rat plasma samples from a discovery PK study. By using a fast gradient and a flowrate of 4 ml/min, the authors were able to assay the samples with chromatographic run times of 30 s/sample. The authors demonstrated that the results obtained by using this higher speed assay were equivalent to those obtained using a more conventional chromatographic procedure. It is likely that additional examples for the use of monolithic columns will be reported as more monolithic column sizes become available.

Jemal (Jemal and Ouyang, 2000; Jemal and Xia, 1999) has discussed the need for chromatographic separation of certain analytes and their metabolites. It is important to understand those times when a metabolite can interfere with the dosed compound when assaying PK samples. Examples of common metabolites that can interfere with the analysis of their "parent compound," include N-oxides and glucuronide metabolites (Jemal and Xia, 1999; Tong *et al.*, 2001; Wainhaus *et al.*, 2002; Yan *et al.*, 2003). For example, Wainhaus (2002) shows an example, Fig. 13, where the signal from an acyl-glucuronide metabolite of the dosed compound would have interfered with the analysis of the dosed compound if the two components were not separated chromatographically.

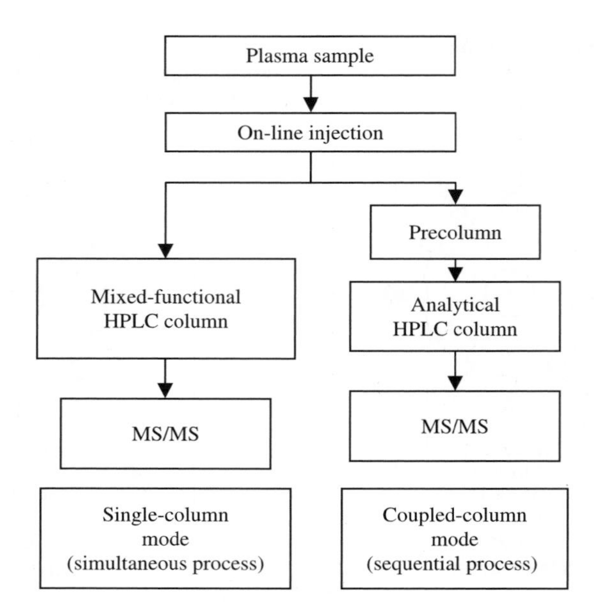

Figure 11.
Schematic comparing direct plasma injection methods using either a single-column or a coupled column approach. Reprinted from Hsieh *et al.* (2002c) copyright 2002, with permission from Russell Publishing

The last two steps in the process for PK assays (Fig. 3) are the data calculations and the report preparation and dissemination procedures. The major vendors of triple quadrupole mass spectrometers have all done a very good job of providing good software tools for chromatographic data integration and assay calculations. Therefore, this step is generally no longer a major hurdle for most PK assays. In order to calculate simple PK parameters such as area under the curve (AUC), half-life, clearance, and volume of distribution, one needs a PK calculation program. Currently, one of the most popular PK programs is included in the WATSON® (InnaPhase Corporation) LIMS program; this program is well-suited for calculating PK parameters in either a drug discovery setting or a good laboratory practices (GLP) environment. Once the PK calculations have been obtained, then various data reporting templates can be used to put the PK results into a format that can be issued electronically to the team members. Using these steps, the data calculation and reporting steps can be performed within a few hours or less.

2.2.1. Enhanced Mass Resolution – A New Analytical Tool for PK Assays

Owing to a technological advance, another dimension for analyte specificity has been added to certain triple quadrupole mass spectrometers; this new capability is higher mass resolution. Until recently, all triple quadrupole mass spectrometers could only provide "unit mass resolution" capability. Unit mass resolution for small molecules (with

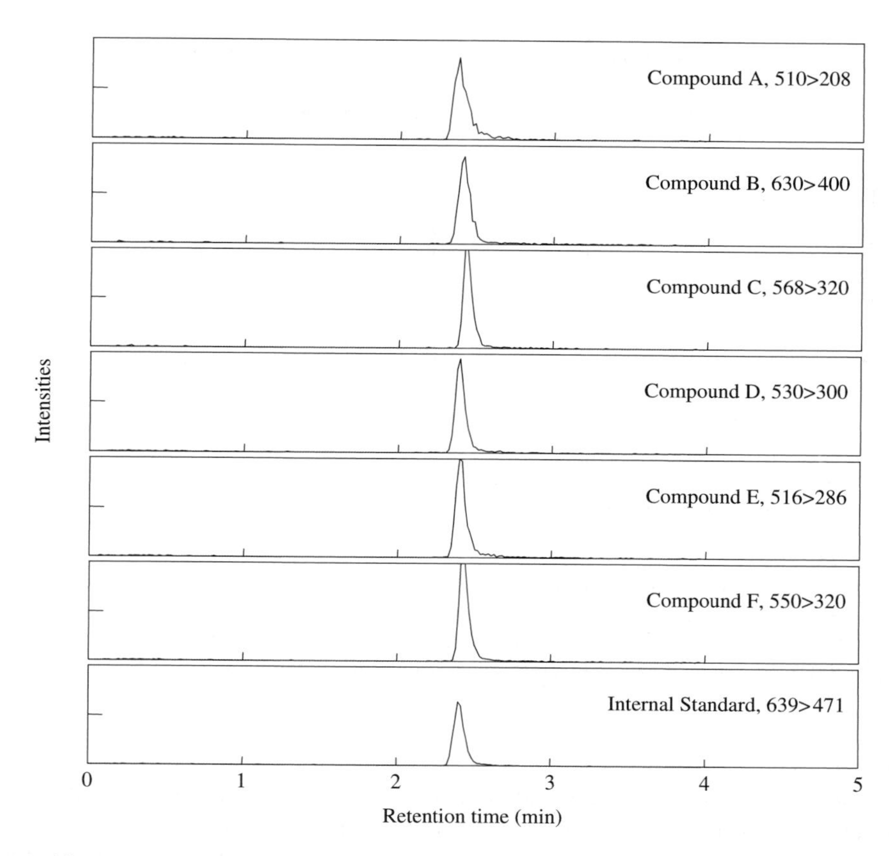

Figure 12.
Direct plasma injection results showing the LC–MS/MS chromatograms for six NCEs spiked into blank rat plasma at a concentration of 500 ng/ml. Reprinted from Hsieh *et al.* (2002c) copyright 2002, with permission from Russell Publishing

molecular weights in the range of 300–800 Da) means simply that the mass spectrometer could distinguish two compounds that differed by at least one mass unit (e.g. *m/z* 520 vs. *m/z* 521). While HPLC–MS/MS systems operated in the selected reaction monitoring (SRM) mode are very selective analytical tools when the mass resolution is set to the typical unit mass resolution setting, there are times when higher mass resolution could provide the additional selectivity that is needed to develop the analytical method quickly. These new triple quadrupole systems are now referred to as having enhanced mass resolution.

Several recent reports have discussed the utility of these enhanced mass resolution triple quadrupole mass spectrometers (Jemal and Ouyang, 2003; Paul *et al.*, 2003; Xu *et al.*, 2003; Yang *et al.*, 2002). In these reports, the typical definition for unit mass resolution was a mass peak width of 0.7 Da full-width at half-maximum (FWHM) and the enhanced mass resolution instruments were set at 0.1–0.2 Da

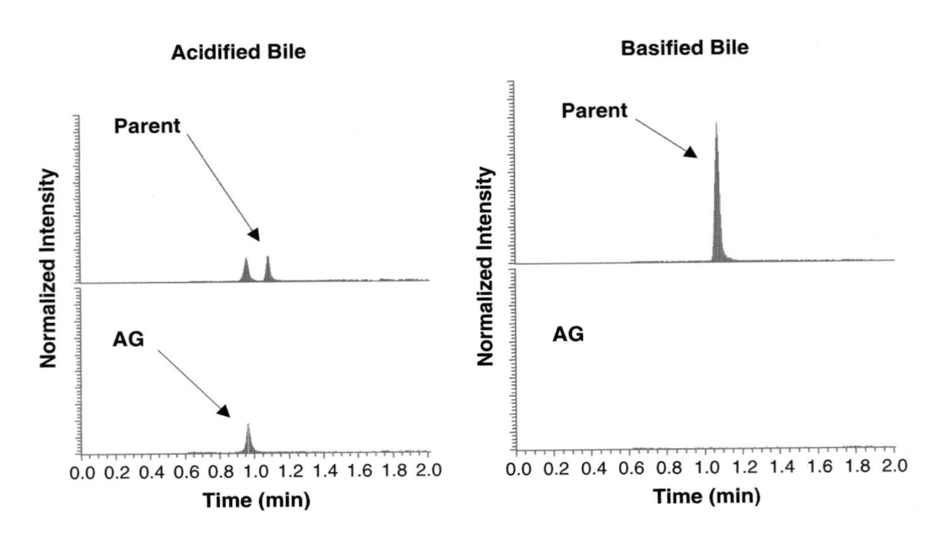

Figure 13.
LC–MS/MS chromatograms showing the results of the analysis of either acidified (left) or basified (right) bile samples from a study of a compound that formed an acyl-glucuronide (AG) of the parent (dosed compound) as a metabolite. Adapted from Wainhaus *et al.* (2002) copyright 2002, with permission from Russell Publishing

FWHM. Figure 14 shows a comparison of the mass spectra of two compounds as recorded in the unit mass resolution mode (upper traces) or the enhanced mass resolution mode (lower traces); it is apparent that the analytical selectivity of the enhanced mass resolution mode is better than the unit mass resolution mode (Yang *et al.*, 2002). With previous triple quadrupole mass spectrometers, if one attempted to set the mass at an enhanced mass resolution of 0.1 or 0.2 Da FWHM, the signal would typically be greatly reduced so that it would be unsuitable for most applications. With the new enhanced mass resolution triple quadrupole mass spectrometers, the signal reduction is typically only a factor of three in going from unit mass resolution (0.7 Da FWHM) to the enhanced mass resolution settings (0.1–0.2 Da FWHM). For example, as shown in Fig. 15, the peak heights of the SRM chromatograms for two compounds were reduced less than threefold in each case (Yang *et al.*, 2002). What is not apparent in this figure is that the signal to noise ratio (S/N) may well increase under the enhanced mass resolution setting even though the signal has been reduced; this outcome can be tested by assaying samples at levels close to the LOQ for the analyte of interest.

Yang *et al.* (2002) compared the results obtained from the enhanced mass resolution triple quadrupole mass spectrometer for an assay of a drug and its metabolite in human plasma over the range of 0.025–25 ng/ml. Standards, QC samples, and blanks were assayed in an overnight test under unit mass resolution (Q1 and Q3 both set to 0.7 Da FWHM) and also at enhanced mass resolution settings (Q1 was set to 0.1 Da FWHM, while Q3 was set to 0.5 Da FWHM). The results of this

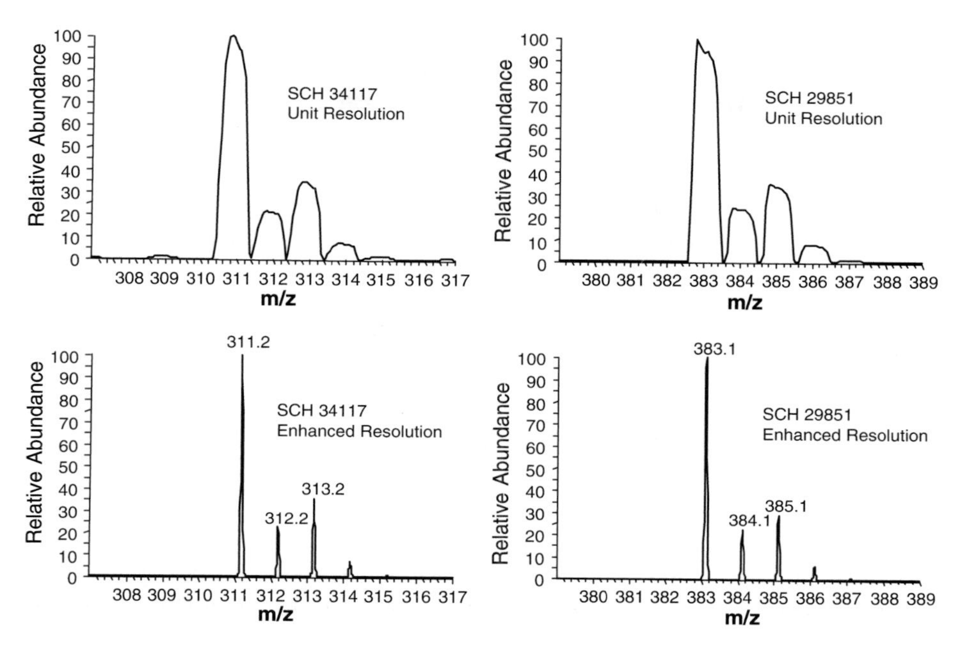

Figure 14.
Mass spectra (Q1 partial scan) of SCH 34117 and SCH 29851 analyzed using either unit (upper traces) or enhanced mass resolution (lower traces). Reprinted from Yang *et al.* (2002) copyright 2002, with permission from John Wiley and Sons Ltd

test were that the assay performance under either the unit mass resolution setting or the enhanced mass resolution setting both met current guidelines for acceptance as a validated assay (Yang *et al.*, 2002).

Jemal and Ouyang (2003) show examples of how the enhanced mass resolution triple quadrupole mass spectrometer provided improved S/N results for some compounds at the LOQ level. For example, Fig. 16 shows a comparison of SRM chromatograms for nefazodone in plasma at unit mass resolution (upper trace) and higher mass resolution (lower trace); this example is for human plasma spiked at a low level, but it is clear that the enhanced mass resolution provided better analytical results in this case. In a second example with the same compound spiked into human urine, the same type of results can be seen; as shown in Fig. 17, the upper trace shows the SRM chromatogram with a interference peak adjacent to the peak for nefazodone, while the lower trace, using enhanced mass resolution, shows only one peak from the target analyte.

Xu *et al.* (2003) recently discussed the utility of the enhanced mass resolution mass spectrometers for bioanalytical applications in a drug discovery environment. In this report, the authors also showed examples of how the enhanced mass resolution could be used to improve the LOQ of an assay. For example, Fig. 18 shows a comparison of SRM chromatograms for discovery compound A in mouse plasma at unit

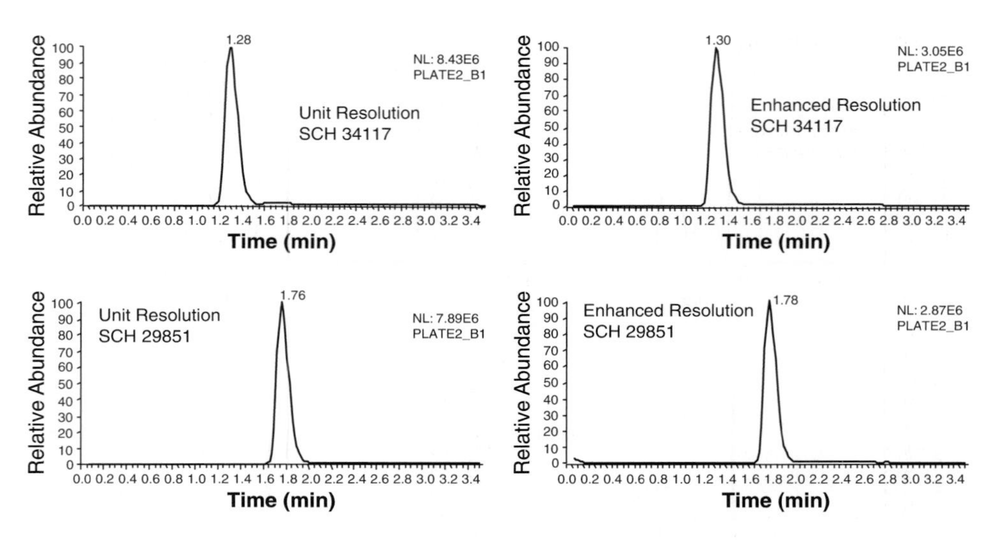

Figure 15.
LC–MS/MS SRM chromatograms of an extracted plasma standard at the ULOQ for the assay (25 ng/ml) obtained using either unit (left traces) or enhanced mass resolution (right traces). Reprinted from Yang *et al.* (2002) copyright 2002, with permission from John Wiley and Sons Ltd

mass resolution in blank mouse plasma (upper traces) vs. mouse plasma spiked at 0.1 ng/ml, the planned LOQ for the assay. It is clear from this figure that the S/N ratio (measured as 3) would be considered unacceptable to most analysts, even for a drug discovery assay. Figure 19 shows a comparison of SRM chromatograms for the same two samples but assayed using the enhanced mass resolution mode; the upper traces are from the blank mouse plasma and it is evident that the background noise level has been reduced and the lower trace shows the SRM chromatograms for the mouse plasma sample that was spiked at 0.1 ng/ml. It is clear from this figure that the S/N ratio (measured as 7) is a significant improvement over what was obtained at unit mass resolution – this S/N would be considered acceptable to many analysts working in a drug discovery setting. It is also worth noting that while the absolute signal decreased for the analyte when the mass resolution was enhanced, the S/N ratio increased – this is because the noise level was reduced more than the signal was reduced.

2.2.2. CARRS – An Example of an Integrated PK Assay Strategy

Korfmacher *et al.* (2001) described a new assay strategy for screening large numbers of compounds in a systematic manner, this screen was called cassette-accelerated rapid rat screen (CARRS). The utility of CARRS has been proven by the fact that it has continued to be used in our laboratory. As with any assay system, it has been improved, but the basic strategy has remained the same. In this section, the

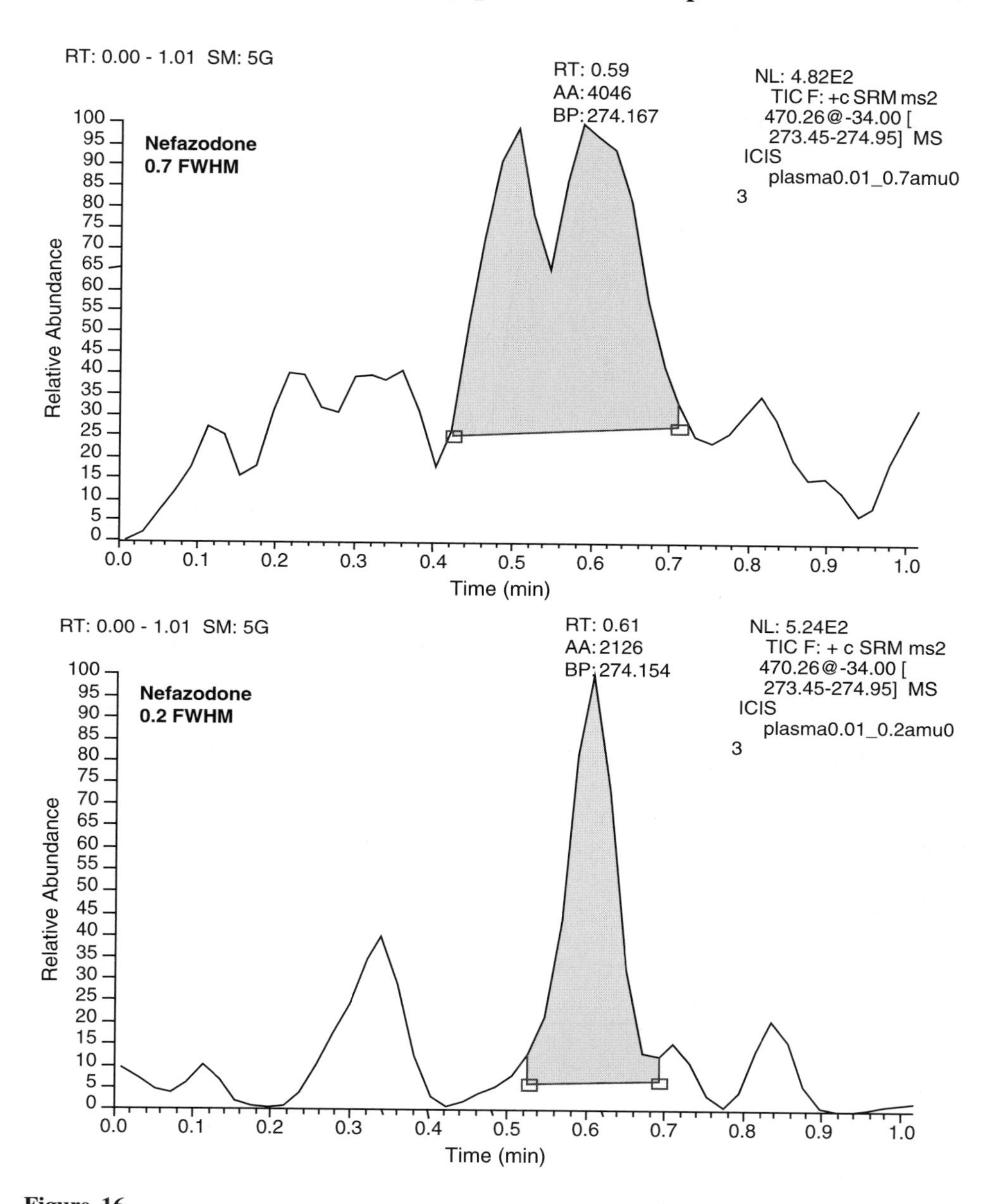

Figure 16.
Comparison of LC–MS/MS SRM chromatograms obtained using either unit mass resolution (0.70 Da FWHM peak width – upper trace) or enhanced mass resolution (0.20 Da FWHM peak width – lower trace) from the analysis of a blank human plasma sample spiked with nefazodone at 30 pg/ml. Reprinted from Jemal and Ouyang (2003) copyright 2003, with permission from John Wiley and Sons Ltd

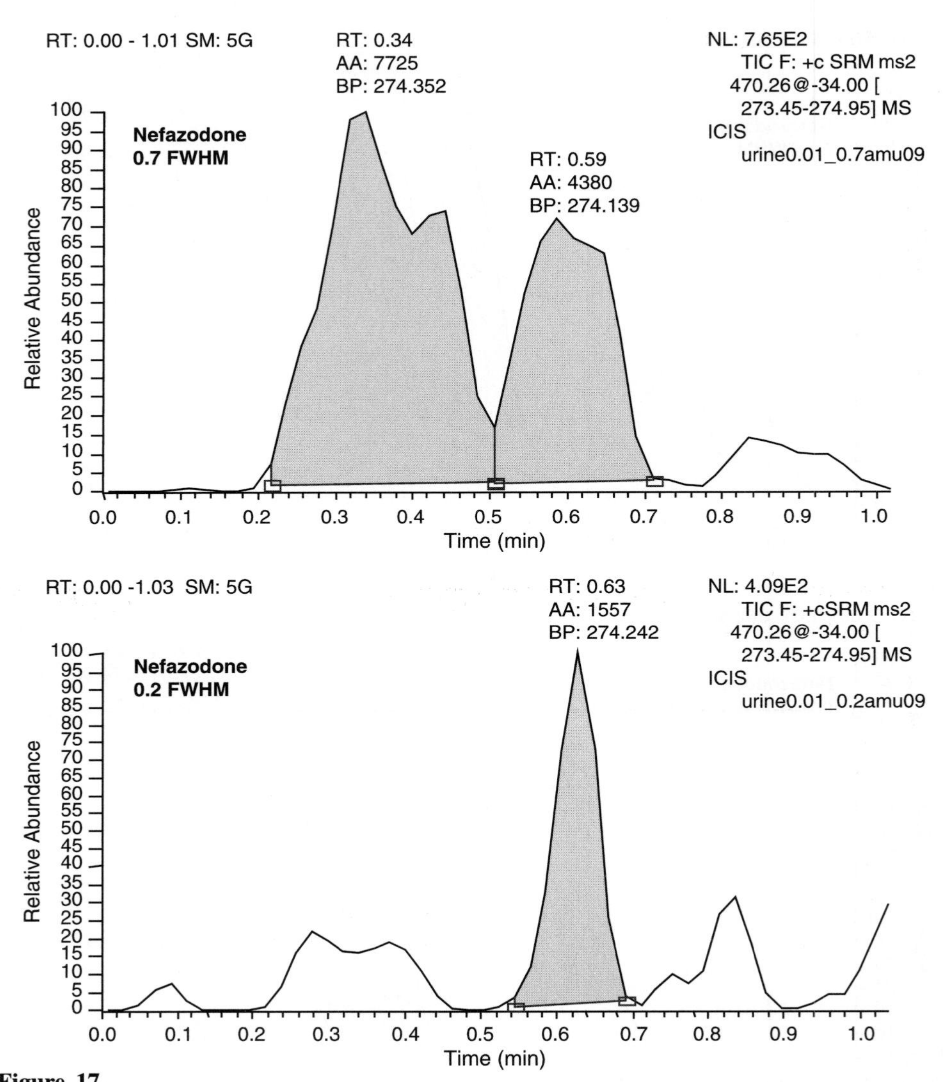

Figure 17.
Comparison of LC–MS/MS SRM chromatograms obtained using either unit mass resolution (0.70 Da FWHM peak width – upper trace) or enhanced mass resolution (0.20 Da FWHM peak width – lower trace) from the analysis of a blank human urine sample spiked with nefazodone at 30 pg/ml. Reprinted from Jemal and Ouyang (2003) copyright 2003, with permission from John Wiley and Sons Ltd

CARRS screen will be described as a model system for setting up an integrated in vivo oral PK screen.

The CARRS system is based on a weekly distribution of compounds to the DMPK discovery dosing group (sufficient amounts for dosing two rats/compound

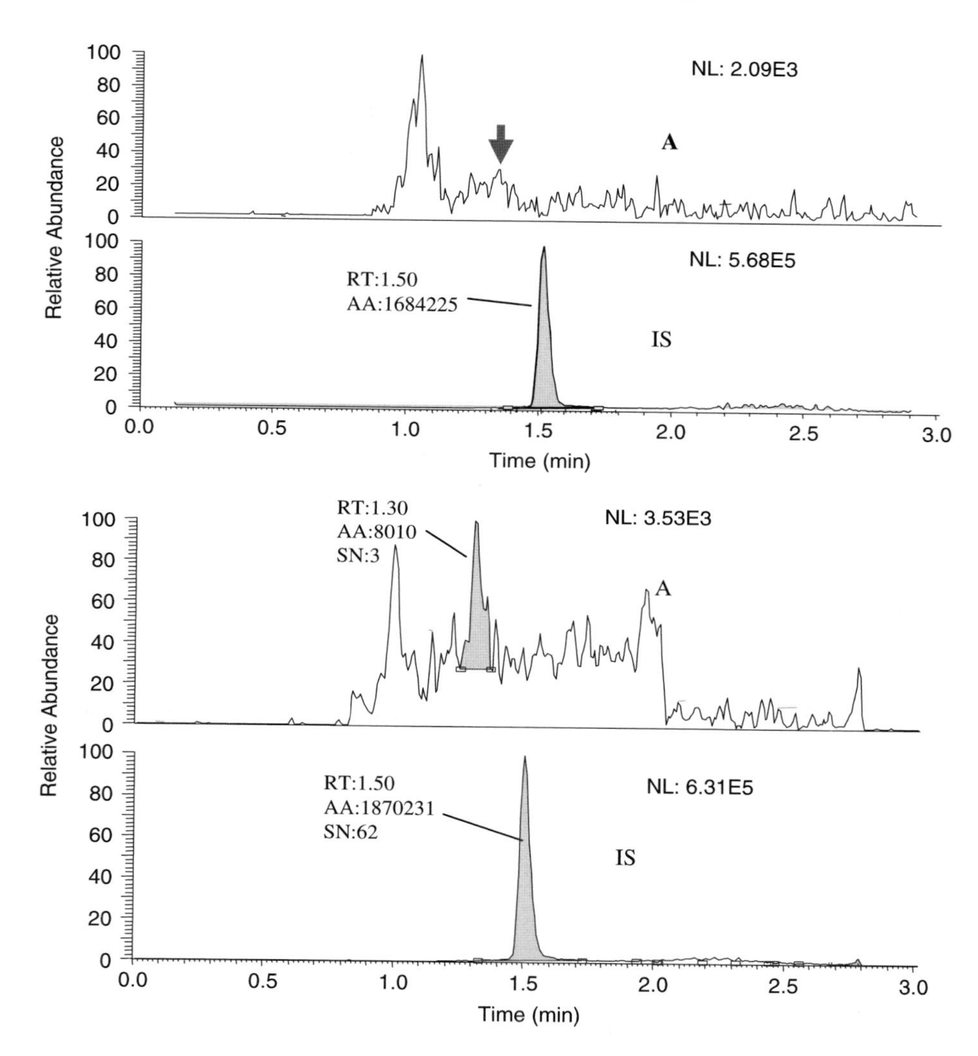

Figure 18.
Comparison of LC–MS/MS SRM chromatograms obtained at unit mass resolution (0.70 Da FWHM peak width) from the analysis of a blank mouse plasma sample spiked with internal standard (IS) only (upper two traces) or spiked with the IS and discovery compound A at 0.1 ng/ml (lower two traces). Reprinted from Xu *et al.* (2003) copyright 2003, with permission from John Wiley and Sons Ltd

at an oral dose of 10 mg/kg) as well as smaller amounts (ca. 2 mg) for the DMPK discovery mass spectrometry group to use for analytical method development and for making standard curves (spiked into drug-free rat plasma). The compounds are delivered as sets (cassettes) of six compounds from one drug discovery program.

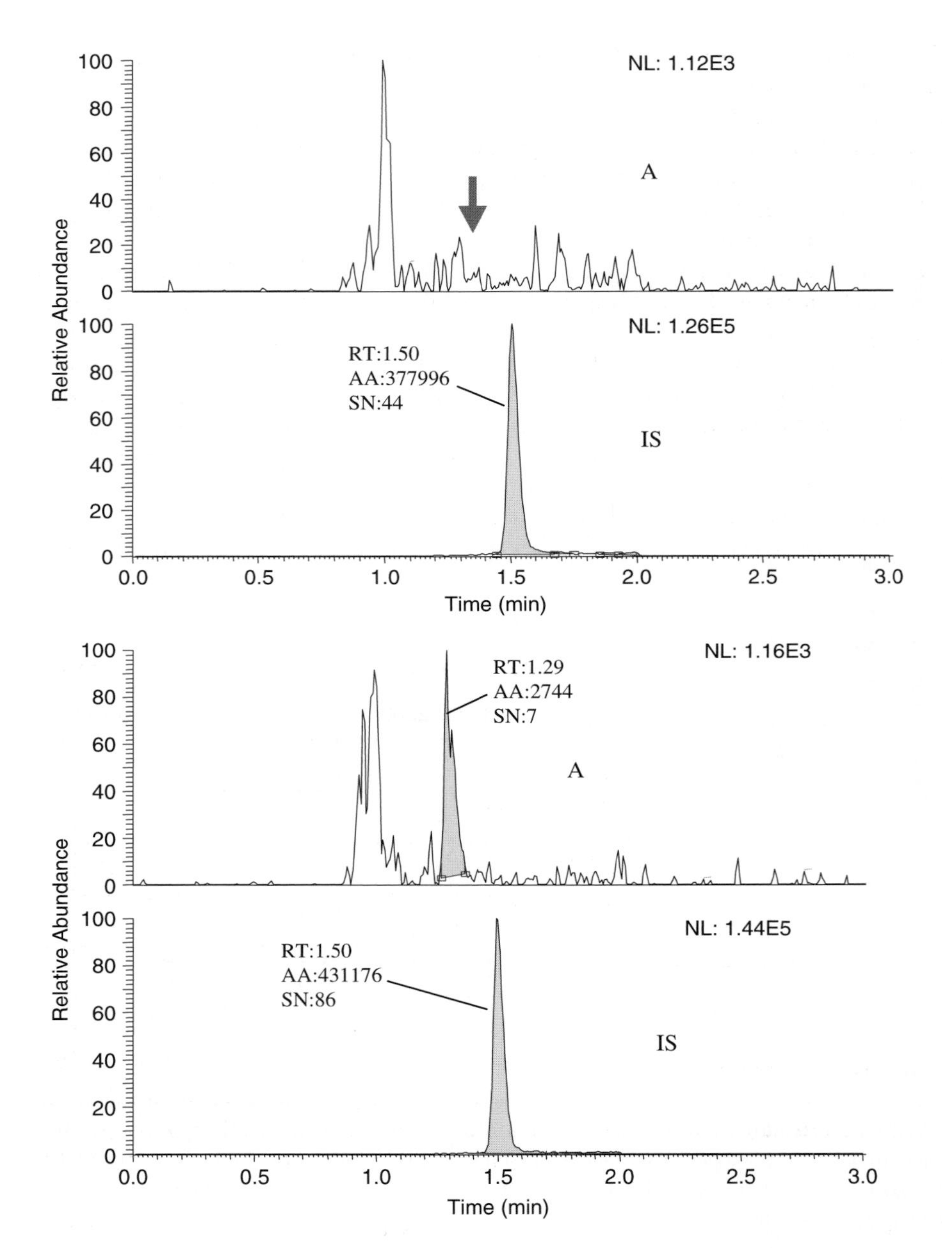

Figure 19.
Comparison of LC–MS/MS SRM chromatograms obtained at enhanced mass resolution (0.20 Da FWHM peak width) from the analysis of a blank mouse plasma sample spiked with internal standard (IS) only (upper two traces) or spiked with the IS and discovery compound A at 0.1 ng/ml (lower two traces). Reprinted from Xu *et al.* (2003) copyright 2003, with permission from John Wiley and Sons Ltd

The CARRS system is modular in that it can be expanded as the needs of the drug discovery teams increase; thus, it started in 2000 with five sets of six compounds (30 total compounds dosed/week) and was expanded over the next few years and is currently set at eight sets of six compounds (48 total compounds dosed/week). This current system allows "slots" for up to eight drug discovery teams to fill each week.

In the week following the delivery of the compound cassettes, the compounds are dosed individually to two rats at an oral dose of 10 mg/kg and blood samples are harvested from the rats at 0.5, 1, 2, 3, 4 and 6 h using serial bleeding techniques that result in about 100 μl plasma samples for each time point from each rat. It is important to note that the dosing is done in a one-in-one manner, i.e., it is not cassette dosing. As described by White and Manitpisitkul (2001), cassette dosing has the risk of drug–drug interactions, which can lead to false-positive and false-negative results. The plasma samples are delivered to the discovery mass spectrometry group in 96-well plates using a standardized template; in this way, the 72 plasma samples from one set of six compounds can be delivered and stored in one 96-well plate.

In the second week, the plasma samples are assayed. The first sample preparation step is to pool equal aliquots from the two rats dosed with a given compound at each

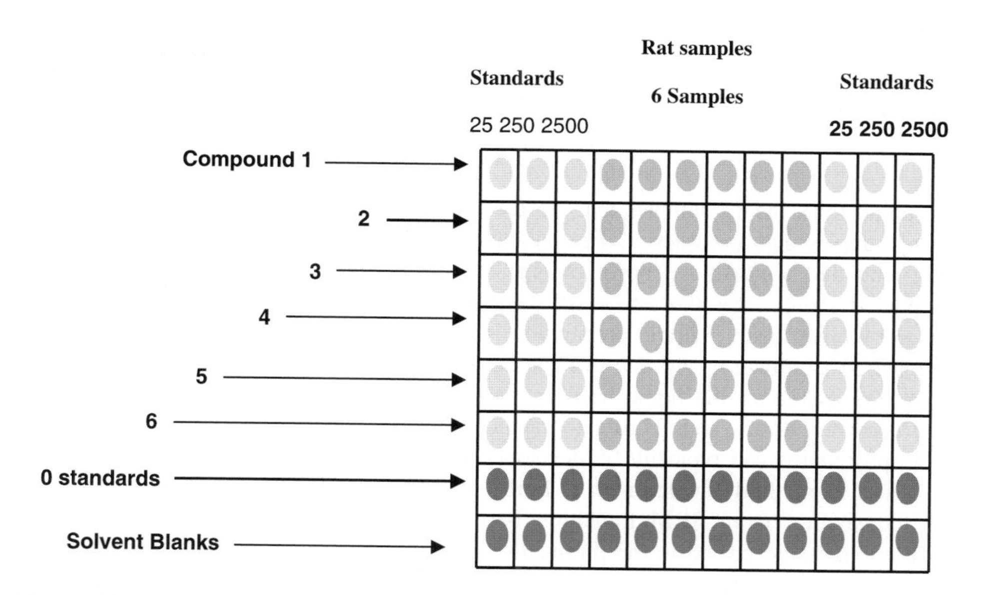

Figure 20.
A schematic diagram of how one 96-well plate was used for holding all the samples and standards required for one set of six compounds as part of the CARRS assay. For each compound, the typical injection sequence is a (solvent) blank and then the 0 standard followed by the first low, medium, and high standards, followed by a blank, followed by the six samples in sampling order, followed by a blank and then the second low, medium and high standard. Reprinted from Korfmacher *et al.* (2001) copyright 2001, with permission from John Wiley and Sons Ltd

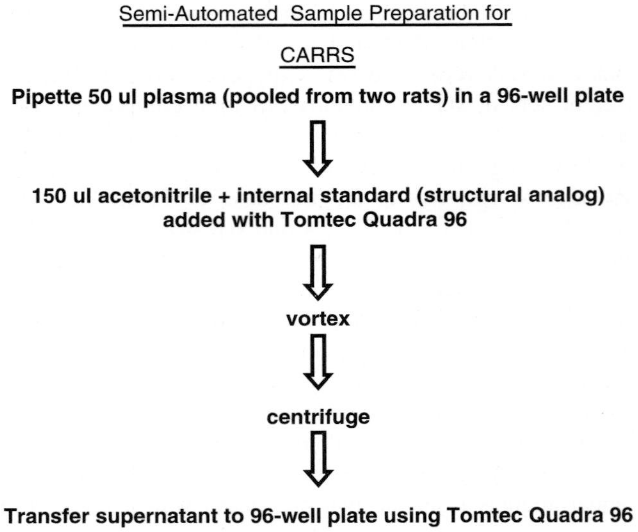

Semi-Automated Sample Preparation for

CARRS

Pipette 50 ul plasma (pooled from two rats) in a 96-well plate

⇩

150 ul acetonitrile + internal standard (structural analog) added with Tomtec Quadra 96

⇩

vortex

⇩

centrifuge

⇩

Transfer supernatant to 96-well plate using Tomtec Quadra 96

Figure 21.
A schematic diagram showing how semi-automated sample preparation can be used as part of the CARRS assay. Reprinted from Korfmacher *et al.* (2001) copyright 2001, with permission from John Wiley and Sons Ltd

time point; thus, the 12 samples that are received for each compound are reduced to six pooled samples that are assayed. The pooled samples are added to a fresh 96-well plate along with the standards as shown in Fig. 20; in this way, all of the standards, blanks, and pooled samples for the set of six compounds in the CARRS assay can be placed into one 96-well plate. Sample preparation is performed using a semi-automated protein precipitation as shown in Fig. 21; this can be accomplished easily by utilizing the Tomtec Quadra 96® robot for the liquid transfer steps. Because all six compounds are from the same discovery team, their chromatographic properties tend to be similar and generic HPLC procedures can typically be developed for each project area set of samples.

The MS method development is fairly straightforward for most compounds, typically consisting of an SRM transition based on the $[M+H]^+$ going to a prominent product ion. As described in the publication (Korfmacher *et al.*, 2001), three standards are selected with a 10-fold increase between standards. Once the assay is complete, the sample results for a compound are compared to the three standards (low, middle, and high). If the sample results are zero, the low and middle standards are selected. If the sample results for a compound fall within one of the two 10-fold ranges, then those two standards are selected for the data calculations. If the samples encompass the 100-fold range between the low and high standards, then these will be selected for the calculation. The assay results can be calculated using an Excel® spreadsheet template because of the use of the selected two-point standard curve based on a simple linear regression calculation.

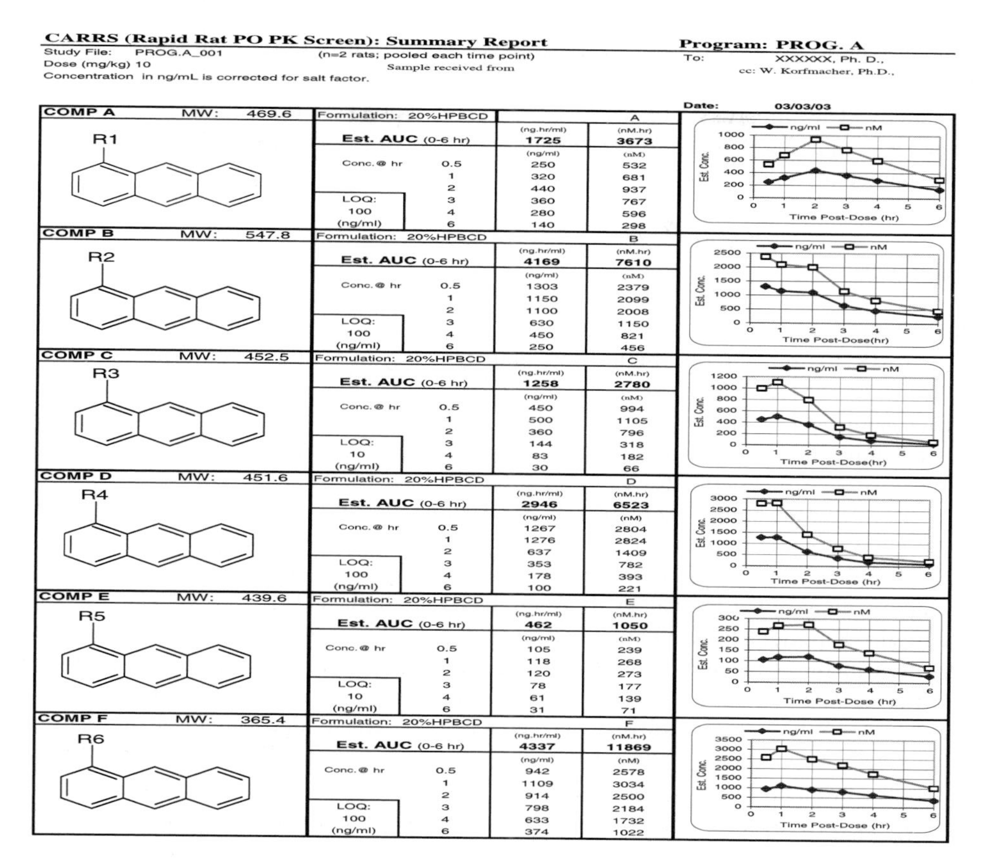

CARRS (Rapid Rat PO PK Screen): Summary Report **Program: PROG. A**

Study File: PROG.A_001 (n=2 rats; pooled each time point) To: XXXXXX, Ph. D.,
Dose (mg/kg) 10 Sample received from cc: W. Korfmacher, Ph.D.,
Concentration in ng/mL is corrected for salt factor.

Date: 03/03/03

Compound	MW	Formulation	LOQ (ng/ml)		(ng/ml)	(nM)
COMP A (R1)	469.6	20%HPBCD	100	Est. AUC (0-6 hr)	1725 (ng.hr/ml)	3673 (nM.hr)
				Conc. @ hr 0.5	250	532
				1	320	681
				2	440	937
				3	360	767
				4	280	596
				6	140	298
COMP B (R2)	547.8	20%HPBCD	100	Est. AUC (0-6 hr)	4169 (ng.hr/ml)	7610 (nM.hr)
				Conc. @ hr 0.5	1303	2379
				1	1150	2099
				2	1100	2008
				3	630	1150
				4	450	821
				6	250	456
COMP C (R3)	452.5	20%HPBCD	10	Est. AUC (0-6 hr)	1258 (ng.hr/ml)	2780 (nM.hr)
				Conc. @ hr 0.5	450	994
				1	500	1105
				2	360	796
				3	144	318
				4	83	182
				6	30	66
COMP D (R4)	451.6	20%HPBCD	100	Est. AUC (0-6 hr)	2946 (ng.hr/ml)	6523 (nM.hr)
				Conc. @ hr 0.5	1267	2804
				1	1276	2824
				2	637	1409
				3	353	782
				4	178	393
				6	100	221
COMP E (R5)	439.6	20%HPBCD	10	Est. AUC (0-6 hr)	462 (ng.hr/ml)	1050 (nM.hr)
				Conc. @ hr 0.5	105	239
				1	118	268
				2	120	273
				3	78	177
				4	61	139
				6	31	71
COMP F (R6)	365.4	20%HPBCD	100	Est. AUC (0-6 hr)	4337 (ng.hr/ml)	11869 (nM.hr)
				Conc. @ hr 0.5	942	2578
				1	1109	3034
				2	914	2500
				3	798	2184
				4	633	1732
				6	374	1022

Figure 22.
An example of the format that has been used to issue CARRS data. The form is based on an Excel® temple that calculates the AUC values and plots the concentration vs. time data automatically

As this is a simple oral PK screen, the only PK parameter that can be easily calculated is AUC (0–6). Once the sample concentrations are entered, the spreadsheet automatically calculates the AUC (0–6). As shown in Fig. 22, the Excel® spreadsheet summarizes the data for the six compounds in a convenient one-page report that can be sent via E-mail to the discovery team. The summary results are normally sent out to the teams by the end of the second week; thus, the cycle time for this PK screen is 2 weeks. In summary, the CARRS assay is an excellent example of an integrated strategy for providing high throughput PK data in a drug discovery setting.

2.3. Conclusions

While the sample load for most drug metabolism scientists working in the drug discovery arena continues to increase, the combination of new technology and new strategies have allowed us to keep up with the demand. It is likely that newer LC–MS/MS systems will continue to become even better in terms of analyte sensitivity and specificity. The importance of keeping abreast of these new developments in order to seize new opportunities for increasing throughput while still producing high quality data cannot be understated.

References

Ackermann, B.L., Berna, M.J., Murphy, A.T., 2002a. Recent advances in use of LC/MS/MS for quantitative high-throughput bioanalytical support of drug discovery. Curr. Top. Med. Chem. 2(1), 53–66.

Ackermann, B.L., Murphy, A.T., Berna, M.J., 2002b. The resurgence of column switching techniques to facilitate rapid LC/MS/MS based bioanalysis in drug discovery. Am. Pharm. Rev. 5(1), 54–63.

Barbarin, N., Mawhinney, D.B., Black, R., Henion, J., 2003. High-throughput selected reaction monitoring liquid chromatography–mass spectrometry determination of methylphenidate and its major metabolite, ritalinic acid, in rat plasma employing monolithic columns. J. Chromatogr. B Anal. Technol. Biomed. Life Sci. 783(1), 73–83.

Bonfiglio, R., King, R.C., Olah, T.V., Merkle, K., 1999. The effects of sample preparation methods on the variability of the electrospray ionization response for model drug compounds. Rapid Commun. Mass Spectrom. 13(12), 1175–1185.

Cox, K.A., White, R.E., Korfmacher, W.A., 2002. Rapid determination of pharmacokinetic properties of new chemical entities: in vivo approaches. Comb. Chem. High Throughput Screen 5(1), 29–37.

Dams, R., Benijts, T., Gunther, W., Lambert, W., De Leenheer, A., 2002. Sonic spray ionization technology: performance study and application to a LC/MS analysis on a monolithic silica column for heroin impurity profiling. Anal. Chem. 74(13), 3206–3212.

Eddershaw, P.J., Beresford, A.P., Bayliss, M.K., 2000. ADME/PK as part of a rational approach to drug discovery. Drug Discov. Today 5(9), 409–414.

Hsieh, Y., Brisson, J.M., Ng, K., Korfmacher, W.A., 2002a. Direct simultaneous determination of drug discovery compounds in monkey plasma using mixed-function column liquid chromatography/tandem mass spectrometry. J. Pharm. Biomed. Anal. 27(1–2), 285–293.

Hsieh, Y., Brisson, J.M., Ng, K., White, R.E., Korfmacher, W.A., 2001a. Direct simultaneous analysis of plasma samples for a drug discovery compound and its hydroxyl metabolite using mixed-function column liquid chromatography–tandem mass spectrometry. Analyst 126(12), 2139–2143.

Hsieh, Y., Brisson, J.M., Wang, G., Ng, K., Korfmacher, W.A., 2003a. Simultaneous fast HPLC–MS/MS analysis of drug candidates and hydroxyl metabolites in plasma. J. Pharm. Biomed. Anal. 33(2), 251–261.

Hsieh, Y., Bryant, M.S., Brisson, J.M., Ng, K., Korfmacher, W.A., 2002b. Direct cocktail analysis of drug discovery compounds in pooled plasma samples using liquid chromatography–tandem mass spectrometry. J. Chromatogr. B Anal. Technol. Biomed. Life Sci. 767(2), 353–362.

Hsieh, Y., Bryant, M.S., Gruela, G., Brisson, J.M., Korfmacher, W.A., 2000. Direct analysis of plasma samples for drug discovery compounds using mixed-function column liquid chromatography tandem mass spectrometry. Rapid Commun. Mass Spectrom. 14(15), 1384–1390.

Hsieh, Y., Chintala, M., Mei, H., Agans, J., Brisson, J.M., Ng, K. *et al.*, 2001b. Quantitative screening and matrix effect studies of drug discovery compounds in monkey plasma using fast-gradient liquid chromatography/tandem mass spectrometry. Rapid Commun. Mass Spectrom. 15(24), 2481–2487.

Hsieh, Y., Ng, K., Korfmacher, W., 2002c. Development and application of single-column direct plasma injection procedures for drug candidate assays using HPLC–MS/MS. Am. Pharm. Rev. 5(4), 88–93.

Hsieh, Y., Wang, G., Wang, Y., Chackalamannil, S., Brisson, J.M., Ng, K. *et al.*, 2002d. Simultaneous determination of a drug candidate and its metabolite in rat plasma samples using ultrafast monolithic column high-performance liquid chromatography/tandem mass spectrometry. Rapid Commun. Mass Spectrom. 16(10), 944–950.

Hsieh, Y., Wang, G., Wang, Y., Chackalamannil, S., Korfmacher, W.A., 2003a. Direct plasma analysis of drug compounds using monolithic column liquid chromatography and tandem mass spectrometry. Anal. Chem. 75(8), 1812–1818.

Jemal, M., Ouyang, Z., 2000. The need for chromatographic and mass resolution in liquid chromatography/tandem mass spectrometric methods used for quantitation of lactones and corresponding hydroxy acids in biological samples. Rapid Commun. Mass Spectrom. 14(19), 1757–1765.

Jemal, M., Ouyang, Z., 2003. Enhanced resolution triple-quadrupole mass spectrometry for fast quantitative bioanalysis using liquid chromatography/tandem mass spectrometry: Investigations of parameters that affect ruggedness. Rapid Commun. Mass Spectrom. 17(1), 24–38.

Jemal, M., Schuster, A., Whigan, D.B., 2003. Liquid chromatography/tandem mass spectrometry methods for quantitation of mevalonic acid in human plasma and urine: method validation, demonstration of using a surrogate analyte, and demonstration of unacceptable matrix effect in spite of use of a stable isotope analog internal standard. Rapid Commun. Mass Spectrom. 17(15), 1723–1734.

Jemal, M., Xia, Y. Q., 1999. The need for adequate chromatographic separation in the quantitative determination of drugs in biological samples by high performance liquid chromatography with tandem mass spectrometry. Rapid Commun. Mass Spectrom. 13(2), 97–106.

Kaitin, K. (2003). Total Cost to Develop a New Prescription Drug. Tufts Center for the Study of Drug Development. Boston, MA.

King, R., Bonfiglio, R., Fernandez-Metzler, C., Miller-Stein, C., Olah, T., 2000. Mechanistic investigation of ionization suppression in electrospray ionization. J. Am. Soc. Mass Spectrom. 11(11), 942–950.

Korfmacher, W.A., 2003. Lead optimization strategies as part of a drug metabolism environment. Curr. Opin. Drug Discov. Dev. 6(4), 481–485.

Korfmacher, W.A., Cox, K.A., Bryant, M.S., Veals, J., Ng, K., Lin, C.C., 1997. HPLC-API/MS/MS: A powerful tool for integrating drug metabolism into the drug discovery process. Drug Discov. Today 2, 532–537.

Korfmacher, W.A., Cox, K.A., Ng, K.J., Veals, J., Hsieh, Y., Wainhaus, S. *et al.*, 2001. Cassette-accelerated rapid rat screen: a systematic procedure for the dosing and liquid chromatography/atmospheric pressure ionization tandem mass spectrometric analysis of new chemical entities as part of new drug discovery. Rapid Commun. Mass Spectrom. 15(5), 335–340.

Kostiainen, R., Kotiaho, T., Kuuranne, T., Auriola, S., 2003. Liquid chromatography/ atmospheric pressure ionization–mass spectrometry in drug metabolism studies. J. Mass Spectrom. 38(4), 357–372.

Lutz, E.S., Markling, M.E., Masimirembwa, C.M., 2002. Monolithic silica rod liquid chromatography with ultraviolet or fluorescence detection for metabolite analysis of cytochrome P450 marker reactions. J. Chromatogr. B Anal. Technol. Biomed. Life Sci. 780(2), 205–215.

Mallet, C.R., Mazzeo, J.R., Neue, U., 2001. Evaluation of several solid phase extraction liquid chromatography/tandem mass spectrometry on-line configurations for high-throughput analysis of acidic and basic drugs in rat plasma. Rapid Commun. Mass Spectrom. 15(13), 1075–1083.

Matuszewski, B.K., Constanzer, M.L., Chavez-Eng, C.M., 1998. Matrix effect in quantitative LC/MS/MS analyses of biological fluids: a method for determination of finasteride in human plasma at picogram per milliliter concentrations. Anal. Chem. 70(5), 882–889.

McCalley, D.V., 2002. Comparison of conventional microparticulate and a monolithic reversed-phase column for high-efficiency fast liquid chromatography of basic compounds. J. Chromatogr. A 965(1–2), 51–64.

Mei, H., Hsieh, Y., Nardo, C., Xu, X., Wang, S., Ng, K. *et al.*, 2003. Investigation of matrix effects in bioanalytical high-performance liquid chromatography/tandem mass spectrometric assays: application to drug discovery. Rapid Commun. Mass Spectrom. 17(1), 97–103.

Miller-Stein, C., Bonfiglio, R., Olah, T.V., King, R.C., 2000. Rapid method development of quantitative LC–MS/MS assays for drug discovery. Am. Pharm. Rev. 3, 54–61.

Muller, C., Schafer, P., Stortzel, M., Vogt, S., Weinmann, W., 2002. Ion suppression effects in liquid chromatography-electrospray-ionisation transport-region collision induced dissociation mass spectrometry with different serum extraction methods for systematic toxicological analysis with mass spectra libraries. J. Chromatogr. B Anal. Technol. Biomed. Life Sci. 773(1), 47–52.

Newton, C.G., Lockey, P.M., 2003. The importance of early pharmacokinetics. Curr. Drug Discov. April, 33–36.

Paul, G., Winnik, W., Hughes, N., Schweingruber, H., Heller, R., Schoen, A., 2003. Accurate mass measurement at enhanced mass-resolution on a triple quadrupole mass-spectrometer for the identification of a reaction impurity and collisionally-induced fragment ions of cabergoline. Rapid Commun. Mass Spectrom. 17(6), 561–568.

Peng, S.X., Barbone, A.G., Ritchie, D.M., 2003. High-throughput cytochrome p450 inhibition assays by ultrafast gradient liquid chromatography with tandem mass spectrometry using monolithic columns. Rapid Commun. Mass Spectrom. 17(6), 509–518.

Prokai, L., Prokai-Tatrai, K., 1999. Metabolism-based drug design and drug targeting. Pharm. Sci. Technol. Today 2(11), 457–462.

Riley, R.J., Martin, I.J., Cooper, A.E., 2002. The influence of DMPK as an integrated partner in modern drug discovery. Curr. Drug Metab. 3(5), 527–550.

Roberts, S.A., 2001. High-throughput screening approaches for investigating drug metabolism and pharmacokinetics. Xenobiotica 31(8–9), 557–589.

Schuhmacher, J., Zimmer, D., Tesche, F., Pickard, V., 2003. Matrix effects during analysis of plasma samples by electrospray and atmospheric pressure chemical ionization mass spectrometry: practical approaches to their elimination. Rapid Commun. Mass Spectrom. 17(17), 1950–1957.

Shou, W.Z., Naidong, W., 2003. Post-column infusion study of the 'dosing vehicle effect' in the liquid chromatography/tandem mass spectrometric analysis of discovery pharmacokinetic samples. Rapid Commun. Mass Spectrom. 17(6), 589–597.

Sinko, P.J., 1999. Drug selection in early drug development: screening for acceptable pharmacokinetic properties using combined in vitro and computational approaches. Curr. Opin. Drug Discov. Devel. 2(1), 42–48.

Smith, J.H., McNair, H.M., 2003. Fast HPLC with a silica-based monolithic ODS column. J. Chromatogr. Sci. 41(4), 209–214.

Tanaka, N., Kobayashi, H., Nakanishi, K., Minakuchi, H., Ishizuka, N., 2001. Monolithic LC columns. Anal. Chem. 73(15), 420A–429A.

Thompson, T.N., 2000. Early ADME in support of drug discovery: the role of metabolic stability studies. Curr. Drug Metab. 1(3), 215–241.

Thompson, T.N., 2001. Optimization of metabolic stability as a goal of modern drug design. Med. Res. Rev. 21(5), 412–449.

Tiller, P.R., Romanyshyn, L.A., 2002. Implications of matrix effects in ultra-fast gradient or fast isocratic liquid chromatography with mass spectrometry in drug discovery. Rapid Commun. Mass Spectrom. 16(2), 92–98.

Tong, W., Chowdhury, S.K., Chen, J.C., Zhong, R., Alton, K.B., Patrick, J.E., 2001. Fragmentation of N-oxides (deoxygenation) in atmospheric pressure ionization: investigation of the activation process. Rapid Commun. Mass Spectrom. 15(22), 2085–2090.

Tong, X.S., Wang, J., Zheng, S., Pivnichny, J.V., Griffin, P.R., Shen, X. *et al.*, 2002. Effect of signal interference from dosing excipients on pharmacokinetic screening of drug candidates by liquid chromatography/mass spectrometry. Anal. Chem. 74(24), 6305–6313.

van Nederkassel, A.M., Aerts, A., Dierick, A., Massart, D.L., Vander Heyden, Y., 2003. Fast separations on monolithic silica columns: method transfer, robustness and column ageing for some case studies. J. Pharm. Biomed. Anal. 32(2), 233–249.

Wainhaus, S.B., White, R.E., Dunn-Meynell, K., Grotz, D.E., Weston, D.J., Veals, J. *et al.*, 2002. Semi-quantitation of acyl glucuronides in early drug discovery by LC–MS/MS. Am. Pharm. Rev. 5(2), 86–93.

White, R.E., Manitpisitkul, P., 2001. Pharmacokinetic theory of cassette dosing in drug discovery screening. Drug Metab. Dispos. 29(7), 957–966.

Wu, J.T., Zeng, H., Deng, Y., Unger, S.E., 2001. High-speed liquid chromatography/tandem mass spectrometry using a monolithic column for high-throughput bioanalysis. Rapid Commun. Mass Spectrom. 15(13), 1113–1119.

Xu, X., Veals, J., Korfmacher, W.A., 2003. Comparison of conventional and enhanced mass resolution triple-quadrupole mass spectrometers for discovery bioanalytical applications. Rapid Commun. Mass Spectrom. 17(8), 832–837.

Yan, Z., Caldwell, G.W., 2001. Metabolism profiling, and cytochrome P450 inhibition & induction in drug discovery. Curr. Top. Med. Chem. 1(5), 403–425.

Yan, Z., Caldwell, G.W., Jones, W.J., Masucci, J.A., 2003. Cone voltage induced in-source dissociation of glucuronides in electrospray and implications in biological analyses. Rapid Commun. Mass Spectrom. 17(13), 1433–1442.

Yang, L., Amad, M., Winnik, W.M., Schoen, A.E., Schweingruber, H., Mylchreest, I. *et al.*, 2002. Investigation of an enhanced resolution triple quadrupole mass spectrometer for high-throughput liquid chromatography/tandem mass spectrometry assays. Rapid Commun. Mass Spectrom. 16(21), 2060–2066.

Zeng, H., Wu, J.T., Unger, S.E., 2002. The investigation and the use of high flow column-switching LC/MS/MS as a high-throughput approach for direct plasma sample analysis of single and multiple components in pharmacokinetic studies. J. Pharm. Biomed. Anal. 27(6) 967–982.

Zheng, J.J., Lynch, E.D., Unger, S.E., 2002. Comparison of SPE and fast LC to eliminate mass spectrometric matrix effects from microsomal incubation products. J. Pharm. Biomed. Anal. 28(2), 279–285.

Identification and Quantification of Drugs, Metabolites and Metabolizing
Enzymes by LC–MS
Swapan K. Chowdhury, editor.

Chapter 3

COMMONLY ENCOUNTERED ANALYTICAL PROBLEMS AND THEIR SOLUTIONS IN LIQUID CHROMATOGRAPHY/ TANDEM MASS SPECTROMETRY (LC/MS/MS) METHODS USED IN DRUG DEVELOPMENT

Tapan K. Majumdar

3.1. Introduction

High-throughput (HT) and parallel approaches for the analysis of new chemical entities are being increasingly adapted in recent years in pharmaceutical research and development (Majumdar *et al.*, 2001, 2004a, b; Bakhtiar *et al.*, 2002; Deng *et al.*, 2002; Hsieh *et al.*, 2002; O'Connor *et al.*, 2002). The increased selectivity, sensitivity, and speed available in LC/MS/MS make it the method of choice for pharmaceutical research and development. Compatibility of atmospheric pressure ionization (API) LC/MS/MS (Fenn *et al.*, 1989; Bruins *et al.*, 1991, 1998; Thomson, 1998; Niessen, 1999) to wide mass and concentration ranges rendered it a very useful tool for sample analysis in drug discovery, pre-clinical and clinical safety, and efficacy studies of new compounds. Because of the rapid pace of drug development, the methods used for the analysis of clinical and toxicokinetic samples require rapid turn around. With appropriate optimization of the different components of an LC/MS or LC–MS/MS system, a vast majority of analytical problems can be solved. In simple cases, such as molecular weight determination, proper calibration of the mass analyzer alone is sufficient. However, when quantitative analysis of trace concentrations of drugs and metabolites is needed in a biomatrix, many variables can dictate the success of the method. These variables include the stability of the samples, cleanliness of the material to be injected, LC separation, number of samples to be analyzed, number of compounds per analysis, lower limit of quantitation (LLOQ), chromatographic resolution, mass resolution, and time allowed for the assay.

New technologies for sample preparation and chromatographic separation are now available for rapid bioassays. The problems and solutions displayed in this presentation are typical during the development of the method for HT analysis in complex biomatrices. A typical HT setup for preparation and analysis of samples from biomatrices is shown in Fig. 1.

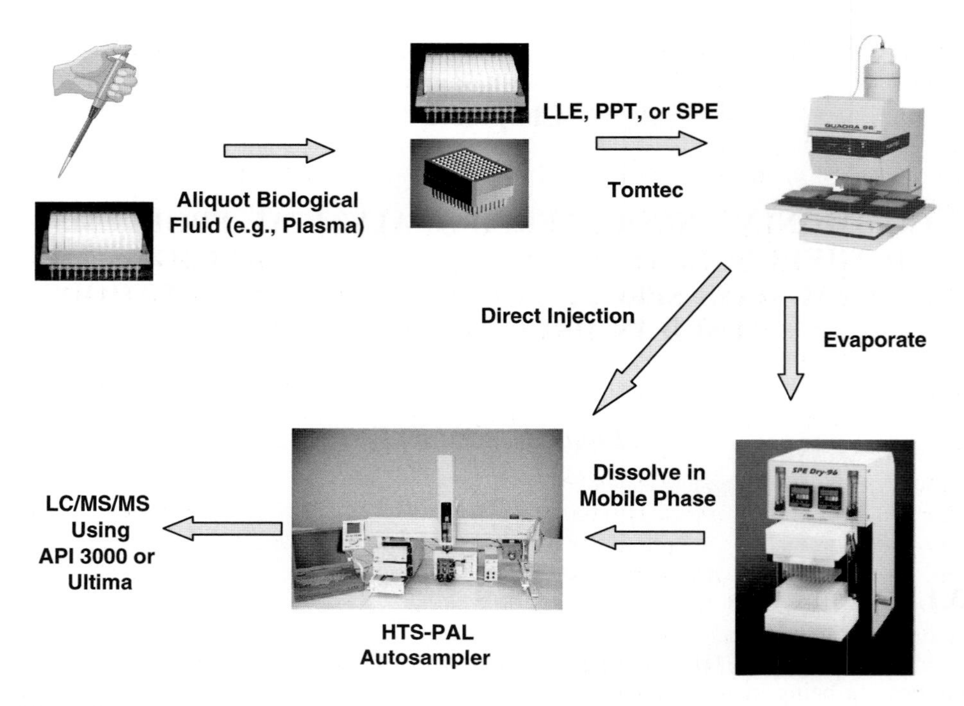

Figure 1.
Typical setup for HT sample analysis in biomatrices

All the traditional methods such as (a) dilute and inject, (b) protein precipitation (PPT), (c) liquid–liquid extraction (LLE), and (d) solid-phase extraction (SPE) can be performed using this setup for rapid sample preparation. In this chapter, problems are discussed in the sequence they are encountered in the bioanalytical laboratories involved in drug development. For example, problems with sample collection and stability, liquid chromatography, and finally mass spectrometry.

3.2. Commonly Encountered Analytical Problems and Solutions

3.2.1. Problems with Sample Collection and Analyte Stability

Analytical problems often begin with sampling and the stability of compounds and metabolites during storage. To avoid problems at this stage, the analytical chemist must devise appropriate sampling procedure. The choice must be made as to whether whole blood, plasma, serum, or urine is to be analyzed. Although plasma and serum are the most commonly sampled body fluids, for some compounds it is more appropriate to monitor their concentrations in whole blood because of time or concentration-dependent binding of the analytes to erythrocytes. For example, analysis of whole blood is needed for the assessment of chlorthalidone or cyclosporin exposure (Brignol *et al.*, 2001).

Upon collection of blood, serum can be collected as a supernatant after clotting. Serum contains no fibrinogen, which is a protein needed in the clotting process. The blood coagulation process takes ~ 30 min at room temperature and the analytical chemist must ascertain the stability of the analytes for this process at room temperature. Alternatively, blood can be collected in containers having an anticoagulant and centrifuged for the collection of plasma as the supernatant. In the separation of plasma, the temperature can have a significant effect on the stability of the analytes.

The type of sampling tube and the nature of the anticoagulant (citrate, Li-heparin, Na-heparin, or ethylenediamine tetra-acetic acid (EDTA)) can have a remarkable effect on the quantitation of analytes (Mei *et al.*, 2003; Suttnar *et al.*, 2001). Li-heparin was found to produce a significant matrix effect (Mei *et al.*, 2003). EDTA plasma was found to exert a higher matrix effect than citrate plasma (Suttnar *et al.*, 2001). Certain plastic tubes used for sample collection were found to introduce matrix interference (Mei *et al.*, 2003). Use of the same brand plastic tubes for sampling, processing, and storage of standard and quality control (QC) samples are recommended (Mei *et al.*, 2003). For some compounds a stabilizer or an enzyme inhibitor may be added to the sampling tube to stop or minimize degradation of the analytes. For example, esters and lactones are unstable due to the presence of esterase in the blood and esterase inhibitors such as sodium fluoride type anticoagulant must be added to the sampling tubes (Jacob and Herbert, 1977).

The typical temperatures used for long-term storage of samples are −20 and −80°C. The long-term storage temperature is dependent on the stability of the analytes and is usually established during the method validation. Low temperatures stop or minimize enzyme (e.g., esterase) activities, thus improve compound stability. For example, theophyline is stable in blood for 1 day at room temperature and 1 month at −20°C (Jonkman *et al.*, 1982). Samples should be frozen in dry ice in sealed containers during shipment to ensure stability. If the samples are to be stored for an appreciable length of time after collection and prior to analysis, long-term storage stability data must be established at the storage temperature for the duration of storage.

Room temperature and freeze–thaw stability of analytes must be established to ensure stability of analytes during sample processing. Thawing of frozen samples must be done with care since localized concentration effects may occur during this process. Frozen samples should be thawed gradually at room temperature without the application of heat. Heating during this process may promote the degradation of analytes. Thawed samples should be vortex-mixed for at least 10 s to ensure sample homogeneity. Centrifugation of the thawed plasma samples is important prior to direct injection to an online extraction column or pre-concentration column, since fibrinogen may have produced fibrin, which could clog the column or pipette tips (such as in Tomtec Quadra-96 sample handler) or the tubing system. Thawed serum samples can be processed directly without centrifugation. Sample tubes must be labeled accurately to avoid any confusion later.

3.2.2. Problem with Positive Controls

Finding quantifiable concentrations of analytes in the method blanks as well as study control samples (especially from toxicokinetic studies) is a common problem in the bioanalytical laboratory. This phenomenon raises questions about the validity of the bioanalytical method and may invalidate expensive toxicological studies. Positive control occurs due to sample contamination in the bioanalytical laboratory, during sample collection, or due to mis-dosing of study animals or subjects. Contaminations in the bioanalytical laboratory can be minimized by careful handling of the samples, calibration standards, and QC samples. Pipeting must be done with extra care to avoid cross contamination. For semi-automated 96-well sample handlers such as Tomtec Quadra-96, caution must be exercised to avoid the dripping of liquids from the pipette heads during the robotic movements. Injector wash procedure must be optimized to minimize contamination due to carryover as discussed in Section 3.2.3. Extra care must be taken in handling the sample tubes during sampling as well as during analysis in the laboratory to avoid any extraneous contamination with analytes from samples or from standards. Positive controls due to contamination outside the laboratory such as during sample collection or mis-dosing of subjects can be confirmed by reanalysis of the contaminated controls. Presence of metabolites at appropriate ratios in the positive controls can also be used as an indication of contamination due to mis-dosing of the control subjects.

3.2.3. Carryover Problem

Carryover is a common problem in liquid chromatographic methods. This problem occurs due to the retention of analytes by adsorption at the active surfaces of the autoinjector system, solvent lines, or in the analytical column. The carryover is also dependent on the types of analytes. This problem is most commonly visible after the injection of analytes at high concentrations. Analyte peaks observed in a solvent blank injected subsequently are due to carryover of analytes from the prior injection. Most carryover problems in the autoinjector system can be minimized by the appropriate choice of injector wash solutions and wash methods in the autoinjector system. The carryover in solvent lines can be minimized by the choice of mobile phase and use of polyetheretherketone (PEEK) tubing instead of metal tubing. Only limitation of PEEK tubing is that it cannot stand high pressure. Carryover problem in the column can be minimized by the appropriate choice of mobile phase and/or stationary phase. Limiting the injection of samples containing very high concentration of analytes for compounds having severe carryover problem is also recommended. An example of carryover problem is shown for AEE788 (an inhibitor of receptor tyrosine kinase) and its metabolite AQM674 (Figs. 2–4). Figure 2 shows the LC/MS/MS chromatograms for AEE788 (1000 ng/ml), the IS, and AQM674 (1000 ng/ml) in human serum extract. Immediately following the injection of this sample a solvent wash was injected through the autosampler. The chromatograms of the solvent wash are shown in Fig. 3. Carryover peaks observed in Fig. 3 was ~50 times higher than the method LLOQ (0.5 ng/ml) for both AEE788 and AQM674. The injector wash solutions during this method were 100% acetonitrile (wash 1) and 50% methanol in water (wash 2).

Troubleshooting of the method indicated that the carryover was from the autoinjector. The carryover problem was solved by the addition of 0.04% trifluoroacetic

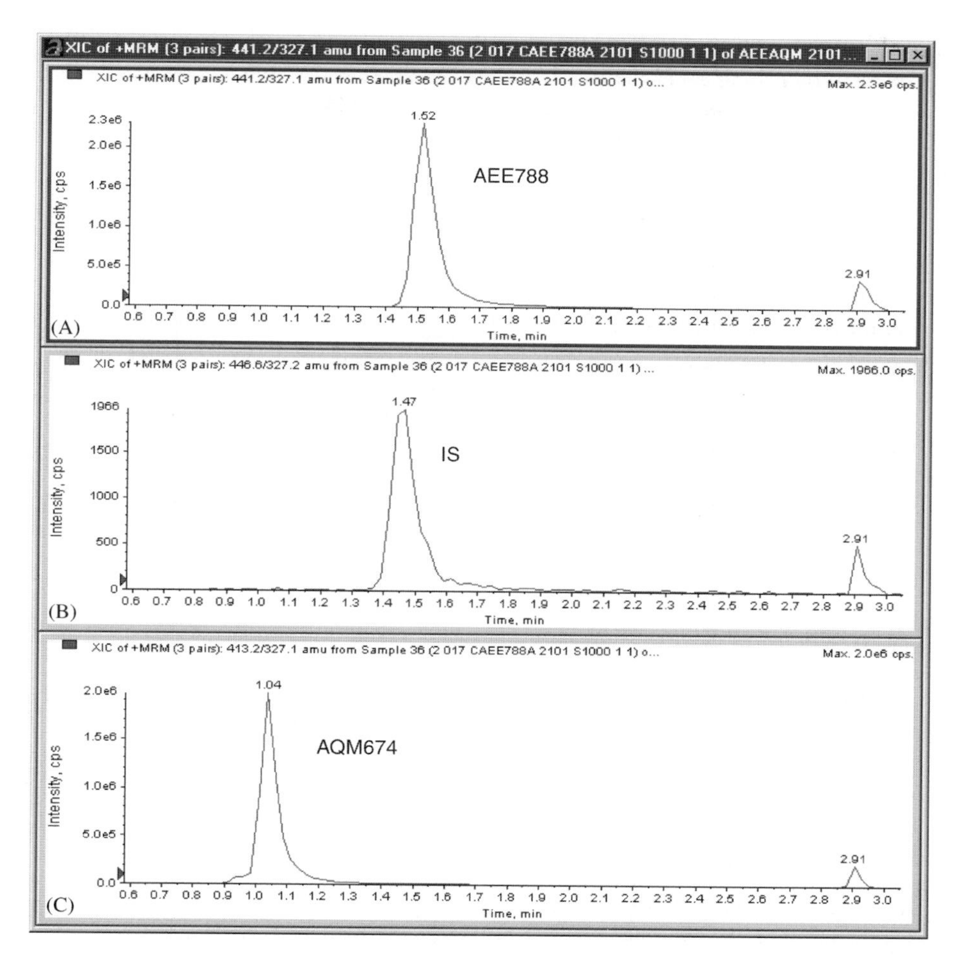

Figure 2.
LC–APCI/MS/MS ion chromatograms of AEE788, the IS, and the metabolite AQM674. Ion Chromatograms for AEE788 (1000 ng/mL) (A), the IS (100 ng/mL) (B), AQM674 (1000 ng/mL) (C). Retention time (min) is shown on the horizontal axis and the peak intensity is shown in the vertical axis

acid to the injector wash 2. With the new injector wash solution the process of injecting the 1000 ng/ml sample following the solvent wash was repeated. Figure 4 shows that the solvent wash is free from carryover peaks.

Another example of a severe carryover problem that was observed and described involves Novartis compound LAG078, a lipid modulator (Majumdar *et al.*, 2004a, b). It was noted during the method development that LAG078 was strongly retained in the injector as well as in the column after every injection of LAG078 solution (50% methanol in water) having concentrations higher than 5 ng/ml (Majumdar *et al.*, 2004a, b). This method needed a LLOQ of 0.05 ng/ml due to low-level dosing

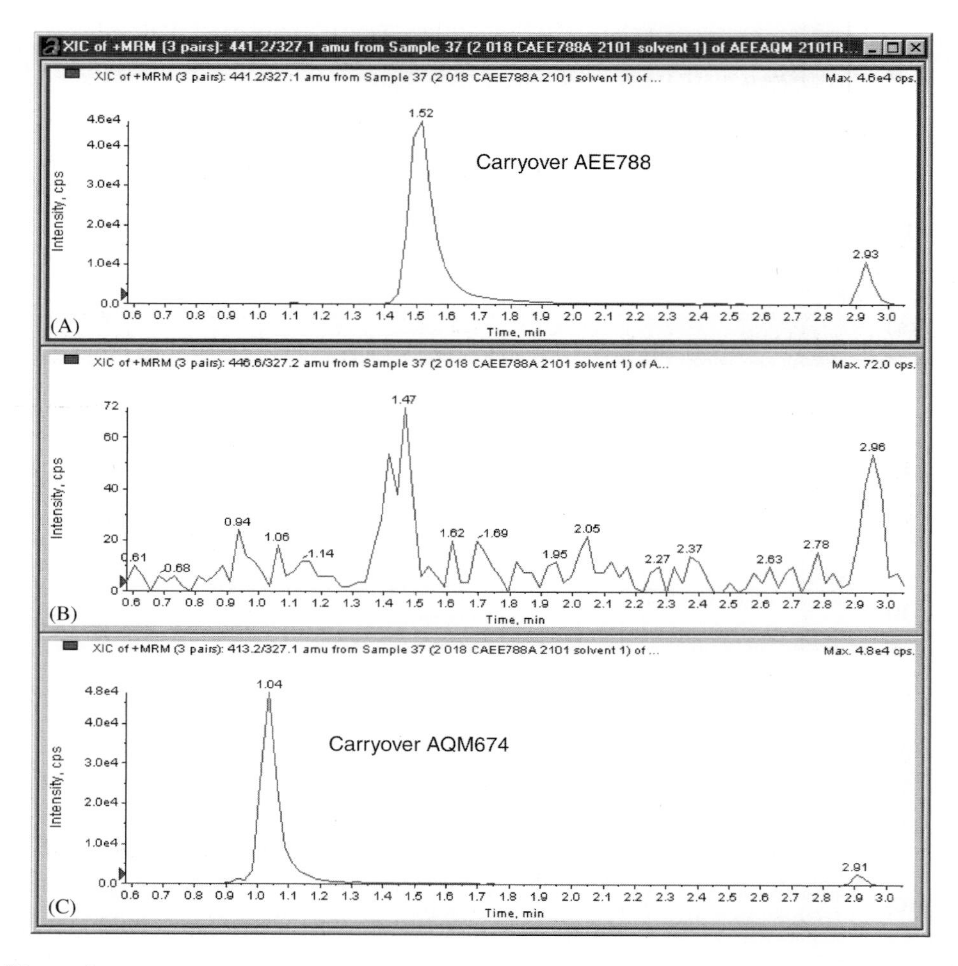

Figure 3.
LC–APCI/MS/MS ion chromatograms from the solvent wash. The carryover peaks of AEE788 (A), the IS (B), AQM674 (C). Retention time (min) is shown on the horizontal axis and the peak intensity is shown in the vertical axis

of the compound. The LAG078 peak from the injector carryover was observed in thesubsequent solvent blank. The problem was affecting the specificity of the method by adding LAG078 contamination in the low concentration (≤ 5 ng/ml) samples. The carryover problem was eliminated by adding ammonium hydroxide solution in the injector wash (Majumdar *et al.*, 2004). The autoinjector syringe, and the injector valve were washed sequentially three times with wash 1 (acetonitrile:methanol:ammonium hydroxide (80:19:1, by vol.)) and wash 2 (methanol:water (50:50, v/v)). The injector wash procedure eliminated the carryover of LAG078 in the injector to the extent that no LAG078 peak was observed in the solvent wash injected following the injection

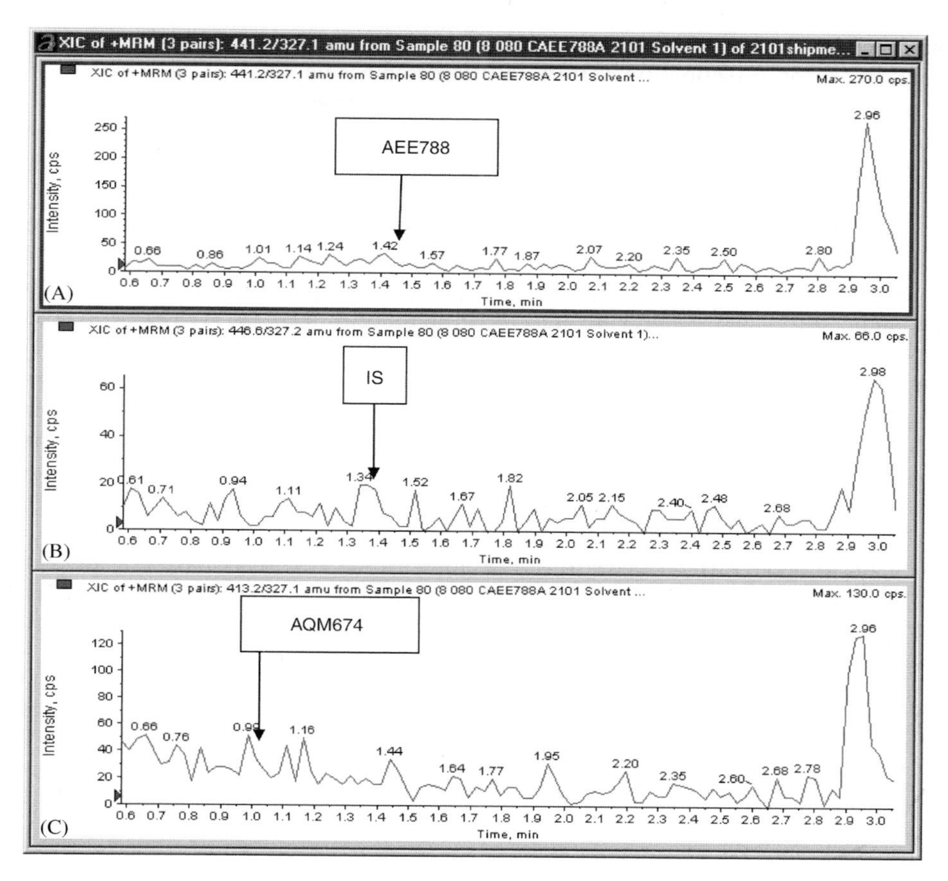

Figure 4.
LC–APCI/MS/MS ion- chromatograms from the solvent wash after adding 0.04% TFA acid to injector wash 2. (A), (B) and (C) are the MS/MS ion chromatograms of AEE788, the IS, and AQM674, respectively. Retention time (min) is shown on the horizontal axis and the peak intensity is shown in the vertical axis

of the 50 ng/ml standard (upper limit of quantification). The column carryover of LAG078 was solved by the use of ammonium hydroxide in the mobile phase at pH 10. Most silica-based LC columns have an upper pH limit of 8 for their operation. For LAG078, a polyvinyl alcohol-based rigid polymer (stable at pH 13) column was used (Majumdar *et al.*, 2004a, b).

3.2.4. Problem with Retention Time Shift in HPLC

Shifting of peaks in HPLC is a traditional problem and could be due to a number of reasons. Unstable pressure in the LC pumps resulting from the pump malfunction or due to trapped air bubbles in the system are the most common contributors of retention time shift. Malfunctioning pumps can be repaired and priming the pumps and solvent lines will remove trapped air bubble. Use of an online degasser and/or the use of degassed solvents will ensure that air bubbles will not appear in the LC system. If retention time shift still exists then appropriate buffering of the mobile phase will mitigate the problem. An example of retention time shift is shown in Fig. 5. In this case, the chromatographic separation of a compound and the internal standard (IS) was performed using a C_{18} (2.1 × 50 mm) column at 45°C. The mobile phase was an isocratic flow of aqueous methanol (45%) and water with 0.01% trifluoroacetic acid (TFA) (55%) at a rate of 0.16 ml/min. The problem was that the retention time of the compound and the IS were changing randomly within a time window of ~ 0.5 min for each peak (Fig. 5). Although the relative retention time for each compound was the same, this random time shift was undesirable and made it difficult to set a narrow time window for data acquisition and automatic peak integration. This problem with random shifting of the chromatographic retention time was resolved by buffering the mobile phase with ammonium acetate. The retention times became stable (Fig. 5) when the mobile was changed to an isocratic flow of methanol (65%) and water (35%) containing 0.04% TFA and 20 mM ammonium acetate. However, buffering the mobile phase also resulted in a slight decrease in the ion intensities, which is discussed in the next section.

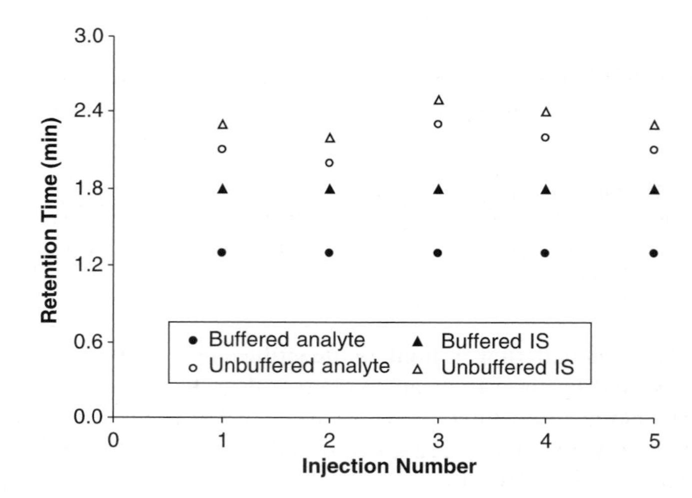

Figure 5.
Effect of buffering of the mobile phase on the retention time of analytes. The analytes and IS showed unstable retention time in unbuffered mobile phase (45% methanol in water with 0.01% TFA). The retention time became stable when the mobile phase was changed to 65% menthol, 0.04% TFA and 20 mM ammonium acetate in water

3.2.5. Problems Related to the Mobile Phase Composition and Ionization of Analytes

A further complicating factor is that the ionization efficiencies of analytes in the API interface is dependent on the composition of the LC mobile phase in both positive and negative ionization modes. A high percentage of organic solvents (e.g., methanol, acetonitrile, etc.) in the mobile phase facilitates the desolvation process in the API interface and increases the ionization efficiencies. For acidic and basic analytes, the pH of the mobile phase plays a significant role in the ionization process (Majumdar *et al.*, 2004a, b; Jemal *et al.*, 1998; Zhao *et al.*, 2002; Law, 2004). For acidic compounds, it was shown that ion intensities gradually decreased in the electrospray ionization (ESI) interface with the increase in the concentration of acidic modifier (e.g., formic acid) in the mobile phase (Jemal *et al.*, 1998). This phenomenon can be explained by the fact that the equilibrium in the ionization of acid (i.e., $AH \leftrightarrow A^- + H^+$) was shifted to the left due to contributing protons from the acid in the mobile phase. As a result a decrease in the number of ionized form of the analyte molecule occurred. On the other hand, increase in the mobile phase pH increases the ion intensities of acidic analytes (Majumdar *et al.*, 2004a, b). Concentration of buffers in the mobile phase also has marked effect on the intensities of analytes in both negative and positive ESI mode. Intensities of acidic analytes were found to decrease gradually with the increase of salt concentration (e.g., ammonium formate) in the mobile phase (Jemal *et al.*, 1998). For basic analytes, the ion intensities in positive ESI mode was decreased with the increase of TFA and adding 20 mM ammonium acetate in the mobile phase mentioned in Section 3.2.4.

3.2.6. Problems with Ion Suppression and Matrix Effect in the API Interface

This is a common problem in API mass spectrometry. There are differences in opinion as to the amount of ion suppression that is acceptable for an analytical method. In some laboratories, ion suppression in an analytical method is not acceptable but in other laboratories it is acceptable if there is no significant effect on the validity of the analytical data with appropriate QC. Ion suppression at the API interface due to co-eluting substances in the matrix such as polar inorganics, fatty acids, amino acids, amines, triglycerides, plasticizers, and other organic substances has been documented (Mei *et al.*, 2003; Matuszewski *et al.*, 1998a, b; Bonfiglio *et al.*, 1999; King *et al.*, 2000; Hsieh *et al.*, 2001; Muller *et al.*, 2002; Avery, 2003). ESI is more likely to suffer from ion suppression than atmospheric pressure chemical ionization (Matuszewski *et al.*, 1998a, b; Bruins *et al.*, 1987; Ikonomou *et al.*, 1991; Kebarle and Teng, 1993; Banks *et al.*, 1994; Kostiainen and Bruins, 1994; Kebarle and Yeunghaw, 1997, p. 3; Tiller and Romanyshyn, 2002). The phenomenon of ion suppression results in reduction of signal intensity. As a result, the LLOQ for highly sensitive bioanalytical methods could be difficult to achieve. The problem can be further complicated by differential ion suppression in intra-patient samples and the standards, which are typically made in blank biomatrix. Differences in ion suppression between the analytes and chemically different ISs may also be problematic (Avery, 2003). This problem can be mitigated by the use of stable-isotope labeled analogues as ISs. The extent of ion suppression is dependent on the methods for sample preparation (King *et al.*, 2000) and chromatographic separation. The extracts produced by protein precipitation method (O'Conor *et al.*, 2002) are most likely to cause ion suppression in ESI due to the dirty nature of the extract (Matuszewski *et al.*, 1998a, b). The extracts obtained from SPE or LLE are much cleaner.

Ion suppression can be assessed at the retention time of the analytes (Matuszewski *et al.*, 1998a, b; Bonfiglio *et al.*, 1999; King *et al.*, 2000; Hsieh *et al.*, 2001). The most widely used method consists of post-column addition of analytes to the LC-eluent, which is described in Fig. 6. Using the setup (Fig. 6), an analyte (compound A) and the IS were infused (flow rate 10 μl/min) using a syringe pump through a "T" connection between column eluent (flow rate 200 μl/min) and the ESI interface of the mass spectrometer. An extract from blank rat serum was injected through the column while the infusion pump was on and the eluent from the analytical column mixed with the infused analytes entered the ESI source of the mass spectrometer. The infusion chromatograms of compound A and the IS are shown in Fig. 7. The LC–ESI/MS/MS chromatograms in Fig. 8 shows 1.78 and 1.52 min as the respective retention times for compound A and the IS. Ion suppressions are shown as lower intensities in the infusion chromatograms (Fig. 7c and d). It is clear from Fig. 7 that ion suppression did not occur at the retention time of compound A and the IS. Therefore, the method is assumed to be free from ion suppression in the ESI interface. If the ion suppressions occurred during the retention times of the analytes,

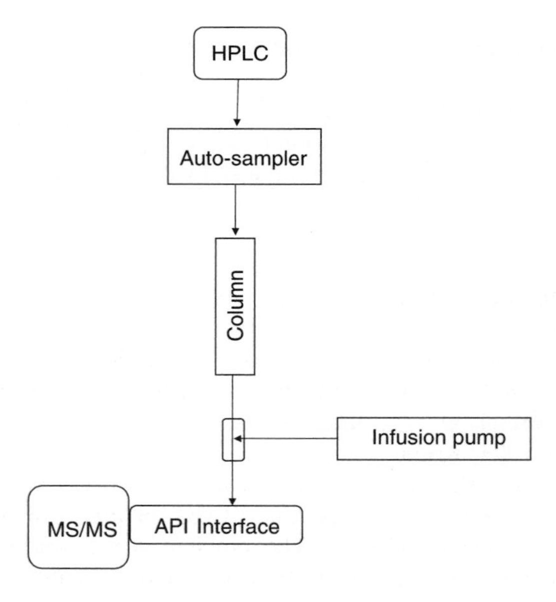

Figure 6.
Post-column infusion system for the determination of ion suppression in the API interface

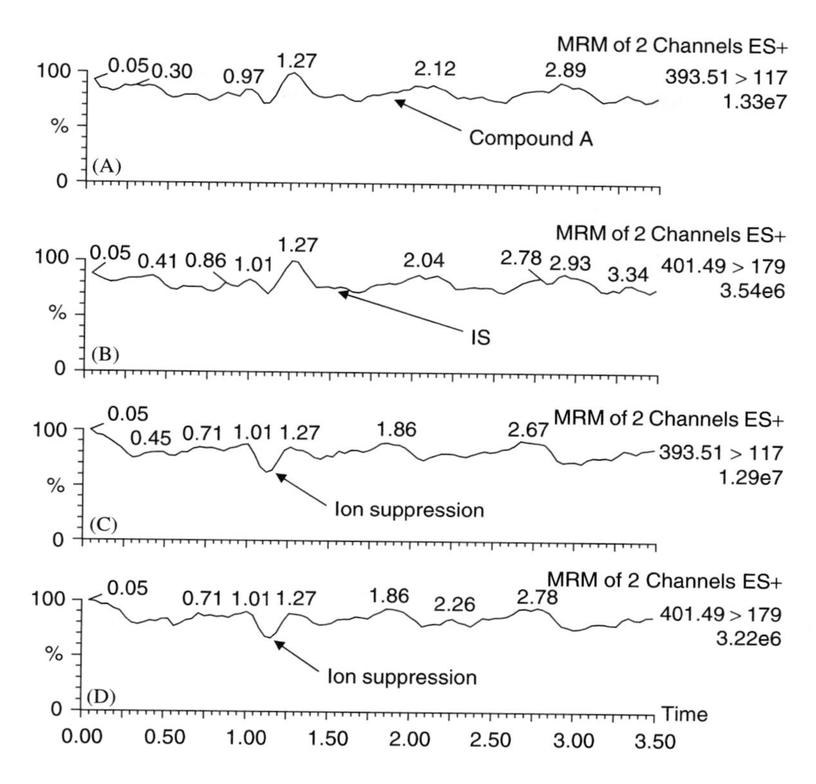

Figure 7.
Infusion LC–ESI/MS/MS chromatograms of compound A and IS: (A) LC–ESI/MS/MS of compound A without injection of matrix (rat serum) blank; (B) LS–ESI/MS/MS of IS without injection of matrix blank; (C) LC–ESI/MS/MS of compound A with the injection of matrix blank; (D) LC–ESI/MS/MS of IS with the injecion of the matrix blank. Retention time (min) is shown on the horizontal axis and peak intensity is shown in the vertical axis

then a change in the LC method may be required to move the peaks away from the ion suppression times.

Another way of minimizing the ion suppression is the use of an optimized sample preparation method in order to obtain a cleaner extract. Finally, an optimized chromatographic method can also yield sharp peaks with minimal ion suppression and without matrix interference.

Matrix effect, which is intrinsically linked to ion suppression, can be further quantified by the comparison of extraction recoveries from spiking solution (50% methanol in water), pre-spiked (i.e., spiked into the matrix prior to extraction), and post-spiked (i.e., spiked into the extract from the matrix) analytes in the biomatrix. An example is shown in Table 1, where blank human serum was pre-spiked with the analyte AQM674 (Section 3) in triplicate and extracted with blank human serum.

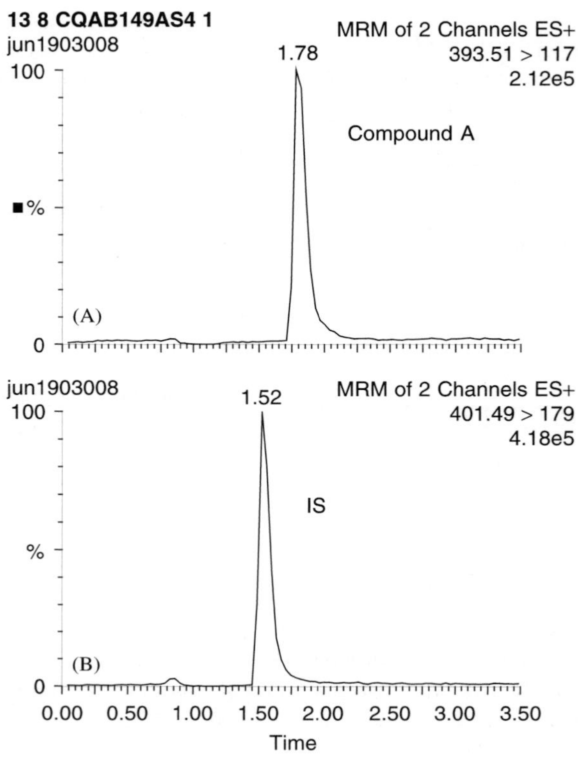

Figure 8.
LC–ESI/MS/MS ion chromatograms of blank rat serum spiked with compound A and the IS. Ion chromatograms for compound A (A), the IS (B). Retention time (min) is shown on the horizontal axis and the peak intensity is shown in the vertical axis

The extracts together with triplicate of spiking solution were dried under nitrogen gas flow. All the tubes were reconstituted with same volume of a solvent (e.g., 50% methanol in water). The reconstituted extracts were then injected into the LC–ESI/MS/MS system and the signal intensities are shown in Table 1. The ratio of the peak intensities in the post-spiked samples and the spiking solution indicates matrix effect only. Typical recoveries are $100 \pm 20\%$. Table 1 shows a recovery of 94.6% indicating a small matrix effect. The ratio of the peak intensities in the pre-spiked and the analytes in the spiking solution provides method recoveries in which both extraction loss and matrix effect are taken into account. Table 1 shows a recovery of 47% indicating a substantial extraction loss and possibly some matrix effect. The ratio of the peak intensities in the pre- and the post-spiked samples is a measure of the extraction loss only. From the data in Table 1, a recovery of 49.7% can be calculated (not shown in Table 1), which indicates a significant extraction loss. Therefore, based on the data provided in Table 1, the method suffered from significant extraction loss and there was a little matrix effect.

Table 1.
Evaluation of matrix effect for AQM674 by comparing LC–MS/MS signal intensities of the spiking solution and adding the same solution to human serum prior to extraction (pre-spiking) and to the extract of the same serum (post-spiking)

Compound	Pre-spiking (0.5 ng/ml)	Post-spiking (0.5 ng/ml)	Spiking solution (0.5 ng/ml)
AQM674	2882	6545	6121
	3427	7342	7650
	3379	5569	6772
Mean peak area	3229	6485	6858
± S.D.	302	888	101
C.V.(%)	9.4	13.7	1.5
Recovery (%)	47.0	94.6	100

3.2.7. Problems with Stable Isotope Labeled IS

A stable isotope labeled IS is very important for a robust quantitative method in the bioanalytical laboratories. The IS follows the same extraction loss, ionization efficiency, and MS/MS reactions as the analyte thereby providing constant ratio of analyte/IS at a specific concentration of analyte. However, high isotopic purity of the stable isotope labeled IS is very important for a quantitative method. The IS should not have significant contribution to the MS/MS channel of the analyte. Typically, the contribution should be <20% of the LLOQ of the analyte. An example is shown in Fig. 9 in which the stable isotope labeled IS ($[C^{13}]_6C_{16}H_{18}FNO_7S$) from MS/MS channel (B) contributed to the MS/MS channel (A) of the analyte LAG078 ($C_{22}H_{18}FNO_7S$) at the same retention time. This contribution was due to a small amount (0.9%) of unlabeled LAG078 present in the IS as isotopic impurity. It is important to keep this contribution to a minimum (<20% of LLOQ) so that the LLOQ of the method was not affected significantly. In this method, the plasma concentration of the IS was 2 ng/ml. Ion chromatograms of LAG078 at the LLOQ (0.01 ng/ml) and the IS (2 ng/ml) are shown in Fig. 10. From Figs. 9 and 10, it is clear that the contribution of the IS was very low compared to the signal intensity of LAG078 at the LLOQ level.

There are also cases, where the main analyte can contribute significantly to the stable isotope-labeled IS peak due to the naturally abundant isotopic patterns in the molecules. During quantitative analysis, the IS concentration is kept constant and the analyte concentration is varied within the range of the calibration curve. If the analyte at low concentration has significant contribution to the IS peak, the contribution can get worse with increase of analyte concentration in a calibration curve leading to gradual lowering of the analyte/IS ratio. This phenomenon can lead to inaccurate quantitation. A typical example is shown in Figs. 11 and 12. Figure 11 shows the

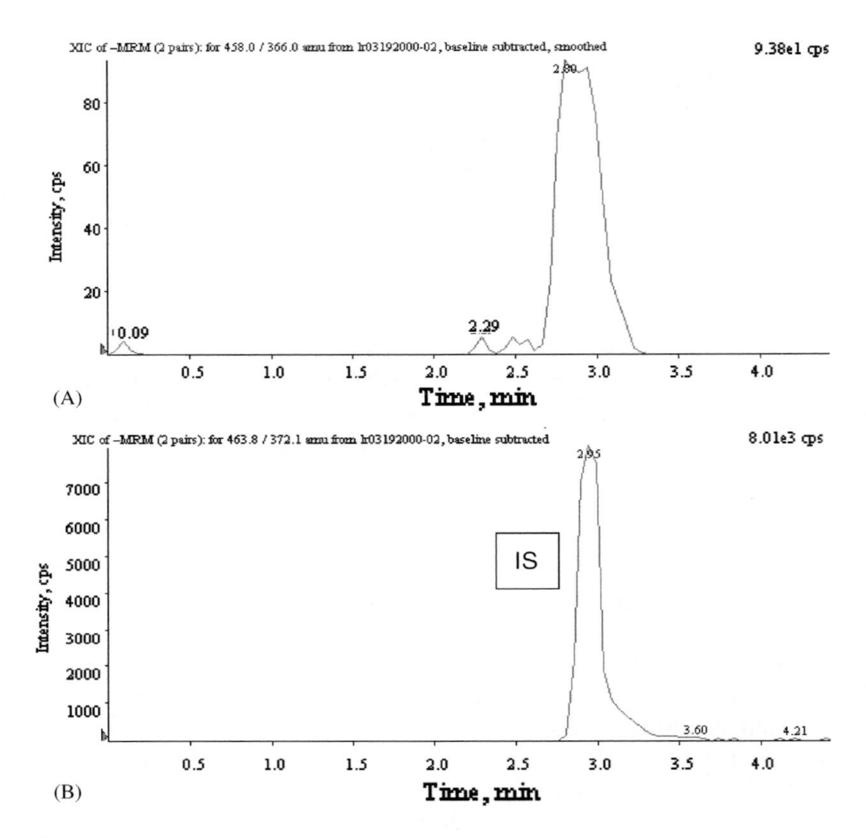

Figure 9.
LC–ESI/MS/MS ion chromatograms of blank human plasma spiked with the internal standard (2 ng/ml). Ion chromatograms for LAG078 ($C_{22}H_{18}FNO_7S$) (A), the IS ($^{13}C_6C_{16}H_{18}FNO_7S$) (B). Retention time (min) is shown on the horizontal axis and peak intensity is shown in the vertical axis

ion chromatograms of sample, where blank dog serum was spiked with the stable isotope-labeled IS ($C_{21}[H^2]_4$ $H_{21}NO_4S$) at a final concentration of 25 ng/ml. The figure clearly shows that the IS has no contribution to the analyte peak (Fig. 11, panel A). Figure 12 shows the ion chromatogram of the analyte LBL752 ($C_{21}H_{25}NO_4S$) spiked in blank dog serum at a final concentration of 2500 ng/ml (mid level in the calibration curve). No IS was spiked in this sample. A huge peak is clearly visible at the retention time of the IS in the MS/MS channel of the IS (Fig. 12, panel B). This contribution occurred due to the natural abundance of the S^{36} isotope present in the analyte molecule. The take-home lesson here is that the molecular weight of the stable isotope-labeled IS should not coincide with the significant naturally abundant isotopes present in the analyte molecule.

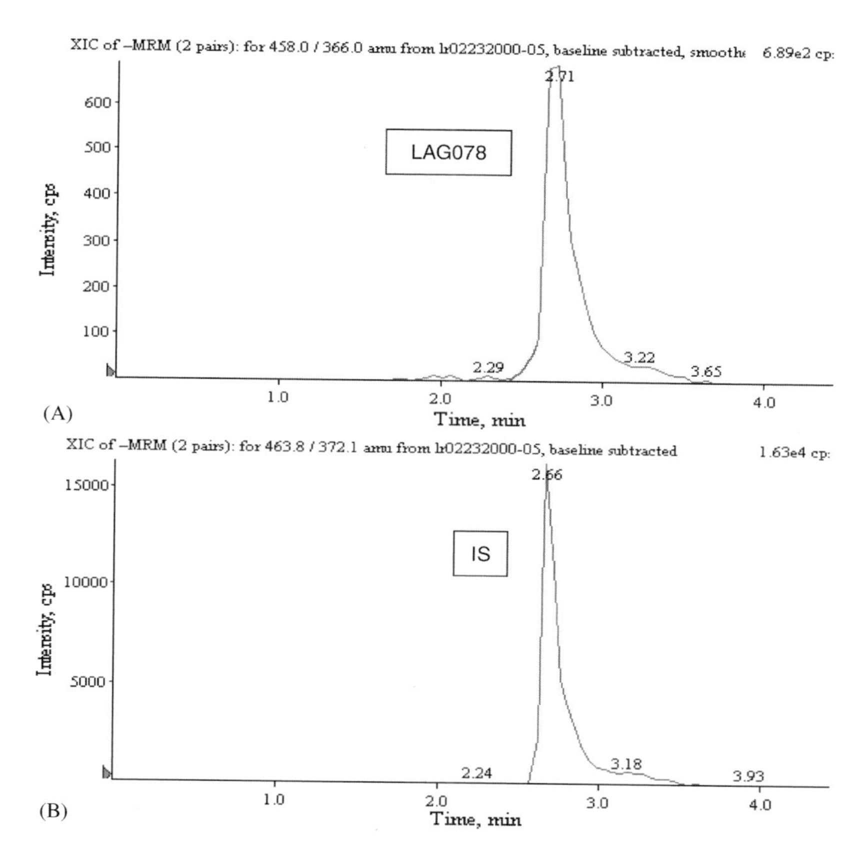

Figure 10.
LC–ESI/MS/MS ion chromatograms of blank human plasma spiked with LAG078 ($C_{22}H_{18}FNO_7S$) at a final concentration of 0.01 ng/ml (LLOQ) (A) and the IS ($^{13}C_6C_{16}H_{18}FNO_7S$; 2 ng/ml) (B). Retention time (min) is shown on the horizontal axis and peak intensity is shown in the vertical axis

3.2.8. Problem of Matrix Interference Due to the Presence of Metabolites or Other Endogenous Compounds

This problem may be observed during the quantification of parent drug and active metabolites in biomatrices in subject samples (Majumdar *et al.*, 2004a, b; Matuszewski *et al.*, 1998a, b; Jemal and Xia, 1999). Method validation is usually done by spiking standard and QC spiking solutions in blank matrices that are usually pre-screened for endogenous metabolites and related interferences. However, in subject samples these interferences may be present, thus creating analytical problems as co-eluting peaks (Majumdar *et al.*, 2004a, b; Matuszewski *et al.*, 1998a, b; Jemal and Xia, 1999). Some of these interferences may be isobaric to the analytes of interest and would not be resolved using MS/MS. Conjugated metabolites (Section 3.2.10) are a potential

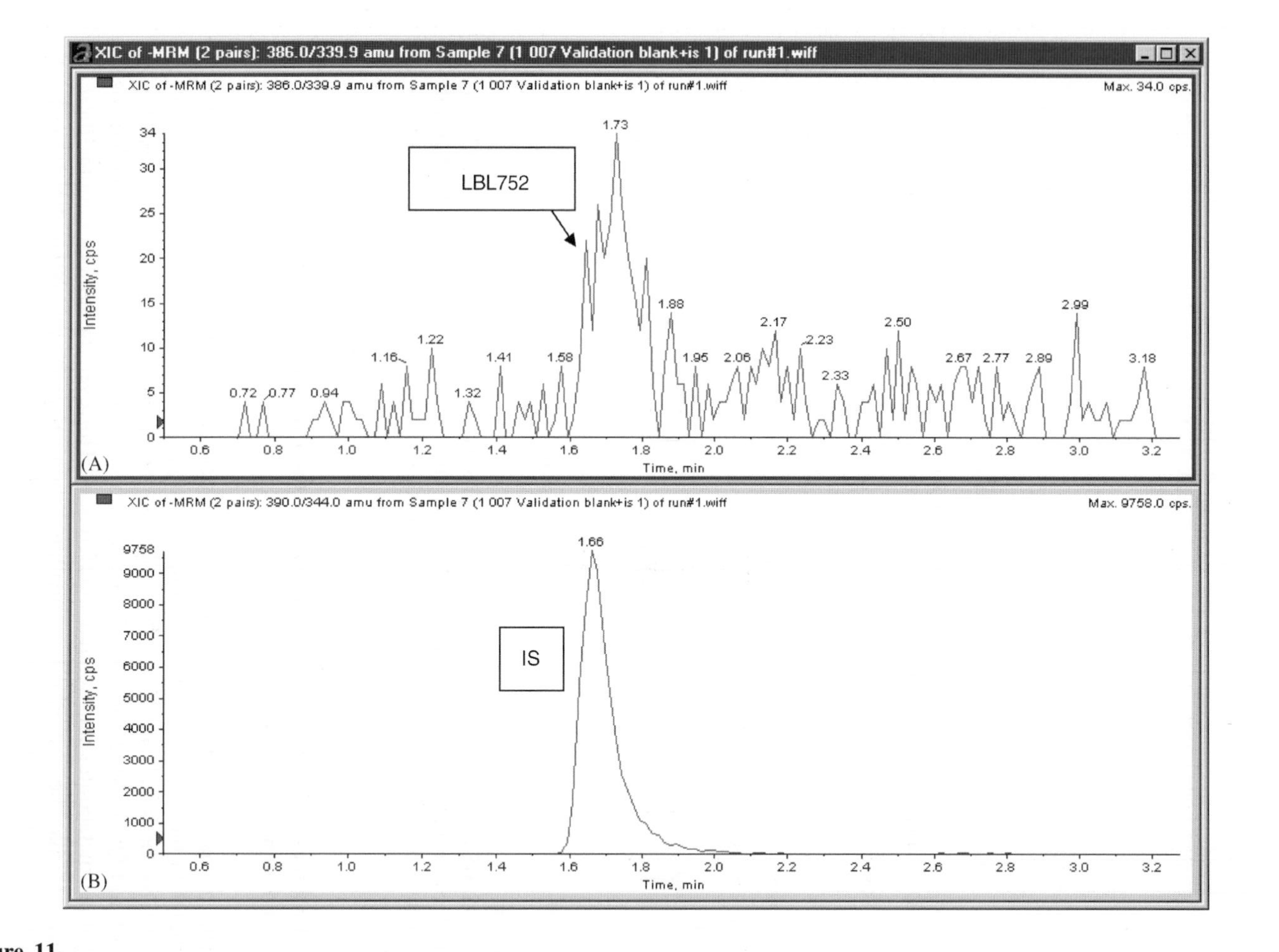

Figure 11.
LC–ESI/MS/MS ion chromatograms in blank dog serum spiked with the internal standard (25 ng/ml). Ion chromatograms for LBL752 ($C_{21}H_{25}NO_4S$) (A), the IS ($C_{212}H_4H_{21}NO_4S$) (B). Retention time (min) is shown on the horizontal axis and peak intensity is shown in the vertical axis

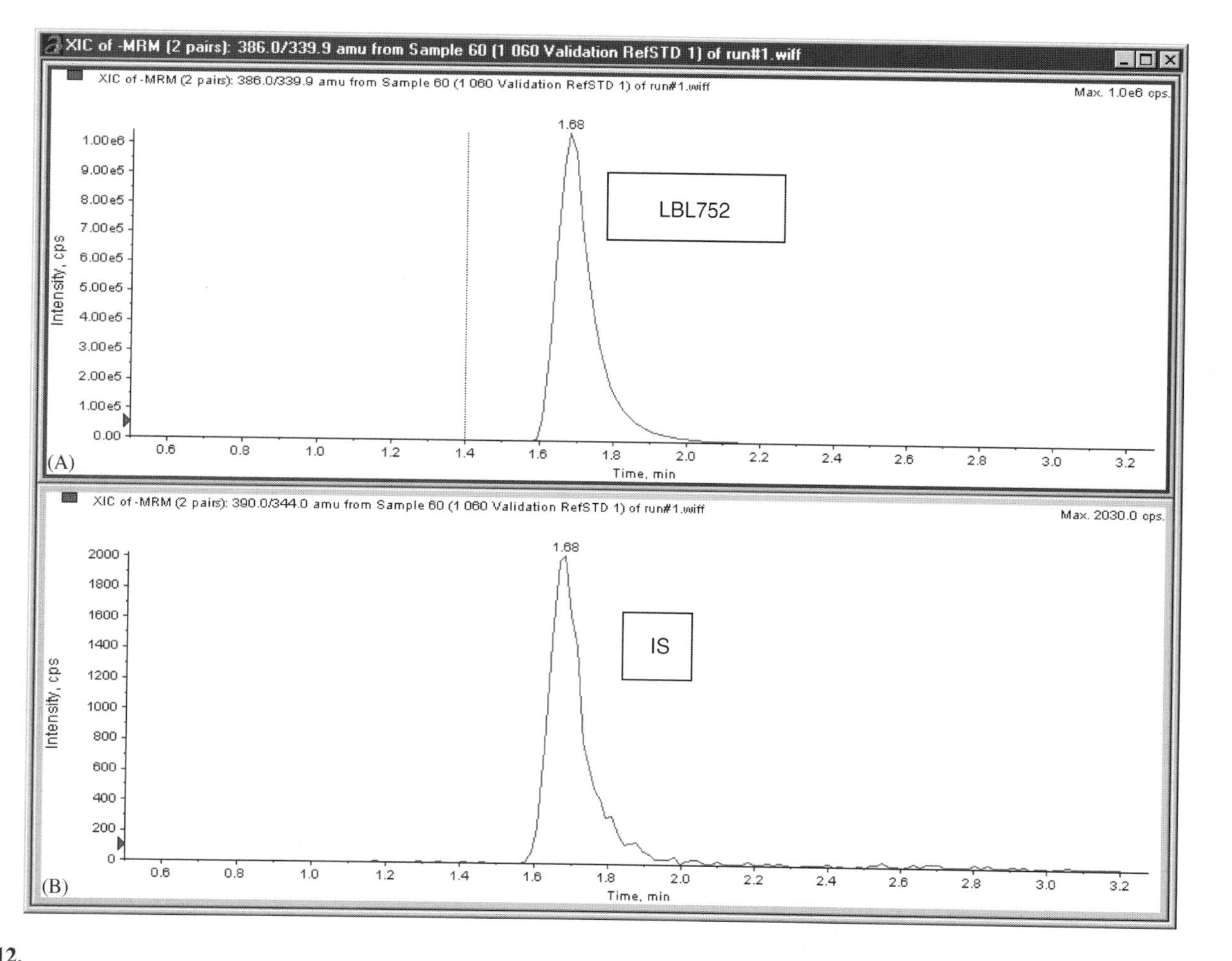

Figure 12.
LC–ESI/MS/MS ion chromatograms in blank dog serum spiked with LBL752 ($C_{21}H_{25}NO_4S$) at a final concentration of 2500 ng/ml (LLOQ) (A) and the IS ($C_{212}H_4H_{21}NO_4S$; 0 ng/ml) (B). Retention time (min) is shown on the horizontal axis and peak intensity is shown in the vertical axis

source of this type of interference and may become more problematic with very fast chromatographic methods. The only way to minimize these co-eluting interferences is by modifying the chromatographic method to achieve adequate separation (Majumdar *et al.*, 2004a, b; Matuszewski *et al.*, 1998a, b; Jemal and Xia, 1999). A typical problem was described for Novartis compound MMI270B, an inhibitor of matrix metalloproteases (Majumdar *et al.*, 2004a, b). The method of quantitation of MMI270B in human plasma was originally validated where the reverse-phase chromatographic separation was performed using a gradient of acetonitrile and water containing 0.05% acetic acid. The method showed excellent specificity for MMI270B in six different lots of human plasma. However, the method failed to quantify MMI270B accurately in serum of patients dosed with the compound due to co-eluted matrix interference (Fig. 13). The interference had several co-eluting components having same precursor and product ion combination as that of MMI270B and could not be resolved using MS/MS techniques. The interference could be due to some endogenous compounds or isomers of MMI270B produced in the patient after dosing with the compound. Similar analytical problems with interfering metabolite peaks were reported in the literature (Matuszewski *et al.*, 1998a, b; Jemal and Xia, 1999). The LC conditions used in the procedure were optimized for separation of the interference peaks and a typical ion chromatogram of patient sample is shown in Fig. 14 with the modified

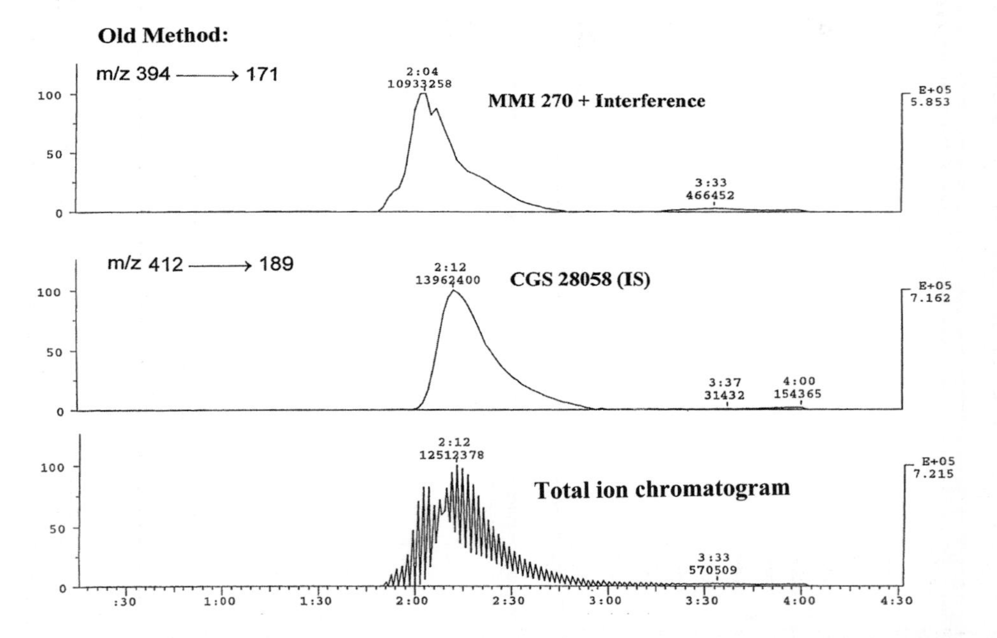

Figure 13.
LC–ESI/MS/MS of MMI270 using a C18 (1 × 50 mm) column. Separation was performed using an isocratic flow (30 µl/min) of 80% acetonitrile in water having 0.05% acetic acid. The extract was from the plasma sample of a patient dosed with 150 mg MMI270

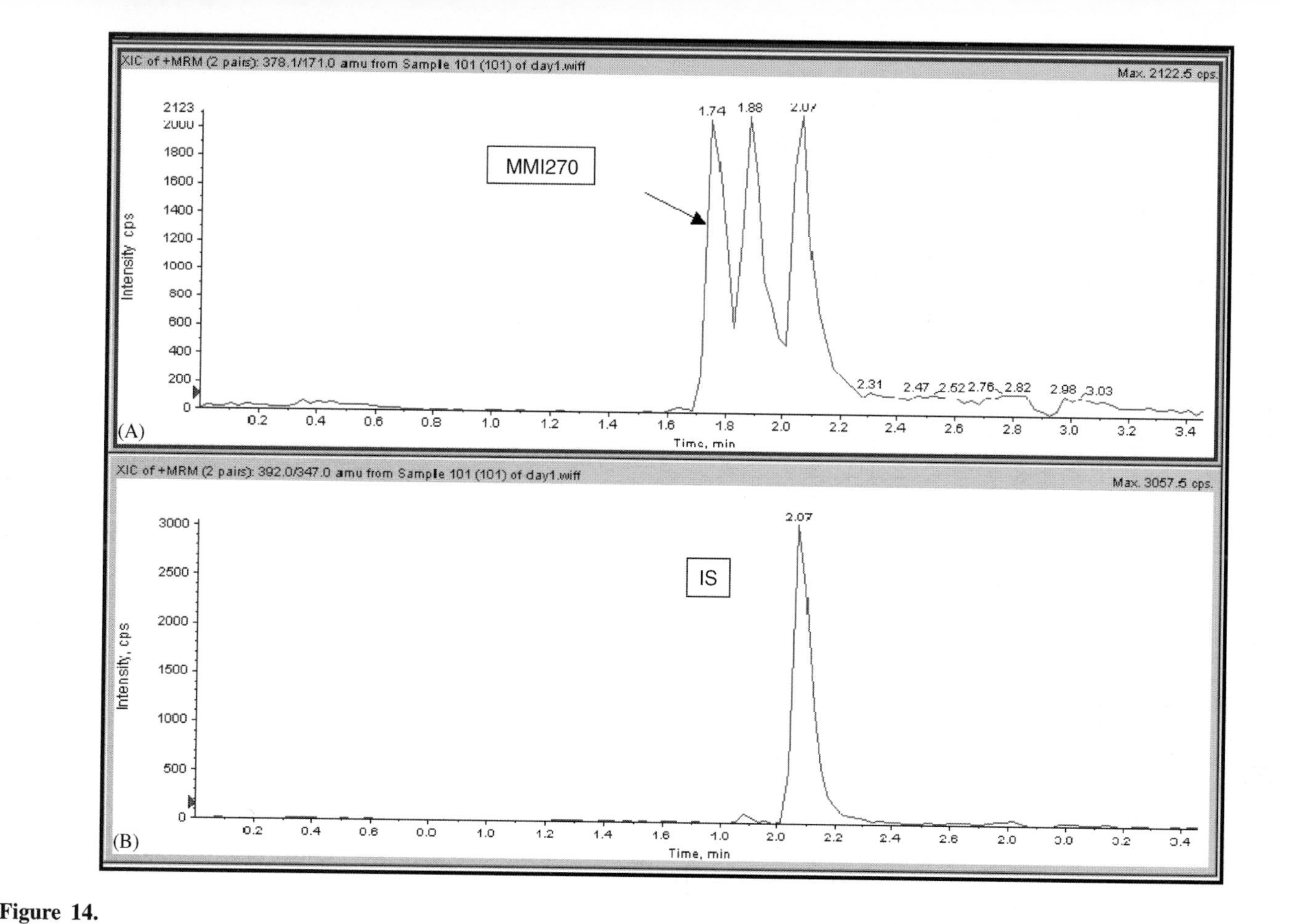

Figure 14.
LC–ESI/MS/MS of MMI270 in the patient plasma sample shown in Fig. 13. HPLC Separation was performed using a C_{18} (30 × 4.6 mm) column using a gradient of 0.01% TFA in water (Solvent A) and 0.01% TFA in methanol (Solvent B). The solvent flow rate was 1 ml/min at room temperature

LC method. The figure clearly shows two additional peaks of comparable intensity of the analyte MMI270B. Chromatographic conditions described in the modified method (Majumdar *et al.*, 2004a, b) separated the interference peaks while maintaining the retention time of MMI270B < 2 min.

3.2.9. The Use of Gas-Phase Cluster Ions for Method Sensitivity Problem

Some bioanalytical problems related to low sensitivity may be solved by the use of gas-phase ion chemistry of the analytes of interest at the API interface. Novartis compound NKP608 (Fig. 15a) and its metabolites CGP 81006 (N-oxide; Fig. 15b) and CGP 60922 (2-hydroxy quinoline derivative; Fig. 15c) provide such an example. Determination of NKP608 and CGP 81006 in human plasma was originally performed simultaneously using a GC/MS method (unpublished Novartis method). The fundamental problem with the GC/MS method was that a fraction of the N-oxide metabolite CGP 81006 was always converted to the parent drug NKP608 in both the heated GC injector and oven leading to an inaccurate quantitation of both compounds. To circumvent this problem a LC–MS/MS method based on ESI was developed in which the reverse-phase separation of the analytes on a C_{18} (50×1.0 mm) column using an isocratic flow (25 µl/min) of acetonitrile:water (75:25, v/v) having 0.05% acetic acid. The ion chromatograms of the analytes obtained by the method are shown in Fig. 16.

Subsequently, CGP 60922 (2-hydroxy quinoline derivative; Fig. 15c) was introduced as a second pharmacologically active metabolite and needed to be quantified. The method described above failed to quantify the 2-hydroxy metabolite CGP 60922 for three reasons. First, CGP 60922 had very low sensitivity compared to CGP 81006 and was almost undetectable below the concentration of 10 ng/ml in the above method. Second, the two metabolites CGP 81006 (Fig. 15b) and CGP 60922 (Fig. 15c) have the same molecular formula $C_{31}H_{24}ClF_6N_3O_3$ (molecular weight 636.0) and produced

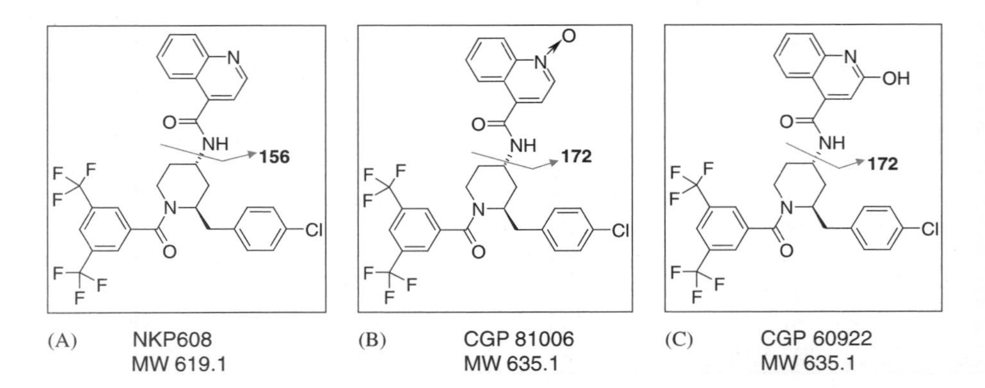

Figure 15.
Structures of (A) NKP608, (B) CGP 81006, and (C) CGP 60922. The arrow refers to the resulting product ion monitored during the MRM experiment

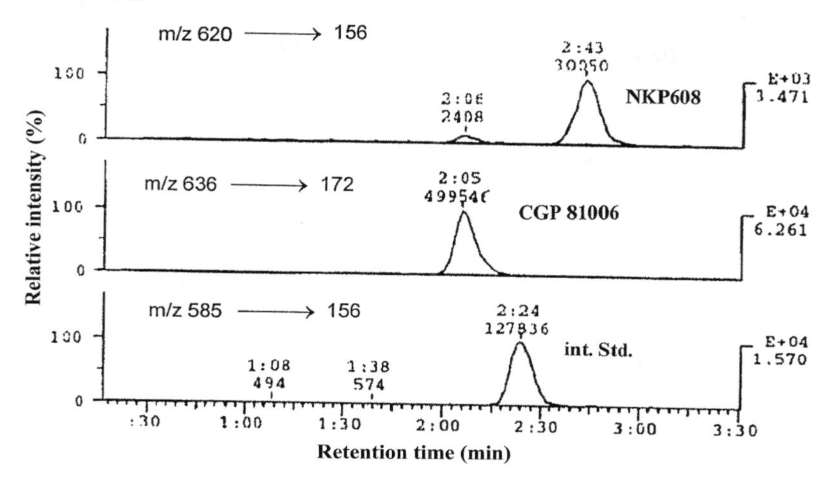

Figure 16.
LC–ESI/MS/MS ion chromatograms of NKP608, CGP 81006, and the IS. The HPLC separation of the analytes were performed on a C_{18} (50 × 1.0 mm i.d., 5-μm particle size) column using an isocratic flow (25 μl/min) of acetonitrile:water (75:25, v/v) having 0.05% acetic acid. Retention time (min) is shown on the horizontal axis and the peak intensity is shown in the vertical axis

product ions having same *m/z* 172 during MS/MS fragmentation process. Therefore, these two compounds were not distinguishable by MS or MS/MS. Third, these two metabolites were co-eluting in the HPLC condition described above (peak CGP 81006 in Fig. 16). Therefore a new chromatographic method was used in which separations were performed using a C_{18} (50 × 4.6 mm i.d., 3-μm particle size) column. The mobile phase was a mixture of methanol:water (73:27, v/v) containing 50 mM ammonium acetate. The separation was performed under isocratic flow of the mobile phase at 1 ml/min. Ionization was performed using a Finnigan APCI interface.

The new method addresses all three problems discussed above. Since metabolites CGP 81006 and CGP 60922 could not be resolved by MS/MS, they were separated chromatographically. It was observed during method development that the hydroxy metabolite CGP 60922 formed intense proton-bound methanol adduct ion (Majumdar *et al.*, 1992; Koenig *et al.*, 1993) with *m/z* 668.1 ($[M+H+CH_3OH]^+$) in the APCI source of the mass spectrometer. MS/MS product ions spectrum of *m/z* 668 at collision energy 43 eV produced ions with *m/z* 636 ($[M+H]^+$) and 172 ($[C_{10}H_6NO_2]^+$) as the major ions (Fig. 17). The sensitivity problem with CGP 60922 was solved by using the $[M+H+CH_3OH]^+$ ion of this molecule as a precursor ion for the MS/MS reaction (Fig. 17). By using the protonated methanol adduct ion a LLOQ of 0.1 ng/ml was achieved for CGP 60922 (Fig. 18). Use of gas-phase cluster ions with ammonia has also been reported for the quantitative analysis of RAD001 (Brignol *et al.*, 2001). When using adduct ions other than $[M+H]^+$, extra care must be taken

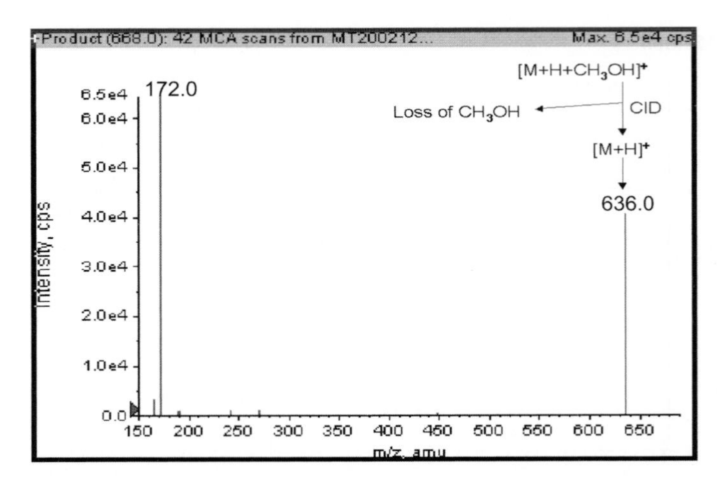

Figure 17.
Full-scan APCI/MS/MS spectrum of the proton-bound menthanol adduct ion $[M + H + CH_3OH]^+$ *m/z* 668 formed from CGP 60922 (collision energy 43 eV and collision gas pressure at 2.5 mTorr) showing the product ion at *m/z* 172.0 and the $[M + H]^+$ ion at *m/z* 636. Mass to charge ratio (*m/z*) is shown on the horizontal axis and the ion intensity is shown in the vertical axis

to ensure that the mobile phase and the ionization source, provide suitable condition to reproducibly form the adduct ions.

3.2.10. Problem with Post-Column Ion Source-Induced Changes of Pro-Drugs and Metabolites

Post-column changes of pro-drugs and metabolites have been documented for LC/MS/MS methods (Matuszewski *et al.*, 1998a, b; Jemal and Xia, 1999; Jemal *et al.*, 2000). These changes commonly observed with lactones, N-oxides, glucuronides, amides, etc. Conversion of analytes in the heated APCI capillary of Finnigan 7000 instruments is common. However, it can also occur in other source designs, depending on the compound stability. Ion source lactonization of simvastatin is a very good example (Jemal *et al.*, 2000). Post-column conversion of N-oxide metabolite CGP 81006 into the parent drug NKP608 in the heated capillary of Finnigan 700 APCI source is shown in Fig. 18. The small peak (ahead of NKP608 peak) in the MS/MS channel of NKP608 (trace C, Fig. 18) has the retention time as CGP 81006 (trace D; Fig. 18). This small peak is clearly due to in-source conversion of CGP 81006 (the N-oxide metabolite) into the parent drug NKP608. This type of source-induced changes in the APCI source of mass spectrometer is very common (Jemal and Xia, 1999; Jemal *et al.*, 2000). The source-induced conversion of metabolite was not a problem in this method due to base-line separation of the peaks in HPLC (trace C, Fig. 18). There is also a big peak ahead of CGP 60922 in trace C (Fig. 18). This

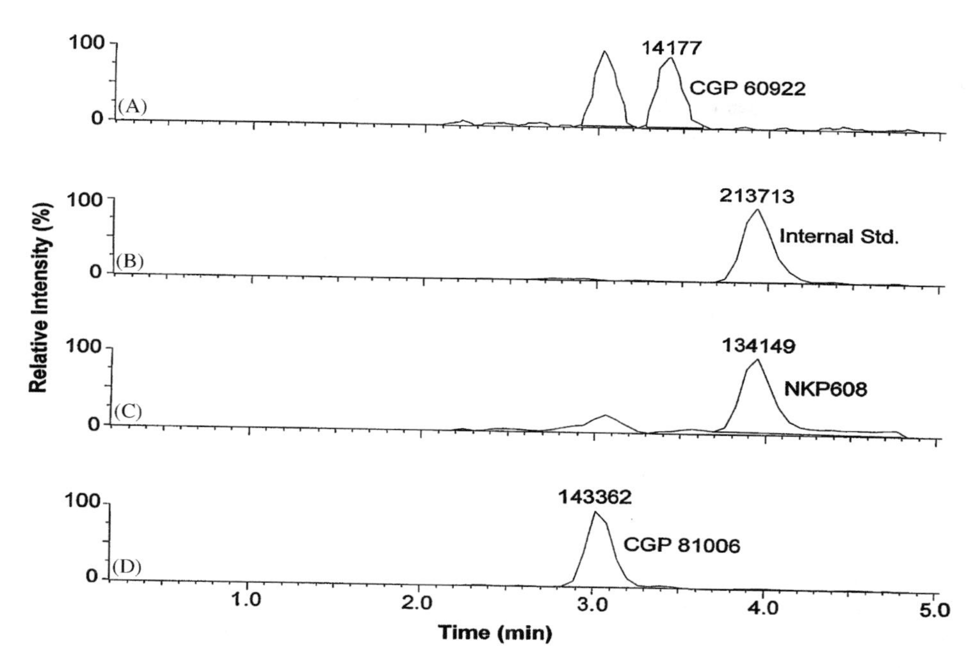

Figure 18.
LC–ESI/MS/MS ion-chromatograms in blank human plasma spiked with CGP 60922 (trace A), the IS (trace B) NKP608 (trace C), and CGP 81006 (trace D). Plasma concentration of each of the analytes was 0.1 ng/ml (LLOQ) and that of the IS was 0.25 ng/ml. Retention time (min) is shown on the horizontal axis and the peak intensity is shown in the vertical axis

peak has the same retention time as CGP 81006 (trace D, Fig. 18). This big peak is due to the contribution of CGP 81006 in the MS/MS channel of CGP 60922. This issue has been discussed in Section 2.9.

3.2.11. Problems with Both Mass Spectrometric and Chromatographic Resolutions in Accurate Quantification of Parent Compound and Metabolites

The need for chromatographic resolution in some bioanalytical methods has been discussed in Section 3.2.8. There are difficult assays in the bioanalytical laboratories, where the chromatographic separation or mass separation using a tandem quadrupole mass spectrometer alone may not be sufficient to accurately quantify the analytes. This type of situation may occur when the subject is dosed with ^{14}C-labeled drugs. Such an example is shown in Figs. 19 and 20. The parent compound ILO (under development at Novartis for the treatment of Schizophrenia) and two metabolites P88 and P95 are shown in Fig. 19. The ^{14}C-labeled ILO and metabolites P88 and P95 have the same mass of 428.2 (two mass unit higher than the parent drug ILO). The

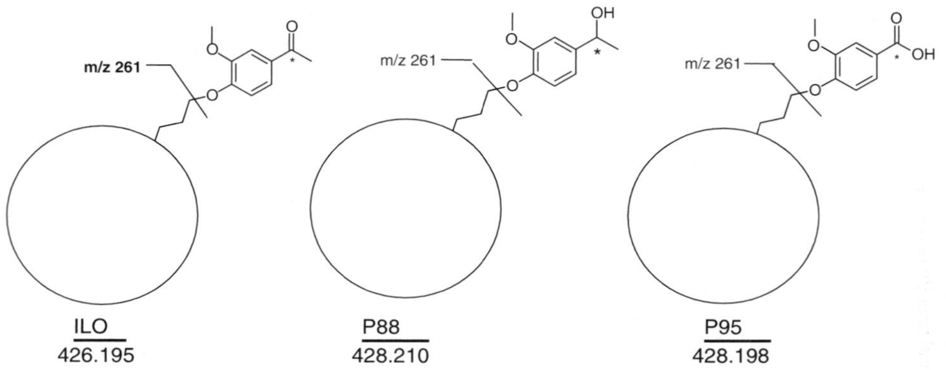

m/z 261
m/z 261
m/z 261

ILO
426.195

P88
428.210

P95
428.198

Figure 19.
Partial structure of ILO ^{14}C-labeled ILO, and the metabolites P88 and P95. The arrow refers to the resulting product ion monitored during the MRM experiment

problem was complicated due to the same precursor and MS/MS fragment ions for all three analytes. The metabolites were co-eluting in a single peak and the ^{14}C-labeled drug was interfering with the quantification of the parent drug (due to the use of low-resolution window in the first quadrupole Q1 of the instrument). This situation required both chromatographic resolution and mass resolution for accurate quantification of all the analytes. The top two panels in Fig. 20 show four peaks after both mass and chromatographic resolutions.

3.2.12. Problem with Quantifying an Unstable and Light-Sensitive Compound at Ultra-Trace Level Concentrations

Lowering of the quantitation limit for an unstable compound is always challenging in a bioanalytical method. However, adding extra steps to typical bioanalytical procedures might be necessary to complete projects with unstable and low-dose compounds. One such example is Iralukast, a peptide-leukotriene antagonist for the treatment of asthma (Majumdar *et al.*, 2000). The analytical challenge was to develop a sensitive LC/MS/MS method with a LLOQ of 10 pg/ml for the analysis of iralukast when administered at low doses during the clinical trials. Several issues had to be addressed in order to devise a LC/MS/MS assay for the above compound. First, iralukast is light sensitive and unstable at the room temperature under acidic conditions. Second, an ultra-low LLOQ of 10 pg/ml was needed to support low dose clinical trials. Third, positive ESI of iralukast did not yield sufficient sensitivity, which was required for the studies in humans. Consequently, LC/MS/MS conditions were optimized for negative ion mode of detection. Fourth, sample preparation steps proved to be critical to reduce the possibility of micro-bore HPLC column (50 mm $\times$ 1.0 mm i.d.) obstruction, chromatographic deterioration, and matrix-mediated electrospray ion suppression. The sample preparation procedure was critical for this method. The

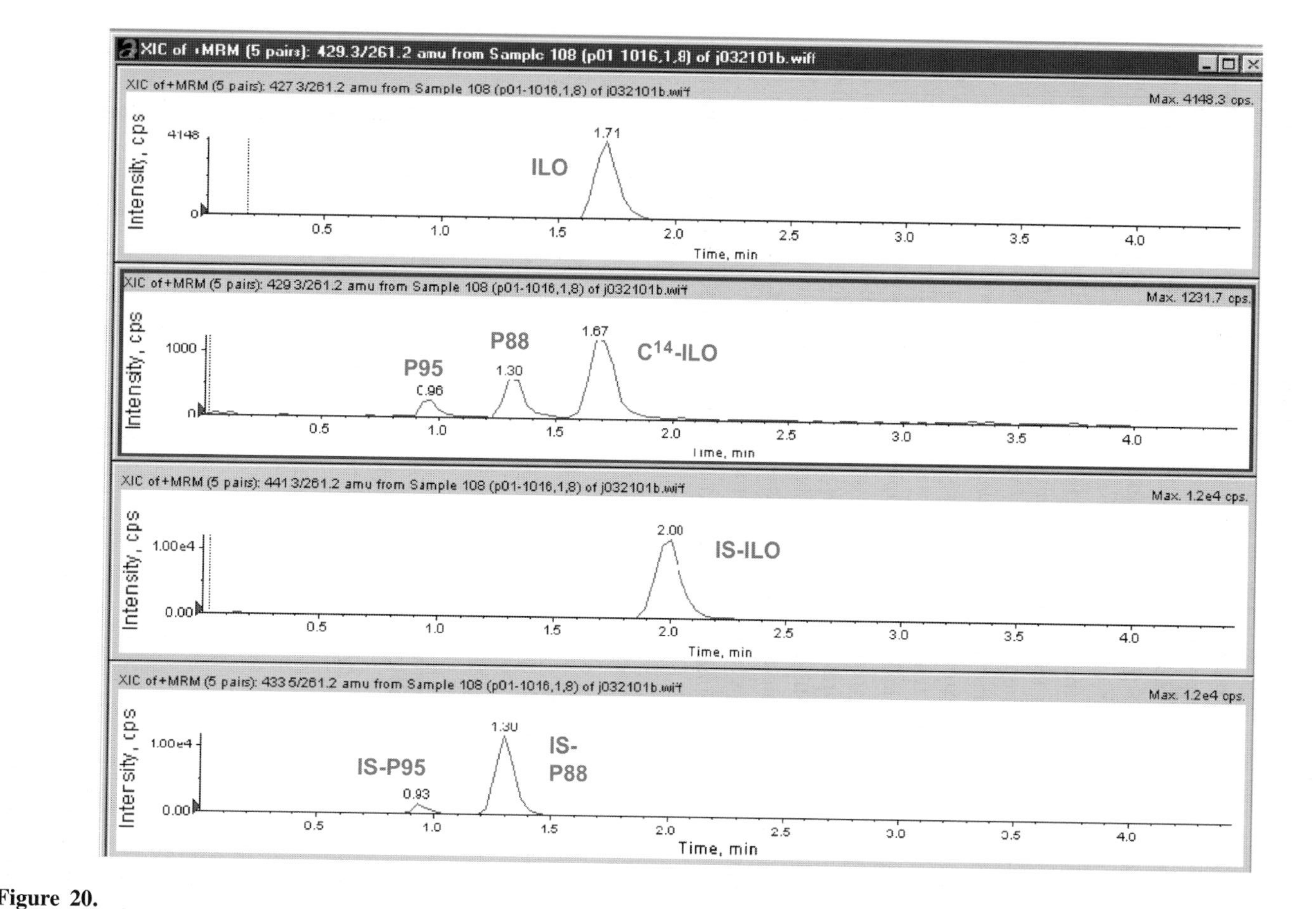

Figure 20.
LC–ESI/MS/MS ion-chromatograms plasma of subject dosed with ^{14}C-labeled ILO: All the analytes and ISs were separated chromatographically and mass resolved using MS/MS. Retention time (min) is shown on the horizontal axis and the peak intensity is shown in the vertical axis

extract obtained from a LLE was further concentrated using a preparative HPLC column-switching system equipped with a fraction collector (Fig. 21). Chromatographic separations were performed at 40°C on a silica analytical column (column B in Fig. 21: 5 μm, 100 mm × 3 mm i.d.) and a silica guard column (column A in Fig. 21: 5 μm, 20 mm × 4 mm i.d.) connected through a 6-port, 2-position valve, using hexane:dichloromethane:isopropanol:acetic acid (60:23:12:5, by vol.) as the mobile phase at a flow rate of 0.3 ml/min. A 0.1 ml aliquot of the sample extract was injected onto column A. After ~ 1.5 min, when Iralukast and the IS were transferred onto column B, the 6-port valve was switched to back flush particles and polar components from column A, using the mobile phase at a flow rate of 2.0 ml/min. After ~ 3.8 min, the 6-port valve was returned to the original position and pump B was stopped. A fraction with retention time from 3.6 to 4.4 min was collected directly into 0.7 ml amber autosampler vials and evaporated under vacuum at room temperature.

Iralukast is light sensitive and unstable at the room temperature under acidic conditions. Therefore, the fraction collector was kept in a refrigerator at 4°C, and an additional pump (pump C in Fig. 21) was employed to deliver pyridine:hexane (30:70, v/v) at a flow rate of 0.1 ml/min during the run via a "T" placed between column-B and the fraction collector. The residue was reconstituted in 20 μl of ethanol:0.2 N-ammonium hydroxide (40:60, v/v) and a 7 μl aliquot were injected for LC/MS/MS analysis. While the validated method addressed the analytical challenges, its major drawback was limited sample throughput capability.

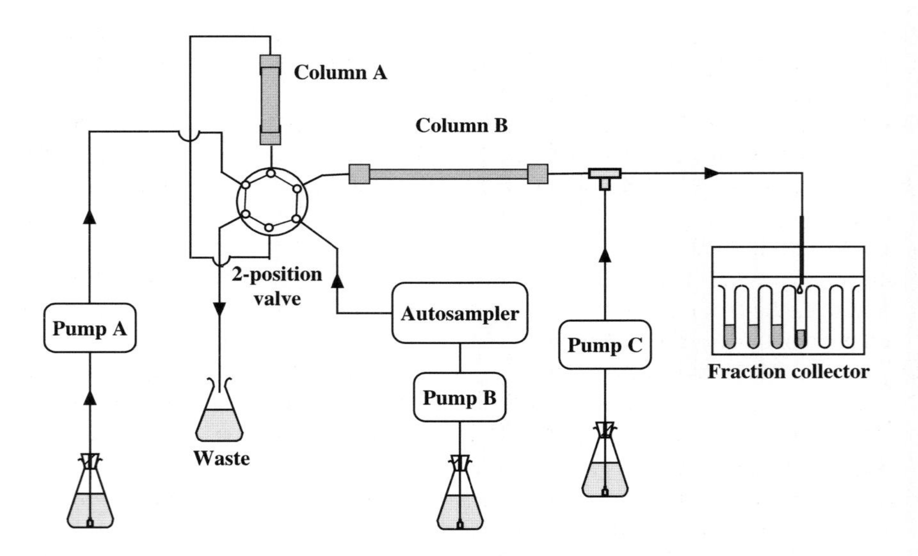

Figure 21.
Instrumental set-up for sample preparatin. Iralukast is light sensitive and unstable at the room temperature under acidic conditions. Therefore, an additional pump (pump C) was employed to deliver pyridine:hexane (30:70, v/v) at a flow-rate of 0.1 ml/min during the run via a 'T' placed between column B and the fraction collector

3.3. Conclusions

The methods discussed were used for HT analysis of drugs and metabolites in biological matrices. In spite of the high sensitivity and specificity achieved by the MS/MS technique, the HPLC separation and the sample extraction procedure are critical. Optimization of the conditions e.g., LC, MS interface, and MS/MS parameters could alleviate many of the problems. In some cases the chemist has to utilize lessons of drug metabolism, gas-phase ion chemistry, and chromatography to solve the problems. The techniques described herein do not encompass all that are available to a bioanalytical chemist. However, some problems faced in our facilities as well those published in the literature are discussed here.

Acknowledgments

The author acknowledges the contributions of Ms. S. Wu, Ms. C. Chen, Ms. P. Nouri, Ms. S. Vedananda, Mr. D. Winn, Mr. L. Ramos, Mr. T. Bedman, Dr. Tapan Ray, and the Isotope Laboratories at Novartis. The author also expresses his gratitude to Dr. F. L. S. Tse, Dr. M. Hayes, and Dr. L. McMahon for their review and comments on the chapters.

References

Avery, M.J., 2003. Quantitative characterization of differential ion suppression on liquid chromatography/atmospheric pressure ionization mass spectrometric bioanalytical methods. Rapid Commun. Mass Spectrom. 17, 197–201.

Bakhtiar, R., Ramos, L., Tse, F.L.S., 2002. High-throughput mass spectrometric analysis of xenobiotics in biological fluids. J. Liq. Chrom. Rel. Tech. 25, 507–540.

Banks, J.F., Shen, S., Whitehouse, C.M., Fenn, J.B., 1994. Ultrasonically assisted electrospray ionization for LC/MS determination of nucleosides from a transfer RNA digest. Anal. Chem. 66, 406–414.

Bonfiglio, R., King, R.C., Olah, T.V., Merkle, K., 1999. The effects of sample preparation methods on the variability of the electrospray ionization response for model drug compounds. Rapid Commun. Mass Spectrom. 13, 1175–1185.

Brignol, N., McMahon, L.M., Luo, S., Tse, F.L.S., 2001. High-throughput semi-automated 96-well liquid/liquid extraction and liquid chromatography/mass spectrometric analysis of everolimus (RAD 001) and cyclosporin a (CsA) in whole blood. Rapid Commun. Mass Spectrom. 15, 898–907.

Bruins, A.P., 1991. Mass spectrometry with ion sources operating at atmospheric pressure. Mass Spectrom. Rev. 10, 53–77.

Bruins, A.P., 1998. Mechanistic aspects of electrospray ionization. J. Chromatogr. A 794, 345–357.

Bruins, A.P., Covey, T.R., Henion, J.D., 1987. Ion spray interface for combined liquid chromatography/atmospheric pressure ionization mass spectrometry. Anal. Chem. 59, 2642–2646.

Deng, Y., Wu, J.T., Lloyd, T.L., Chi, C.L., Olah, T.V., Unger, S.E., 2002. High-speed gradient parallel liquid chromatography/tandem mass spectrometry with fully automated sample preparation for bioanalysis: 30 seconds per sample from plasma. Rapid Commun. Mass Spectrom. 16, 1116–1123.

Fenn, J.B., Mann, M., Meng, C.K., Wong, S.F., Whitehouse, C.M. 1989. Electrospray ionization for mass spectrometry of large biomolecules. Science 246, 64–71.

Hsieh, Y., Chintala, M., Mei, H., Agans, J., Brisson, J., Ng, K., Korfmacher, W.A., 2001. Quantitative screening and matrix effect studies of drug discovery compounds in monkey plasma using fast-gradient liquid chromatography/tandem mass spectrometry. Rapid Commun. Mass Spectrom. 15, 2481–2487.

Hsieh, Y., Wang, G., Wang, Y., Chackalamannil, S., Brisson, J.M., Ng, K., Korfmacher, W.A., 2002. Simultaneous determination of a drug candidate and its metabolite in rat plasma samples using ultrafast monolithic column high-performance liquid chromatography/tandem mass spectrometry. Rapid Commun. Mass Spectrom. 16, 944–950.

Ikonomou, M.G., Blades, A.T., Kebarle, P., 1991. Electrospray-ion spray: a comparison of mechanisms and performance. Anal. Chem. 63, 1989–1998.

Jacob, E., Herbert, V., 1977. Effect of sodium fluoride anticoagulant on measurement of transcobalamins in plasma and serum. J. Lab. Clin. Med. 90, 949–950.

Jemal, M., Ouyang, Z., Powell, M.L., 2000. Direct-injection LC–MS–MS method for high-throughput simultaneous quantitation of simvastatin and simvastatin acid in human plasma. J. Pharm. Biomed. Anal. 23, 323–340.

Jemal, M., Ouyang, Z., Teitz, D.S., 1998. High performance liquid chromatography mobile phase composition optimization for the quantitative determination of a carboxylic acid compound in human plasma by negative ion electrospray high performance liquid chromatography tandem mass spectrometry. Rapid Commun. Mass Spectrom. 12, 429–434.

Jemal, M., Xia, Y.Q., 1999. The need for adequate chromatographic separation in the quantitative determination of drugs in biological samples by high performance liquid chromatography with tandem mass spectrometry. Rapid. Commun. Mass Spectrom. 13, 97–106.

Jonkman, J.H.G., Franke, J.P., Schoenmaker, R., De Zeeuw, R.A., 1982. Influence of sample preparation and storage on theophylline concentrations in biological fluids. Int. J. Pharm. 10, 177–180.

Kebarle, P., Teng, L., 1993. From ions in solution to ions in the gas phase – the mechanism of electrospray mass spectrometry. Anal. Chem. 65, 972–986A.

Kebarle, P., Yeunghaw, H., 1997. On the mechanism of electrospray mass spectrometry. In: Cole, R.B. (Ed.), Electrospray Ionization Mass Spectrometry. Wiley, New York, pp. 3–63.

King, R., Bonfiglio, R., Fernandez-Merzler, C., Miller-Stein, D., Olah, T., 2000. Mechanistic investigation of ionization suppression in electrospray ionization. J. Am. Soc. Mass Spectrom. 11, 942–950.

Koenig, S., Kofel, P., Reinhard, H., Saegesser, P., Schlungger, U.P., 1993. Reaction of protonated acetaldehyde with methanol in the gas phase. Org. Mass Spectrom. 28, 1101.

Kostiainen, R., Bruins, A.P., 1994. Effect of multiple sprayers on dynamic range and flow rate limitations in electrospray and ionspray mass spectrometry. Rapid Commun. Mass Spectrom. 8, 549–558.

Law, B., 2004. The effect of eluent pH and compound acid–base character on the design of generic-gradient reversed-phase high-performance liquid chromatography (RP-HPLC) methods for use in drug discovery. J. Pharm. Biomed. Anal. 34, 215–219.

Majumdar, T.K., Bakhtiar, R., Melamed, D., Tse, F.L.S., 2000. Trace-level quantitation of iralukast in human plasma by microbore liquid chromatography/tandem mass spectrometry. Rapid Commun. Mass Spectrom. 14, 476–481.

Majumdar, T.K., Bakhtiar, R., Wu, S.; Winn, D., Tse, F., 2001. Troubleshooting LC–MS/MS methods for the bioanalysis of drugs: some typical problems and solutions. Adv. Mass Spectrom. 15, 681–682.

Majumdar, T.K., Clairet, F.C., Tabet, J.C., Cooks, R.G., 1992. Epimer distinction and structural effects on gas-phase acidities of alcohols measured using the Kinetic Method. J. Am. Chem. Soc. 114, 2897–2903.

Majumdar, T.K., Vedananda, S., Tse, F.L.S., 2004a. Rapid analysis of MMI270B, an inhibitor of matrix metalloproteases in human plasma by liquid chromatography–tandem mass spectrometry: matrix interference in patient samples. Biomed. Chromatogr. 18, 77–85.

Majumdar, T.K., Wu, S., Vedananda, S., Tse, F.L.S., 2004b. Determination of LAG078, a lipid-lowering compound, in dog plasma using liquid chromatography–tandem mass spectrometry: application in a toxicokinetic study. J. Pharm. Biomed. Anal. 35, 853–866.

Matuszewski, B.K., Chavez-Eng, C.M., Constanzer, M.L., 1998a. Development of high-performance liquid chromatography–tandem mass spectrometric methods for the determination of a new oxytocin receptor antagonist (L-368,899) extracted from human plasma and urine: a case of lack of specificity due to the presence of metabolites. J. Chromatogr. B 716, 195–208.

Matuszewski, B.K., Constanzer, M.L., Chavez-Eng, C.M., 1998b. Matrix effect in quantitative LC/MS/MS analyses of biological fluids: a method for determination of finasteride in human plasma at picogram per milliliter concentrations. Anal. Chem. 70, 882–889.

Mei, H., Hsieh, Y., Nardo, C., Xu, X., Wang, S., Ng, K., Korfmacher, W.A., 2003. Investigation of matrix effects in bioanalytical high-performance liquid chromatography/tandem mass spectrometric assays: application to drug discovery. Rapid Commun. Mass Spectrom. 17, 97–103.

Muller, C., Schafer, P., Stortzel, M., Vogt, S., Weinmann, W., 2002. Ion suppression effects in liquid chromatography–electrospray-ionisation transport-region collision induced dissociation mass spectrometry with different serum extraction methods for systematic toxicological analysis with mass spectra libraries. J. Chromatogr. B 773, 47–52.

Niessen, W.M.A., 1999. State-of-the-art in liquid chromatography–mass spectrometry. J. Chromatogr. A 856, 179–197.

O'Connor, D., Clarke, D.E., Morrison, D., Watt, A.P., 2002. Determination of drug concentrations in plasma by a highly automated, generic and flexible protein precipitation and liquid chromatography/tandem mass spectrometry method applicable to the drug discovery environment. Rapid Commun. Mass Spectrom. 16, 1065–1071.

Suttnar, J., Masova, L., Dyr, J.E., 2001. Influence of citrate and EDTA anticoagulants on plasma malondialdehyde concentrations estimated by high-performance liquid chromatography. J. Chromatogr. B 751, 193–197.

Thomson, B.A., 1998. Atmospheric pressure ionization and liquid chromatography/mass spectrometry – Together at last. J. Am. Soc. Mass Spectrom. 9, 187–193.

Tiller, P., Romanyshyn, L., 2002. Implications of matrix effects in ultra-fast gradient or fast isocratic liquid chromatography with mass spectrometry in drug discovery. Rapid. Commun. Mass Spectrom. 16, 92–98.

Zhao, J.J., Yang, A.Y., Rogers, J.D., 2002. Effects of liquid chromatography mobile phase buffer contents on the ionization and fragmentation of analytes in liquid chromatographic/ionspray tandem mass spectrometric determination. J. Mass. Spectrom. 37, 421–433.

Identification and Quantification of Drugs, Metabolites and Metabolizing
Enzymes by LC–MS
Swapan K. Chowdhury, editor.

Chapter 4

PITFALLS IN QUANTITATIVE LC–MS/MS: METABOLITE CONTRIBUTION TO MEASURED DRUG CONCENTRATION

Mohammed Jemal

4.1. Introduction

High-performance liquid chromatography (HPLC) coupled with atmospheric pressure ionization (API) tandem mass spectrometry (LC–MS/MS) has become the technique of choice for quantification of small-molecule drugs and metabolites in biological matrices such as plasma, serum, blood, urine, and in vitro biological samples (Bakhtiar *et al.*, 2002; Jemal, 2000; Naidong, 2003; Tiller *et al.*, 2003). In order to achieve enhanced sensitivity and selectivity (specificity), the LC–MS/MS acquisition is conducted via selected reaction monitoring (SRM), simultaneously obtaining the SRM transitions of the analyte and its internal standard. In spite of the great specificity provided by the SRM-based LC–MS/MS analysis, the use of this technique for quantification of drugs in post-dose (incurred) biological samples could be hampered by interferences arising from the presence of some metabolites, which originate from the in vivo biotransformation of the administered drug. In this chapter, two kinds of potential sources of interference and approaches to overcome their untoward consequences are discussed. The first kind is due to inadequate chromatographic or mass resolution between a drug and its metabolite. The second type is due to instability of the metabolite during the multiple sample handling and cleanup steps involved prior to the LC–MS/MS analysis, thereby generating the drug due to degradation of the metabolite.

4.2. Analytical Pitfalls Due to Metabolite In-Source Conversion

In the period immediately following the introduction of the LC–MS/MS technique for quantitative bioanalysis, the thinking among some bioanalysts was that LC–MS/MS bioanalysis could be achieved with little or no chromatographic resolving capability. However, as the technique began to be used more widely, it was realized that some chromatographic resolving power was indeed required to achieve the specificity needed for analysis of post-dose samples. An article that systematically demonstrated such a need was published in 1999 (Jemal and Xia, 1999). In this article, the authors discussed the behavior of drugs of different functional groups, which, when administered to animals or humans, gave metabolites that could cause interference with the quantification of the drug. The interference problem arose from the fact that the metabolite produced, via in-source conversion, a molecular ion that was identical to the parent ion of the drug. The presence of a prodrug, due to incomplete in vivo conversion to the intended drug, also caused a similar interference in the analysis of the drug. The consequence of such interference is that the SRM transition adopted for the quantitative determination of the drug will respond not only to the drug but also to the metabolite. Thus, in the absence of adequate chromatography to separate the drug from the metabolite, the LC–MS/MS method will not be specific to the drug, the analyte of interest.

The drugs and their metabolites used in the aforementioned article (Jemal and Xia, 1999) to demonstrate the interference caused by the metabolites are shown in Fig. 1. The partial chemical structures of the compounds shown are adequate for the discussion of the mass spectrometric behavior of each compound, as it relates to the production of the same SRM transitions by the drug and its metabolite. Compounds **I**, **III**, **V**, and **IX** are drugs, which have methyloxime, lactone, carboxylic acid, and sulfhydryl (thiol) functional groups, respectively. The corresponding metabolites, compounds **II**, **IV**, **VI**, and **X**, have isomeric methyloxime, hydroxy acid, acylglucuronide, and disulfide functional groups. Compound **VII** is a phenolic drug and compound **VIII** is its prodrug. The demonstration of the interference of a metabolite or a prodrug with the accurate quantification of a drug is represented below using the data obtained for three of the five drugs (Jemal and Xia, 1999).

The positive ion electrospray mass and MS/MS spectra of **III**, a lactone drug, and its hydroxy acid metabolite **IV** are shown in Fig. 2. The mass spectrum in Fig. 2A shows that **III** gave an intense $[\mathbf{III}+\mathrm{H}]^+$ ion at *m/z* 363, and the MS/MS spectrum in Fig. 2B using *m/z* 363 as the precursor ion shows an intense product ion at *m/z* 285. Therefore, it would be reasonable to use an SRM transition of *m/z* 363 to *m/z* 285 for the LC–MS/MS quantitative determination of **III** in a biological sample. On the other hand, the mass spectrum in Fig. 2C shows that **IV** gave, in addition to an intense $[\mathbf{IV}+\mathrm{H}]^+$ ion at *m/z* 381, a fragment ion arising from in-source lactonization at *m/z* 363, which is the same as the precursor ion in the SRM transition utilized for the proposed LC–MS/MS method of **III**. Thus, unless the LC–MS/MS method can chromatographically separate **III** from **IV**, the method developed and validated for **III** would be interfered with by the presence of **IV**. This is illustrated in Fig. 3. The SRM chromatograms of **III** and **IV** shown in Figs. 3A and B were obtained by conducting, separately, the SRM transition of *m/z* 363 to *m/z* 285, with an isocratic mobile phase of 25% formic acid (0.1%) and 75% methanol at a flow rate of 0.8 ml/min. Compounds **III** and **IV** were not resolved under these LC conditions, with retention factor (k) values of 0.94 and 0.85 for **III** and **IV**, respectively. Figure 3C shows that **III** and **IV** were chromatographically resolved by using an isocratic mobile phase of 45% formic acid (0.1%) and 55% methanol, with k values of 9.0 and 8.0 for **III** and **IV**, respectively, and a resolution of 1.1. This demonstrates an example where a k value of ≥ 8 was required to achieve the needed chromatographic resolution, due

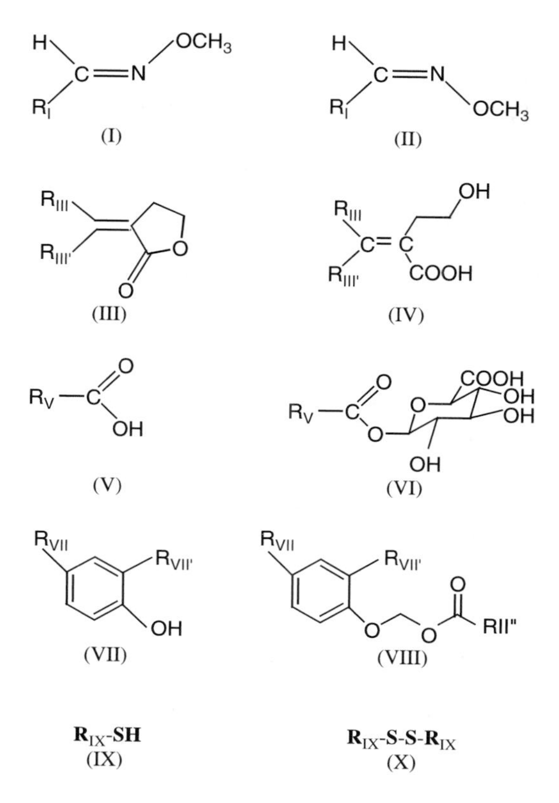

Figure 1.
Representation of the essential functional groups of the different classes of the compounds used to demonstrate potential interference due to in-source conversion. Jemal and Xia (1999) Copyright 1999. John Wiley & Sons Limited. Reproduced with permission

to the small separation factor (α) value obtained under the conditions used. It was of interest to note that the positive ion electrospray mass spectra of both **III** and **IV** showed the $[M+NH_4]^+$ (at *m/z* 380 and *m/z* 398, respectively) adduct to be the predominant molecular ion, with little response at $[M+H]^+$, when a mobile phase of ammonium acetate (10 mM, pH 3) and methanol was used instead of the 0.1% formic acid and methanol. More interestingly, under these conditions, there was no ion common to the two compounds arising from the in-source lactonization of **IV**. This illustrates an example where the judicious choice of the mobile phase would alleviate the problem of interference arising from the presence of a common ion.

The negative ion electrospray mass and MS/MS spectra of **IX**, a thiol drug, and of its disulfide **X**, are shown in Fig. 4. The mass spectrum in Fig. 4A shows that **IX** gave an intense $[\mathbf{IX}-H]^-$ ion at *m/z* 407, and the MS/MS spectrum in Fig. 4B using *m/z* 407 as the precursor ion shows an intense product ion at *m/z* 280. Thus,

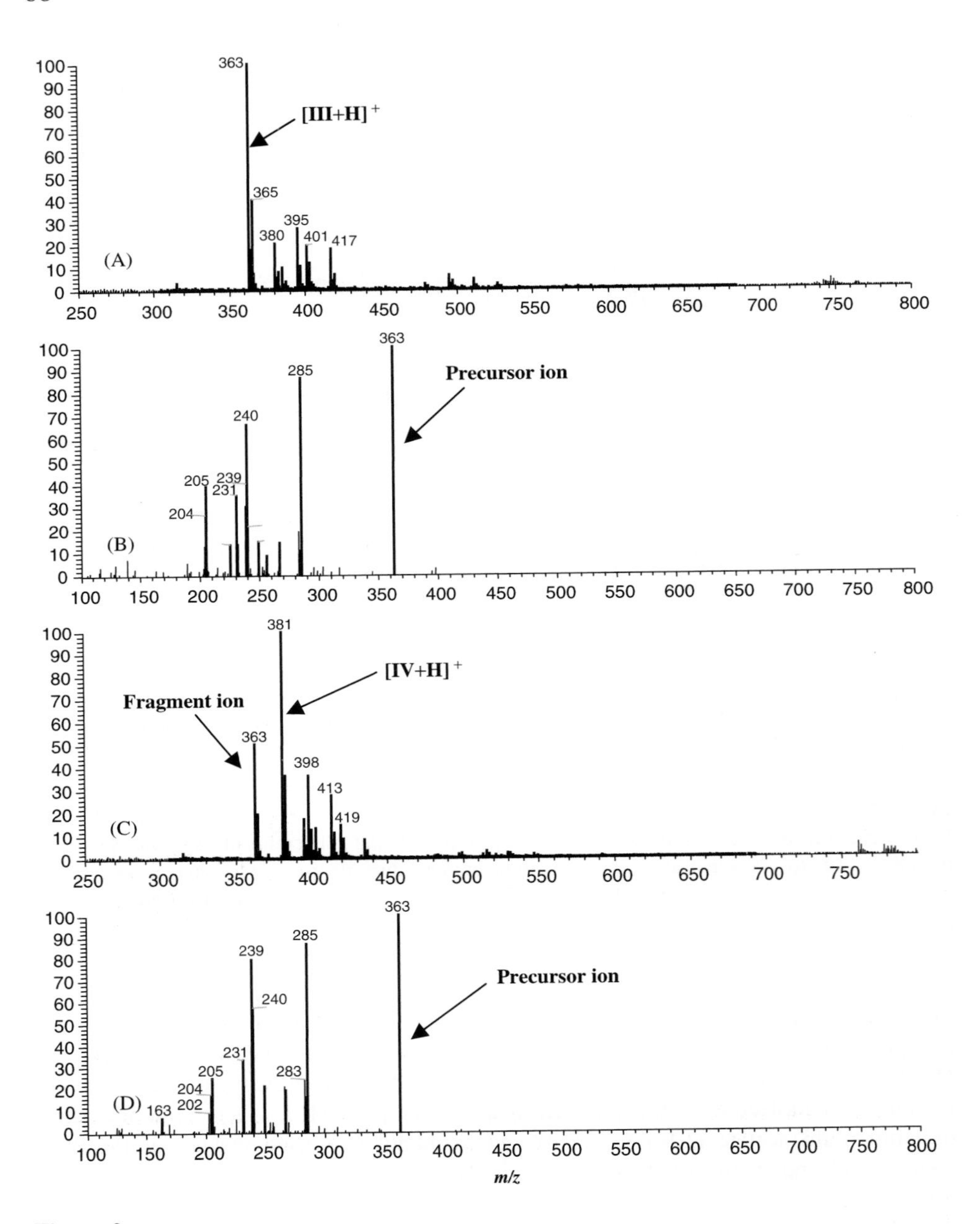

Figure 2.
Positive ion electrospray mass spectrum (A) and MS/MS spectrum using precursor ion *m/z* 363 (B) for lactone drug **III**; and positive ion electrospray mass spectrum (C) and MS/MS spectrum using precursor ion *m/z* 363 (D) for **IV**, the hydroxy acid metabolite. Jemal and Xia (1999) Copyright 1999. John Wiley & Sons Limited. Reproduced with permission

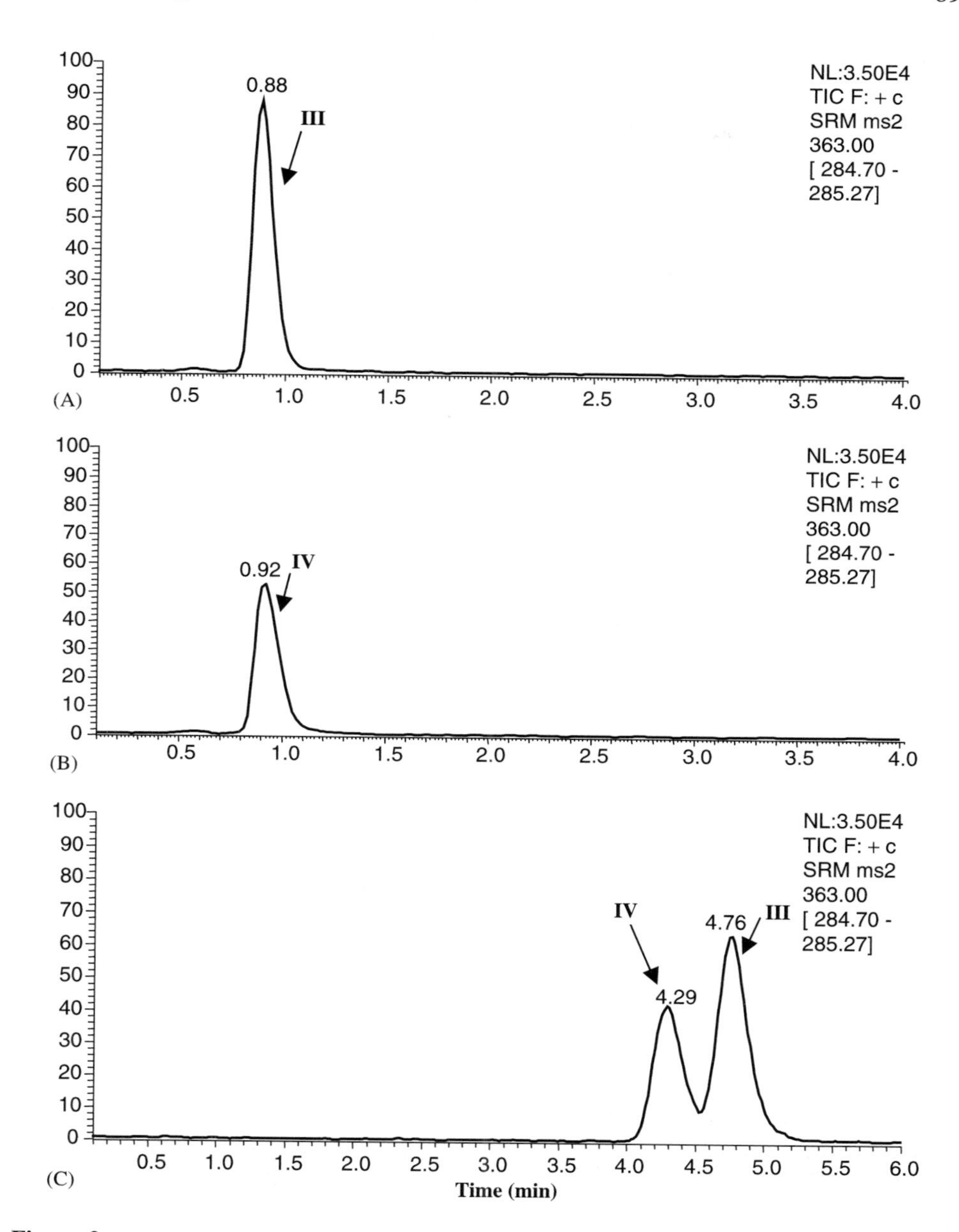

Figure 3.
SRM (m/z 363→m/z 285) chromatograms of **III** (A) and **IV** (B) in a mobile phase of 0.1% formic acid and methanol (25/75, v/v); and that of **III** and **IV** (C) in a mobile phase of 0.1% formic acid and methanol (45/55, v/v). Jemal and Xia (1999) Copyright 1999. John Wiley & Sons Limited. Reproduced with permission

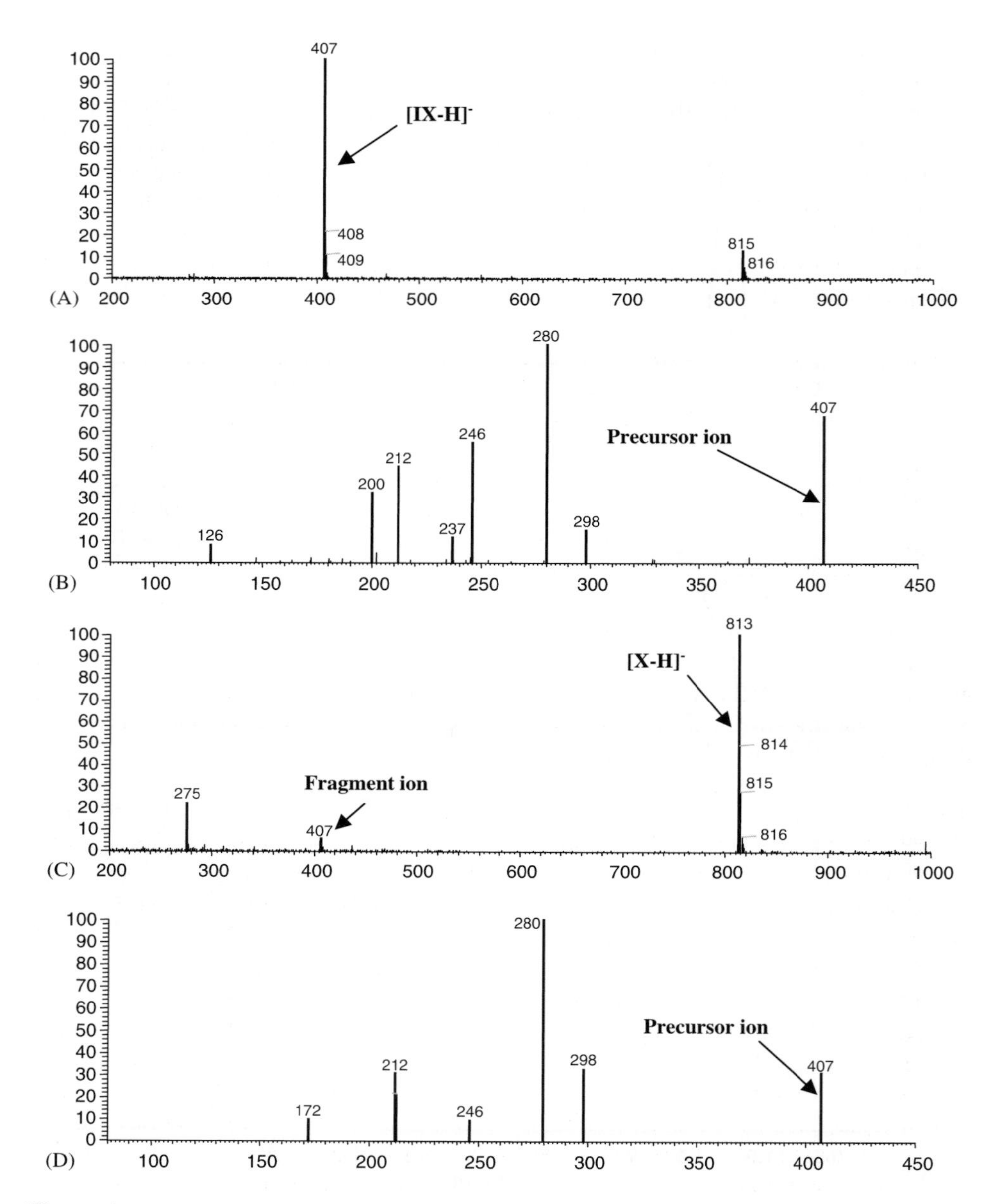

Figure 4.
Negative ion electrospray mass spectrum (A) and MS/MS spectrum, using precursor ion *m/z* 407 (B) for thiol drug **IX**; and negative ion electrospray mass spectrum (C) and MS/MS spectrum using precursor ion *m/z* 407 (D) for **X**, the disulfide. Jemal and Xia (1999) Copyright 1999. John Wiley & Sons Limited. Reproduced with permission

it would be reasonable to develop and validate a SRM LC–MS/MS method using the transition of *m/z* 407 to *m/z* 280 for the quantitative determination of **IX** in a biological sample. On the other hand, the mass spectrum in Fig. 4C shows that **X** gave, in addition to an intense [**X** − H]$^-$ ion at *m/z* 813, a fragment ion arising from in-source conversion at *m/z* 407, which is the same as the precursor ion in the SRM transition utilized for the proposed LC–MS/MS method for **IX**. Thus, unless the LC–MS/MS method can chromatographically separate **IX** from **X**, the method developed and validated for **IX** would be interfered with by the presence of **X**. This is illustrated in Fig. 5. Figures 5A and B show the SRM chromatograms of **IX** and **X** obtained by conducting the SRM transition of *m/z* 407 to *m/z* 280, with an isocratic mobile phase of 45% ammonium acetate (5 mM, pH 5.5) and 55% acetonitrile at a flow rate of 0.8 ml/min. Compounds **IX** and **X** were not resolved under these LC conditions, with *k* values of 0.43 and 0.56 for **IX** and **X**, respectively. Figures 5C and D show that **IX** and **X** were well resolved by using an isocratic mobile phase of 55% formic acid (0.1%) and 45% acetonitrile, with *k* values of 0.68 and 1.3 for **IX** and **X**, respectively, and a resolution of 1.7. This demonstrates an example where a baseline separation can be achieved with a *k* value of only 1.3. It should be noted that **IX** gave a signal at *m/z* 815 due to [2**IX** − H]$^-$. However, this ion should not interfere with a SRM transition for **X** utilizing *m/z* 813 as the precursor ion provided that unit-mass resolution is maintained in the precursor ion analyzer.

The negative ion electrospray mass and MS/MS spectra of **VII**, a phenolic drug, and of its prodrug **VIII**, are shown in Fig. 6. The mass spectrum in Fig. 6A shows that **VII** gave an intense [**VII** − H]$^-$ ion at *m/z* 369, and the MS/MS spectrum in Fig. 6B using the *m/z* 369 as the precursor ion gave an intense product ion at *m/z* 229. Thus, it would be reasonable to develop and validate a SRM LC–MS/MS method using the transition of *m/z* 369 to *m/z* 229 for the quantitative determination of **VII** in a biological sample. On the other hand, the mass spectrum in Fig. 6C shows that **VIII** gave, in addition to an intense ion at *m/z* 646, a fragment ion arising from in-source conversion at *m/z* 369, which is the same as the precursor ion in the SRM transition utilized for the proposed LC–MS/MS method of **VII**. Thus, unless the LC–MS/MS method can chromatographically separate **VII** from **VIII**, the method developed and validated for **VII** would be interfered with by the presence of **VIII**. It is interesting to note that **VIII**, a quaternary ammonium compound, did yield a negative ion at *m/z* 646, which is attributed to the formation of an adduct consisting of **VIII** and two units of acetate ion to obtain [**VIII**$^+$ + 2CH_3COO^-]$^-$. It should also be noted that **VII**, a phenolic drug, gave a negative ion at [M − H]$^-$ in an acidic mobile phase in which the compound is expected to be in the non-ionized form.

Other published examples of interference due to in-source conversion, in the absence of adequate chromatographic separation, caused by a metabolite or a prodrug in the accurate quantification of the drug include the analysis of morphine in the presence of its morphine-3-glucuronide and morphine-6-glucuronide metabolites (Naidong *et al.*, 1999); analysis of a drug in the presence of its *N*-oxide metabolite (Ayrton *et al.*, 1999; Ramanathan *et al.*, 2000); analysis of fosinoprilat in the presence of its prodrug fosinopril (Jemal *et al.*, 2000a); analysis of simvastatin acid in the presence of its prodrug simvastatin (Jemal *et al.*, 2000b); analysis of hydromorphone in the

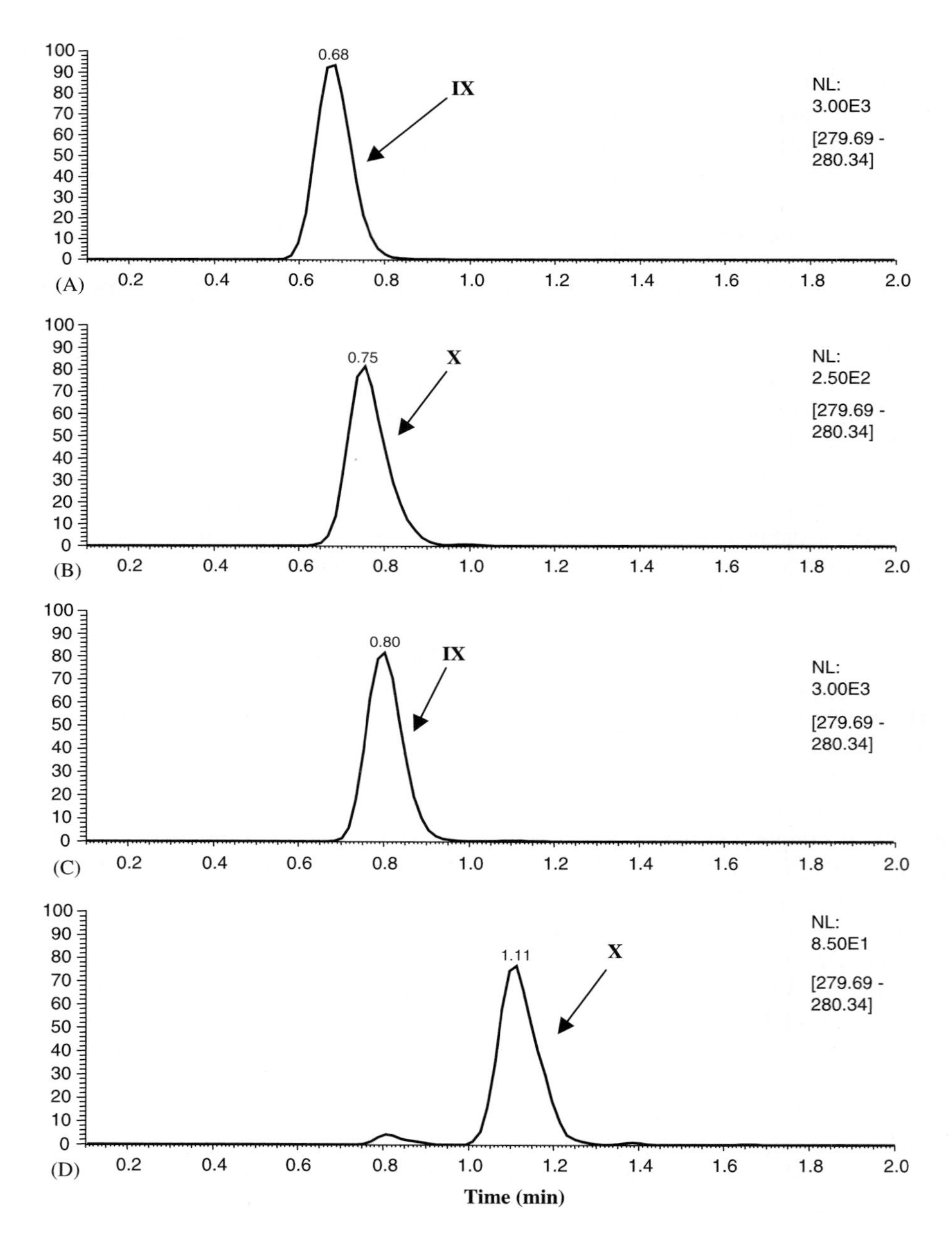

Figure 5.
SRM (m/z 407 → m/z 280) chromatograms of **IX** (A) and **X** (B) in a mobile phase of ammonium acetate (5 mM, pH 5.5) and acetonitrile (45/55, v/v); and those of **IX** (C) and **X** (D) in a mobile phase of ammonium acetate (5 mM, pH 5.5) and acetonitrile (55/45, v/v). Jemal and Xia (1999) Copyright 1999. John Wiley & Sons Limited. Reproduced with permission

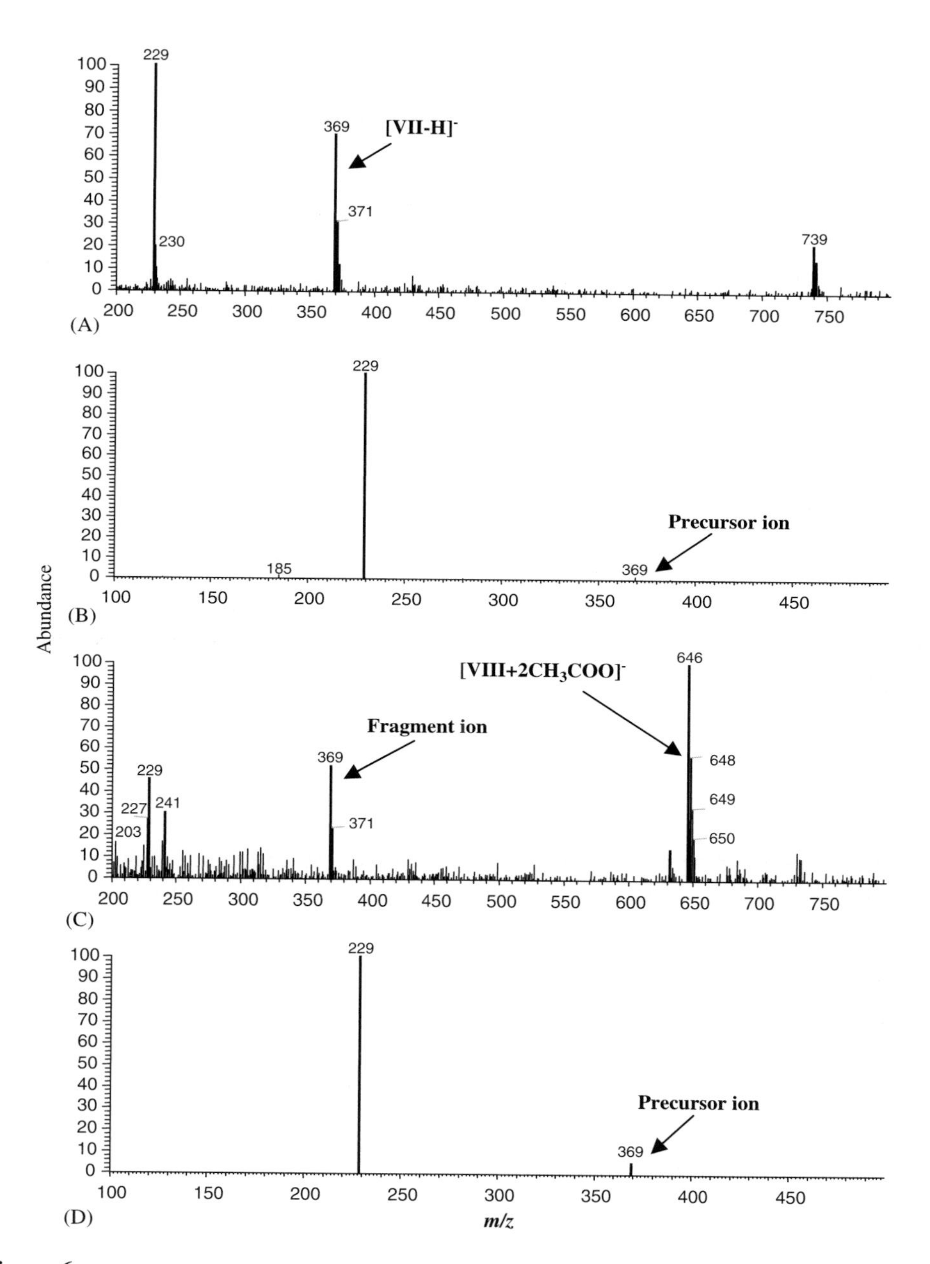

Figure 6.
Negative ion electrospray mass spectrum (A) and MS/MS spectrum using precursor ion *m/z* 369 (B) for phenolic drug **VII**; and negative ion electrospray mass spectrum (C) and MS/MS spectrum using precursor ion *m/z* 369 (D) for **VIII**, the prodrug. Jemal and Xia (1999) Copyright 1999. John Wiley & Sons Limited. Reproduced with permission

presence of its hydromorphone-3-glucuronide metabolite (Naidong *et al.*, 2000); analysis of a carboxylic acid containing compound in the presence of its acylglucuronide metabolite (Romanyshyn *et al.*, 2000); and analysis of a primary amine containing drug in the presence of its carbamoyl glucuronide (Liu and Pereira, 2002). The in-source dissociation behavior of more than 100 *O*- and *N*-glucuronides in electrospray ionization, operated in both positive and negative modes, has recently been reported (Yan *et al.*, 2003). It was found that the source voltage was the most critical parameter contributing to in-source fragmentation of the glucuronides, and that the degree of in-source dissociation was different for the different glucuronides.

The metabolites discussed above as contributing to the SRM signal of the drug due to their in-source conversion, such as metabolites **IV**, **VI**, and **X** in Fig. 1, can be classified as non-isomeric metabolites. However, post-dose samples may also contain isomeric metabolites, such as metabolite **II** shown in Fig. 1, which is a *Z*-isomeric metabolite of **I** (Xia *et al.*, 1999), and also epimeric metabolites (Testa *et al.*, 1993). Isomeric metabolites will obviously interfere with the SRM transition used for the quantitation of the drug.

4.3. Analytical Pitfalls Due to the Simultaneous Formation of $[M+H]^+$ and $[M+NH_4]^+$ Ions from a Drug as Well as Its Metabolite

There are special situations where a chromatographic separation between a drug and its non-isomeric metabolite may be needed even in the absence of in-source conversion in order to accurately quantitate the drug. One such situation involves a lactone drug and its hydroxy acid metabolite (or vice versa). The unique feature of a lactone and the corresponding hydroxy acid is the 18-mass-unit difference between the two compounds, which coincidentally is only one mass unit different from the 17-mass-unit difference that exists between the $[M+H]^+$ and $[M+NH_4]^+$ ions of any analyte. This has implications for the development of LC–MS/MS methods for the accurate quantitation of a lactone drug in the presence of its hydroxy acid metabolite, or vice versa. This is demonstrated, as summarized below from a previously reported work (Jemal and Ouyang, 2000), by using simvastatin, a prodrug, and simvastatin acid, its active metabolite (Morris *et al.*, 1993), as model lactone and hydroxy acid compounds, respectively. The chemical structures of the two compounds are shown in Fig. 7.

Under the mobile phase and mass spectrometric conditions used (Jemal and Ouyang, 2000), both simvastatin and simvastatin acid gave adequately large responses for both $[M+H]^+$ and $[M+NH_4]^+$ ions. Four SRM chromatograms, representing two lactone channels and two acid channels, were obtained at different mass resolution settings for simvastatin as well as simvastatin acid samples prepared separately. The four channels were the $[M+H]^+$ lactone channel (*m/z* 419→*m/z* 285), the $[M+NH_4]^+$ lactone channel (*m/z* 436→*m/z* 285), the $[M+H]^+$ acid channel (*m/z* 437→*m/z* 303), and the $[M+NH_4]^+$ acid channel (*m/z* 454→*m/z* 303).

Four SRM chromatograms, representing the two lactone channels and the two acid channels, obtained at a unit-mass resolution (Q1 FWHM = 0.65 u and Q3

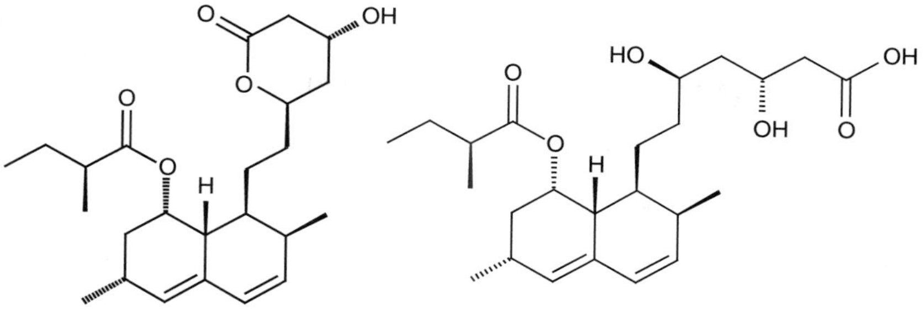

Simvastatin
$C_{25}H_{38}O_5$
Molecular weight: 418.58
Monoisotopic exact mass: 418.27

Simvastatin acid
$C_{25}H_{40}O_6$
Molecular weight: 436.59
Monoisotopic exact mass: 436.28

Figure 7.
Chemical structures of simvastatin and simvastatin acid. Jemal and Ouyang (2000) Copyright 2000. John Wiley & Sons Limited. Reproduced with permission

FWHM = 0.65 u, where FWHM = full-width at half-maximum) for a simvastatin sample are shown in Fig. 8. The four corresponding SRM chromatograms for a simvastatin acid sample are shown in Fig. 9. Looking at the $[M+H]^+$ lactone channel (*m/z* 419→*m/z* 285), the simvastatin sample gave a large peak at the retention time of simvastatin (1.89 min) but did not give a peak at the retention time of simvastatin acid (1.31 min), as shown in Fig. 8A. The simvastatin acid sample gave a small peak at the retention time of simvastatin (Fig. 9A). This peak, which is ~2.0% of the lactone peak seen in the simvastatin sample (Fig. 8A), is due to a small amount of the simvastatin present in the simvastatin acid sample. The simvastatin acid sample gave little or no peak at the retention time of the acid, which indicates that there is practically no in-source lactonization of the acid (Fig. 9A). The combined results of the simvastatin and simvastatin acid samples show that the $[M+H]^+$ lactone channel is selective for the lactone and does not respond to the acid. Therefore, this lactone channel could be used for the accurate quantitation of the lactone in the presence of the acid even in the absence of chromatographic separation between the two compounds. Chromatographic separation would have been required if in-source lactonization of the acid had occurred. It should be noted that under a different set of chromatographic and mass spectrometric conditions (Jemal *et al.*, 2000b), a substantial in-source lactonization was obtained for simvastatin acid.

Looking at the $[M+NH_4]^+$ lactone channel (*m/z* 436→*m/z* 285), the simvastatin sample gave a very large peak at the retention time of simvastatin, but did not give a peak at the retention time of simvastatin acid (Fig. 8B). The simvastatin acid sample gave a small peak at the retention time of simvastatin (Fig. 9B). This peak is due to a small amount of the simvastatin present in the simvastatin acid sample. The simvastatin acid sample gave little or no response at the retention time of the acid, which indicates that there is practically no in-source lactonization. The combined results of the simvastatin and simvastatin acid samples show that the $[M+NH_4]^+$ lactone

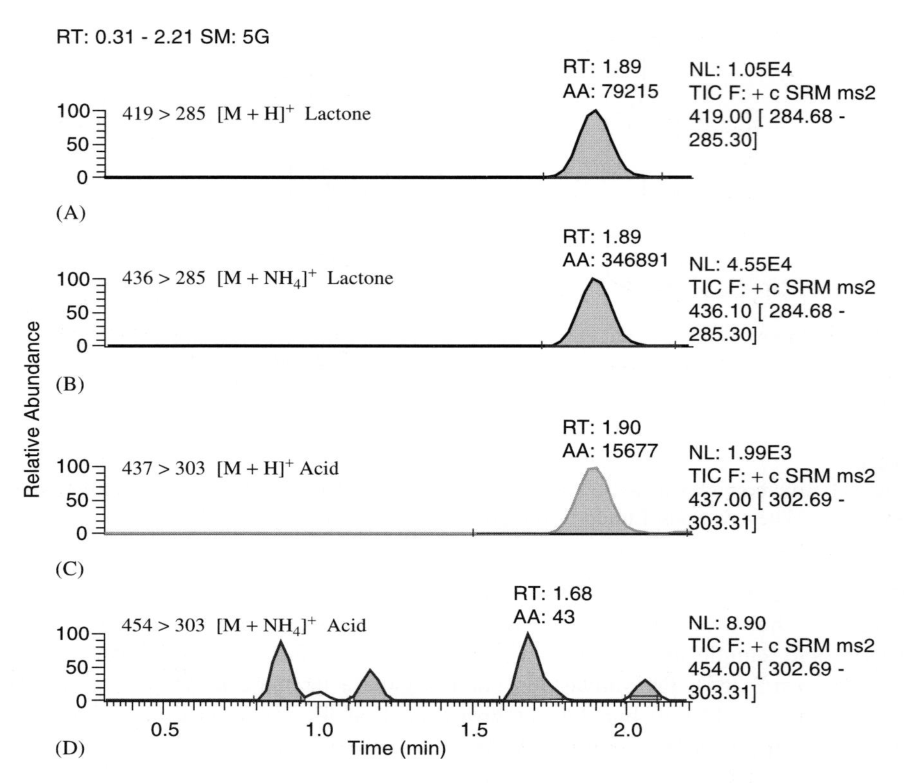

Figure 8.
SRM chromatograms obtained from a simvastatin sample (500 ng/ml), with Q1 and Q3 at a unit-mass resolution: (A) the $[M+H]^+$ lactone channel (*m/z* 419→*m/z* 285); (B) the $[M+NH_4]^+$ lactone channel (*m/z* 436→*m/z* 285); (C) the $[M+H]^+$ acid channel (*m/z* 437→*m/z* 303); and (D) the $[M+NH_4]^+$ acid channel (*m/z* 454→*m/z* 303). Jemal and Ouyang (2000) Copyright 2000. John Wiley & Sons Limited. Reproduced with permission

channel is selective for the lactone and does not respond to the acid. Therefore, this lactone channel could be used for the accurate quantitation of the lactone in the presence of the acid even in the absence of chromatographic separation between the two compounds. Chromatographic separation would have been required if in-source lactonization of the acid had occurred.

Looking at the $[M+H]^+$ acid channel (*m/z* 437→*m/z* 303), the simvastatin acid sample gave a peak at the retention time of simvastatin acid, but did not give a peak at the retention time of simvastatin lactone (Fig. 9C). The simvastatin sample did not give a peak at the retention time of the acid, which indicates the absence of any acid in the lactone sample. However, the simvastatin sample gave a sizable peak at the retention time of simvastatin (Fig. 8C). This peak is ~82% of the acid peak seen for the simvastatin acid sample (Fig. 9C). The peak is due to the A + 1 isotopic

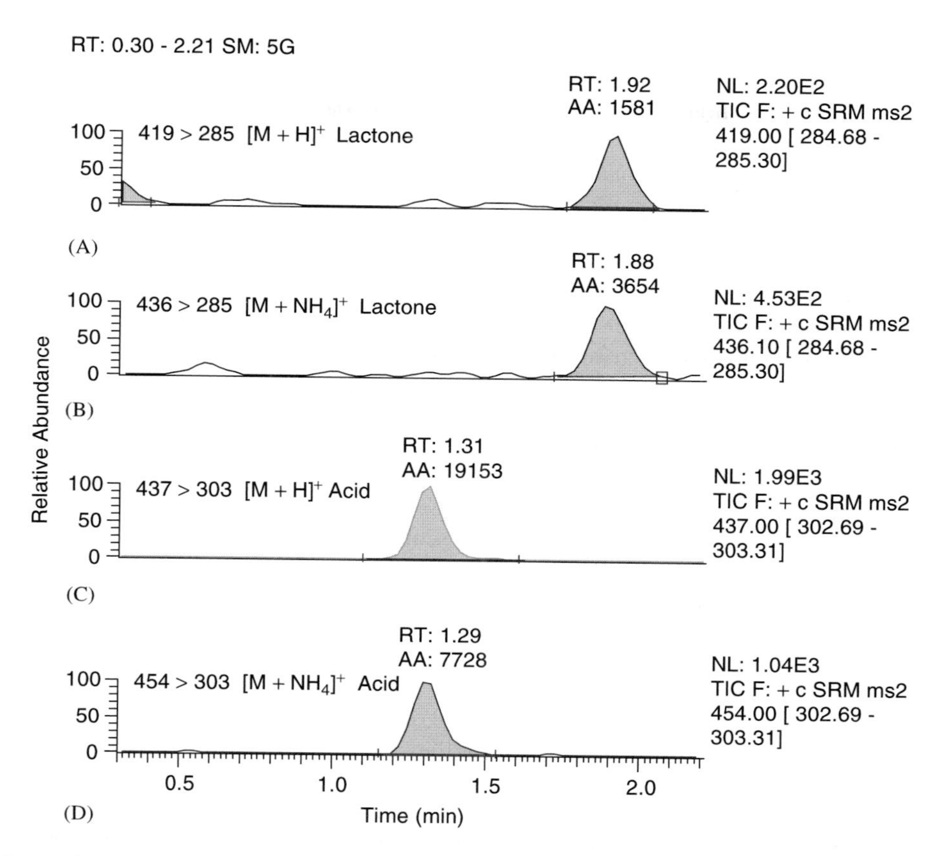

Figure 9.
SRM chromatograms obtained from a simvastatin acid sample (500 ng/ml), with Q1 and Q3 at a unit-mass resolution: (A) the $[M+H]^+$ lactone channel (*m/z* 419→*m/z* 285); (B) the $[M+NH_4]^+$ lactone channel (*m/z* 436→*m/z* 285); (C) the $[M+H]^+$ acid channel (*m/z* 437→*m/z* 303); and (D) the $[M+NH_4]^+$ acid channel (*m/z* 454→*m/z* 303). Jemal and Ouyang (2000) Copyright 2000. John Wiley & Sons Limited. Reproduced with permission

signal from the $[M+NH_4]^+$ ion of simvastatin, which also generates a product ion at *m/z* 303. In-source hydrolysis of the lactone would also give a peak at the same lactone retention time. However, the results from the $[M+NH_4]^+$ acid channel (see below) show that there is no in-source hydrolysis. The combined results of the simvastatin and simvastatin acid samples show that, even in the absence of in-source hydrolysis, the $[M+H]^+$ acid channel is not selective for the acid in the absence of a chromatographic separation between the two compounds.

Looking at the $[M+NH_4]^+$ acid channel (*m/z* 454→*m/z* 303), the simvastatin acid sample gave a peak at the retention time of simvastatin acid, but did not give a peak at the retention time of simvastatin (Fig. 9D). The simvastatin sample did not give a peak at the retention time of simvastatin acid (Fig. 8D), which indicates the absence of any acid in the lactone sample. The simvastatin sample did not give a peak at the retention time of simvastatin (Fig. 8D), which indicates the absence of in-source hydrolysis. The combined results of the simvastatin and simvastatin acid samples show that the $[M+NH_4]^+$ acid channel is selective for the acid. Therefore, this acid channel could be used for the accurate quantitation of the acid in the presence of the lactone, even without chromatographic separation between the two compounds. Chromatographic separation would have been required if in-source hydrolysis of the lactone had occurred.

As a contrast to the results obtained under a unit-mass resolution setting, the four SRM chromatograms, representing the two lactone channels and the two acid channels, obtained at a two-mass resolution setting (Q1 FWHM = 1.51 u and Q3 FWHM = 0.65 u) for a simvastatin sample are shown in Fig. 10. The four corresponding SRM chromatograms for a simvastatin acid sample are shown in Fig. 11. Looking at the $[M+H]^+$ lactone channel (*m/z* 419→*m/z* 285), the results of the simvastatin sample (Fig. 10A) and simvastatin acid sample (Fig. 11A) were similar to those seen in the $[M+H]^+$ lactone channel under a unit-mass resolution setting except for the larger response due to the wider Q1 mass window. Typically, changing the Q1 resolution from a unit-mass resolution to a two-mass resolution resulted in three-fold increase in response.

Looking at the $[M+NH_4]^+$ lactone channel (*m/z* 436→*m/z* 285), the simvastatin sample result (Fig. 10B) was similar to that seen in the $[M+NH_4]^+$ lactone channel under a unit-mass resolution setting except for the larger response due to the wider Q1 mass window. However, the simvastatin acid sample gave a minor peak at the retention time of simvastatin acid (Fig. 11B), which was not seen under a unit-mass resolution setting. The peak is 0.15% of the lactone peak obtained from the lactone sample (Fig. 10B). This peak could not be due to in-source lactonization since such a peak was not seen with Q1 at a unit-mass resolution setting. This peak is attributed to the contribution from the $[M+H]^+$ ion of the acid (*m/z* 437) due to inadequate mass resolution of *m/z* 436 from *m/z* 437, which also generates a product ion at *m/z* 285.

Looking at the $[M+H]^+$ acid channel (*m/z* 437→*m/z* 303), the simvastatin acid sample result (Fig. 11C) was similar to that seen in the $[M+H]^+$ acid channel under a unit-mass resolution setting except for the larger response due to the wider Q1 mass window. However, the simvastatin sample peak obtained at the retention time of simvastatin (Fig. 10C) was quantitatively different from that obtained under a unit-mass resolution setting. This peak was 91% of the acid peak in the simvastatin acid sample (Fig.11C). This is larger than the corresponding value of 82% obtained under a unit-mass resolution. This peak is attributed to a combination of the response from the simvastatin $[M+NH_4]^+$ ion (due to inadequate mass resolution of *m/z* 437 from *m/z* 436) and the response from the A + 1 isotopic peak arising from the same simvastatin $[M+NH_4]^+$ ion.

Looking at the $[M+NH_4]^+$ acid channel (*m/z* 454→*m/z* 303), the results of the simvastatin sample (Fig. 10D) and simvastatin acid sample (Fig. 11D) were similar to those seen in the $[M+NH_4]^+$ acid channel under a unit-mass resolution setting except for the larger response due to the wider Q1 mass window.

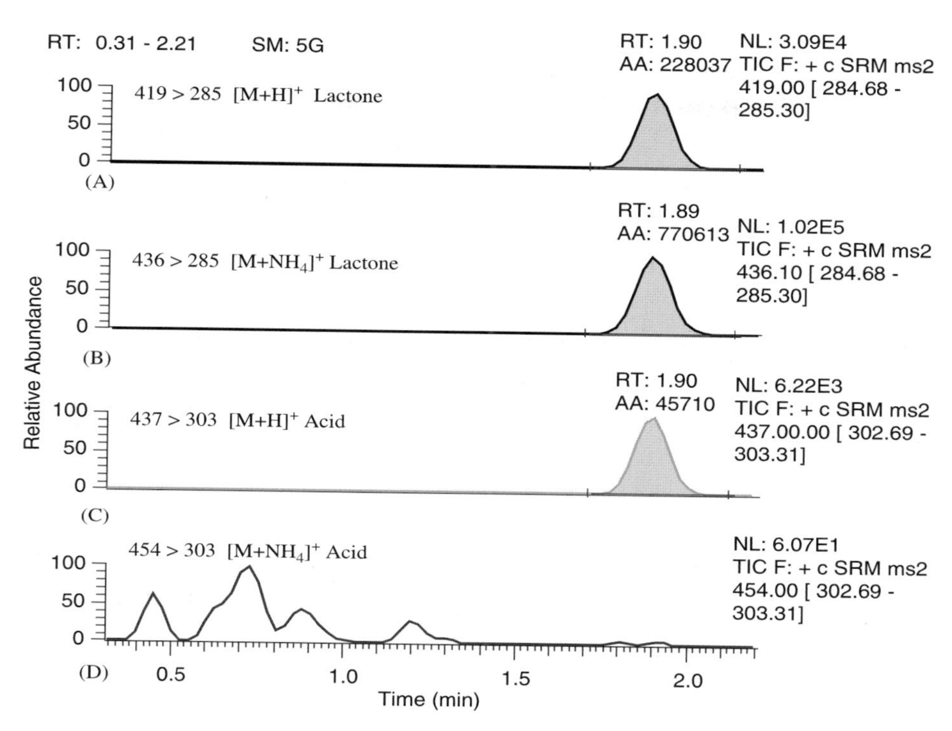

Figure 10.
SRM chromatograms obtained from a simvastatin sample (500 ng/ml), with Q1 at a two-mass resolution and Q3 at unit-mass resolution: (A) the $[M+H]^+$ lactone channel (*m/z* 419→*m/z* 285); (B) the $[M+NH_4]^+$ lactone channel (*m/z* 436→*m/z* 285); (C) the $[M+H]^+$ acid channel (*m/z* 437→*m/z* 303); and (D) the $[M+NH_4]^+$ acid channel (*m/z* 454→*m/z* 303). Jemal and Ouyang (2000) Copyright 2000. John Wiley & Sons Limited. Reproduced with permission

Carrying out the same set of experiments under four-mass resolution conditions (Q1 FWHM = 3.11 u, Q3 FWHM = 0.65 u) confirmed the conclusions arrived at from the two-mass resolution experiments. The following conclusions and observations can be made from the results of the above experiments. Between the two lactone channels that can be used for the quantitation of the lactone, the $[M+H]^+$ lactone channel is the more specific of the two. With this channel, chromatographic separation of the lactone from the hydroxy acid will be needed only if there is in-source lactonization, no matter what the Q1 mass resolution is. On the other hand, the $[M+NH_4]^+$ lactone channel will require chromatographic separation of the lactone from the hydroxy acid when a wide Q1 mass window is used even if there is no in-source lactonization. This is the consequence of $[M+NH_4]^+$ of the lactone being only one mass unit lower than the $[M+H]^+$ of the acid. Between the two acid channels that can be used for the quantitation of the hydroxy acid, the $[M+NH_4]^+$ acid

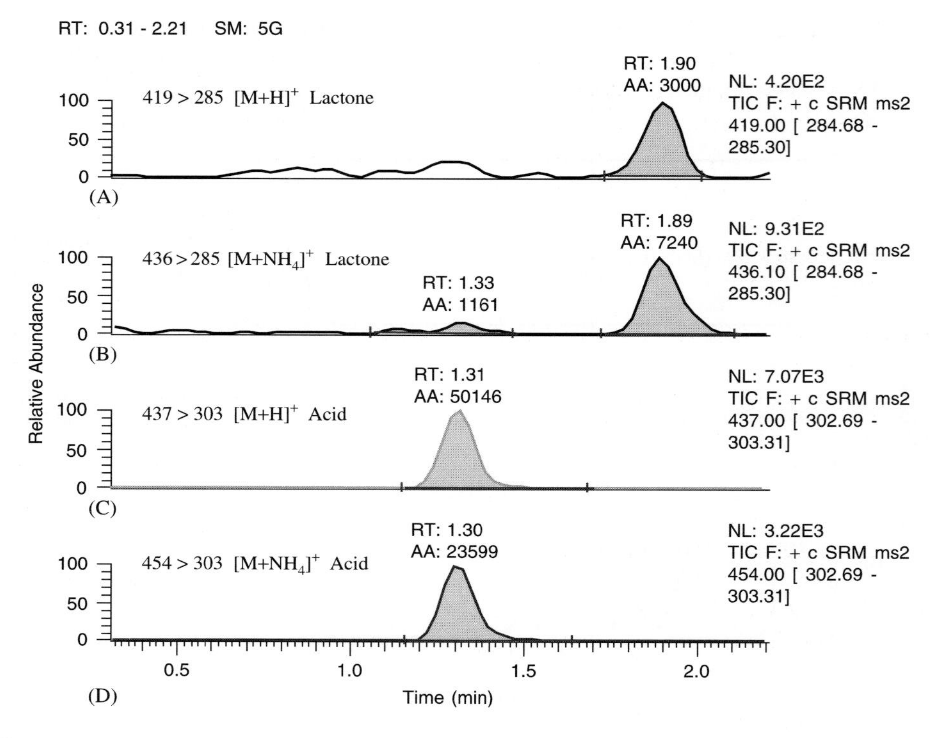

Figure 11.
SRM chromatograms obtained from a simvastatin acid sample (500 ng/ml), with Q1 at a two-mass resolution and Q3 at unit-mass resolution: (A) the $[M+H]^+$ lactone channel (*m/z* 419→*m/z* 285); (B) the $[M+NH_4]^+$ lactone channel (*m/z* 436→*m/z* 285); (C) the $[M+H]^+$ acid channel (*m/z* 437→*m/z* 303); and (D) the $[M+NH_4]^+$ acid channel (*m/z* 454→*m/z* 303). Jemal and Ouyang (2000) Copyright 2000. John Wiley & Sons Limited. Reproduced with permission

channel is by far the more specific of the two. With this channel, chromatographic separation of the hydroxy acid from the lactone will be needed only if there is in-source hydrolysis, no matter what the Q1 mass resolution is. On the other hand, the $[M+H]^+$ acid channel will require chromatographic separation of the hydroxy acid from the lactone no matter what the Q1 mass resolution is or whether or not there is in-source hydrolysis. This is the consequence of $[M+H]^+$ of the hydroxy acid being only one mass unit higher than the $[M+NH_4]^+$ of the lactone. When working with lactone drugs or hydroxy acid drugs, it should be noted that the $[M+NH_4]^+$ ions may be formed, in addition to the $[M+H]^+$ ions, even when the LC mobile phase does not contain known sources of the ammonium ion. It should also be noted that the ratio of the $[M+NH_4]^+$ ion to the $[M+H]^+$ ion depends not only on the LC mobile phase but also on the mass spectrometer source parameters.

Another noteworthy situation involves a primary amide drug ($RCONH_2$) and its acid metabolite (RCOOH), where there is only one-mass-unit difference between the two compounds. Thus, unless there is chromatographic separation between the two, the $[M+H]^+$ channel of the acid metabolite will be interfered with by the amide drug because of the A + 1 isotopic contribution from the $[M+H]^+$ ion of the drug. The same type interference will also occur when the $[M+NH_4]^+$ channel is used instead of the $[M+H]^+$ channel. When a wide resolution setting (larger than unit-mass resolution) is used, there is the additional problem of the $[M+H]^+$ channel of the amide drug being interfered with by the acid metabolite because of inadequate mass resolution between the $[M+H]^+$ channels of the two compounds. Same kind of problem occurs when $[M+NH_4]^+$ channel is used. Other noteworthy drug–metabolite pairs that can present analytical pitfalls involve a primary alcohol drug ($RCH_2(OH)$) and its aldehyde metabolite (RCHO) or vice versa, and a secondary alcohol drug (RCH(OH)R′) and its ketone metabolite (RCOR′) or vice versa. In both cases, there is only a two-mass difference between a drug and the corresponding metabolite. Hence, there is a potential for drug–metabolite interference due to A + 2 isotopic contribution or when inadequate mass resolution is used. Another situation of interest is where a lactone drug or metabolite, in the presence of ammonium hydroxide that may be used in any step of sample preparation, may produce the acid, amide, or lactam form of the compound prior to or during LC–MS/MS. The potential interference of one form over another should be evaluated.

4.4. Analytical Pitfalls Due to Conversion of Metabolite to the Parent Drug During Sample Collection, Handling, Extraction, and Analysis

Another metabolite-related analytical problem is due to the fact that an isomeric or non-isomeric metabolite present in the post-dose sample may be unstable and hence may revert back to the drug during the multiple steps of sample collection and preparation that precede the introduction of the processed sample into the LC–MS/MS system. If such a conversion occurs, the measured concentration of the drug would be artificially inflated.

Acylglucuronide metabolites tend to be unstable, especially under alkaline conditions and elevated temperatures, and degrade to generate the parent drug (Khan *et al.*, 1998; Shipkova *et al.*, 2003). The mildly acidic pH of 3–5 tend to be the most desirable pH region for minimizing the hydrolysis of acylglucuronides in biological samples or during the numerous steps involved during sample analysis, including the solution used in the final step, namely, the injection or reconstitution solution. It should be emphasized that different acylglucuronides have been shown to have different rates of hydrolysis (Shipkova *et al.*, 2003). Another class of metabolites that are prone to degradation to generate the parent drug are the lactone metabolites of hydroxy acid drugs. Systematic studies of the effects of pH and temperature on the stability of the lactone metabolites of hydroxy acid drugs show that the sample pH should be adjusted to 3–5 in order to minimize the hydrolysis of the metabolite back to the drug or vice versa (Jemal *et al.*, 1999a,b, 2000b; Kantola *et al.*, 1998; Kaufman, 1990; Kearney *et al.*, 1993; Morris *et al.*, 1993; Won, 1994). Other metabolite functional groups that could undergo conversion to generate the parent drug include *O*-methyloximes (Xia *et al.*, 1999) and carbon–carbon double bonds (Wang *et al.*, 2003). Such compounds may undergo *E* to *Z* isomerization (or vice versa) due to exposure to light or undesirable pH conditions. Epimerization is another potential source of metabolite interference (Testa *et al.*, 1993; Won, 1994). A sample that contains a thiol drug and its disulfide metabolite may also cause an analytical challenge due to the potential for the conversion of the thiol to the disulfide or vice versa (Gilbert, 1995).

4.5. Bioanalytical Method Design for Drug Quantification in the Presence of a Metabolite that May Degrade to Generate the Drug

During method development for the quantification of a drug in a post-dose sample, which contains the drug and its potentially unstable metabolite, conditions must be optimized to minimize such interconversion. However, even the optimal conditions adopted may not totally prevent conversion of the metabolite to the drug. It is thus essential to appropriately design method development in order to minimize the adverse effect of such a conversion on the accuracy and precision of the method. The significance of proper method design has been systematically illustrated (Jemal and Xia, 2000) using two compounds that can potentially interconvert, namely, a hydroxy acid compound (pravastatin, analyte 1, Fig. 12) and the corresponding lactone compound (pravastatin lactone, analyte 2, Fig. 12). The important feature of the method design is the use of the appropriate composition (in terms of the ratio of the concentration of one analyte to that of the other) of quality control (QC) samples vis-à-vis the composition of calibration standards. This is illustrated by the work (Jemal and Xia, 2000) presented below.

The calibration standard curve in human plasma ranged from 5 to 500 ng/ml for both pravastatin and pravastatin lactone with each standard concentration point containing equal amounts of the two analytes. The QC samples, prepared in human plasma, contained only pravastatin, or only pravastatin lactone, or both in varying ratios of pravastatin to pravastatin lactone: 1:1, 1:3, 1:10, 3:1, and 10:1. The composition of the standard and QC samples are summarized in Table 1. Each concentration point of the calibration standard curve was run in duplicate and each QC sample was run in five replicates.

The standard and QC samples were analyzed by direct injection of the plasma samples without prior extraction (Jemal and Xia, 2000) after they were subjected to three different interconversion conditions. Under condition 1, all the standard and QC samples were prepared in an ice-bath and analyzed immediately after preparation. For analysis, a portion of internal standard working solution was added to each calibration standard and QC sample, which lowered the pH of the samples to 4.2, where interconversion was minimal. Under condition 2, all the standard and QC samples were analyzed after keeping them as is (i.e. at pH 7.3, without adjusting the pH) at

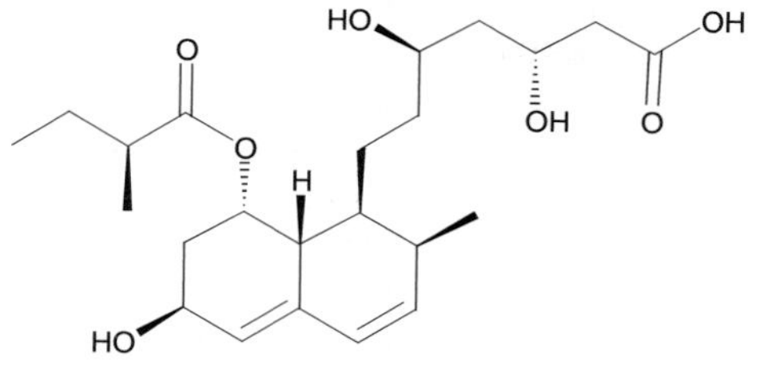

Pravastatin

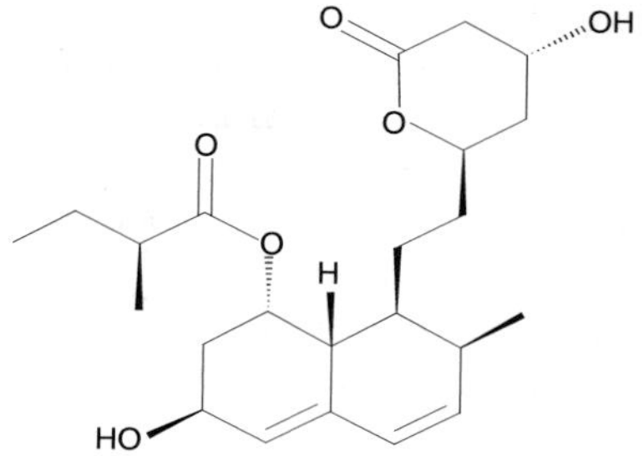

Pravastatin lactone

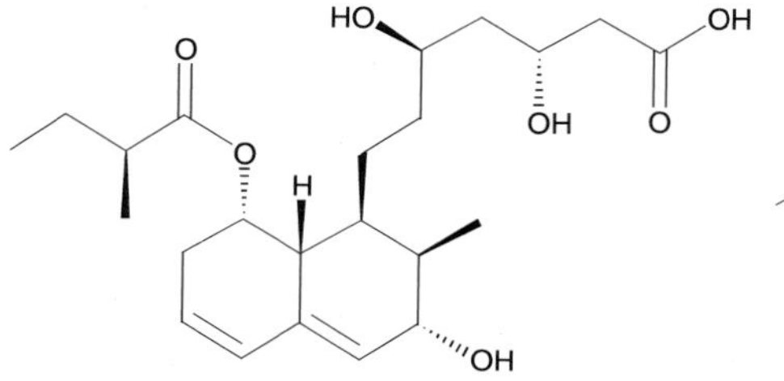

SQ-31906

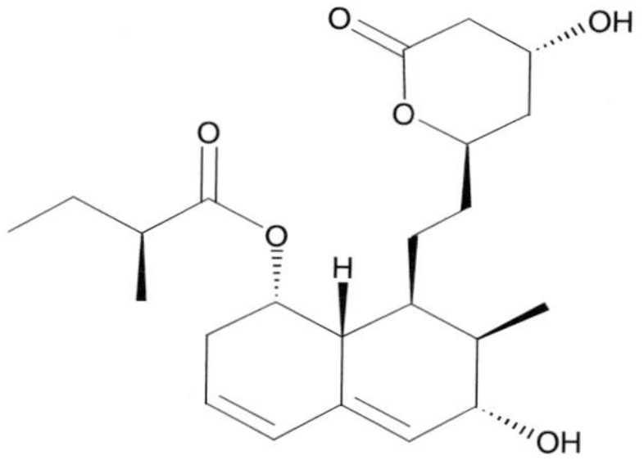

SQ-31906 lactone

Figure 12.
Chemical structures of pravastatin and pravastatin lactone. Reproduced from Jemal and Xia (2000) Copyright 2000, with permission from Elsevier

room temperature for 4 h to promote the conversion of one analyte to another. For analysis, a portion of internal standard working solution was added to each calibration standard and QC sample, which lowered the pH of the samples to 4.2. Under condition 3, all the standard and QC samples were adjusted to pH 1.8 by adding the hydrochloric acid/sodium chloride solution to each calibration standard and QC sample. The samples were then kept at room temperature for 2 h to promote analyte conversion. For analysis, a portion of internal standard working solution was added to each sample, which raised the pH to 4.2. The processed samples from conditions 1–3 were kept at 4°C on the HPLC autosampler during analysis.

The behavior of pravastatin-only and pravastatin lactone-only QC samples under the three conditions of sample preparation, shown in Figs. 13 and 14, was used to gage the degree of conversion of the two analytes, namely, no conversion, one-way conversion, or two-way conversion (interconversion). Under condition 1, there was no conversion of pravastatin to pravastatin lactone or vice versa. Under condition 2, pravastatin lactone was hydrolyzed to produce pravastatin, but there was no conversion from pravastatin to pravastatin lactone. Under condition 3, there were several conversion pathways. Pravastatin not only underwent intramolecular cyclization to yield pravastatin lactone, but also underwent isomerization of 6-α-hydroxyl to

Table 1.
Composition of calibration standards and QC samples

	Pravastatin (ng/ml)	Pravastatin lactone (ng/ml)
Standard 1 (1:1)	5.00	5.00
Standard 2 (1:1)	10.0	10.0
Standard 3 (1:1)	25.0	25.0
Standard 4 (1:1)	50.0	50.0
Standard 5 (1:1)	100	100
Standard 6 (1:1)	200	200
Standard 7 (1:1)	300	300
Standard 8 (1:1)	400	400
Standard 9 (1:1)	500	500
QC1 (1:1)	15.0	15.0
QC2 (1:1)	200	200
QC3 (1:1)	420	420
QC4 (1:3)	5.00	15.0
QC5 (1:10)	20.0	200
QC6 (1:10)	42.0	420
QC7 (0:15)	0	15.0
QC8 (0:200)	0	200
QC9 (0:420)	0	420
QC10 (3:1)	15.0	5.00
QC11 (10:1)	200	20.0
QC12 (10:1)	420	42.0
QC13 (15:0)	15.0	0
QC14 (200:0)	200	0
QC15 (420:0)	420	0

Note: The numbers within parentheses show the ratios of the pravastatin concentrations to the pravastatin lactone concentrations. Reproduced from Jemal and Xia (2000) Copyright 2000, with permission from Elsevier.

3-α-hydroxyl to produce positional isomers of pravastatin (SQ-31906) and pravastatin lactone (SQ-31906 lactone). Meanwhile, pravastatin lactone produced not only pravastatin but also SQ-31906 and SQ-31906 lactone. Thus, the same number of products were obtained whether the starting material was pravastatin or pravastatin lactone.

After having defined the three conditions of sample preparation by the behavior of the pravastatin-only and pravastatin lactone-only QC samples, the accuracy and precision obtained for the standard curve and the different sets of QC samples could be interpreted intelligently. The results of analysis of samples from condition 1 are shown in Table 2. The performance of the standard curve for both pravastatin and

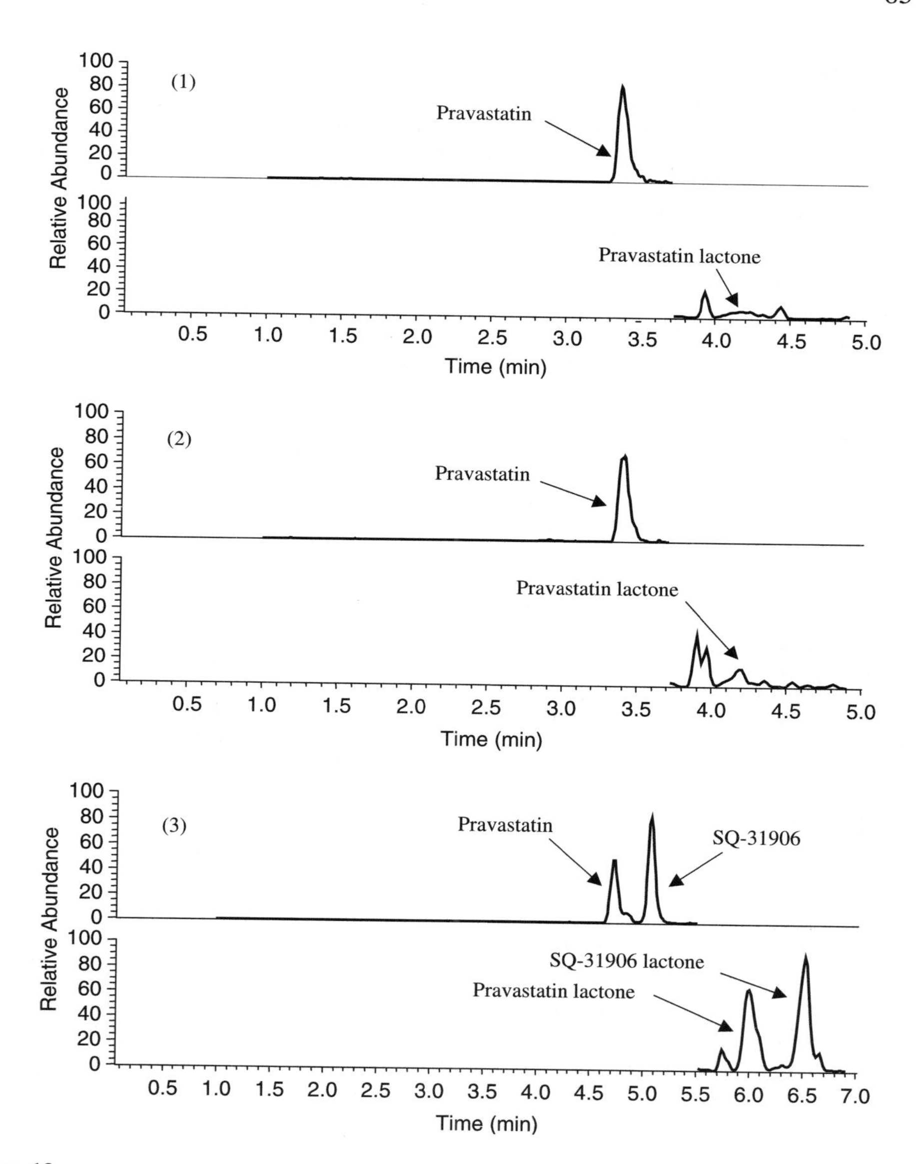

Figure 13.
SRM chromatograms of QC14 (pravastatin-only QC sample at 200 ng/ml) under conditions (1), (2), and (3). A longer analytical column was used for (3). Reproduced from Jemal and Xia (2000) Copyright 2000, with permission from Elsevier

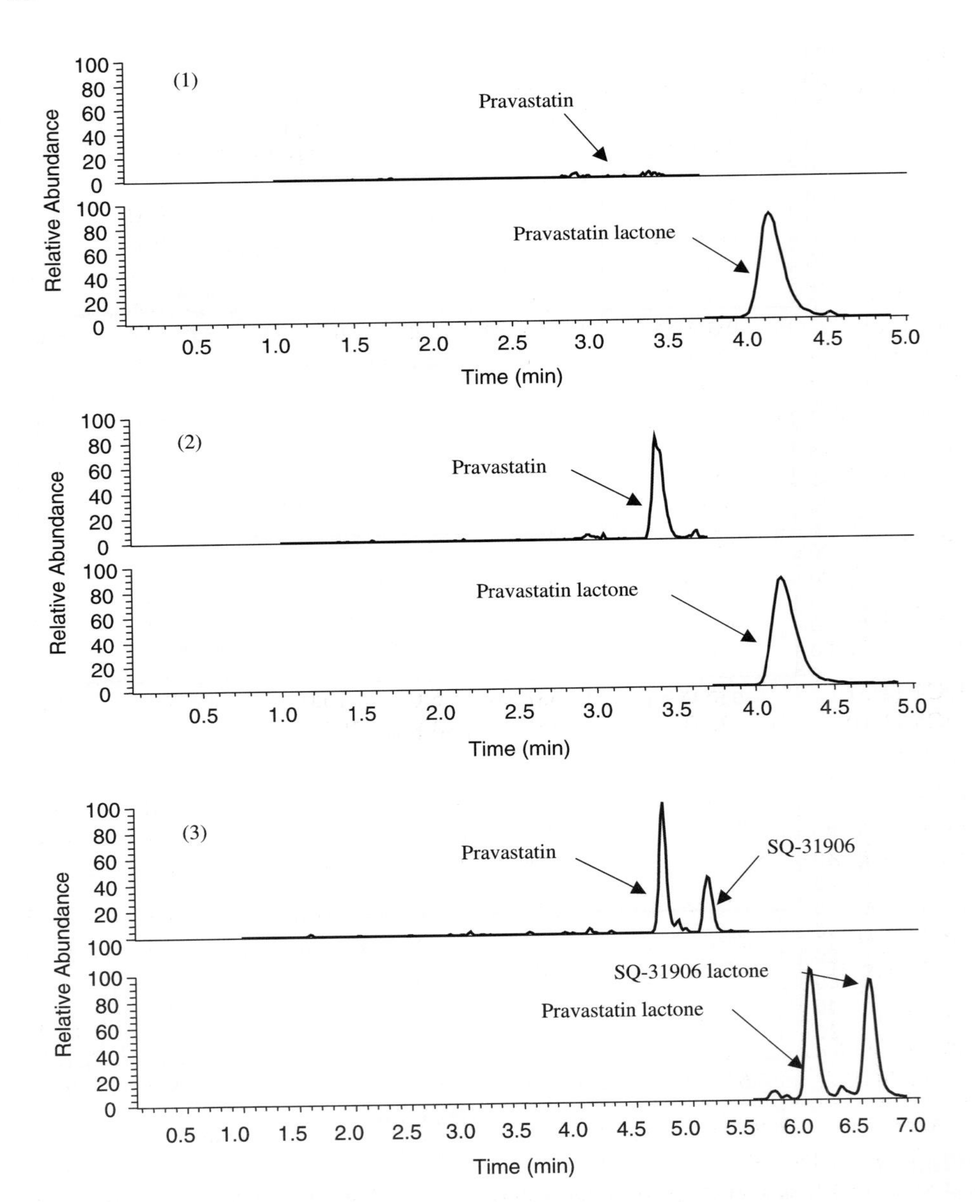

Figure 14.
SRM chromatograms of QC8 (pravastatin lactone-only QC sample at 200 ng/ml) under conditions (1), (2), and (3). A longer analytical column was used for (3). Reproduced from Jemal and Xia (2000) Copyright 2000, with permission from Elsevier

Table 2.
Accuracy and precision under condition 1

	Pravastatin			Pravastatin lactone		
	Nominal conc. (ng/ml)	RSD (%)	Dev. (%)	Nominal conc. (ng/ml)	RSD (%)	Dev. (%)
Standard 1 (1:1)	5.00	7.3	−1.7	5.00	7.9	+0.1
Standard 2 (1:1)	10.0	5.2	+2.5	10.0	1.3	−0.5
Standard 3 (1:1)	25.0	6.4	+1.2	25.0	5.2	−2.1
Standard 4 (1:1)	50.0	6.6	+3.4	50.0	2.8	+5.2
Standard 5 (1:1)	100	5.5	−0.1	100	1.9	+2.6
Standard 6 (1:1)	200	3.0	−2.1	200	1.7	+0.3
Standard 7 (1:1)	300	0.5	−1.5	300	3.4	−1.9
Standard 8 (1:1)	400	1.4	−0.7	400	1.7	−2.8
Standard 9 (1:1)	500	0.5	−0.9	500	0.7	−0.9
QC1 (1:1)	15.0	5.9	+1.3	15.0	4.7	−2.2
QC2 (1:1)	200	2.8	−7.1	200	1.6	−7.9
QC3 (1:1)	420	0.6	−8.6	420	3.9	−5.8
QC4 (1:3)	5.00	3.0	−6.6	15.0	4.3	−1.7
QC5 (1:10)	20.0	3.9	−4.8	200	3.3	−5.3
QC6 (1:10)	42.0	5.1	−2.6	420	2.0	−6.3
QC7 (0:15)	0	N/A	N/A	15.0	5.8	−3.5
QC8 (0:200)	0	N/A	N/A	200	3.7	−6.2
QC9 (0:420)	0	N/A	N/A	420	1.8	−1.3
QC10 (3:1)	15.0	3.1	−7.8	5.00	8.7	+4.3
QC11 (10:1)	200	2.4	−7.4	20.0	6.0	+3.2
QC12 (10:1)	420	5.0	−4.8	42.0	2.4	−0.3
QC13 (15:0)	15.0	8.2	−2.6		N/A	N/A
QC14 (200:0)	200	3.0	−8.0	0	N/A	N/A
QC15 (420:0)	420	2.3	−6.5	0	N/A	N/A

Note: Dev. = deviation of the mean concentration from the nominal concentration ($n = 2$ for standards; $n = 5$ for QC samples). N/A = not applicable, because the measured concentration was below LLQ. The numbers within parentheses show the ratios of the pravastatin concentrations to the pravastatin lactone concentrations. Reproduced from Jemal and Xia (2000) Copyright 2000, with permission from Elsevier

pravastatin lactone was excellent. The accuracy and precision for both pravastatin and pravastatin lactone were excellent in all QC samples.

Table 3 summarizes the results of analysis of samples from condition 2. The performance of the standard curve for both pravastatin and pravastatin lactone was

Table 3.
Accuracy and precision under condition 2

	Pravastatin			Pravastatin lactone		
	Nominal conc. (ng/ml)	RSD (%)	Dev. (%)	Nominal conc. (ng/ml)	RSD (%)	Dev. (%)
Standard 1 (1:1)	5.00	2.0	−7.6	5.00	4.6	+4.7
Standard 2 (1:1)	10.0	0.2	+10.7	10.0	11.0	−6.3
Standard 3 (1:1)	25.0	3.3	+9.6	25.0	5.7	−6.4
Standard 4 (1:1)	50.0	1.1	+3.4	50.0	4.0	−3.5
Standard 5 (1:1)	100	1.9	+5.4	100	6.2	−4.0
Standard 6 (1:1)	200	2.1	−2.8	200	1.8	+5.0
Standard 7 (1:1)	300	2.5	−4.9	300	0.2	+5.0
Standard 8 (1:1)	400	3.7	−8.9	400	3.2	+2.4
Standard 9 (1:1)	500	1.5	−4.8	500	0.2	+3.2
QC1 (1:1)	15.0	7.0	+2.7	15.0	6.7	+3.5
QC2 (1:1)	200	5.8	+4.8	200	2.6	−6.3
QC3 (1:1)	420	0.7	−10.6	420	2.1	−4.2
QC4 (1:3)	5.00	6.0	+27.6	15.0	3.5	+5.0
QC5 (1:10)	20.0	7.7	+158.2	200	2.9	−6.4
QC6 (1:10)	42.0	4.4	+65.4	420	2.3	+2.1
QC7 (0:15)	0	8.8	*4.72	15.0	6.7	−5.6
QC8 (0:200)	0	5.1	*58.4	200	2.9	−6.4
QC9 (0:420)	0	5.6	*74.7	420	5.9	−6.1
QC10 (3:1)	15.0	1.3	−20.2	5.00	5.0	+7.2
QC11 (10:1)	200	1.1	−19.4	20.0	6.3	+6.5
QC12 (10:1)	420	2.5	−17.9	42.0	2.0	+6.4
QC13 (15:0)	15.0	3.5	−19.2	0	N/A	N/A
QC14 (200:0)	200	1.7	−19.8	0	N/A	N/A
QC15 (420:0)	420	1.4	−18.4	0	N/A	N/A

Note: Dev. = deviation of the mean concentration from the nominal concentration (n = 2 for standards; n = 5 for QC samples). N/A = not applicable, because the measured concentration was below LLQ. The numbers within parentheses show the ratios of the pravastatin concentrations to the pravastatin lactone concentrations. * = measured concentration in ng/ml. Reproduced from Jemal and Xia (2000) Copyright 2000, with permission from Elsevier

excellent. The accuracy and precision for both pravastatin and pravastatin lactone were excellent in the QC samples in which the ratio of pravastatin to pravastatin lactone was identical to that in the calibration standards (ratio of 1:1). The accuracy for pravastatin was unacceptable for all QC samples in which the ratio of pravastatin to pravastatin lactone was different from the 1:1 ratio in the calibration standards. The lactone-only and 1:10 QC samples showed positive deviations, and the 10:1 and pravastatin-only QC samples showed negative deviations. On the other hand, the accuracy and precision for pravastatin lactone were excellent in all QC samples, including those in which the ratio of pravastatin to pravastatin lactone was different from the 1:1 ratio in the calibration standards. Thus, under condition 2, where the conversion was one-way, with pravastatin lactone, as substrate analyte, producing pravastatin, as product analyte, acceptable accuracy for the product analyte was obtained only in QC samples in which the ratio of the two analytes was identical to the ratio in the calibration standards. On the other hand, acceptable accuracy and precision for the substrate analyte were obtained for all QC samples.

Table 4 summarizes the results of analysis of samples from condition 3, which allowed interconversion between pravastatin and pravastatin lactone in both the standards and QC samples. The performance of the standard curve for both pravastatin and pravastatin lactone was excellent. The accuracy and precision for both pravastatin and pravastatin lactone were excellent in the QC samples in which the ratio of pravastatin to pravastatin lactone was identical to that in the calibration standards (ratio of 1:1). The accuracy and precision for pravastatin were excellent in the pravastatin-only and 10:1 QC samples. In these QC samples, the effect of interconversion under condition 3 was a net decrease in the concentration of pravastatin. Thus, in these QC samples, pravastatin is a substrate analyte; therefore, no adverse effect is expected on the accuracy of the substrate analyte, as discussed above under condition 2. However, the accuracy for pravastatin was unacceptable in the pravastatin lactone-only and 1:10 QC samples. The measured concentrations of these QC samples showed positive deviations from the nominal concentrations. In these QC samples, the effect of interconversion under condition 3 was a net increase in the concentration of pravastatin. Thus, in these QC samples, pravastatin is a product analyte; therefore, an adverse effect is expected on the accuracy of the product analyte, as discussed above under condition 2. The accuracy results of pravastatin lactone in the QC samples in which the pravastatin to pravastatin lactone ratios were different from the 1:1 ratio in the calibration standards were the reverse of those of pravastatin. Thus, the accuracy for pravastatin lactone was excellent for the pravastatin lactone-only and 1:10 QC samples. However, the accuracy for pravastatin lactone was unacceptable for the pravastatin-only and 10:1 QC samples. The measured concentrations of these QC samples showed positive deviations from the nominal concentrations.

The above results illustrate that QC samples that contain the two analytes in the same ratio as in the calibration standards give acceptable accuracy for both analytes under all conditions, including those conditions that allow the conversion of one analyte to the other. On the other hand, QC samples that contain the analytes in ratios different from that in the calibration standards give acceptable accuracy for both analytes only under conditions where there is no interconversion during analysis.

Table 4.
Accuracy and precision under condition 3

	Pravastatin			Pravastatin lactone		
	Nominal conc. (ng/ml)	RSD (%)	Dev. (%)	Nominal conc. (ng/ml)	RSD (%)	Dev. (%)
Standard 1 (1:1)	5.00	11.8	+1.2	5.00	0.4	+3.3
Standard 2 (1:1)	10.0	10.7	−3.2	10.0	1.0	−3.5
Standard 3 (1:1)	25.0	6.1	−3.1	25.0	1.0	−9.5
Standard 4 (1:1)	50.0	0.4	+8.5	50.0	1.1	+0.4
Standard 5 (1:1)	100	0.0	+6.4	100	2.9	+3.3
Standard 6 (1:1)	200	1.9	−1.8	200	2.9	+0.2
Standard 7 (1:1)	300	0.7	−0.8	300	0.4	+5.1
Standard 8 (1:1)	400	5.1	−0.8	400	0.0	+5.0
Standard 9 (1:1)	500	1.3	−6.3	500	0.1	−4.4
QC1 (1:1)	15.0	8.2	−3.6	15.0	2.5	−3.4
QC2 (1:1)	200	2.7	+8.4	200	1.2	+10.2
QC3 (1:1)	420	5.9	+5.3	420	1.8	+7.2
QC4 (1:3)	5.00	24.2	+11.9	15.0	2.9	+0.9
QC5 (1:10)	20.0	8.0	+87.8	200	2.5	−1.1
QC6 (1:10)	42.0	7.6	+89.6	420	1.7	+4.1
QC7 (0:15)	0	N/A	N/A	15.0	3.7	−3.9
QC8 (0:200)	0	12.2	*20.4	200	2.5	−1.1
QC9 (0:420)	0	5.0	*41.3	420	1.7	+6.4
QC10 (3:1)	15.0	5.2	−9.6	5.00	N/A	N/A
QC11 (10:1)	200	2.3	+1.2	20.0	3.6	+79.5
QC12 (10:1)	420	4.3	−4.1	42.0	2.6	+73.4
QC13 (15:0)	15.0	5.7	−10.0	0		N/A
QC14 (200:0)	200	6.8	+3.6	0	6.7	*7.91
QC15 (420:0)	420	3.9	−2.4	0	2.7	*28.1

Note: Dev. = deviation of the mean concentration from the nominal concentration (n = 2 for standards; n = 5 for QC samples). N/A = not applicable, because the measured concentration was below LLQ. The numbers within parentheses show the ratios of the pravastatin concentrations to the pravastatin lactone concentrations. * = measured concentration in ng/ml. Reproduced from Jemal and Xia (2000) Copyright 2000, with permission from Elsevier

Under conditions that allow conversion between the analytes, such QC samples will fail the accuracy criteria for at least one of the analytes. Thus, a method validated for the quantitation of the two analytes using calibration standards with 1:1 analyte concentrations and QC samples with the same 1:1 analyte concentrations can be used for accurate measurement of the analytes in post-dose samples only if such samples contain the two analytes in the 1:1 ratio. On the other hand, analyte concentration ratios in post-dose samples, even from the same clinical or non-clinical study, normally vary from sample to sample due to, among other things, sampling time following the administration of the drug. Therefore, in a method to be used for quantitating interconverting analytes in post-dose samples, where analyte concentration ratios (composition) normally vary from sample to sample, QC samples that cover the entire spectrum of the composition of the post-dose samples should be incorporated during validation of the method.

In summary, the recommended first step when developing a method for quantitation of two analytes that can potentially undergo interconversion is to select conditions that will eliminate or minimize the interconversion. When optimizing the conditions for the elimination or minimization of analyte interconversion, the conditions adopted should be more favorable (in terms of eliminating conversion) toward the analyte that is expected to be the major component in the post-dose samples. The second step is to judiciously select the composition of the QC samples vis-à-vis the composition of the calibration standards. The following recommended set of calibration standards and QC samples are expected to meet the requirements for most of post-dose samples: 1:1 calibration standards throughout the curve range; 1:1 low-QC samples, 1:1 mid-QC samples; 1:1 high-QC samples; analyte 1-only high-QC samples; 10:1 mid-QC samples; analyte 2-only high-QC samples; and 1:10 mid-QC samples.

4.6. Strategy of Using Post-Dose Samples to Establish Validity of Bioanalytical LC–MS/MS Methods

The common practice of validating an LC–MS/MS method for the quantitation of a drug in a biological matrix is based on the performance of the method using standard curves and QC samples prepared by spiking the drug into a blank biological matrix (Shah *et al.*, 2000). However, as described above, post-dose biological samples could contain, in addition to the administered drug, isomeric and non-isomeric metabolites that could affect the accuracy and precision of the drug measurement. Ideally, the QC samples used to validate a bioanalytical method developed for the quantification of the drug should thus contain the metabolites in addition to the drug. However, this is not achievable in practice since, at the time of the drug method validation, the metabolites may not be known or their reference standards may not be available. Under this circumstance, a practical strategy is to challenge with post-dose samples, as soon as feasible, the method already validated using QC samples containing only the drug. Such a strategy has been developed (Jemal *et al.*, 2002) to address the metabolite-related analytical pitfalls discussed earlier in this chapter. The strategy, presented below, is illustrated using a carboxylic acid drug (RCOOH, Fig. 15), which

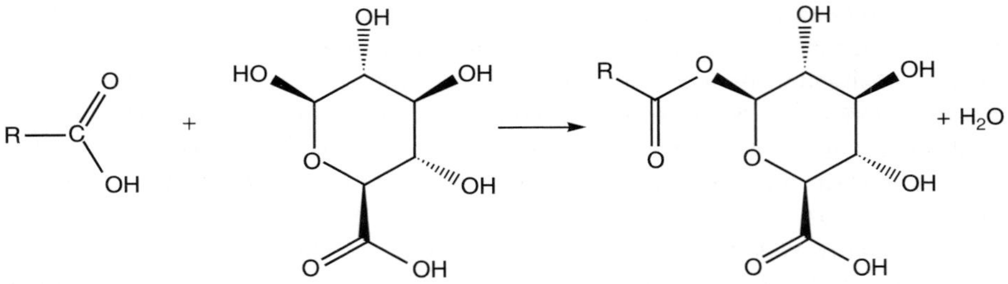

RCOOH monoisotopic mass 516.19 + Glucuronic acid monoisotopic mass 194.04 → ACG monoisotopic mass 692.22 + H_2O

Figure 15.
Chemical structures (full or partial) of the drug (RCOOH), glucuronic acid, and the acylglucuronide of RCOOH (ACG). Jemal *et al.* (2002) Copyright 2002, John Wiley & Sons Limited. Reproduced with permission

produces an acylglucuronide metabolite (ACG, Fig. 15) that is potentially capable of presenting metabolite-related analytical challenges.

An LC–MS/MS method (method 1), developed for the quantitation of RCOOH in human plasma, was fully validated using standards and QC samples prepared by spiking specified amounts of RCOOH into matrix-free human plasma. The method was then used to analyze post-dose plasma samples within a day or so of receipt of the samples from the different panels of the ascending-dose FIH study. Subsequently, portions of a variety of post-dose samples from the high-dose panels, including different subjects and different post-dosing time points, were pooled to obtain a large amount of sample to carry out the different experiments described below.

The first objective was to use the pooled post-dose sample to unequivocally show that the RCOOH metabolite, ACG, did not contribute to the RCOOH SRM channel used in method 1. Three variations of method 1 (methods 2, 3, and 4) were used for this purpose. The chromatographic conditions used for methods 2 were identical to those used for method 1, with same analytical column and acetonitrile/aqueous mobile phase. The difference between the two was only in the SRM transitions monitored (presented below). The chromatographic conditions for methods 3 and 4 differed from those of methods 1 and 2 in the composition of the mobile phase and the analytical column used. The same analytical column was used for methods 3 and 4, but the mobile phases were different for the two methods. For method 1, the SRM scheme employed involved transitions of the $[M+H]^+$ precursor ion to selected product ions: *m/z* 517→*m/z* 186 for RCOOH. The mass spectrometric conditions of methods 2–4 were identical to those of method 1 except for the additional SRM transitions used in methods 2–4. In addition to the RCOOH SRM channel used in method 1, the following additional SRM transitions were incorporated in methods 2–4: *m/z* 693→*m/z* 186 for ACG; and *m/z* 693→*m/z* 517 for ACG. Since there was no reference standard for the ACG, the transitions were selected based on the expected mass spectrometric behavior of an acylglucuronide (Jemal and Xia, 1999; Romanyshyn *et al.*, 2000).

The pooled post-dose sample was first analyzed using method 2. As shown in Fig. 16, the post-dose sample showed peaks not only in the RCOOH *m/z* 517→*m/z* 186 channel (0.46 min), as expected, but also in the ACG *m/z* 693→*m/z* 186 channel (0.45 min). On the other hand, a QC sample analyzed under the same conditions showed a peak only in the RCOOH *m/z* 517→*m/z* 186 channel at the 0.46 min retention time of RCOOH (SRM chromatograms not shown). Because there was no chromatographic separation between the RCOOH and the ACG peaks of the post-dose sample (0.46 min versus 0.45 min), there was a concern that the ACG might contribute to the *m/z* 517→*m/z* 186 channel, and hence cause the overestimation of the concentration of the RCOOH. From Fig. 16, it is interesting to note that the post-dose sample showed little or no peak in the second ACG channel (*m/z* 693→*m/z* 517). This indicates that the *m/z* 517 product ions generated, if any, from the *m/z* 693 precursor ions could not withstand the collision cell conditions and hence underwent further fragmentation to produce the *m/z* 186 ions.

In order to unequivocally establish whether or not the ACG contributed to the *m/z* 517→*m/z* 186 channel, method 2 was modified by changing the chromatographic conditions to obtain method 3. The pooled post-dose sample SRM chromatograms obtained using method 3 are shown in Fig. 17. As with method 2, there was a peak in the ACG *m/z* 693→*m/z* 186 channel, which in this case eluted well before the RCOOH peak seen in the *m/z* 517→*m/z* 186 channel. On the other hand, the *m/z* 517→*m/z* 186 channel did not show a peak at the retention of the ACG. This unequivocally established that the *m/z* 693 ion from the ACG did not undergo in-source fragmentation to generate the *m/z* 517 ion in the source. Therefore, the originally validated method 1 was valid for the quantitation of RCOOH in post-dose samples in spite of the fact that method 1 did not achieve chromatographic separation between the RCOOH and the ACG.

It was desirable to achieve a chromatographic separation of RCOOH from ACG under chromatographic conditions (especially the LC mobile phase) different from those of method 3. The main objective was to ascertain that the absence of the in-source generation of *m/z* 517 from *m/z* 693 still holds true under the new chromatographic conditions. This was needed since the absence/presence of in-source fragmentation is dependent not only on the in-source parameters but also on the LC mobile phase used (Jemal and Ouyang, 2000). The pooled post-dose sample SRM chromatograms obtained using method 4, with a mobile phase different from that of method 3, are shown in Fig. 18. As with method 3, the ACG was chromatographically separated from RCOOH, and there was no peak in the *m/z* 517→*m/z* 186 channel at the retention time of ACG. Thus, the results of both methods 3 and 4 established that method 1 was valid for analysis of post-dose samples. As a means of a further confirmation of this finding, the concentration of RCOOH in the pooled post-dose sample determined using method 4 was found to be equal to the concentration determined using method 1.

The second objective was to use the pooled post-dose sample for undertaking bench-top stability investigation in order to show that ACG did not degrade to generate the parent drug, RCOOH, during the multiple steps preceding measurement by LC–MS/MS. Stability study at room temperature showed that the pooled

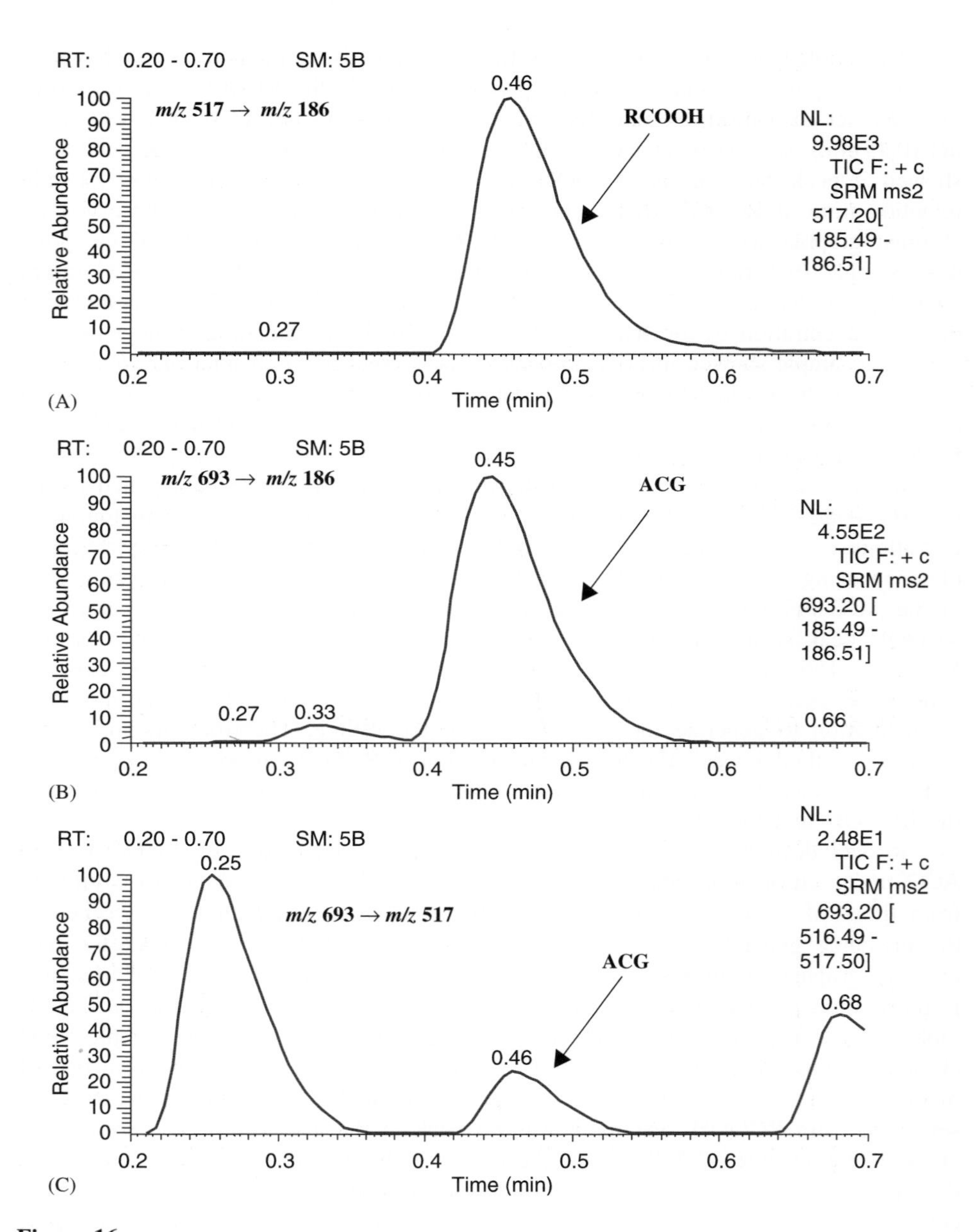

Figure 16.
SRM chromatograms obtained using method 2 from a post-dose sample: (A) the RCOOH channel (*m/z* 517→*m/z* 186); (B) the first ACG channel (*m/z* 693→*m/z* 186); and (C) the second ACG channel (*m/z* 693→*m/z* 517). Jemal *et al.* (2002) Copyright 2002, John Wiley & Sons Limited. Reproduced with permission

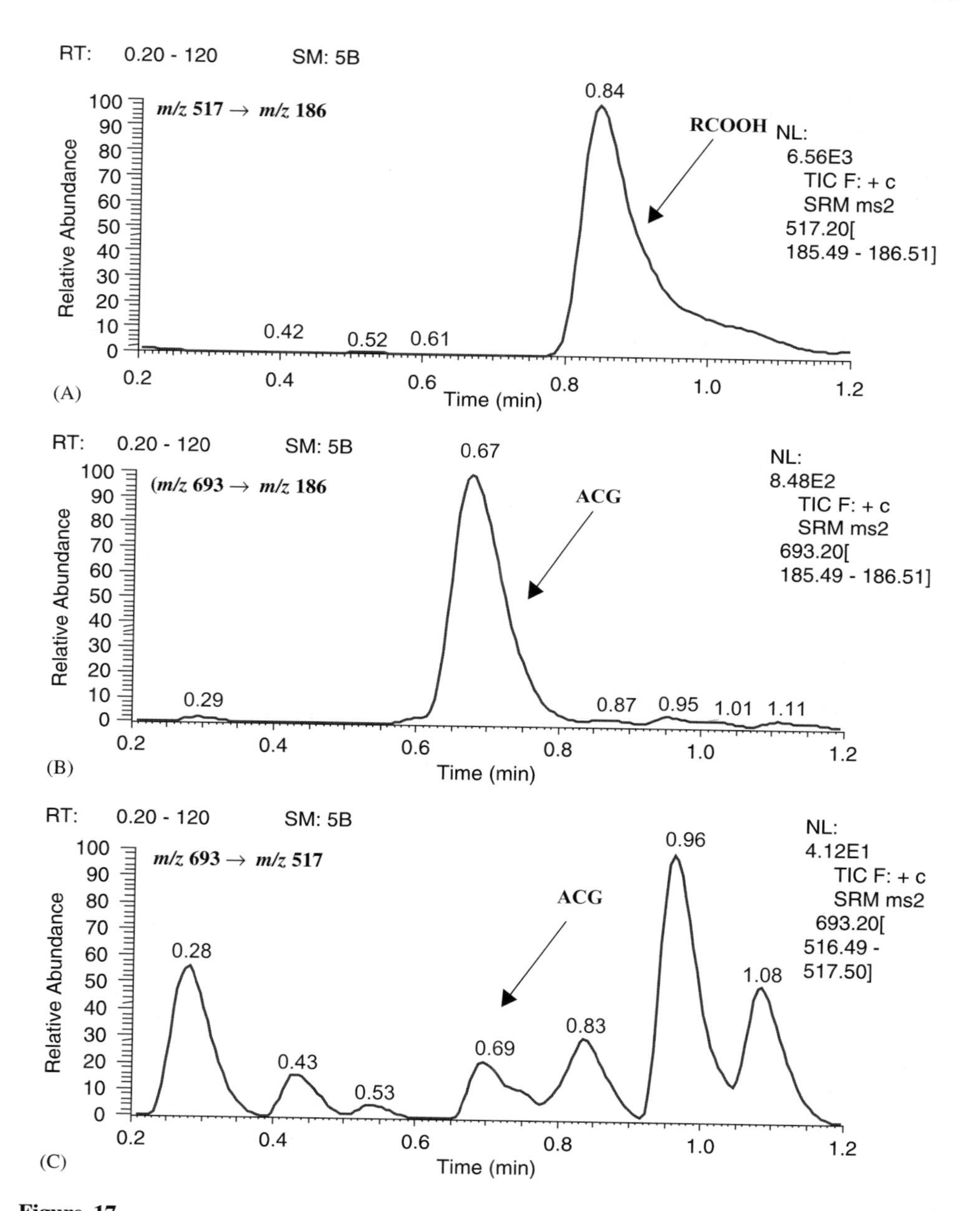

Figure 17.
SRM chromatograms obtained using method 3 from a post-dose sample: (A) the RCOOH channel (*m/z* 517→*m/z* 186); (B) the first ACG channel (*m/z* 693→*m/z* 186); and (C) the second ACG channel (*m/z* 693→*m/z* 517). Jemal *et al.* (2002) Copyright 2002, John Wiley & Sons Limited. Reproduced with permission

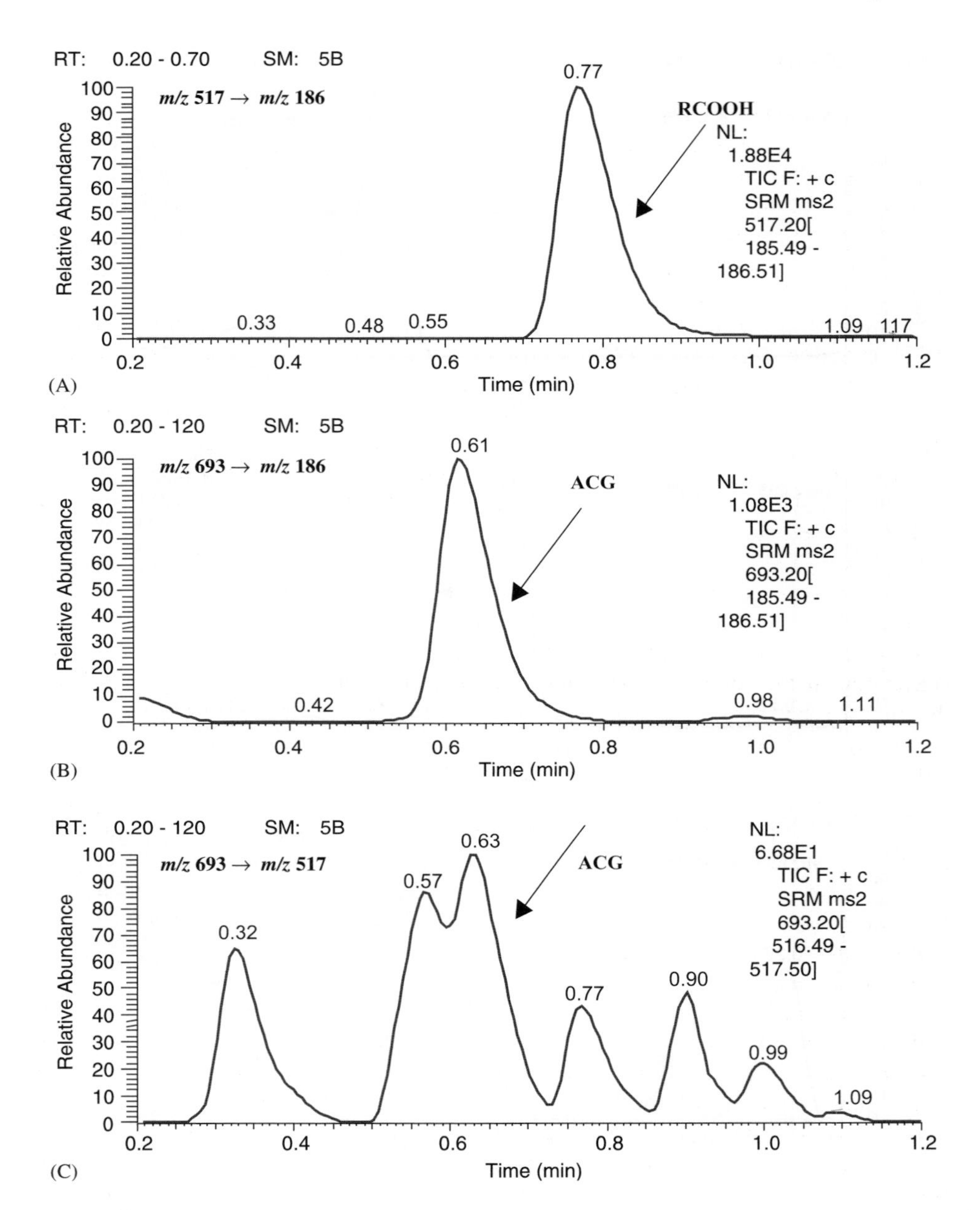

Figure 18.
SRM chromatograms obtained using method 4 from a post-dose sample: (A) the RCOOH channel (*m/z* 517→*m/z* 186); (B) the first ACG channel (*m/z* 693→*m/z* 186); and (C) the second ACG channel (*m/z* 693→*m/z* 517). Jemal *et al.* (2002) Copyright 2002, John Wiley & Sons Limited. Reproduced with permission

post-dose sample was stable as is or after-processing (in the injection vial), as evidenced by the constancy of the measured concentrations of RCOOH at different time points for at least 24 h. Thus, the sample handling conditions adopted in method 1, such as maintaining room temperature (as opposed to ice/water temperature or 4°C) during the plasma sample aliquotting and for the autosampler, and keeping the plasma pH as is (as opposed to adjusting it to 3–5, where ACG is expected to be more stable), were acceptable. When the pooled post-dose sample was subsequently subjected to a harsher condition by raising the pH of the plasma to 12 with sodium hydroxide and then keeping the sample at room temperature for 30 min, the ACG was no longer present. However, the measured concentration of RCOOH did not increase under the harsher treatment, as was also the case with a QC sample treated under identical conditions. The reason for the apparent non-increase in the RCOOH concentrations in the stressed post-dose sample is probably due to the relatively low concentration of ACG compared to the concentration of RCOOH. Thus, the concentration of RCOOH produced as a result of the hydrolysis of ACG would be within the experimental error of measurement of the concentration of the total RCOOH.

Similar strategies can be applied to other LC–MS/MS methods validated for drugs with different chemical structures and functional groups. Listed in Table 5 are typical classes of drugs and some of their potential non-isomeric metabolites, which can cause problems with validated LC–MS/MS methods due to metabolite-to-drug conversion in the source of the mass spectrometer or in the various steps preceding the introduction of the sample into the LC–MS/MS system. The suggested SRM transitions in Table 5 can be used to challenge LC–MS/MS methods with pooled post-dose samples following the strategy described above.

With isomeric metabolites (other than enantiomeric metabolites), such as epimers and *E* or *Z* isomers, there are no additional SRM channels that can distinguish the metabolites from the drug. Thus, when changing the LC conditions in an attempt to chromatographically resolve the drug from an isomeric metabolite that may or may not be present in the post-dose sample, one has to rely only on the SRM transition that is common to both the drug and the putative isomeric metabolite. It is impossible to know the chromatographic conditions or chromatographic run times required to achieve the chromatographic separation of the drug from the metabolite in the absence of an authentic reference standard for the metabolite. The ease of chromatographic separation between a drug and its metabolite depends on the nature of the drug and the metabolite (Jemal and Xia, 1999; Romanyshyn *et al.*, 2001; Yan *et al.*, 2003). It should be noted that for the same LC column material, linear velocity, and mobile phase composition (i.e. maintaining the same separation factor and retention factor), the chromatographic run time required to achieve a given chromatographic resolution can be varied by as much as 10-fold by simply changing the column length and particle size (Neue, 1997). Therefore, it is meaningless to even suggest recommended run times without taking into consideration the nature of the analyte and the LC parameters used. However, it can generally be recommended that the post-dose sample be run under at least two sets of LC conditions obtained by changing the mobile phase and/or the LC column. For each set of conditions, it is recommended that a retention factor of about 10 be

Table 5.
Putative metabolites of drugs of different chemical structures and the SRM transitions for the metabolites vis-à-vis the SRM transitions of the drug

Drug type	Drug SRM	Metabolite	Metabolite SRM
Carboxylic acid	$[M+H]^+ \rightarrow P^+$	Acylglucuronide	(a) $[M+H+176]^+ \rightarrow [M+H]^+$ (b) $[M+H+176]^+ \rightarrow P^+$
γ or δ hydroxy carboxylic acid	$[M+H]^+ \rightarrow P^+$	Lactone	(a) $[M+H-18]^+ \rightarrow [M+H]^+$ (b) $[M+H-18]^+ \rightarrow P^+$
Lactone	$[M+H]^+ \rightarrow P^+$	Hydroxy acid	(a) $[M+H+18]^+ \rightarrow [M+H]^+$ (b) $[M+H+18]^+ \rightarrow P^+$
Alcohol or phenol	$[M+H]^+ \rightarrow P^+$	*O*-Glucuronide	(a) $[M+H+176]^+ \rightarrow [M+H]^+$ (b) $[M+H+176]^+ \rightarrow P^+$
Alcohol or phenol	$[M+H]^+ \rightarrow P^+$	*O*-Sulfate	(a) $[M+H+80]^+ \rightarrow [M+H]^+$ (b) $[M+H+80]^+ \rightarrow P^+$
Amine	$[M+H]^+ \rightarrow P^+$	*N*-Glucuronide	(a) $[M+M+176]^+ \rightarrow [M+H]^+$ (b) $[M+H+176]^+ \rightarrow P^+$
Amine	$[M+H]^+ \rightarrow P^+$	*N*-Oxide	(a) $[M+H+16]^+ \rightarrow [M+H]^+$ (b) $[M+H+16]^+ \rightarrow P^+$
Thiol (sulfhydryl)	$[M+H]^+ \rightarrow P^+$	Disulfide	(a) $[M+M-1]^+ \rightarrow [M+H]^+$ (b) $[M+M-1]^+ \rightarrow P^+$
Sulfide	$[M+H]^+ \rightarrow P^+$	*S*-Oxide	(a) $[M+H+16]^+ \rightarrow [M+H]^+$ (b) $[M+H+16]^+ \rightarrow P^+$

Note: The SRM transitions shown are for electrospray ionization in the positive ion mode. M is the mono-isotopic mass of the drug. P is the product ion in the SRM transition used for quantitation of the drug. For each drug type, the fragmentation exhibited by the metabolite SRM transition designated as (a) can potentially take place within the source of the mass spectrometer as well. If such in-source fragmentation occurs and there is no chromatographic separation between the drug and the metabolite, the concentration of the drug determined by using the $[M+H]^+ \rightarrow P^+$ transition would be inflated. A similar list of SRM transitions can be prepared for electrospray negative ionization, and for atmospheric pressure chemical ionization in the positive or negative ion mode. Jemal *et al.* (2002) Copyright 2002, John Wiley & Sons Limited. Reproduced with permission

achieved for the drug. It should be noted that the strategy proposed in this paper does not address the problem caused by the presence of an enantiomeric metabolite in the post-dose sample.

A structural analog of a drug used as the internal standard can potentially cause a problem in LC–MS/MS bioanalysis (Matuszewski *et al.*, 1998). Bioanalysts should be aware that a drug may produce a metabolite that has the same SRM transition as the internal standard. Such a metabolite, if not separated chromatographically from the internal standard, will interfere with the SRM signal of the internal standard. Drug functional groups that can produce potentially interfering metabolites include: (a) methoxy, *N*-methyl, and methyl ester groups, which are capable of undergoing demethylation to produce metabolites having [M − 14] mass; (b) thiolic and catechol groups, which are capable of undergoing methylation to produce metabolites having [M + 14] mass; (c) CH_3 and CH_2 and CH containing groups, which are capable of undergoing oxidation to produce hydroxylated metabolites having [M + 16] mass; (d) amine and sulfide groups, which are capable of undergoing oxidation to produce *N*- and *S*-oxide metabolites having [M + 16] mass; and (e) CH_3 and CH_2 containing groups, which are capable of undergoing oxidation to produce aldehyde and ketone metabolites having [M + 14] mass. Other possible metabolite masses include [M − 2] due to two-electron oxidation and [M + 2] due to two-electron reduction. The recommendation is to select the internal standard judiciously and then establish the acceptability of the internal standard when post-dose samples are available. For this, the post-dose sample is to be analyzed following the validated method procedure without adding the internal standard. The absence of interfering signal at the retention time and signal channel of the internal standard will establish the acceptability of the internal standard.

4.7. Summary and Conclusions

Post-dose biological samples normally contain metabolites in addition to the drug. It is therefore very important to consider the potential interference such metabolites cause in the quantification of the drug in post-dose samples. Conjugate and other non-isomeric metabolites may contribute to the SRM channel used for the drug due to in-source conversion of the metabolite unless there is adequate chromatographic separation between the two compounds, thereby artificially inflating the measured concentration of the drug. It is important to point out that in-source conversion is a function of the mobile phase and source parameters. There may also be non-isomeric metabolites, such as the lactone metabolite of a hydroxy acid drug, which may require chromatographic separation from the drug even in the absence of in-source generation of the drug entity. This is a consequence of the $[M+NH_4]^+$ ion of the lactone metabolite being lower than the $[M+H]^+$ ion of the hydroxy acid by only one mass unit. This becomes a problem when the drug and the metabolite simultaneously form the $[M+H]^+$ $[M+NH_4]^+$ ions. In this regard, it should be realized that $[M+NH_4]^+$ ions may be formed even when no source of the ammonium is added to the mobile phase. Some metabolites, such as the acid metabolite of a primary amide drug, the secondary alcohol metabolite of a ketone drug, and the primary alcohol metabolite of an aldehyde drug, will cause interference with the drug channel, in the absence of chromatographic separation, only when larger than unit-mass resolution setting is used. Isomeric metabolites will, on the other hand, directly contribute to the SRM channel of the drug in the absence of chromatographic separation from the drug. The mass spectral interference encountered due to the presence of metabolites can always be overcome by achieving chromatographic separation between the drug and the metabolite. However, it is difficult to predict the retention factor required to achieve the chromatographic separation.

Aside from the mass spectral interference summarized above, isomeric as well as non-isomeric metabolites may adversely affect the accurate quantification of the drug if such metabolites are unstable and convert to the parent drug during the multiple steps preceding the LC–MS/MS analysis. This type of metabolite-related analytical problem could be avoided by using conditions that stabilize the metabolite, such as appropriate sample pH and temperature, during the different steps of sample collection, storage, handling, extraction, and analysis. For most metabolites that are normally considered to be prone to degradation, such as lactones, hydroxy acids, and acylglucuronides, buffering the sample to pH 3–5 and keeping the sample at a low temperature of about 4°C provide adequate conditions for stability. However, even under the selected optimum conditions, there may still be slight conversion of the metabolite to the drug or vice versa. The untoward effect of such an occurrence could be avoided by the judicious choice of the composition of the QC samples (in terms of the concentration ratio of the drug to the metabolite) vis-à-vis the composition of the calibration standards.

Under the normal course of drug development, the metabolites are either unidentified or their reference standards are unavailable at the time the method for the drug quantification is developed and validated in anticipation of a non-clinical or clinical study. Thus, a priori investigation of the effect of the metabolites on the performance of the parent drug method cannot be undertaken. Under this circumstance, the prudent approach to the drug method development is to select experimental conditions that avoid problems from mass spectral interference and bench-top conversion of anticipated metabolites. This is not easy in the absence of the authentic metabolite compounds. But bioanalysts have to do the best they can to build into method development some means of overcoming the undesirable effects of potential metabolites. Subsequently, as soon as feasible, the validated method should be challenged with a pooled post-dose sample using a strategy similar to the one described above in this chapter.

References

Ayrton, J., Clare, R.A., Dear, G.J., Mallet, D.N., Plumb, R.S., 1999. Ultra-high flow rate capillary liquid chromatography with mass spectrometric detection for the direct analysis of pharmaceuticals in plasma at sub-nanogram per milliliter concentrations. Rapid Commun. Mass Spectrom. 13, 1657–1662.

Bakhtiar, R., Ramos, L., Tse, F.L.S., 2002. High-throughput mass spectrometric analysis of xenobiotics in biological fluids. J. Liq. Chromatogr. Rel. Technol. 25, 507–540.

Gilbert, H.F., 1995. Thiol/disulfide exchange equilibria and disulfide bond stability. Methods Enzymo. 251, 8–29.

Jemal, M., 2000. High-throughput quantitative bioanalysis by LC/MS/MS. Biomed. Chromatogr. 14, 422–429.

Jemal, M., Huang, M., Mao, Y., Whigan, D., Schuster, A., 2000a. Liquid chromatography/electrospray tandem mass spectrometry method for the quantitation of fosinoprilat in human serum using automated 96-well solid-phase extraction for sample preparation. Rapid Commun. Mass Spectrom. 14, 1023–1028.

Jemal, M., Ouyang, Z., 2000. The need for chromatographic and mass resolution in liquid chromatography/tandem mass spectrometric methods used for quantitation of lactones and corresponding hydroxy acids in biological samples. Rapid Commun. Mass Spectrom. 14, 1757–1765.

Jemal, M., Ouyang, Z., Chen, B.H., Teitz, D., 1999a. Quantitation of the acid and lactone forms of atorvastatin and its biotransformation products in human serum by high-performance liquid chromatography with electrospray tandem mass spectrometry. Rapid Commun. Mass Spectrom. 13, 1003–1015.

Jemal, M., Ouyang, Z., Powell, M., 2000b. Direct-injection LC-MS-MS method for high-throughput simultaneous quantitation of simvastatin and simvastatin acid in human plasma. J. Pharmaceut. Biomed. Anal. 23, 323–340.

Jemal, M., Ouyang, Z., Powell, M.L., 2002. A strategy for a post-method-validation use of incurred biolgical samples for establishing the acceptability of a liquid chromatography/tandem mass-spectrometric method for quantitation of drugs in biolgical samples. Rapid Commun. Mass Spectrom. 16, 1538–1547.

Jemal, M., Rao, S., Salahudeen, I., Chen, B.H., Kates, R., 1999b. Quantitation of cerivastatin and its seven acid and lactone biotransformation products in human serum by liquid chromatography-electrospray tandem mass spectrometry. J. Chromatogr. B 736, 19–41.

Jemal, M., Xia, Y.Q., 1999. The need for adequate chromatographic separation in the quantitative determination of drugs in biological samples by high performance liquid chromatography with tandem mass spectrometry. Rapid Commun. Mass Spectrom. 13, 97–106.

Jemal, M., Xia, Y.Q., 2000. Bioanalytical method validation design for the simultaneous quantitation of analytes that may undergo interconversion during analysis. J. Pharmaceut. Biomed. Anal. 22, 813–827.

Kantola, T., Kivisto, K.T., Neuvonen, P.J., 1998. Effect of itraconazole on the pharmacokinetics of atorvastatin. Clin. Pharmacol. Ther. 64, 58–65.

Kaufman, M.J., 1990. Rate and equilibrium constants for acid-catalyzed lactone hydrolysis of HMG-CoA reductase inhibitors. Int. J. Pharm. 66, 97–106.

Kearney, A.S., Crawford, L.F., Mehta, S.C., Radebaugh, G.W., 1993. The interconversion kinetics, equilibrium, and solubilities of the lactone and hydroxyacid forms of the HMG-CoA reductase inhibitor, CI-981. Pharmaceut. Res. 10, 1461–1465.

Khan, S., Teitz, D.S., Jemal, M., 1998. Kinetic analysis by HPLC-electrospray mass spectrometry of the pH-dependent acyl migration and solvolysis as the decomposition pathways of ifetroban 1-O-acyl glucuronide. Anal. Chem. 70, 1622–1628.

Liu, D.Q., Pereira, T., 2002. Interference of a carbamoyl glucuronide metabolite in quantitative liquid chromatography/tandem mass spectrometry. Rapid Commun. Mass Spectrom. 16, 142–146.

Matuszewski, B.K., Chavez-Eng, C.M., Constanzer, M.L., 1998. Development of high-performace liquid chromatography-tandem mass spectrometric methods for the determination of a new oxytocin receptor antagonist (L-368,899) extracted from human plasma and urine: a case of lack of specificity due to the presence of metabolites. J. Chromatogr. B 716, 195–208.

Morris, M.J., Gilbert, J.D., Hsieh, J.Y.K., Matuszewski, B.K., Ramjit, H.G., Bayne, W.F., 1993. Determination of the HMG-CoA reductase inhibitors simvastatin, lovastatin, and pravastatin in plasma by gas chromatography/chemical ionization mass spectrometry. Biol. Mass Spectrom. 22, 1–8.

Naidong, W.J., 2003. Bioanalytical liquid chromatography tandem mass spectrometry methods on underivatized silica columns with aqueous/organic mobile phases. J. Chromatogr. B 796, 209–224.

Naidong, W., Jiang, X., Newland, K., Coe, R., Lin, P., Lee, J., 2000. Development and validation of a sensitive method for hydromorphone in human plasma by normal phase liquid chromatography-tandem mass spectrometry. J. Pharmaceut. Biomed. Anal. 23, 697–704.

Naidong. W., Lee, J.W., Jiang, X., Wehling, M., Hulse, J.D., Lin, P.P., 1999. Simultaneous assay of morphine, morphine-3-glucuronide and morphine-6-glucuronide in human plasma using normal-phase liquid chromatography-tandem mass spectrometry with a silica column and an aqueous organic mobile phase. J. Chromatogr. B 735, 255–269.

Neue, U.D., 1997. HPLC Columns: Theory, Technology, and Practice Wiley-VCH, New York, pp. 41–53.

Ramanathan, R., Su, A.D., Alvarez, N., Blumenkranz, N., Chowdhury, S.K., Alton, K., Patrick, J., 2000. Liquid chromatography/mass spectrometry methods for distinguishing N-oxides from hydroxylated compounds. Anal. Chem. 72, 1352–1359.

Romanyshyn, L., Tiller, P.R., Alvaro, R., Pereira, A., Hop, C.E.C.A., 2001. Ultra-fast gradient vs. fast isocratic chromatography in bioanalytical quantification by liquid chromatography/tandem mass spectrometry. Rapid Commun. Mass Spectrom. 15, 313–319.

Romanyshyn, L., Tiller, P.R., & Hop, C. E. C. A., 2000. Bioanalytical applications of 'fast chromatography' to high-throughput liquid chromatography/tandem mass spectrometric quantitation. Rapid Commun. Mass Spectrom. 14, 1662–1668.

Shah, V.P., Midha, K.K., Findlay, J.W.A., Hill, H.M., Hulse, J.D., McGilveray, I.J., McKay, G., Miller. K.J., Patnaik, R.N., Powell, M.L., Tonelli, A., Viswanathan, C.T., Yacobi, A., 2000. Bioanalytical method validation - a revisit with a decade of progress. Pharmaceut. Res. 17, 1551–1557.

Shipkova, M., Armstrong, V.W., Oellerich, M., Wieland, E., 2003. Acyl glucuronide drug metabolites: toxicological and analytical implications. Ther. Drug Monit. 25, 1–16.

Testa, B., Carrupt, P.A., Gal, J., 1993. The so-called "interconversion" of stereoisomeric drugs: an attempt at clarification. Chirality, 5, 105–111.

Tiller, P.R., Romanyshyn, L.A., Neue, U.D., 2003. Fast LC/MS in the analysis of small molecules. Anal. and Bioanal. Chem. 377, 788–802.

Wang, C.J., Pao, L.H., Hsiong, C.H., Wu, C.Y., Whang-Peng, J.J.K., Hu, O.Y.P., 2003. Novel inhibition of *cis/trans* retinoic acid interconversion in biological fluids – an accurate method for determination of *trans* and 13-*cis* retinoic acid in biological fluids. J. Chromatogr. B 796, 283–291.

Won, C.M., 1994. Epimerization and hydrolysis of dalvastatin, a new hydroxymethylglutaryl coenzyme A (HMG-CoA) reductase inhibitor. Pharmaceut. Res. 11, 165–170.

Xia, Y.Q., Whigan, D.B., Jemal, M., 1999. A simple liquid-liquid extraction with hexane for low-picogram determination of drugs and their metabolites in plasma by high-performance liquid chromatography with positive ion electrospray tandem mass spectrometry. Rapid Commun. Mass Spectrom. 13, 1611–1621.

Yan, Z., Caldwell, G.W., Jones, W.J., Masucci, J.A., 2003. Cone voltage induced in-source dissociation of glucuronides in electrospray and implications in biological analyses. Rapid Commun. Mass Spectrom. 17, 1433–1442.

Identification and Quantification of Drugs, Metabolites and Metabolizing
Enzymes by LC–MS
Swapan K. Chowdhury, editor.

Chapter 5

IN VITRO DMPK SCREENING IN DRUG DISCOVERY, ROLE OF LC–MS/MS

Inhou Chu and Amin A. Nomeir

5.1. Introduction

The goal of drug discovery programs is to identify new chemical entities (NCEs) suitable for development as therapeutic agents. Recent advances in chemical synthesis (parallel and combinatorial chemistry), structural chemistry, molecular biology and robotics have greatly increased the number of compounds requiring evaluation; therefore, increasing the demands for drug metabolism and pharmacokinetics (DMPK) screening (White, 2000). This is particularly true in the early stage of drug discovery, where a large number of NCEs needs to be evaluated so that the selected few can proceed to drug development. While traditional in vivo animal studies can provide direct information regarding how a drug candidate behaves in the integrated biological system, these studies are usually costly and time consuming, therefore can no longer keep up with the increased demands. In vitro models for the assessment of the biopharmaceutical properties such as metabolic stability, permeability across the human colon carcinoma cell line (Caco-2) and inhibition of cytochrome P450 isoforms (CYPs), are faster, easier to perform and require less amount of compound; therefore, amendable to higher-throughput screening. Performance of many of these in vitro screens in a higher-throughput format has been made possible primarily by the availability of liquid chromatography–mass spectrometry (LC–MS) and liquid chromatography–tandem mass spectrometry (LC–MS/MS). In this chapter, two primary human-based in vitro screening systems, namely the CYP enzyme inhibition and Caco-2 permeability will be discussed, along with a secondary screen, the blood–brain-barrier (BBB) penetration model. The role of LC–MS/MS in making these assays possible in drug discovery will be emphasized.

5.2. CYP Inhibition Screening

5.2.1. Background

The majority of drugs are oxidatively metabolized, primarily in the liver and intestine, by a family of heme-containing enzymes namely CYPs (Smith *et al.*, 1997).

Inhibition of one or more of these enzymes by co-administered drugs may alter the pharmacokinetics resulting in a significant drug–drug interaction that could lead to adverse drug effects including fatalities (Honig *et al.*, 1993). Therefore, it is prudent to evaluate the inhibition of these drug-metabolizing enzymes by NCEs in early drug discovery in order to avoid potential inhibitory drug–drug interactions. At the same time, CYP inhibition screening is easy to perform because the assay for each isoform involves the analysis of the same end product regardless of the compound to be evaluated.

Approximately 80% of commercially available drugs are oxidatively metabolized by CYP 3A4 and CYP 2D6 (Fig. 1; adapted from Mizutani, 2003). Consequently, our approach has been an initial screen for these two isoforms, and if there is still an interest in the NCE, it is evaluated for potential inhibition of CYPs 2C9, 2C19 and 1A2. Our current higher-throughput approach has been set up to evaluate both CYP 3A4 and CYP 2D6 in one combined assay, CYPs 2C9 and 2C19 in another combined assay and CYP 1A2 in a separate assay. The potential of drug–drug interaction of a NCE is expressed as IC_{50} values, which is the NCE concentration that reduces the enzyme activity by 50%. The CYP substrates are added at concentrations near the K_m values for the enzymes; therefore, good estimates of the K_i values are obtained (K_i is approximately equal to or one-half of the IC_{50} depending on the type of inhibition). The IC_{50} is derived from a semi-log plot of the NCE concentrations versus percent inhibition as described later.

5.2.2. Types of Inhibition

Two types of inhibition are evaluated. The first is *direct inhibition*, in which the NCE itself reversibly inhibit the enzyme resulting in the alteration of Michaelis–Menten

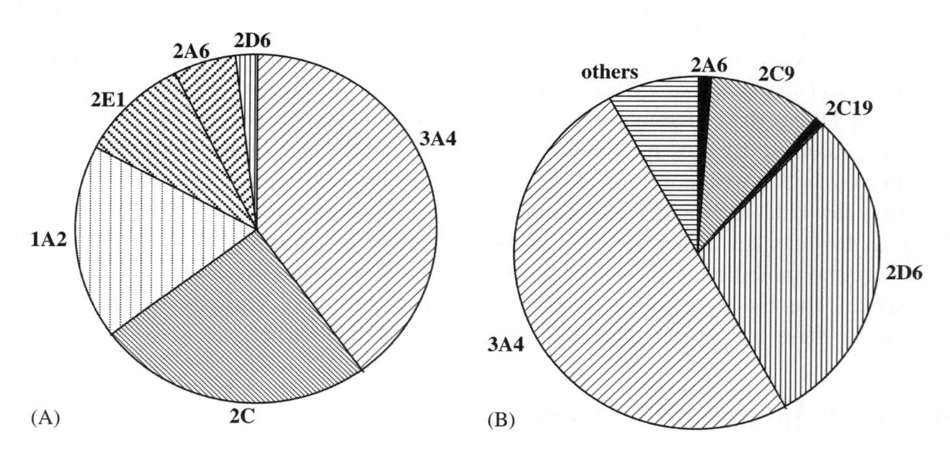

Figure 1.
Distribution of various CYP isozymes in average human liver (A) and relative contribution of each isoform in the metabolism of commercially available drugs (B). (Adapted from Mizutani, 2003)

kinetic parameters K_m and V_{max}. This reversible inhibition could be competitive, non-competitive, uncompetitive or mixed. The other type of inhibition is *metabolism/mechanism-based* inhibition, which is caused by a metabolite of the NCE that is either a more potent direct reversible inhibitor (*metabolism-based*) or a time-dependent irreversible inhibitor (*mechanism-based*) (Silverman, 1988; White, 2000). To distinguish between direct and metabolism/mechanism-based inhibition, co-incubation and pre-incubation experiments are carried out (Newton *et al.*, 1995; Nomeir *et al.*, 2004). In the co-incubation experiment, the NCE is incubated at several concentrations with human liver microsomes, NADPH and probe substrates; and the IC_{50} is determined. In the pre-incubation experiment, the NCE is first incubated with liver microsomes and NADPH for 30 min without the probe substrates, then the probe substrates are added and the reaction is allowed to proceed as with the co-incubation experiment; again, the IC_{50} is determined. Both IC_{50} values from the pre- and co-incubation are required to determine whether or not the NCE is a direct, and/or metabolism/mechanism-based inhibitor. In both cases, the % inhibition is determined based on a blank solvent incubation (with the same solvent concentration used to add the NCE to the incubation mixture). It should be emphasized that the solvent concentration should be kept to a minimum as it may alter the basal enzyme activity.

The NCE is generally classified as a direct inhibitor if the IC_{50} values under both pre- and co-incubation conditions are similar (within threefold) or higher under pre-incubation. If the IC_{50} under the pre-incubation conditions is much lower compared to that after co-incubation (greater inhibition), the NCE is likely to be a metabolism/mechanism-based inhibitor. To distinguish between metabolism-based and mechanism-based inhibition, additional experiments (not discussed in this chapter) are required (Nomeir *et al.*, 2004). Both direct and mechanism-based inhibitors could result in inhibitory drug–drug interactions. In addition, mechanism-based inhibitors carry an additional risk of potential drug idiosyncratic effects. It should be recognized that a compound could be a weak direct inhibitor but a potent metabolism/mechanism-based inhibitor. Therefore, it is important to evaluate both types of inhibition. Usually potent metabolism/mechanism-based inhibitors are excluded from development except under special circumstances such as a first in class drug for previously untreated life-threatening disease.

5.2.3. Experimental Procedure

A schematic representation of the experimental set-up is shown in Fig. 2. Pools of human liver microsomes from 10 to 20 donors are obtained from the International Institute for the Advancement of Medicine (Exton, PA). The microsomes are frozen in small aliquots and stored at −80°C till use. The metabolites dextrorphan and 6*β*-hydroxytestosterone that are formed from dextromethorphan *O*-demethylase (CYP 2D6 reaction, Chen *et al.*, 1990) and testosterone 6*β*-hydroxylase (CYP 3A4 reaction, Waxman *et al.*, 1991), respectively, are quantified by LC–MS/MS.

As mentioned above, for each NCE, two incubation conditions are performed. In the co-incubation experiment, NCEs are added in methanol at 0.3, 3, 10 and 30 μM,

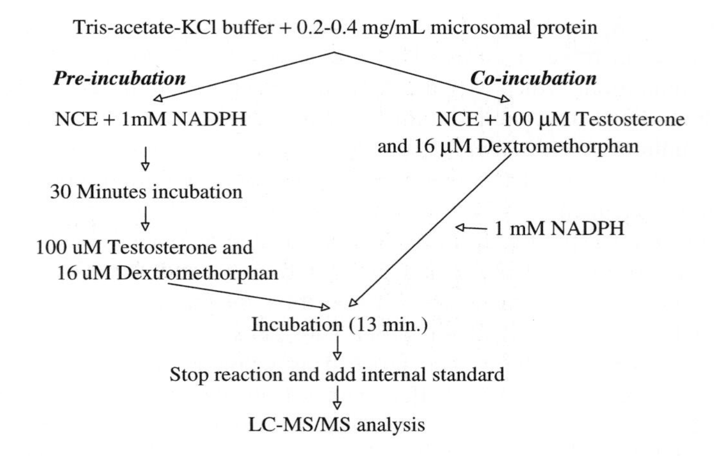

Figure 2.
Schematic representation of CYP 2D6 and CYP 3A4 inhibition assay

dimethyl sulfoxide ((DMSO) is used if the NCE is insoluble in methanol) to separate wells of 96-well plates. Each well contains a mixture of microsomal protein (final concentration, 0.2–0.4 mg/ml) and probe substrates [dextromethorphan (16 μM) and testosterone (100 μM) for CYPs 2D6 and 3A4, respectively] in 50 mM Tris-acetate buffer, pH 7.4, containing 150 mM KCl. The plates are pre-warmed at 37°C for 5 min. The reactions are initiated by the addition of 20 μl of a 10 mM solution of NADPH prepared in buffer (final concentration, 1 mM) followed by a brief shaking. The total volume of the reaction mixture is 200 μl. Following an incubation period of 13 min, the reactions are terminated by the addition of 35% perchloric acid followed by shaking and centrifugation for protein precipitation.

In the pre-incubation experiment, the NCE is incubated with liver microsomes and NADPH at the same concentrations as above for 30 min in the absence of probe substrates. The substrates for CYP 2D6 and CYP 3A4 are then added, and the reaction is allowed to proceed as indicated above. The reactions are terminated after 13 min and the samples are processed for LC–MS/MS analysis as indicated above.

5.2.4. Sample Plate Set-Up

The entire processes of incubation and sample analysis are carried out in the same 96-well plates. Sample arrangement is shown in Fig. 3. For both co- and pre-incubation experiments, the first row of the 96-well plate is reserved for blank, solvent and positive control incubations (prototype direct and mechanism-based inhibitors for each CYP at one concentration each). These serve as quality control. Rows 2–8 and rows 1–8 of a second 96-well plate are for NCEs (for either pre- or co-incubation). Each row is for one NCE at the four concentrations mentioned above, prepared in triplicates. Therefore, every 15 NCEs require four

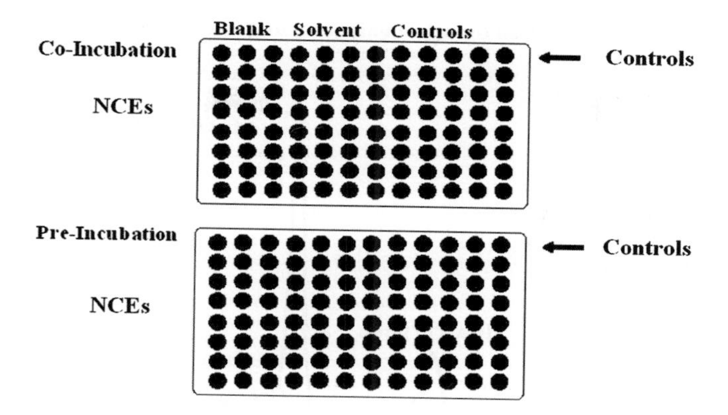

Figure 3.
Schematic representation of the experimental set-up for CYP inhibition

96-well plates (two for co-incubation and two for pre-incubation) which would generate 60 IC_{50} (for both CYP 3A4 and 2D6). A similar protocol is followed for a combined CYP 2C9 and 2C19 assay, while CYP 1A2 is analyzed separately in a similar manner.

5.2.5. Quantification with LC–MS/MS

The quantification of the metabolites for both CYP 3A4 and 2D6 reactions is carried out using a SCIEX API 3000, running in the positive ion multiple reaction monitoring (MRM) mode. The high performance liquid chromatography (HPLC) gradient system used consists of two solvent mixtures. Solvent mixture A (SMA) is 94.9% H_2O, 5% MeOH and 0.1% formic acid; solvent mixture B is 94.9% MeOH, 5% H_2O and 0.1% formic acid. The analytical column used is a Develosil Combi-RP5, 3.5 × 20 mm (Phenomenex Inc., Torrance, CA), which has been proven to be rugged and reproducible at the elevated flow rate (1.1 ml/min) that is used to speed up the analysis. The HPLC gradient starts with 65% SMA, which is maintained for 0.2 min, then switched to 100% SMB and held for another 0.2 min. At 0.4 min, the solvent is changed back to the initial conditions. A representative ion chromatogram for the simultaneous analysis of both 6β-hydroxytestosterone and dextrorphan is shown in Fig. 4.

5.2.6. IC_{50} Determination

The % inhibition of enzyme activity by an NCE at each concentration is calculated from the ratio of the metabolite generated in the presence and absence of the NCE under investigation. The concentrations of both 6β-hydroxytestosterone and dextrorphan in the reaction mixture are calculated based on standard curves that contain both metabolites. The IC_{50} values are determined by semi-log plotting of NCE

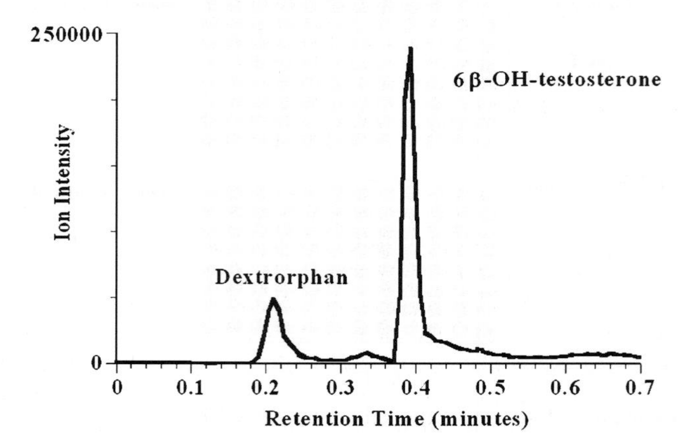

Figure 4.
Positive MRM ion chromatogram of 6β-hydroxytestosterone and dextrorphan

concentrations on the *x*-axis versus % inhibition on the *y*-axis. The % inhibition is calculated as follows:

$$\% \text{ inhibition} = \left[1 - \left\{ \frac{[\text{metabolite}]_{NCE}}{[\text{metabolite}]_{\text{control}}} \right\}\right] \times 100\%$$

For direct inhibition, NCEs with IC_{50} values >10 μM are considered to be less likely to cause inhibitory drug–drug interactions. NCEs with IC_{50} values <1 μM are considered potent inhibitors and are likely to produce drug–drug interactions. These potent inhibitors are usually excluded from consideration except in rare occasions. The fate of NCEs with IC_{50} values between 1 and 10 μM is determined by additional factors such as potency, therapy area, the stage of the discovery program, pharmacokinetics and projected efficacious plasma concentrations.

5.2.7. Throughput

The overall LC–MS/MS cycle time for this assay is 1.3 min. Usually eight 96-well plates containing 30 NCEs (for both CYP 3A4 and 2D6 pre- and co-incubation experiments) are assayed overnight. Since for each NCE, four IC_{50} values are determined, a total of 120 IC_{50} values could be generated daily. Compared to HPLC-UV assays, which was used when we started the CYP inhibition screening, the current LC–MS/MS assay provides more than 100-fold increase in throughput demonstrating the power of LC–MS/MS technology. Such increase in throughput is allowing the utilization of CYP inhibition assay as one of the first-line DMPK screening tools in several drug discovery programs.

5.3. Caco-2 Permeability

5.3.1. Background

Oral dosing is the most common route of drug administration for systemic drug delivery. It is the safest and least invasive as it utilizes the natural body systems of nutrients absorption. As a result, for the majority of drug discovery programs, the objective is to select NCEs that possess a good potential for oral absorption in humans. In vivo animal models are the most reliable methods to evaluate oral pharmacokinetics, which include absorption, as the entire organism provides an integrated system of absorption, distribution, metabolism and excretion. However, in vivo evaluation in animals of a large number of NCEs is not only costly but also time consuming and requires relatively large amounts of compounds. Also, animal models provide the net pharmacokinetic outcome of all processes involved, consequently, it is difficult to ascertain if the problem is due to oral absorption and/or other factors. Utilizing physicochemical properties such as pK_a, molecular weight, $\log P$, the numbers of hydrogen bond donors and acceptors, etc., could provide a good initial assessment of transcellular passive diffusion mode of absorption as reported by Lipiniski *et al.* (1997). However, other modes of absorption including transporter-mediated would be difficult to determine based on physicochemical properties (Brayden, 1997). Thus, in vitro models that can assess the potential for oral absorption of NCE have become major components of drug discovery programs in major pharmaceutical companies (Wilson *et al.*, 1990; Stevenson *et al.*, 1999). Caco-2 monolayer has been the most utilized screening tool for the assessment of potential oral absorption of NCEs in drug discovery. Caco-2 is a human colon carcinoma cell line, which exhibits properties resembling those of the intestinal enterocytes, such as the formation of microvilli, tight junctions and P-glycoprotein (P-GP) expression (Hidalgo *et al.,* 1989; Wilson, 1990). Several studies have shown a good correlation between Caco-2 permeability and oral drug absorption in humans (Gres *et al.,* 1998; Polli and Ginski, 1998); therefore, it has been generally accepted that compounds with good permeability across the Caco-2 monolayer would possess good human intestinal absorption, unless dissolution rate and/or solubility are limiting factors. Early on, we have established a Caco-2 permeability screen utilizing LC–MS/MS as an assay tool (Krishna *et al.*, 2001; Fung *et al.,* 2003). As the demand for Caco-2 screening increased, we recognized that LC–MS/MS analysis has become the bottleneck for higher-throughput. As a result, we introduced a multiplexed ion source (MUX®) coupled with the LC–MS/MS (developed by Waters Corp.) to increase the throughput of the Caco-2 screen. This MUX® system allows the analysis of four samples simultaneously, which greatly reduced sample analysis time.

In addition to transport studies, the Caco-2 system is used to evaluate if the NCE is a substrate for the efflux transporter P-GP. This has been carried out by the performance of bidirectional transport studies. Details of these experiments are described later.

5.3.2. Mechanisms of Absorption

There are four major pathways for a compound to be absorbed from the intestinal lumen (Wilson *et al.,* 1990; Brayden, 1997). Most drugs are absorbed via *transcellular passive diffusion*, which is driven by a concentration gradient. Lipophilic compounds partition into the lipid bilayer of the enterocytes, thereby crossing the cell membrane. The compound must possess some degree of solubility in the intestinal lumen in order to partition into the cell membrane. Low-molecular-weight hydrophilic compounds may be absorbed via the *paracellular route*, by passing through cell–cell junctions. Since the junctions are tight and represent a small surface area compared with the enterocytes, this mode of transport is slow and is primarily for water soluble low-molecular-weight compounds (<350 Da). Compounds can also be absorbed by a transmembrane transporter without requiring energy, i.e. *facilitated diffusion,* which is driven by concentration gradient, or requiring energy, i.e. *active transport*, which could proceed against concentration gradient. Since a compound forms a reversible drug-carrier complex in order to be transported, this process is selective. Also, as the number of carriers available is limited, this process is saturable either by the compound itself (at high concentrations) or other compounds that use the same carrier.

In addition to absorption from the lumen to the blood side of the intestine, P-GP is an important transporter that pumps certain compounds back into the intestinal lumen (efflux) after they are partitioned into the cell membrane. This transporter could hinder absorption and may influence the overall disposition of the compound. It is important to note that a compound could be simultaneously absorbed by several modes with varying degrees.

5.3.3. Caco-2 Screening Assay

The Caco-2 monolayers are first incubated at 37°C for 30 min in a CO_2 incubator with transport media (TM) prior to initiating a transport experiment. The TM for the apical side is made of Hank's balance salt solution (HBSS) containing 10 mM N-(2-hydroxyethyl)piperazine-N′-(2-ethanesulfonic acid) (HEPES) buffer (pH 7.4) and 25 mM D-glucose (Krishna *et al.*, 2001). In the basolateral compartment, 4% bovine serum albumin (BSA) is added to the TM. The addition of BSA has been shown to reduce non-specific binding that is often encountered with highly lipophilic compounds (Krishna *et al.*, 2001). For transport studies, NCEs are placed into the apical compartment of the Transwell® (Fig. 5). The integrity of the monolayer is confirmed by the measurement of the trans-epithelial electrical resistance (TEER), which must be above a certain limit to be used for transport experiments. In addition, with each experiment, two standard compounds, propranolol (a transcellular marker) and mannitol (a paracellular marker), are included for quality control.

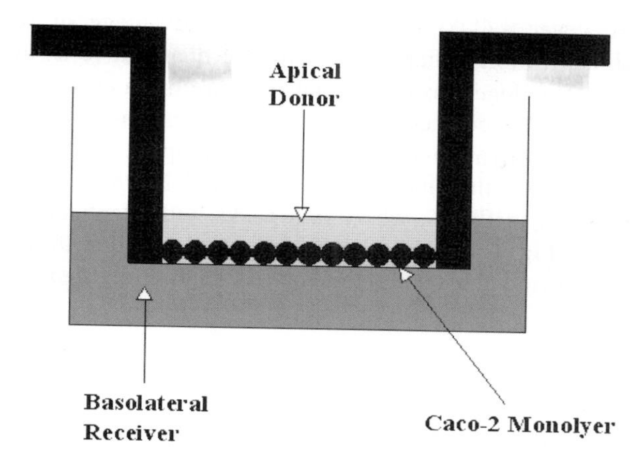

Figure 5.
Transwell for the Caco-2 permeability assay. Compounds are placed in the pical compartment and the concentrations in the apical and basolateral compartments are determined at given time points

The permeability of an NCE is determined in duplicate transwells. Duplicate samples are taken from the apical compartment at 0 min. At 120 min after the initiation of the transport experiment, duplicate samples are taken from both the apical and basolateral compartments. Samples are transferred into 96-well plates and mixed with 3 volumes of acetonitrile containing an internal standard. Sample-acetonitrile mixtures are vortexed, centrifuged and the supernatant is subjected to LC–MS/MS analysis in the same 96-well plate.

5.3.4. Efflux Evaluation

A bidirectional transport system has been set up to determine if the NCE is a substrate for the efflux transporter, P-GP which has been reported to be located on the apical compartment of the Caco-2 monolayer (Hunter *et al.*, 1993). In this assay, the permeability of a NCE is determined for both the apical-to-basolateral (A–B) and the basolateral-to-apical (B–A) directions. If a compound possesses a higher permeability in the B–A compared to the A–B direction, this would suggest that it is a substrate for an efflux transporter, primarily P-GP.

5.3.5. LC–MS or LC–MS/MS?

We had initially evaluated both LC–MS and LC–MS/MS as potential analytical tools for Caco-2 permeability screening. LC–MS offers the advantage of a simpler method development, which is a major advantage when dealing with a large number of compounds. However, LC–MS/MS provides the advantage of selectivity that is necessary to establish a more reliable and sensitive quantitative assay. Since in vitro matrices are devoid of endogenous compounds that are often encountered with in vivo samples, it has been always debated whether the more expensive LC–MS/MS systems are needed for the quantification of such in vitro samples. We have evaluated both techniques, and concluded that the selectivity of LC–MS/MS is required in many cases. This is illustrated in Fig. 6, where two peaks with identical *m/z* were detected in the LC–MS mode (Fig. 6A), while in the LC–MS/MS mode they produced different daughter ions allowing the recognition of the analyte of interest (Fig. 6B).

5.3.6. Multiplexed Ion Source (MUX®)

Figure 7 shows a schematic representation of the LC–MS/MS system used. The mass spectrometer is a Quattro Ultima equipped with a 4-way MUX® interface. This 4-way MUX® ion source consists of four electrospray needles around a rotating disk that contains an orifice allowing independent sampling from each sprayer. Four samples from the same gradient system (if a single HPLC pump system is used) or different gradient systems (if a multiple HPLC pump system is used) are simultaneously introduced into the mass spectrometer. Thus, both method development and sample analysis times are greatly shortened.

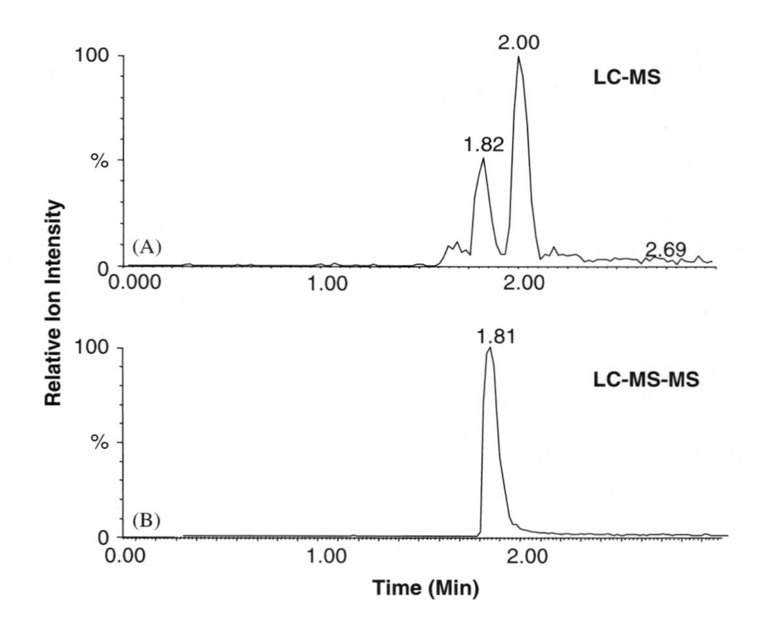

Figure 6.
Ion chromatograms of a basolateral sample of the Caco-2 permeability screen analyzed under (A) LC–MS and (B) LC–MS/MS conditions. (Adapted from Fung *et al.*, 2003)

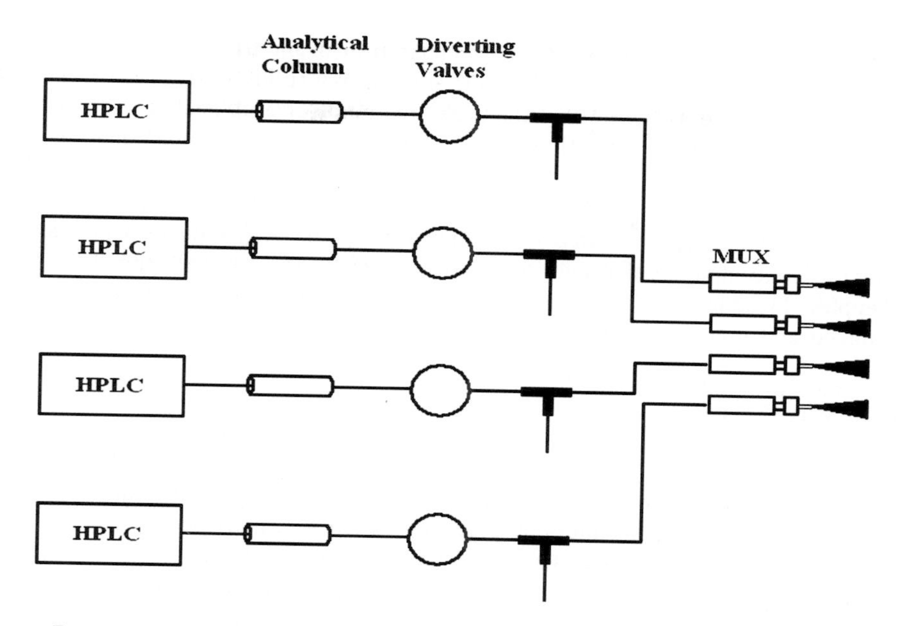

Figure 7.
A schematic representation of the LC–MS/MS system with a 4-sprayer MUX® interface

Several inherent issues needed to be addressed prior to utilizing the MUX® system for sample analysis in the Caco-2 screening assay. Examples include inter-channel variability and inter-channel cross-talk (Morrison *et al.*, 2002). Inter-channel variability is eliminated by setting up the system so that all samples associated with one compound are injected into the same channel; therefore, the response signal is independent of the ionization efficiency of each channel. The inter-channel cross-talk was evaluated in advance by injecting a benchmark compound (propranolol) on one channel and quantifying propranolol from all four channels. The results showed minimal cross-talk (<0.2%). Also, since compounds with different *m/z* are injected into different channels, the possibility of cross-talk is virtually eliminated.

5.3.7. Determination of Apparent Permeability

Caco-2 screening data for NCEs are presented as apparent permeability coefficient (P_{app}) which is calculated as follows:

$$P_{\text{app}} = (C_{B_t}/C_{A_0})V_b/(T \times S)$$

where C_{B_t} is the concentration of the compound in the basolateral side at time t, C_{A_0} the concentration of the compound in the apical side at the beginning of the experiment

(0 min), V_b the volume of the basolateral compartment of the transwell, T the sampling time (120 min) and S the Caco-2 membrane surface area (Eddy *et al.*, 1977). The term (C_{B_t}/C_{A_0}) is calculated as the ratio of the peak area ratios of each analyte to the internal standard, which resulted in the elimination of calibration curve samples that occupied approximately one-third of the total injection time. This approach has been validated prior to use (Fung *et al.*, 2003). The linearity of the response between the basolateral and apical samples is insured by diluting the A_0 and A_{120} samples with transport medium to concentrations that fall within approximately one order of magnitude of the B_t samples. In addition to permeability, the total recovery of the added compound is determined and used in the evaluation of the overall permeability data.

5.3.8. Assay Precision

The precision of the Caco-2 permeability system has been established prior to use as demonstrated by the inter-day precision of propranolol permeability determined on six different days. Excellent inter-day precision (%CV = 12) has been observed (Fung *et al.*, 2003), which demonstrates the reproducibility of both the biological and analytical systems. As a quality control for the assay, the permeability of the two standard compounds, propranolol and mannitol must fall within an acceptable range in order for the results to be accepted.

5.3.9. Throughput

The Caco-2 screening assay has been set up on a 1-week cycle. The cycle starts on Tuesday of every week, when methanol solutions of NCEs are delivered into 96-well plates from Compound Distribution Center. On the same day, the compounds are prepared for an overnight automated method development using the generic MUX® LC–MS/MS system. On Wednesday morning, the compounds that could not be analyzed with the generic system (usually 5–15%) are transferred to a different mass spectrometer (Sciex 3000 or 4000) for method development. On the same day sample lists are generated for each of the two instruments. Also, on Wednesday, the Caco-2 experiment is carried out and the samples are delivered for analysis in 96-well plates. Sample analysis is completed by Friday for both instruments and the analytical data are evaluated for error and submitted electronically to the Caco-2 experimental group. On Monday, the data are processed and the Caco-2 permeability reports are issued to individual discovery teams. Also, the instruments are subjected to regular maintenance to be ready for the next cycle. With this arrangement, approximately 60 compounds can be evaluated weekly.

5.4. Blood–Brain Barrier

5.4.1. Background

In addition to the two primary screening assays discussed above, we also conducted secondary, project-specific assays. One of such screening tool is the BBB permeability assay. By assessing the potential brain uptake of a drug candidate, discovery programs could advance or eliminate the NCE depending on program-specific central nervous system (CNS) activities. For example, for non-CNS disease targets, a drug candidate that shows extensive penetration into CNS may be eliminated because of potential undesirable side effects. On the other hand, for CNS indications, compounds with higher BBB permeability may be more desirable.

The BBB results from high-resistance tight junctions formed by the microvessel endothelial cells in the brain capillaries (Eddy *et al.*, 1977; Rubin and Staddon, 1999). Also, transporters such as P-GP have been reported to reside in the lumen side of the membrane, which provide additional control for the exchange of substances between the blood and the CNS. Due to the presence of various transporters and efflux systems, permeability across the BBB is difficult to assess from physicochemical properties alone.

The traditional methods for the evaluation of brain penetration have been the determination of brain and plasma concentrations in animals following compound administrations, which are time consuming, costly and require large amounts of compounds. We had adopted the format of coupling cultured bovine brain microvessel endothelial cells (BBMEC) with LC–MS/MS for higher-throughput evaluation of potential brain penetration of NCEs (Chu *et al.*, 2002). This assay was set up in order to screen compounds prior to the performance of brain penetration studies in animals.

5.4.2. Cell Culture

The BBMEC are isolated from the gray matter of the bovine cerebrum (Eddy *et al.*, 1977; Chu *et al.*, 2002). Briefly, the gray matter is incubated with protease to separate the microvessels from fat, myelin and other tissues. A subsequent enzyme digestion with collagenase/dispase is carried out to separate the endothelial cells from the basement membrane. Percoll gradient is then used to separate BBMEC from other cells. Finally, the BBMEC are aliquotted, frozen in liquid nitrogen with 20% horse serum and 10% DMSO until use. The entire isolation process takes approximately 12 h. The isolated cells can be used for approximately 3 months after isolation and freezing.

For drug transport studies, the BBMEC are cultured in 12-well transwell plates. The plating culture medium (CM) is made of 1:1 ratio of MEM and F-12 Ham containing 10% horse serum (Chu *et al.*, 2002). The cell culture plates are maintained in an incubator at 37°C with 5% CO_2. Confluent monolayers suitable for drug transport studies are established within 11–13 days.

5.4.3. Transport Studies

Transport experiments are carried similar to those described for the Caco-2 studies, where NCEs are added to the apical compartment of the transwell plates as a solution in DMSO (final concentration of DMSO is 1%). Samples are taken at 0 and 60 min from both the apical and basolateral compartments. The samples are mixed with internal standard in acetonitrile, vortexed, centrifuged and the supernatant is subjected to LC–MS/MS analysis. The BBB permeability and total compound recovery are calculated as described for Caco-2 permeability.

5.4.4. Assay Precision

The precision of this bioanalytical assay as well as the in vitro BBB model has been demonstrated by examining the inter-day precision of propranolol permeability. Six assays were performed at 0.5 μM on six different days, and four assays were carried out at 5 μM. The permeability was independent of the concentration used, and showed excellent inter-day precisions (%CV≤10) at both concentration levels, which demonstrate the reproducibility of both the biological and bioanalytical systems (Table 1). In addition, the reproducibility of the LC–MS/MS assay was established by the preparation of 8-point calibration curves for seven representative compounds (Chu *et al.*, 2002). These calibration standards were injected on three different days. Excellent inter-day reproducibility was observed between the three reconstructed calibration curves as shown for one of these test compounds (Fig. 8). The results also showed good linearity within the concentration range used.

5.4.5. In Vivo–In Vitro Correlation

The relevance of this in vitro cell-based model has been evaluated by comparing the brain/plasma area under the curve (AUC) (0–24 h) ratio in the rat to the in vitro BBB permeability for 29 compounds. A good concordance with a correlation coefficient of 0.87 has been established (Fig. 9), which is comparable to data reported by other investigators (Eddy *et al.*, 1977; Dehouck *et al.*, 1995).

Table 1.
Inter-day precision of the blood–brain barrier model using propranolol

	Apparent permeability of propranolol (nm/s)	
	0.5 μM	5 μM
Day 1	470	507
Day 2	434	—[a]
Day 3	418	445
Day 4	384	468
Day 5	355	439
Day 6	388	—[a]
Average	408	464
%CV	10	6

[a]Permeability was not measured on this day. Adapted from Chu *et al.* (2002)

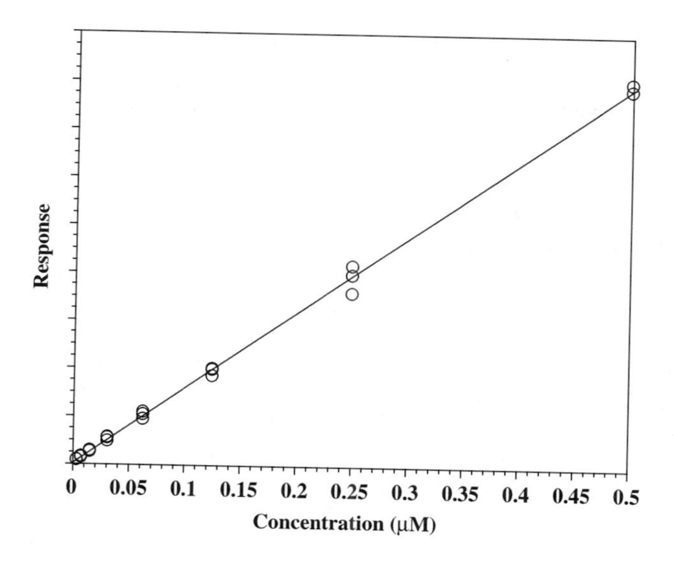

Figure 8.
Reproducibility of LC–MS/MS response. Triplicate injections of calibration curve standards of a test compound performed on three different days

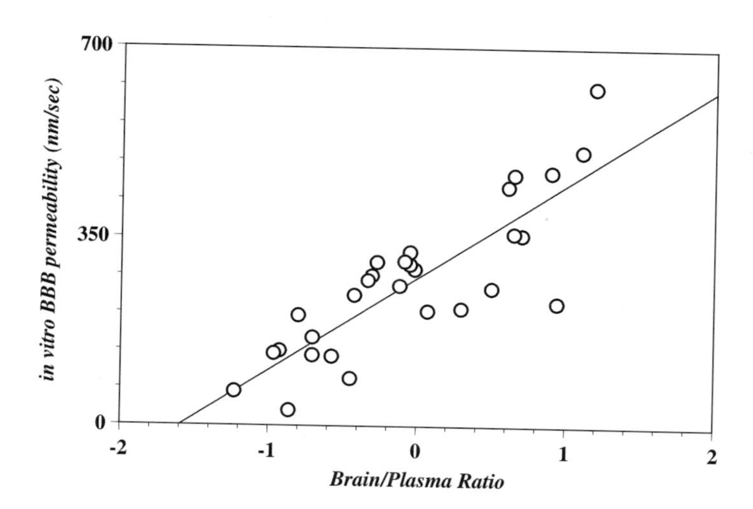

Figure 9.
Brain/plasma ratio versus in vitro BBB permeability of 29 compounds. (Adapted from Chu *et al.*, 2002)

5.5. Conclusion

The fate of an NCE in modern drug discovery is typically determined by the results of several screens performed in the drug discovery stage. These include drug metabolism and pharmacokinetics screens such as in vitro metabolic stability, permeability across the Caco-2 monolayers, pharmacokinetics in animals and weather or not, and to what extent the NCE inhibits various human CYPs. In recent years, identification of drug candidates has became more challenging because the number of NCEs has increased substantially creating the need for higher-throughput screens. The specificity and speed provided by the triple quadrupole mass spectrometer have allowed HPLC-co-eluting compounds to be analytically resolved by their different fragment ions. This unique feature of LC–MS/MS paved the way for rapid, generic fast gradient LC–MS/MS assays to be developed and utilized across the industry accelerating sample analysis.

When implementing these higher-throughput assay systems into a drug discovery decision tree, it is important to understand the relevance, precision, accuracy and shortcomings of each assay. Also, setting up the sequence of these screens in the decision tree should be done in a way to maximize the benefit by eliminating as many compounds with undesirable DMPK attributes as early as possible. This would minimize the cost and speed up the selection of drug candidates. Furthermore, it is crucial to understand how to use these higher-throughput data to advance various discovery programs, for each acceptance/rejection criteria could be different.

The three assays described in this chapter provide discovery scientists with information regarding some DMPK attributes of NCEs without going into lengthy and costly in vivo experiments. Compounds that pass through these assay move forward to pharmacokinetic studies in animals. Such increase in the screening rate of drug candidates using these in vitro higher-throughput assays has been made possible by the use of LC–MS/MS technology.

Acknowledgment

The authors acknowledge the scientific expertise of L. Favreau, C. Li, F. Liu, T. Tang, P. Kumari, T.-T. Liu and A. Soares in various techniques discussed in this chapter.

References

Brayden, D.J., 1997. Human intestinal epithelial cell monolayers as prescreens for oral drug delivery. Pharmaceut. News 4, 11–15.

Chen, Z.R., Somogyi, A.A., Bochner, F., 1990. Simultaneous determination of dextromethorphan and three metabolites in plasma and urine using HPLC with application to their disposition in man. Ther. Drug Monit. 12, 97–104.

Chu, I., Favreau, L., Soares, A., Lin, C.C., Nomeir, A.A., 2000. Validation of higher-throughput high performance liquid chromatography/atmospheric pressure chemical ionization tandem mass spectrometry assays to conduct cytochrome P-450s CYP2D6 and CYP3A4 enzyme inhibition studies in human liver microsomes. Rapid Commun. Mass Spectrom. 14, 207–214.

Chu, I., Liu, F., Soares, T., Kumari, P., White, R., Nomeir, A.A., 2002. Generic fast gradient LC–MS/MS techniques for the assessment of the *in vitro* permeability across the blood–brain barrier in drug discovery. Rapid Commun. Mass Spectrom. 16, 1501–1505.

Dehouck, M.P., Dehouck, B., Schluep, C., Lemaire, M., Cecchelli, R., 1995. Drug transfer across blood–brain barrier. Eur. J. Pharm. Sci. 3, 357–1797.

Eddy, E.P., Maleef, B.E., Hart, T.K., Smith, P.L., 1977. *In vitro* models to predict blood–brain barrier permeability. Adv. Drug Deliv. Rev. 23, 185–194.

Fung, E.N., Chu, I., Li, C.C., Liu, T., Soares, A., Morrison, R., Nomeir, A.A., 2003. Higher-throughput screening for Caco-2 permeability utilizing a multiple sprayer LC–MS/MS system. Rapid Commun. Mass Spectrom. 17, 2147–2152.

Gres, M.C., Julian, B., Bourrie, M., Meunier, V., Roques, C., Berger, M., Boulenc, X., Berger, Y., Fabre, G., 1998. Correlation between oral drug absorption in human, and apparent drug permeability in TC-7 cells. A human epithelial intestinal cell line: comparison with the parental Caco-2 cell line. Pharmaceut. Res. 15, 726–733.

Hidalgo, I.J., Raub, T.J., Borchardt, R.T., 1989. Characterization of human colon carcinoma cell line (Caco-2) as a model system for intestinal epithelial permeability. Gastroenterology 96, 736–749.

Honig, P.K., Wortham, D.C., Zamani, K., Conner, D.P., Mullin, J.C., Cantilena, L.R., 1993. Terfenadine-ketoconazole interaction. Pharmacokinetic and electro-cardiographic consequences. JAMA 269, 1513–1518.

Hunter, J., Jepson, M.A., Tsuruo, T., Simmons, N.L., Hirst, B.H., 1993. Functional expression of P-glycoprotein in apical membranes of human intestinal Caco-2 cells. Kinetics of vinblastine secretion and interaction with modulators. J. Biol. Chem. 268, 14991–14997.

Krishna, G., Chen, K.J., Lin, C.C., Nomeir, A.A., 2001. Permeability of lipophilic compounds in drug discovery using in-vitro human absorption model, Caco-2. Int. J. Pharm. 222, 77–89.

Lipiniski, C.A., Lombardo, F., Dominy, B.W., Feeney, P.J., 1997. Experimental and computational approaches to estimate solubility and permeability in drug discovery and development settings. Adv. Drug Deliv. Rev. 23, 3–25.

Mizutani, T., 2003. PM frequencies of major CYPs in Asians and Caucasians. Drug Metab. Rev. 35, 99–106.

Morrison, D., Davies, A.E., Watt, A.P., 2002. An evaluation of four channel multiplexed electrospray tandem mass spectrometry for higher throughput quantitative analysis. Anal. Chem. 74, 1896–1902.

Newton, D.J., Wang, R.W., Lu, A.Y.H., 1995. Cytochrome P450 inhibitors. Evaluation of specificities in the in vitro metabolism of therapeutic agents by human liver microsomes. Drug Metab. Dispos. 23, 154–158.

Nomeir, A.A., Palamanda, J.R., Favreau, L., 2004. Optimization in drug discovery: In vitro methods. In: Yan, Z., Caldwell, G.W. (Eds.), Identification of CYP Mechanism-based Inhibitors. Humana Press, Totowa, NJ, pp. 245–261.

Polli, J.E., Ginski, M.J., 1998. Human drug absorption kinetics and comparison to Caco-2 monolayer permeabilities. Pharmaceut. Res. 15, 47–52.

Rubin, L.L., Staddon, J.M., 1999. The cell biology of blood–brain barrier. Annu. Rev. Neurosci. 22, 11–28.

Silverman, R.B., 1988. Mechanism-based Enzyme Inactivation: Chemistry and Enzymology. CRC Press, Raton Boca, FL.

Smith, D.A., Ackland, M.J., Jones, B.C., 1997. Properties of cytochrome P450 isoenzymes and their substrates. Part 2: Properties of cytochrome P450 substrates. Drug Discov. Today 2, 479–486.

Stevenson, C.L., Augustijns, P.F., Hendren, R.W., 1999. Use of Caco-2 cells and LC/MS/MS to screen a peptide combinatorial library for permeable structures. Int. J. Pharm. 177, 103–115.

Tamvakopoulos, C.S., Colwell, L.F., Barakat, K., Fenyk-Melody, J., Griffin, P.R., Nargund, R., Palucki, B., Sebhat, I., Shen, X., Stearns, R.A., 2000. Determination of brain and plasma drug concentrations by liquid chromatography/tandem mass spectro-metry. Rapid Commun. Mass Spectrom. 14, 1729–1735.

Waxman, D.J., Lapenson, D.P., Aoyama, T., Gelboin, H.V., Gonzalez, F.J., Korzekwa, K., 1991. Steroid hormone hydroxylase specificities of eleven cDNA-expressed human cytochrome P450s. Arch. Biochem. Biophys. 290, 160–166.

White, R.E., 2000. High-throughput screening in drug metabolism and pharmacokinetic support of drug discovery. Ann. Rev. Pharmacol. Toxicol. 40, 133–157.

Wilson, G., 1990. Cell culture techniques for the study of drug transport. Eur. J. Drug Metabol. Pharmacokinet. 15, 159–163.

Wilson, G., Hassan, I.F., Dix, C.J., Williamson, I., Shah, R., Mackay, M., Artursson, P., 1990. Transport and permeability properties of human Caco-2 cells: an in vitro model of the intestinal epithelial cell barrier. J. Control. Rel. 11, 25–40.

Identification and Quantification of Drugs, Metabolites and Metabolizing
Enzymes by LC–MS
Swapan K. Chowdhury, editor.

Chapter 6

METABOLITE IDENTIFICATION BY LC–MS: APPLICATIONS IN DRUG DISCOVERY AND DEVELOPMENT

Cornelis E.C.A. Hop and Chandra Prakash

6.1. Introduction

Recent data indicate that the discovery and development of a new drug costs around 1 billion dollars and it may take approximately 10 years for the drug to reach the marketplace (Fig. 1). Considering these staggering numbers, it is critical that efforts are made to reduce attrition of drug candidates during the various stages of drug discovery and development. One of the sources of attrition can be inappropriate drug disposition characteristics. Indeed, data from a joint meeting from the Pharmaceutical Manufacturers Association (PMA) and the Food and Drug Administration (FDA) in 1991 showed that about 40% of all lead candidates failed in their development due to poor pharmacokinetics (Baillie and Pearson, 2000). Thus, it is critical that attention is paid to the disposition characteristics of a drug candidate early on. This information sheds light on the absorption, distribution, metabolism and excretion (ADME) of potential drugs (Lin and Lu, 1997; Borchardt *et al.*, 1998; Eddershaw and Dickins, 1999; Woolf, 1999; White, 2000; Riley *et al.*, 2002; Smith *et al.*, 2002) and these studies are performed (a) to support the selection of more efficacious and safer drugs for development, (b) to help understand pharmacological and toxicological observations and (c) to determine dose levels and dose regimens. Up to 1985, the disposition of a drug candidate was examined predominantly once the compound reached phase I or subsequent clinical trials. However, around 1990, most major drug companies started to explore the drug disposition characteristics of lead candidates during the discovery stage of drug development. The ADME data was being used to guide synthetic chemistry efforts in order to come up with candidates with more appropriate drug disposition characteristics. This approach appears to have been successful because informal data suggest that the attrition due to inappropriate ADME characteristics has been reduced significantly.

Currently, a wide variety of in vitro and in vivo screens are in place to obtain valuable information about ADME parameters for lead candidates. The in vitro studies include (a) metabolic stability in liver microsomes, hepatocytes or with recombinant cytochrome P450 enzymes, (b) metabolite formation in liver microsomes, hepatocytes

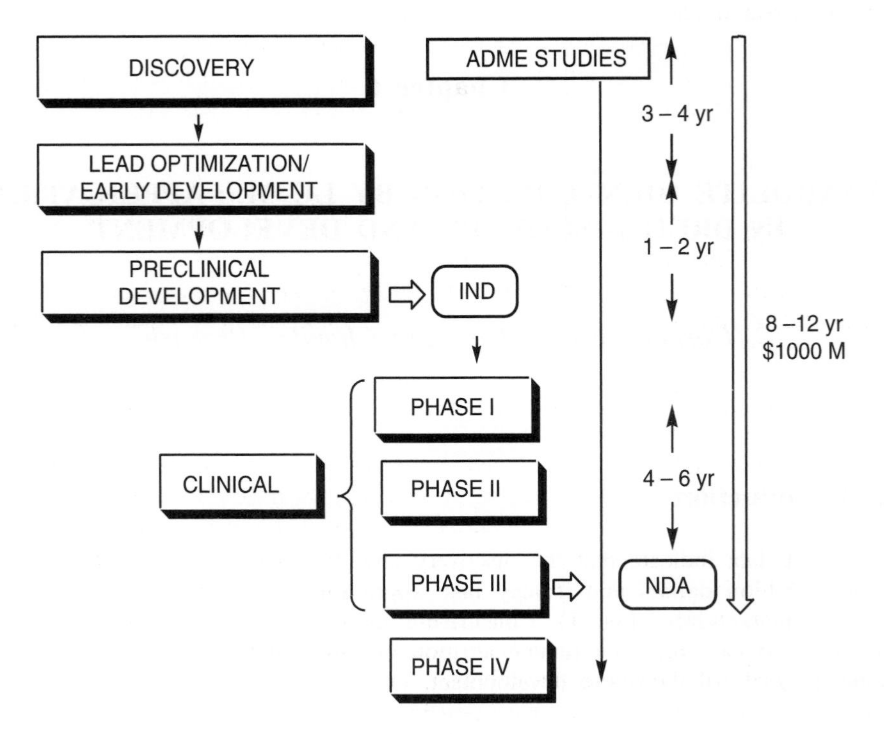

Figure 1.
Generic timeline for the various stages of drug discovery and development

or with recombinant cytochrome P450 enzymes, (c) absorption/transport studies in Caco-2 cells or cell lines over expressing various transporters, (d) cytochrome P450 inhibition, (e) cytochrome P450 induction, (f) plasma protein binding and (g) red blood cell partitioning. The in vivo studies include (a) pharmacokinetic studies via various routes of administration (oral, intravenous, subcutaneous, etc.), (b) tissue distribution (e.g., brain penetration) and (c) metabolite identification in various biological fluids (plasma, bile, urine, etc.). It has been shown that many of these studies benefit from analysis by liquid chromatography interfaced with mass spectrometry (LC–MS) (Niessen, 1998; Willoughby *et al.*, 1998; Lee and Kerns, 1999; Lee, 2002).

In the recent past, analysis of biological samples was a bottleneck. In some cases GC–MS, with or without derivatization of the analyte, was possible, but in many cases more traditional analytical techniques were employed, such as radioimmunoassays or HPLC with UV, fluorescence or electrochemical detection. With the introduction of atmospheric pressure ionization (API), in particular electrospray ionization (ESI) and atmospheric pressure chemical ionization (APCI), in the late 1980s, it became possible to sample liquids directly by mass spectrometry. Shortly after the introduction of these ionization techniques, they were employed in combination with liquid chromatography for pharmacokinetic studies and metabolite identification. The availability of LC–MS enabled increased and earlier ADME evaluation in drug discovery and development. With minimal effort, it became possible to develop sensitive and selective assays for drug candidates and, if required, their metabolites. Recent developments, such as monolithic columns, ballistic gradients and parallel analytical columns, have increased the throughput of LC–MS for quantitative in vitro and in vivo studies dramatically (Bakhtiar *et al.*, 2002; Cox *et al.*, 2002; Hopfgartner and Bourgogne, 2003). Another area, which benefited greatly from the introduction of LC–MS, is metabolite identification. Both the general approach and specific examples displaying the power of LC–MS for metabolite identification will be presented here. For general reviews describing the use of LC–MS in drug disposition studies see Korfmacher *et al.* (1997), Poon (1997), Brewer and Henion (1998), Baillie and Pearson (2000) and Rossi and Sinz (2002).

6.2. Metabolite Identification

Drugs are metabolized generally to more polar, hydrophilic entities, which can be excreted from the body more easily. The main routes of elimination are bile and urine. Detailed knowledge of the metabolic fate of drugs is important because the metabolites could be (a) pharmacologically active, (b) toxic, (c) involved in drug–drug interactions via inhibition or induction of drug-metabolizing enzymes or (d) competing for plasma protein binding with the parent compound. Examples include phenacetin, which is metabolized to acetaminophen and the latter compound is responsible for the majority of the analgesic activity.

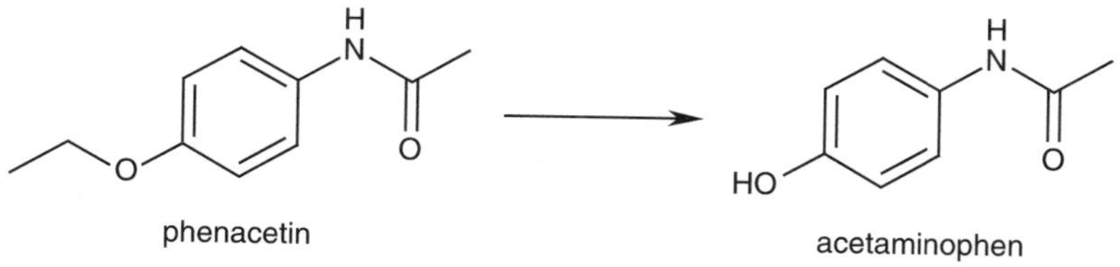

phenacetin → acetaminophen

A few years ago, terfenadine was withdrawn from the market, because it is metabolized by CYP3A4 to fexofenadine and inhibition of its metabolism by CYP3A4 inhibitors, such as erythromycin and ketoconazole, can give rise to elevated levels of terfenadine, which can cause cardiotoxicity. However, the metabolite, fexofenadine, is marketed, because it does not have these interactions and is pharmacologically active.

terfenadine → fexofenadine

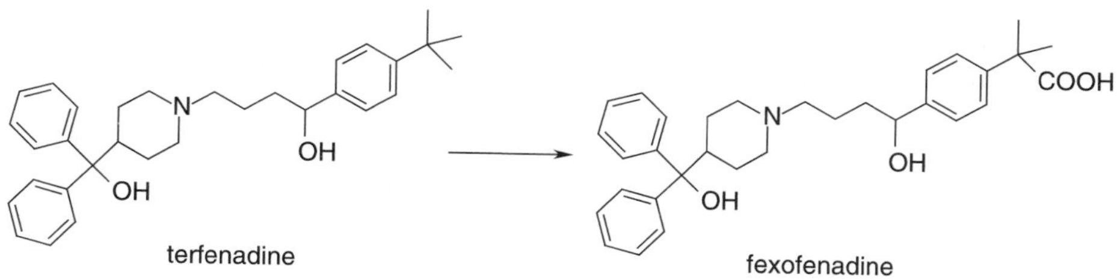

During toxicology studies, detailed knowledge about the metabolism is important, because, for proper assessment of the safety of a drug for human use, it must be shown that the animal species used for safety evaluation are exposed to the same metabolites as humans. If this is not the case, specific toxicology studies with the metabolite(s) not formed in the animal species may be needed.

In the discovery stage, identification of metabolites may reveal metabolic liabilities in molecules, which can subsequently be addressed by medicinal chemists who can block these particular sites of metabolism. This could result in the discovery of drug candidates with superior pharmacokinetic properties. Alternatively extensive metabolism has been used to create prodrugs, which may be better absorbed than the drug itself (e.g., fenofibrate). Thus, details about the metabolism of a drug and the enzymes involved in the metabolism are a necessity throughout drug discovery and development.

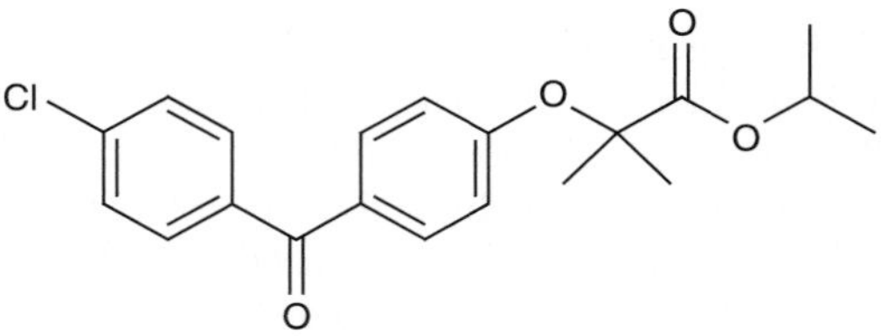

fenofibrate

6.3. Mass Spectrometers For Metabolite Identification

Although impressive metabolite identification data have been obtained with GC–MS, the convenience of interfacing conventional LC columns with atmospheric pressure ionization sources has shifted metabolite identification studies toward employing almost exclusively LC–MS. The two most commonly employed atmospheric pressure ionization techniques are APCI and ESI. In APCI, also called heated nebulizer, the effluent from the LC column is converted into a fine mist via nebulization caused by a high velocity jet of air or nitrogen. The spray travels through a quartz tube heater, which vaporizes the droplets and, consequently, the analyte in the gas stream. A corona discharge from a needle initiates chemical ionization using the vaporized solvent as the reagent gas. Approximately 5–10 kV is applied to the needle, which generates a 2–5 μA corona discharge current. For the generation of positive ions, the needle is at a high positive voltage with respect to counter electrode. A high negative voltage is used for the production of negative ions.

In ESI a dilute solution is sprayed from a fine needle, which carries a high potential (about 3–5 kV). If the needle carries a positive potential, the droplets will have an excess of positive charges, usually protons. Evaporation of the volatile solvent (e.g., H_2O, CH_3OH or CH_3CN) results in increased Coulombic repulsion between the positive charges, which eventually results in fragmentation of the droplet, generating smaller droplets. There are two theories for ion formation in ESI. Iribane and Thomson have suggested that ionized sample molecules are expelled from the droplets. Alternatively, it has been proposed that single, ionized sample molecules remain after continuous solvent evaporation and droplet fragmentation. One of characteristic features of ESI is the mild nature of the ionization process, which usually results in the formation of $[M + H]^+$ or $[M - H]^-$ ions only. The absence of fragment ions necessitates MS/MS experiments to obtain structural information (*vide infra*).

A wide range of mass spectrometers has been used for metabolite identification and several new types have become available recently. The most frequently used instruments are single and triple quadrupole and 3-dimensional ion trap mass spectrometers. Time-of-flight, quadrupole-time-of-flight (Q-TOF), linear ion trap and Fourier-transform mass spectrometers have been used as well. In many ways, the instruments listed above are complementary. Although detailed descriptions of these types of mass spectrometers have been presented elsewhere, it is important to describe specific advantages and disadvantages of these types of mass spectrometers for metabolite identification.

6.3.1. Single and Triple Quadrupole Mass Spectrometer

With single and triple quadrupole mass spectrometers, MS data can be obtained by scanning the first quadrupole and detecting the mass-separated ions. Product ion MS/MS data can be obtained with a triple quadrupole mass spectrometer by mass-selecting the ions of interest with the first quadrupole, fragmenting these ions in the collision cell and mass separating the ions with the third quadrupole for subsequent detection. Triple quadrupole mass spectrometers also allow constant neutral loss and precursor ion MS/MS experiments. A constant neutral loss scan allows detection of all ionized compounds, which lose a specific entity upon collision-induced dissociation. For example, sulfate-conjugated metabolites are known to lose SO_3 (i.e., 80 Da) easily upon collision-induced dissociation and, therefore, an 80 Da constant neutral loss allows detection of all sulfate-conjugated metabolites present in the sample of interest. Precursor ion scans allow detection of all ionized compounds, which give rise to a specific fragment ion upon collision-induced dissociation. For example, glucuronide-conjugated metabolites frequently yield a signal at m/z 175 in the negative ion mode and this “fingerprint” can be used to detect glucuronide-conjugated metabolites. Constant neutral loss and precursor ion scans greatly facilitate the process of finding metabolites, especially if the matrix contains many endogenous components. The various MS/MS scan modes available for a triple quadrupole mass spectrometer are summarized in Fig. 2 and the availability of multiple MS/MS scan modes increases the selectivity of this type of mass spectrometers significantly. Finally, triple quadrupole mass spectrometers can operate in the selected reaction-monitoring (SRM) mode. In this mode, a specific ion is mass-selected with the first quadrupole and fragmented in the collision cell and the third quadrupole is set to transmit a specific fragment ion. This mode results in greatly enhanced sensitivity compared to full scan MS/MS experiments.

6.3.2. 3-Dimensional Ion Trap Mass Spectrometer

In a 3-dimensional ion trap mass spectrometer all ions are in the center of the ion trap and a spectrum is obtained by sequentially ejecting ions out of the trap. For

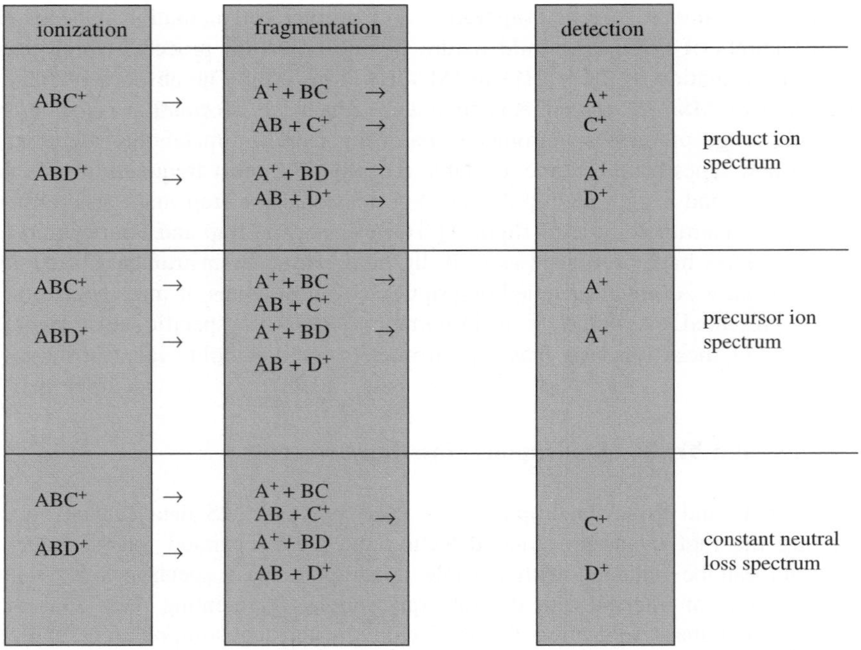

ionization		fragmentation		detection	
ABC^+	→	$A^+ + BC$ $AB + C^+$	→ →	A^+ C^+	product ion spectrum
ABD^+	→	$A^+ + BD$ $AB + D^+$	→ →	A^+ D^+	
ABC^+	→	$A^+ + BC$ $AB + C^+$	→	A^+	precursor ion spectrum
ABD^+	→	$A^+ + BD$ $AB + D^+$	→	A^+	
ABC^+	→	$A^+ + BC$ $AB + C^+$	→	C^+	constant neutral loss spectrum
ABD^+	→	$A^+ + BD$ $AB + D^+$	→	D^+	

Figure 2.
Scan modes feasible with a triple quadrupole mass spectrometer

a product ion MS/MS spectrum, the ions with a lower and higher molecular weight than that of the desired ions are ejected and the remaining ions with the desired *m/z* value are activated, which results in fragmentation. Because of the 3-dimensional nature of the ion trap, MS^n is possible and this can provide more information about the structure of a metabolite. (In the ion trap, MS/MS and MS^n are performed in "time" rather than in "space" as is the case for a triple quadrupole mass spectrometer.) However, constant neutral loss and precursor ion MS/MS data cannot be obtained with an ion trap mass spectrometer. For more details about the use of a 3-dimensional ion trap for metabolite identification see Tiller *et al.* (1998) and Dear *et al.* (1999).

6.3.3. Q-TOF Mass Spectrometer

The Q-TOF mass spectrometer was introduced in 1996 and has found widespread use for metabolite identification. In the MS mode the time-of-flight analyzer is used for mass analysis. For product ion MS/MS spectra, the ions of interested are selected with the first quadrupole and the fragment ions are mass-separated with the time-of-flight analyzer. The advantages of the Q-TOF mass spectrometer are sensitivity, scan speed and mass accuracy. The mass accuracy is usually within 5 ppm and, therefore, it is possible to assign a specific molecular formula to the metabolite under investigation (Hopfgartner *et al.*, 1999; Hop *et al.*, 2001; Qin and Frech, 2001; Tiller *et al.*, 2001).

Thus, each type of mass spectrometers has specific strengths and frequently more than one type is used to come up with more definitive structural data for metabolites.

6.3.4. Common Biotransformations

Subsequent to acquisition, MS spectra can be inspected for the presence of (anticipated) phase I metabolites. Table 1 presents a list of common phase I biotransformations and the corresponding mass changes. The most common metabolic process is oxidation resulting in a hydroxy moiety, which generates an M + 16 Da metabolite. Liver microsomes are frequently used to assess formation of phase I metabolites by a compound. Most phase I biotransformations are mediated by the ubiquitous cytochrome P450 enzymes. Table 2 presents a list of common phase II biotransformations and the

Table 1.
Common phase I biotransformations and the corresponding change in mass of the parent compound

Mass change (Da)	Type of biotransformation
−28	Deethylation ($-C_2H_4$)
−14	Demethylation ($-CH_2$)
−2	Two-electron oxidation ($-H_2$)
+2	Two-electron reduction ($+H_2$)
+14	Addition of oxygen and two-electron oxidation ($+O{-}H_2$)
+16	Addition of oxygen (+O)
+18	Hydration ($+H_2O$)
+30	Addition of two oxygen atoms and two-electron oxidation ($+2O{-}H_2$)
+32	Addition of two oxygen atoms (+2O)

Table 2.
Common phase II conjugation reactions and the corresponding change in mass of the parent compound

Mass change (Da)	Conjugate
+42	Acetyl
+80	Sulfate
+57	Glycine
+107	Taurine
+176	Glucuronide
+307 or 305	Glutathione

corresponding mass changes. Phase II biotransformations involve conjugation of the compound, with the objective of making the drug more hydrophilic. These phase II biotransformations are mediated by enzymes, such as *N*-acetyl transferases, sulfotransferases, UDP glucuronosyltransferases and glutathione-S-transferases. Both phase I and phase II enzymes are present in hepatocytes. Although MS data are frequently sufficient to identify the type of biotransformation taking place, product ion MS/MS spectra are necessary to identify the location of the structural modification. Interpretation of MS/MS data is time-consuming and requires a high level of expertise. (Several textbooks are available discussing interpretation of MS/MS data in detail.) Reliable software to automate interpretation of product ion MS/MS spectra is not yet available.

6.4. Additional Tools For Metabolite Identification

6.4.1. Chemical Derivatization

Although product ion MS/MS data can provide a substantial amount of structural information for metabolite identification, it is sometimes difficult to differentiate between regioisomers. For example, based on the product ion MS/MS spectra alone, it was not possible to define the site of glucuronidation for two glucuronide conjugates, **1** and **2**, of CP-101,606 (an NMDA receptor antagonist) observed in humans (Johnson *et al.*, 2003). The site of conjugation was established after derivatization of the sample with diazomethane. After treatment of **1** with diazomethane, the full-scan MS spectrum showed an intense protonated molecule at *m/z* 518, 14 Da higher than **1**, suggesting that the phenolic group was substituted with glucuronide. On the other, the full-scan MS spectrum of the methylated product of **2** showed an intense protonated molecule at *m/z* 532, 28 Da higher than **2**, indicating the methylation of both phenolic group as well as the carboxylic acid moiety of the glucuronide and, therefore, **2** was characterized as a benzylic glucuronide.

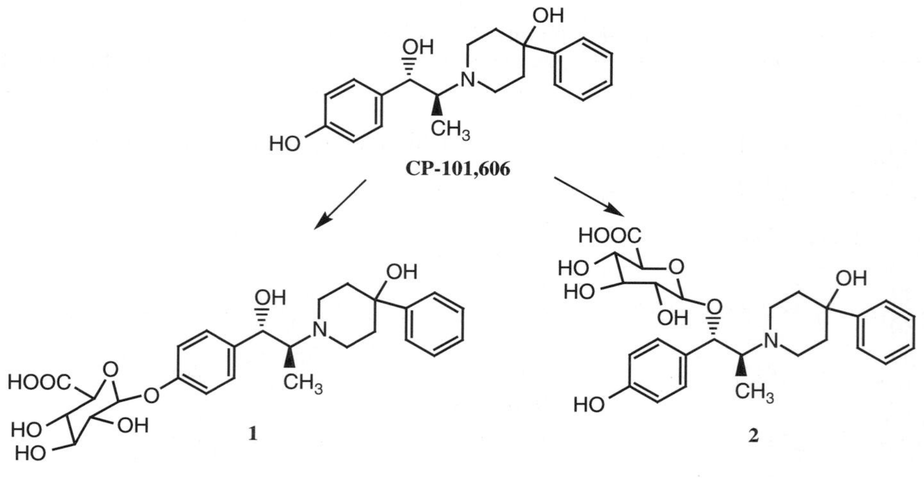

CP-101,606

1

2

A few common derivatization reactions are presented in Table 3 and others can be found in the literature (Knapp, 1979). Sometimes derivatization of a metabolite is employed to create a more hydrophobic, readily ionizable entity. Dalvie *et al.* (1998) dansylated several polar urinary metabolites of a proprietary compound, which were detectable by radiochromatography, but which could not be identified by LC–MS. The presence of the dimethylamino group in the dansyl moiety facilitated protonation and characterization of the metabolites by LC–MS. An additional advantage is that increasing the molecular weight of the analyte frequently reduces the amount of chemical interference.

6.4.2. Radiolabeled Compounds

During the later stages of drug discovery or drug development a radiolabeled (^{3}H and/or ^{14}C) analog is synthesized to obtain more detailed metabolism and disposition data. In vitro studies can be performed using the radiolabeled analog. Upon LC–MS analysis the column eluent is split with a fraction going to the mass spectrometer for detection and the remainder going to an on-line radio-flow detector or a fraction collector for offline liquid scintillation counting (LSC). The radiochromatogram provides the retention times of all metabolites generated (unless the radiolabel is not retained in the metabolite) and, therefore, the inspection of mass spectrometric data for identification of metabolites can be more focused. Only a narrow region of the total-ion-current (TIC) trace, as specified by the radiochromatogram, needs to be examined for drug-related components. A small amount of a radiolabeled compound can be administered to animals or humans as well. LSC of the various biological fluids (usually plasma, bile and urine) helps identify the fluid with the largest amount of radioactivity and this fluid can be subsequently examined by LC–MS for identification of metabolites.

6.4.3. Isotope Patterns

If the parent compounds contains a characteristic (stable) isotope pattern, for example, due to the presence of one or more chlorine or bromine atoms, the metabolites can be identified by looking for this specific isotope pattern (cluster analysis).

Table 3.
Common reagents used to derivatize compounds

Reagent	Derivatization reaction
Anhydrous methanolic HCl	Methylation of carboxylic acids and allylic and benzylic alcohols
$LiAlH_4$ or $NaBH_4$	Reduction of aldehydes to alcohols
$TiCl_3$	Reduction of *N*-oxides
Acetic anhydride	Acetylation of amines and phenols

The presence of the desired isotope pattern increases the confidence that the signal is indeed related to the administered compound. Wienkers *et al.* (1995, 1996) identified metabolites for tirilazad, a potent inhibitor of membrane lipid peroxidation, and they employed a mixture of unlabeled and [2,4,6-$^{13}C_3$, 1,3-$^{15}N_2$-pyrimidine]tirilazad; metabolites were identified by looking for two signals of equal abundance separated by 5 Da.

6.4.4. LC–NMR–MS

Although LC–MS is a sensitive analytical technique for obtaining molecular weights and molecular formulas for metabolites, it provides only limited structural information. Due to the nature of the fragmentation pattern in product ion MS/MS spectra it is only possible to identify the moiety where structural modification has occurred without being able to isolate the specific site. If an authenthic standard is available, a comparison of the HPLC retention times, as well as MS and MS/MS spectra may be sufficient to make more definitive assignments. Nuclear magnetic resonance (NMR) spectroscopy is frequently required to obtain the exact structure of each metabolite. The NMR spectrum can be obtained following isolation of the metabolite of interest. Alternatively, LC–NMR–MS can be employed (Shockcor *et al.*, 1996; Burton *et al.*, 1997; Clayton *et al.*, 1998; Loudon *et al.*, 2000). In the latter case, about 95% of the LC eluent is directed into the NMR spectrometer and the remainder is diverted into the mass spectrometer. Several elegant applications have appeared, especially in combination with generation of metabolites in bioreactors, but NMR spectroscopy still suffers from limited sensitivity compared to mass spectrometry, which makes this technique less practical for the identification of minor metabolites.

6.4.5. H/D Exchange

Frequently, the use of D_2O as the mobile phase for HPLC can be as informative as derivatization procedures. The number of exchangeable hydrogen atoms can provide valuable information about the functional groups present in the analyte. For example, it was feasible to distinguish two possible structures for an oxidative metabolite of a drug candidate containing a thiazolidine moiety (Liu *et al.*, 2001). If hydroxylation of one of the carbon atoms of the thiazolidine group occurred, the H/D shift should be one more than anticipated for the sulfoxide. H/D exchange also facilitates distinguishing *N*-oxides from hydroxylated metabolites. Other examples have been presented by Olsen *et al.* (2000).

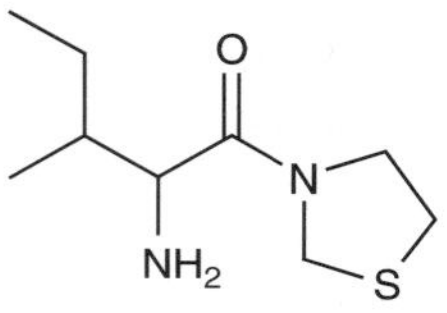

O
N
NH_2
S

6.5. Applications

Several examples are presented illustrating the use of various types of mass spectrometers to facilitate metabolite identification. It is critical to realize that, although LC–MS is a very powerful and useful technique, definitive assignments frequently require the use of additional analytical techniques (such as NMR) and/or the synthesis of authentic standards.

6.5.1. Metabolism of a Potent Neurokinin 1 Receptor Antagonist

Substance P is the most abundant neurokinin in the human central nervous system. The substance P preferring neurokinin-1 (NK1) receptor is highly expressed in brain regions that are critical for regulation of neurochemical responses to stress. NK1 receptor antagonists have been proven in concept to have excellent potential for the treatment of major depression (Kramer *et al.*, 1998) and their side effect profile might allow favorable differentiation against currently available therapies. In addition, NK1 receptor antagonists allow superior and sustained protection from acute and delayed chemotherapy-induced emesis (Navari *et al.*, 1999; Tattersall *et al.*, 2000). Here, the metabolism of a potent NK1 receptor antagonist, [3R,5R,6S]-3-(2-cyclopropyloxy-5-trifluoromethoxyphenyl)-6-phenyl-1-oxa-7-azaspiro[4.5]decane, (compound **A** hereafter; see Fig. 3) in rat plasma as well as in rat liver microsomes and rat hepatocytes is described (Hop *et al.*, 2002).

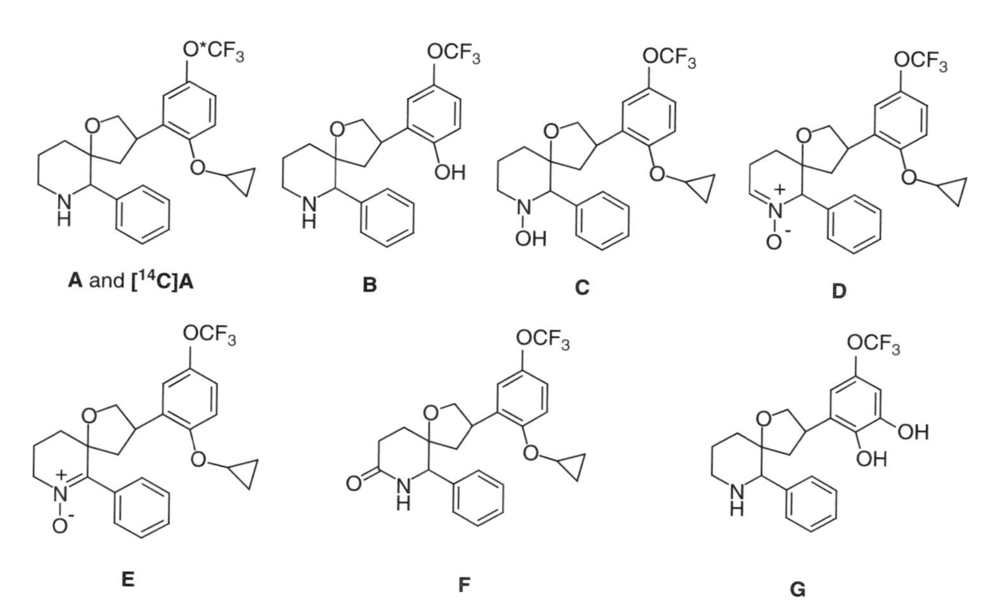

Figure 3.
Structures of compound **A** and its metabolites (* indicates the position of the ^{14}C-label)

Interpretation of product ion spectra of metabolites frequently hinges upon similarity with that of the parent compound. However, sometimes interpretation of the product ion spectrum of the parent compound is ambiguous. Availability of ^{14}C-labeled compound **A** with the label in the trifluoromethyl group facilitated interpretation of the product ion spectrum. Comparison of the product ion spectrum of compound **A** with that of its ^{14}C-labeled analog indicates that the fragments at *m/z* 231, 215, 203, 191 and 175 are associated with the trifluoromethoxy phenyl moiety, whereas the fragments at *m/z* 184, 172, 159, 131, 91 and 56 are associated with the phenyl piperidine moiety. Accurate mass measurements obtained with a Q-TOF mass spectrometer provided additional information. Table 4 summarizes the molecular formula assignments for the major fragment ions of [**A** + H]$^+$; all assignments are within 10 ppm with most of them being within 5 ppm. Based on this information, structural assignments for the most informative fragmentation pathways of [**A** + H]$^+$ ions are feasible and the data are summarized in Fig. 4. Concern about the potential pharmacological activity of anticipated metabolites led to synthesis of authentic standards of the *O*-dealkylated metabolite (**B**), the hydroxylamine metabolite (**C**), the nitrones (**D** and **E**), the lactam metabolite (**F**) and the hydroxylated and *O*-dealkylated metabolite (**G**). Comparison of the product ion spectra of the metabolites **B–G** with that of the parent compound **A** indicates that the product ion spectra are compatible with the assigned structures.

Table 4.
Assignment of the fragment ions observed in the product ion spectrum of protonated compound **A**, $C_{24}H_{26}NO_3F_3$, using the micromass Q–TOF II

Fragmention (Da)	Observed mass (Da)	Formula assignment	Theoretical mass (Da)	Error (ppm)
416	416.1839	$C_{24}H_{25}NO_2F_3$	416.1837	0.5
231	231.0628	$C_{11}H_{10}O_2F_3$	231.0633	−2.2
215	215.0325	$C_{10}H_6O_2F_3$	215.0320	2.4
203[a]	203.0677	$C_{10}H_{10}OF_3$	203.0684	−3.1
203[a]	203.0304	$C_9H_6O_2F_3$	203.0320	−7.8
191	191.0325	$C_8H_6O_2F_3$	191.0320	2.4
184	184.1137	$C_{13}H_{14}N$	184.1126	5.8
175	175.0378	$C_8H_6OF_3$	175.0371	4.4
172	172.1136	$C_{12}H_{14}N$	172.1126	5.4
159	159.1053	$C_{11}H_{13}N$	159.1048	3.4
131	131.0873	$C_{10}H_{11}$	131.0861	9.1
91	91.0526	C_7H_7	91.0548	−23.9 [b]

[a]The signal at *m/z* 203 was composite.

[b]The signal at *m/z* 91 was outside the calibrated mass range resulting in a larger error.

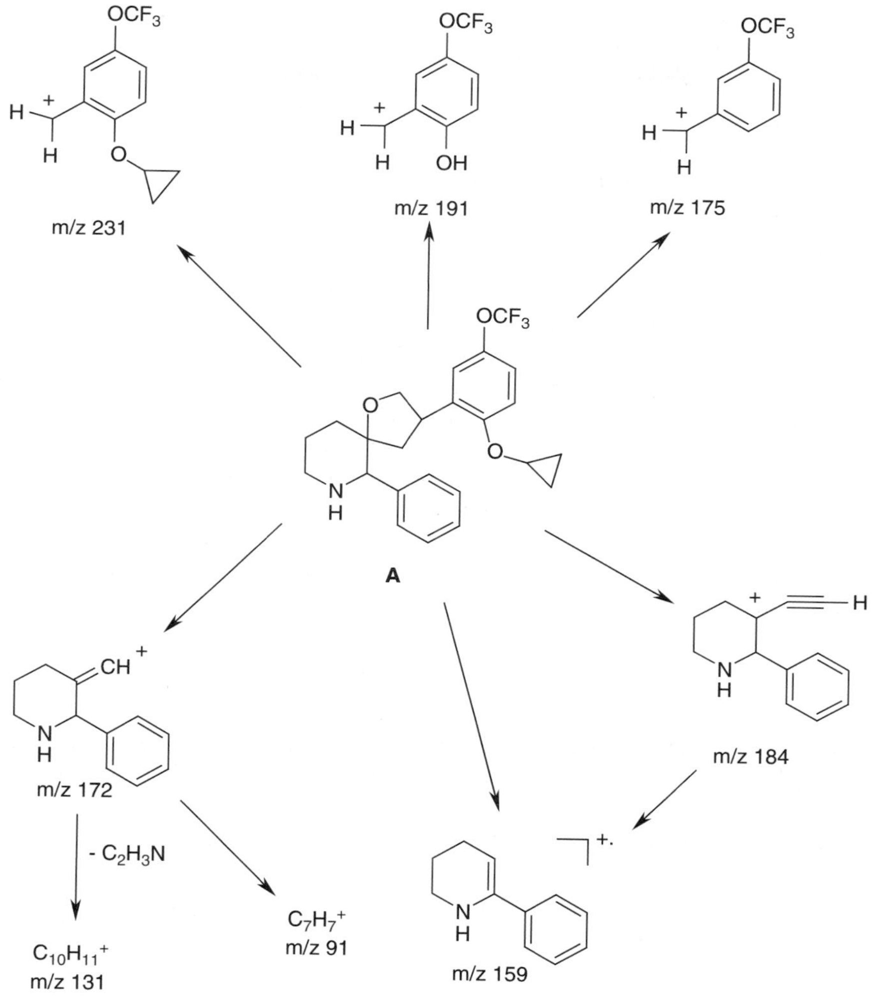

Figure 4.
Tentative structural assignments for the most informative fragmentation pathways observed for protonated compound **A** upon collision-induced dissociation

The turnover of compound **A** in rat hepatocytes was only 30% and, therefore, the most abundant signal (at a retention time of 23.8 min) is the parent compound, **A**. The radiochromatogram is presented in Fig. 5 and the three metabolites eluting at retention times of 27.0, 29.8 and 32.8 min correspond to the two nitrones (**D** and **E**) and the lactam (**F**), respectively, based on comparison of the product ion spectra with those of the authentic standards. In addition, three minor metabolites were observed at retention times of 10.0, 15.2 and 32.8 min. The product ion spectrum

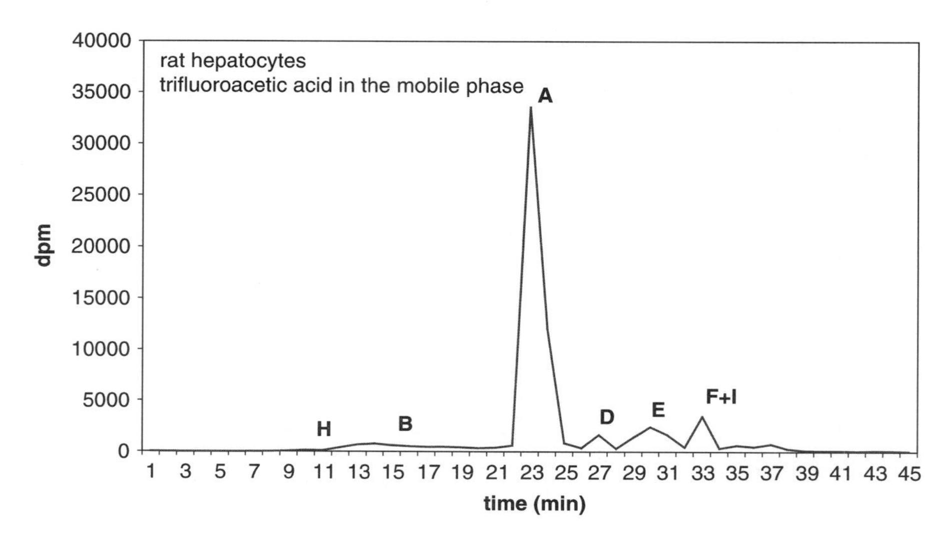

Figure 5.
HPLC-radiochromatogram of ^{14}C-labeled compound **A** and its metabolites generated in rat hepatocytes. The liquid chromatography mobile phase contained trifluoroacetic acid

of the metabolite at 15.2 min indicates that this was the *O*-dealkylated metabolite (**B**). The MS and MS/MS spectra of the metabolite eluting at 10.0 min indicate that the molecular weight of the metabolite is 585 Da and that loss of 176 Da occurs upon collision-induced dissociation. These data suggest that this metabolite (**H**) is a glucuronide of the hydroxylated and *O*-dealkylated metabolite (**G**, retention time = 13.0 min). Definitive structural assignment of the metabolite eluting at 32.8 min (tentatively assigned as metabolite **I**) was not possible. The mass spectrum indicates a molecular weight of 463 Da and signals at *m/z* 175, 191, 203 and 215 in the product ion spectrum imply that 30 Da has been added to the phenyl-piperidine moiety of the molecule (see below for more details).

The radiochromatogram obtained from rat plasma following oral or intravenous dosing with ^{14}C-labeled **A** diluted with unlabeled compound **A** differs significantly from that obtained with hepatocytes (compare Figs. 5 and 6). The signal corresponding with the parent compound was relatively weak and the major component in plasma was a new metabolite (**J**) eluting later (at 35.9 min) than all metabolites observed in rat hepatocytes. The mass spectrum of metabolite **J** contains signals at *m/z* 447, 465, 482 and 487, which correspond to $[\mathbf{J} + H - H_2O]^+$, $[\mathbf{J} + H]^+$, $[\mathbf{J} + NH_4]^+$ and $[\mathbf{J} + Na]^+$ ions. Thus, the molecular weight of metabolite **J** is 464 Da. Note that compounds with a low-proton affinity (due to the absence of a basic nitrogen atom) frequently give rise to $[M + NH_4]^+$ and $[M + Na]^+$ signals, which are more abundant than the $[M + H]^+$ signal. These data were obtained with an acidic mobile phase containing 0.1% trifluoroacetic acid. Without trifluoroacetic

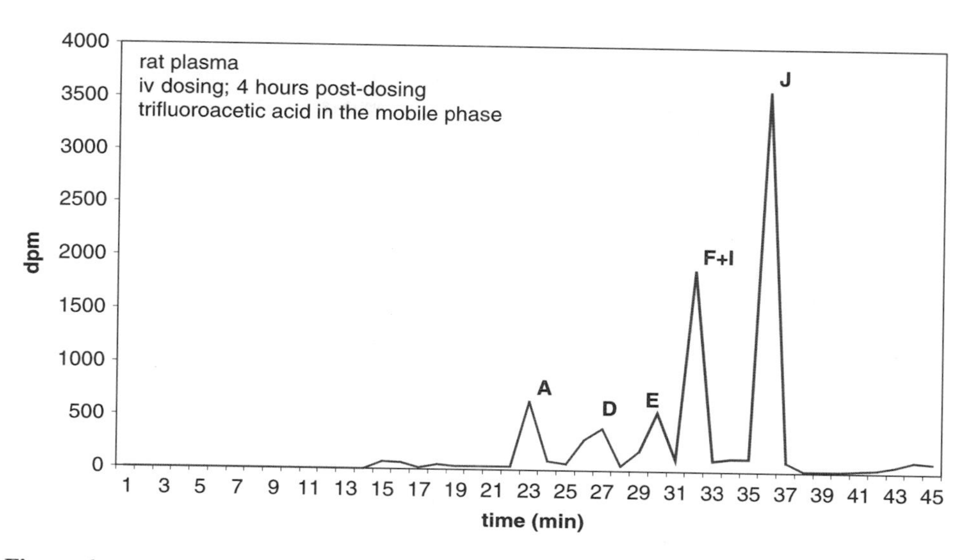

Figure 6.
HPLC-radiochromatogram of ^{14}C-labeled compound **A** and its metabolites present in rat plasma following intravenous administration at 10 mg/kg. The liquid chromatography mobile phase contained trifluoroacetic acid

acid in the mobile phase, the parent compound, **A**, eluted slightly later (at 29.6 min) and metabolite **J** eluted much earlier (at 23.1 min). The major change in the retention time of metabolite **J** suggests that this metabolite contains a carboxylic acid moiety. At low pH, the latter metabolite will be neutral, which results in more interaction with the C18 stationary phase and a longer retention time. Without trifluoroacetic acid in the mobile phase, it was also possible to obtain mass spectrometric data in the negative ion mode; an intense signal at *m*/*z* 463 was obtained for metabolite **J**, which confirms the molecular weight derived from the positive ion mode data.

The positive ion mode product ion spectrum of metabolite **J** contains signals at *m*/*z* 175, 191, 203 and 215, which were also observed in the product ion spectrum of the parent compound. Thus, the metabolic biotransformation must have occurred at the phenyl piperidine moiety of the molecule. The most intense signal, *m*/*z* 105, corresponds to $[C_6H_5{-}C{=}O]^+$ ions, which suggests oxidative ring opening of the piperidine ring. The negative ion product spectrum provides little additional information; the spectrum was dominated by a signal at *m*/*z* 85, which corresponds with $[F_3C{-}O]^-$ anions. Combination of these data with knowledge of biotransformations feasible for piperidine-containing compounds led to the keto acid structure presented in Fig. 7. This was confirmed by the ^1H NMR spectrum of the isolated metabolite **J**. The signals due to the aliphatic protons at the 2 and 6 positions of the piperidine ring in the parent compound disappeared and the splitting patterns of the protons (due to proton–proton interaction) at the 5 position of the piperidine ring and the ortho positions on the phenyl ring changed significantly. The latter aromatic protons were also displaced downfield, which suggests the presence of an adjacent carbonyl moiety and, consequently, opening of the piperidine ring. Thus, the MS and NMR data combined suggest that metabolite **J** has the keto acid structure. The oxidative deamination mechanism presented in Fig. 8 is proposed for formation of metabolite **J** and a similar mechanism has been presented for the metabolism of *N*-deacetyl ketoconazole by Rodriguez *et al.* (1999). Many of the metabolic intermediates presented in Figure 8, such as the secondary hydroxylamine (**C**, minor component) and the nitrones (**D** and **E**), were observed in rat plasma after dosing with compound **A**. A metabolite with a molecular weight of 463 Da (**I**; retention time = 32.8 min) was observed in rat plasma as well as in rat hepatocyte incubations (see above). The product ion spectrum contains signals at *m/z* 175, 191, 203 and 215, which implies that the trifluoromethoxy phenyl moiety is intact and that 30 Da has been added to the phenyl piperidine moiety. The most intense signal in the product ion spectrum is

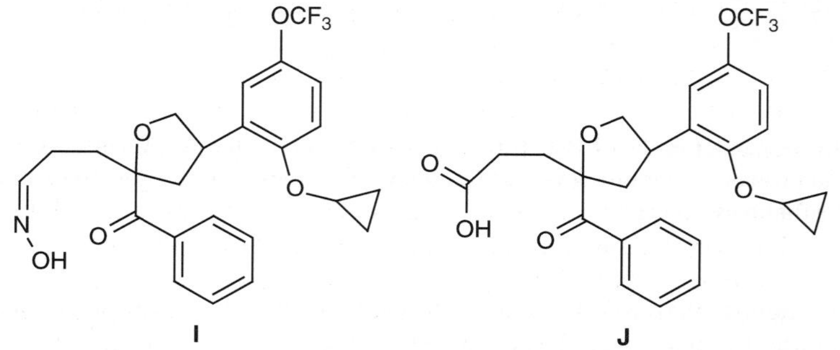

Figure 7.
The structures of metabolites **I** and **J**

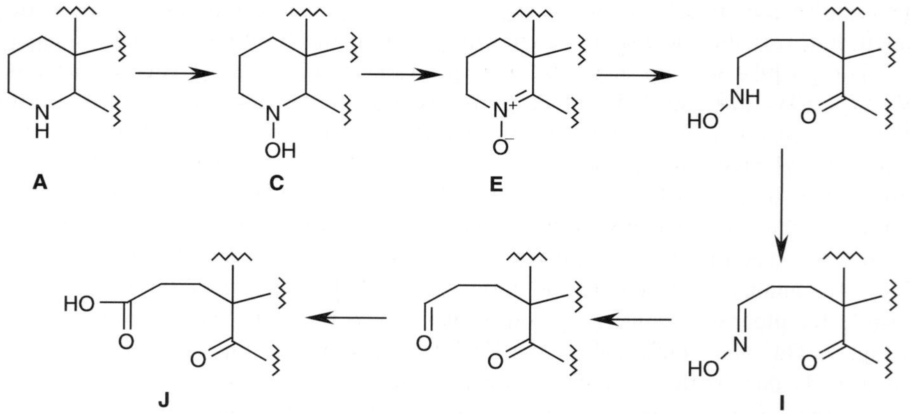

Figure 8.
The proposed mechanism for oxidative deamination of compound **A**

at *m/z* 105, which was also observed for metabolite **J**. Thus, metabolite **I** could be the oxime presented in Fig. 7, which is the analog of the oxime intermediate proposed for the metabolism of *N*-deacetyl ketoconazole (Rodriguez *et al.*, 1999).

Thus, mass spectrometric data were employed to elucidate the metabolism of [3R,5R,6S]-3-(2-cyclopropyloxy-5-trifluoromethoxyphenyl)-6-phenyl-1-oxa-7-azaspiro [4.5]decane and significantly different metabolic profiles were observed for rat hepatocytes and rat plasma with the major circulating metabolite, **J**, being generated by oxidative deamination of the piperidine ring. However, it is not clear whether the high-plasma levels of metabolite **J** are due to rapid formation of metabolite **J** or slow elimination of metabolite **J** from systemic circulation.

6.5.2. Metabolism of a Potent PPARγ Agonist

Diabetes mellitus is a major threat to human health and the number of diabetics continues to increase (Zimmet *et al.*, 2001). The most common form of diabetes is type 2 diabetes, which is characterized by insulin resistance and/or abnormal insulin secretion. Hyperglycemia may lead to nephropathy, neuropathy, retinopathy and atherosclerosis. Control of blood glucose levels can be achieved with oral hypoglycemic agents, such as sulphonylureas, metformin, peroxisome proliferator-activated receptor-γ (PPARγ) agonists, α-glucosidase inhibitors and insulin (Moller, 2001). Upon ligand binding, the nuclear hormone receptor PPARs regulate specific gene expression by binding to peroxisome proliferator responsive elements after heterodimerization with another nuclear receptor, retinoid X receptor (Berger and Moller, 2002). The nuclear hormone receptor PPARγ governs expression of genes involved in the regulation of glucose and lipid metabolism. The ultimate outcome of administration of a PPARγ agonist is an increase in the sensitivity of certain tissues toward insulin, which enhances glucose metabolism and inhibits hepatic gluconeogenesis. Troglitazone, pioglitazone and rosiglitazone, all thiazolidinediones (TZDs), are PPARγ agonists and have shown clinical efficacy, but troglitazone was withdrawn because of idiosyncratic, but severe, hepatotoxicity. As an alternative to TZDs, oxazolidinediones (OZDs) have been identified as PPARγ agonists by Desai *et al.* (2003). Here, the in vitro metabolism of a new PPARγ agonist, compound **K**, is described.

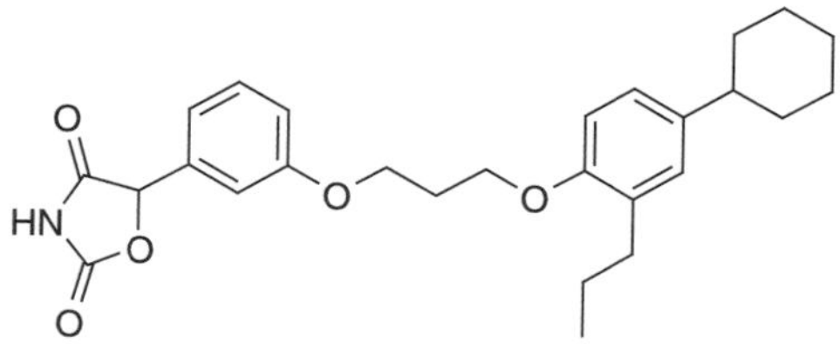

K, 451 Da

Compound **K** ionizes better in the negative than the positive ion mode due to the acidic nature of the OZD moiety. This facilitated metabolite identification because signals due to endogenous material are usually less abundant in the negative ion mode. The product ion spectrum of the [**K** – H]$^-$ ions at *m/z* 450 was dominated by an intense signal at *m/z* 42, which can be ascribed to [N≡C–O]$^-$ from the OZD ring; the abundance of all other fragment ions (*m/z* 191, 203 and 217) was less than 5% of the intensity of the *m/z* 42 signal. Thus, negative ion mode data provided little structural information.

In the positive ion mode, the [**K** + H]$^+$ signal at *m/z* 452 was very small, but the [**K** + NH_4]$^+$ signal at *m/z* 469 was more abundant. Fragmentation of the [**K** + NH_4]$^+$ ions gave rise to several structure characteristic signals, including *m/z* 234, 190, 177 and 83, and the nature of these ions is described in Fig. 9. The intensity of *m/z* 83 ([C_6H_{11}]$^+$), albeit small, defined the cyclohexane ring.

Thus, metabolites of compound **K** were identified in a three-step process:

1. Identification of metabolites via LC–MS in the negative ion mode.
2. Confirmation of metabolites via LC–MS/MS in the negative ion mode using *m/z* 42 precursor ion detection.
3. Characterization of metabolites via LC–MS/MS in the positive ion mode using [**K** + NH_4]$^+$ and/or [**K** + H]$^+$ ions as precursor.

Incubation of compound **K** with rat, dog, monkey and human liver microsomes resulted in a large extent of metabolism with the major pathways being associated with oxidation of the cyclohexyl ring. (Other, mostly minor, metabolic processes include oxidation of the propyl side chain and opening of the OZD ring.) Figure 10 shows the extracted ion chromatograms from rat, dog, monkey and human liver microsomal samples for the *m/z* 466 [**K** + O – H]$^-$ ions. Generally, microsomes from all four species form the same hydroxylated metabolite, but there are striking quantitative differences. In the positive ion mode, the corresponding *m/z* 468 ([**K** + O + H]$^+$) signals were small, but signals at *m/z* 485 ([**K** + O + NH_4]$^+$) and 450 ([**K** + O– H_2O + H]$^+$) were present. The abundance of the *m/z* 450 signal indicates that the protonated hydroxy metabolites are relatively unstable and loss of H_2O prevails, which suggests that hydroxylation of aliphatic carbon atoms, but not aromatic

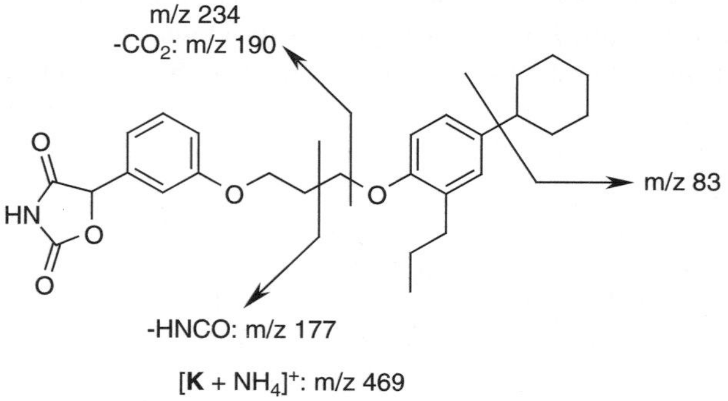

Figure 9.
Tentative structural assignments for the most informative fragmentation pathways observed for [**K** + NH_4]$^+$ upon collision-induced dissociation

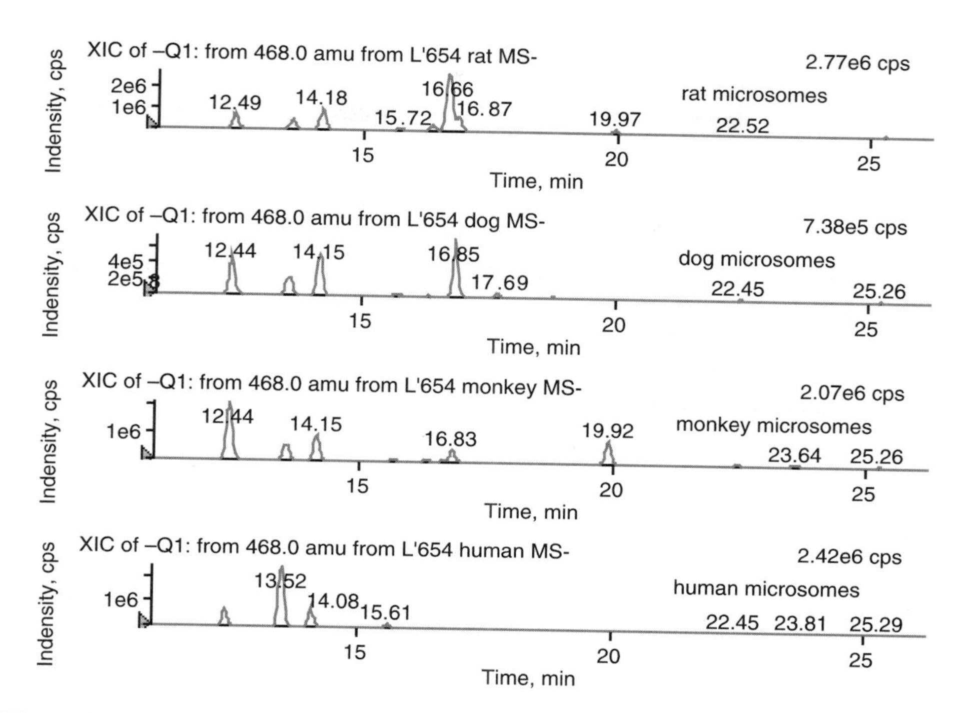

Figure 10.
Extracted ion chromatograms for the *m/z* 466 [**K** + O − H]$^-$ ions from compound **K** incubated with rat, dog, monkey and human liver microsomes

carbon atoms, has taken place. The positive ion mode product ion spectra of the [**K** + O+ NH_4]$^+$ ions from the four hydroxylated metabolites eluting at 12.4, 13.5, 14.2 and 16.8 min had characteristic signals at *m/z* 234, 190, 177 and 81. The signal at *m/z* 81 corresponds with [C_6H_{11} + O − H_2O]$^+$, which is indicative of hydroxylation of the cyclohexane ring in multiple positions. However, LC–MS/MS does not provide regiochemical information regarding hydroxylation of the cyclohexane ring. Relatively, small quantities of several [**K** + 14 Da] metabolites were detected as well and all product ion spectra had a signal at *m/z* 97. The signal at *m/z* 97 corresponds with [C_6H_9O]$^+$, which is indicative of ketone formation in multiple positions on the cyclohexane ring.

Compound **K** was also incubated with recombinant cytochrome P450 2C8, 2C19 and 3A4 enzymes. The turnover was greatest with 2C8 followed by 2C9 and the turnover with 3A4 was small. The respective extracted ion chromatograms for the *m/z* 466 [**K** + O −H]$^-$ ions are presented in Fig. 11 and, again, the profiles, in particular the regions between 10 and 20 min, are quantitatively quite distinct. The hydroxylated metabolites generated by CYP2C8 and CYP2C19 feature prominently in the extracted ion chromatogram generated for human liver microsomes (Figs. 10 and 11). The product ion spectra obtained in the positive ion mode indicate that the major hydroxylated metabolites involve hydroxylation of the cyclohexane ring.

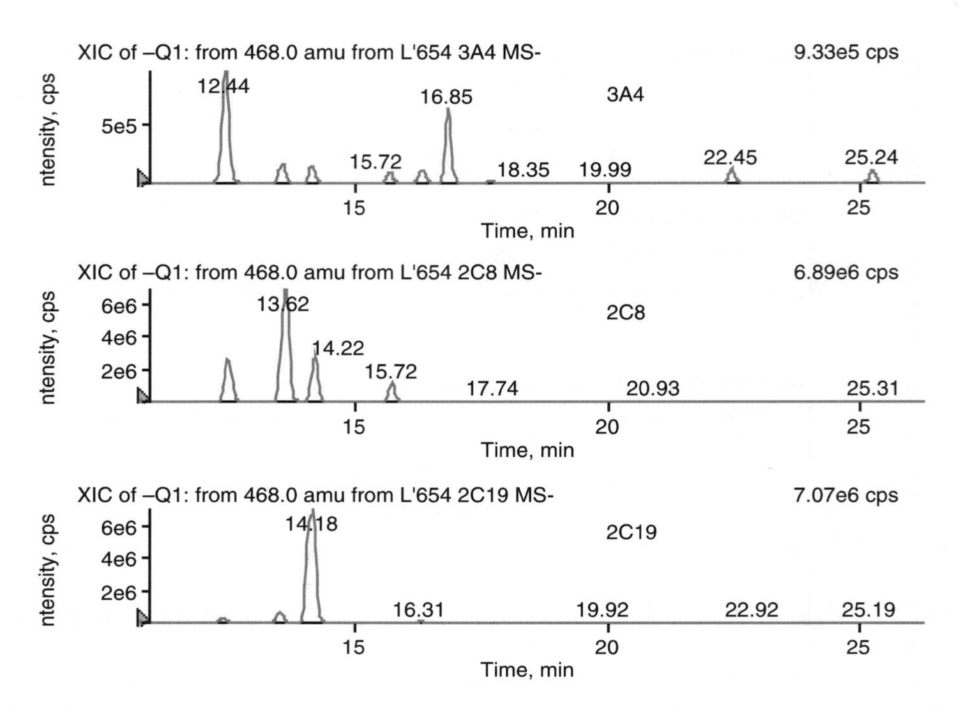

Figure 11.
Extracted ion chromatograms for the *m/z* 466 $[\mathbf{K}+\mathrm{O}-\mathrm{H}]^-$ ions from compound **K** incubated with recombinant cytochrome P450 3A4, 2C8 and 2C19 enzymes

To obtain more information about the regiochemistry and stereochemistry of the hydroxylated metabolites, compound **K** was incubated in a bioreactor (Rushmore *et al.*, 2000) with the two cytochrome P450 enzymes which had the most turnover, i.e. CYP2C8 and CYP2C19, followed by isolation of the most abundant metabolites and analysis by ^{1}H NMR. Two chromatographic fractions each were isolated for CYP2C8 (13.6 and 14.2 min) and CYP2C19 (13.6 and 14.2 min). Chemical shifts and coupling constants in the ^{1}H NMR spectrum indicated that CYP2C8 forms two equatorial hydroxy metabolites and CYP2C19 forms two axial hydroxy metabolites (Fig. 12).

These data indicate that CYP enzymes from different species display subtle quantitative differences in the regioselectivity for hydroxylation of the cyclohexane ring. In addition, the human CYP enzymes 2C8 and 2C19 possess stereoselectivity for hydroxylation of the cyclohexane ring.

6.5.3. Metabolism of an Antipsychotic Drug Ziprasidone

Schizophrenia is a complex disorder characterized by thought disturbances, auditory hallucinations and inappropriate effects (Howard and Seeger, 1993). While the classical antipsychotic drugs of the phenothiazine and butyrophenone classes are

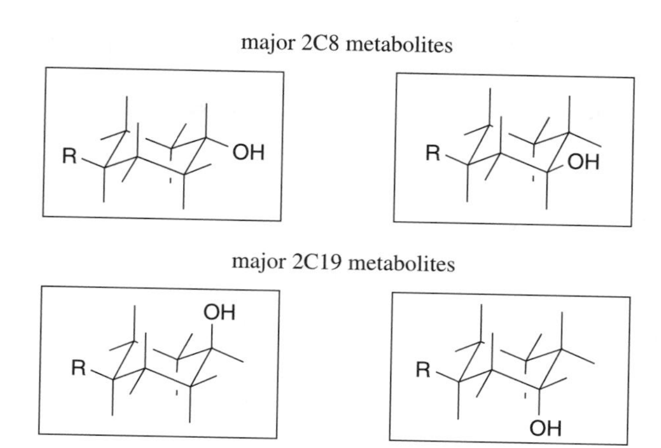

Figure 12.
The structures of major hydroxylated metabolites of compound **K** in recombinant CYP2C8 and CYP2C19

effective in the treatment of schizophrenia, their usage is commonly associated with extrapyrimidal side effects (Tarsey, 1983). In addition, these drugs are not effective in all patients or against both positive and negative symptoms of schizophrenia (Ortiz and Gershon, 1986). Laboratory and clinical findings have suggested that antagonism of serotonin 5-HT_{2C} receptors in the brain limits the undesirable motor side effects associated with dopamine receptor blockade and improves efficacy against the negative symptoms of schizophrenia (Meltzer, 1995). Ziprasidone, a substituted benzisothiazolyl-piperazine analog, was recently approved for the treatment of schizophrenia (Fig. 13). It exhibits potent and highly selective antagonistic activity on the dopamine D_2 and serotonin 5-HT_{2A} receptors. It also has high affinity for the 5-HT_{1A}, 5-HT_{1D} and 5-HT_{2C} receptor subtypes that could contribute to the overall therapeutic effect (Howard *et al.*, 1994). Preclinical ADME studies were needed to support the safety assessment package for registration of this new drug. Here, the strategies for the identification of metabolites of ziprasidone in rats are described.

6.5.3.1. Use of Two Different Radiolabels

The metabolism of tiospirone, a structurally related analog of ziprasidone (Fig. 13), has been described earlier (Mayol *et al.*, 1991). One of its main metabolic pathways, *N*-dealkylation of the butyl group attached to the piperazinyl nitrogen, resulted in the cleavage of the molecule into two major portions. However, these studies were conducted using tiospirone labeled with ^{14}C at the piperazine ring and hence only the metabolites containing the benzisothiazole piperazine (BITP) moiety were monitored. It is desirable to trace both fragments by providing each with a different label. Thus, we studied the metabolism of ziprasidone in preclinical species and humans after administration of a mixture of ziprasidone labeled with ^{14}C at the C-2 of the ethyl group attached to the piperazinyl nitrogen and ziprasidone labeled with ^{3}H at the C-7 position of the benzisothiazole (Prakash *et al.*, 1997). The use of two labels not only facilitated the tracing of metabolites formed through *N*-dealkylative cleavage of ziprasidone, but also aided their identification. A total of 11 peaks were detected in the radiochromatograms in the rat urine (Fig. 14). Metabolites **M1**, **M2** and **M3** were detected only in the ^{3}H radiochromatogram indicating that they were cleavage products associated with the benzisothiazole moiety. On the other hand, metabolite **M4** was detected mainly in the ^{14}C radiochromatogram with a smaller peak in ^{3}H radiochromatogram representing spillover, indicating that it was also a cleaved metabolite but containing the oxindole moiety. The remaining metabolites, **M6**, **M7**, **M8**, **M9** and **M10**, were detected by peaks of similar

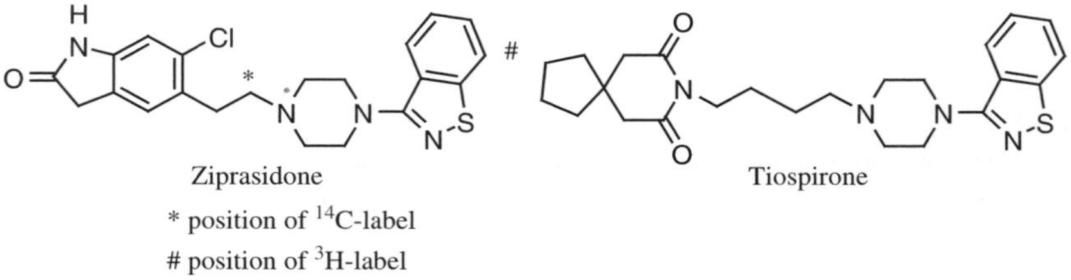

Ziprasidone   Tiospirone

* position of ^{14}C-label

position of ^{3}H-label

Figure 13.
Structures of ziprasidone and a structurally related analog tiospirone

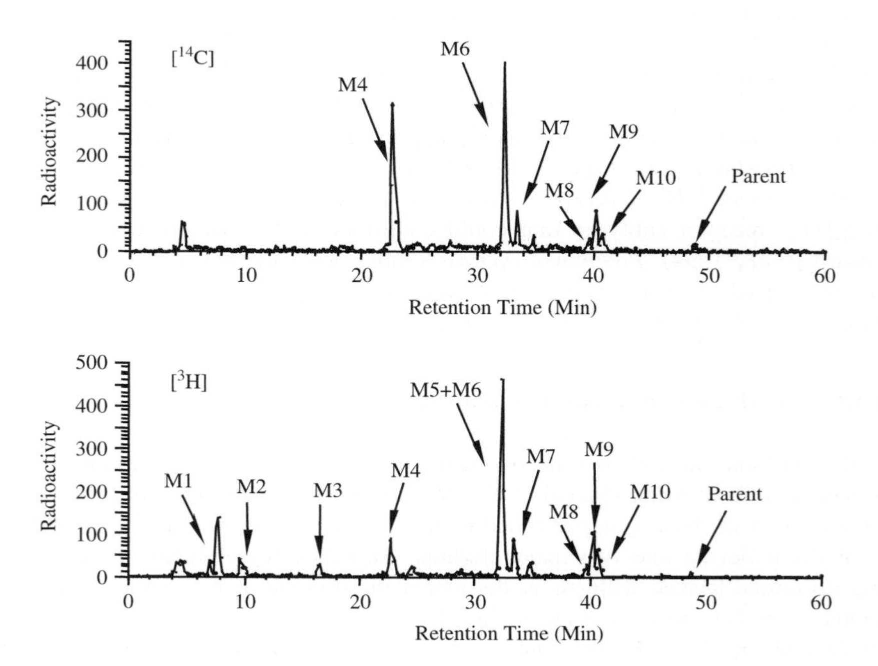

Figure 14.
HPLC-radiochromatograms of ziprasidone metabolites in rat urine

heights in both radiochromatograms, suggesting that they did not undergo cleavage at the piperazinyl nitrogen and, therefore, are expected to be dual labeled.

6.5.3.2. Precursor Ion Spectrum

Ziprasidone generated a strong pseudomolecular ion $[M+H]^+$ at *m/z* 413. The product ion mass spectrum of *m/z* 413 showed signals at *m/z* 220, 194, 177, 166 and 159. The two diagnostic fragment ions at *m/z* 220 and 194 resulted from the cleavage of the C–N (piperazine) bond and corresponded to $[BITP+H]^+$ and $[oxindole\text{-}CH_2CH_2]^+$, respectively. The other fragment ions at *m/z* 177 and 159 were due to losses of 43 ($CH_3N{=}CH_2$) and 35 Da (Cl) from the ions at *m/z* 220 and 194, respectively (Fig. 15). The assignment of these ions was confirmed by the parallel product ion spectrum of *m/z* 415 ($[M + H]^+$, ^{37}Cl), which gave fragment ions at *m/z* 220, 196, 177, 166 and 159. Precursor ion scanning (see Fig. 2) of *m/z* 194 provided molecular ion information for seven metabolites from rat urine (Fig. 16). Product ion spectra of the protonated molecules provided structurally significant fragment ions. Metabolites **M1**, **M2**, **M3**, **M4** and **M5** were identified as BITP–sulfone, BITP–sulfoxide, BITP–sulfone–lactam, chlorooxindole acetic acid and BITP, respectively. Metabolites **M6**, **M8**, **M9** and **M10** were characterized as *S*-methyl-dihydro-ziprasidone-sulfoxide, ziprasidone sulfone, *S*-methyl-dihydro-ziprasidone and ziprasidone sulfoxide, respectively (Fig. 17).

6.5.3.3. Differentiation of Isobaric Metabolites

Metabolite **M9** showed a protonated molecule at *m/z* 429, the same as of **M10**, suggesting that **M9** and **M10** were isobars. The product ion spectrum of **M9** (*m/z* 429) showed fragment ions at *m/z* 280, 263, 219, 194, 150 and 123. The ion at *m/z* 280 corresponds to a charge-initiated fragmentation of the piperazinyl nitrogen benzisothiazole carbon bond with the expulsion of the benzisothiazole +16 Da moiety as a neutral molecule. The ion at *m/z* 150 resulted from the cleavage of the same nitrogen–carbon bond with charge retention on the benzisothiazole +16 Da moiety and suggested modification of the benzisothiazole ring. The assignment of these ions was confirmed by a parallel product ion spectrum of *m/z* 431 ($[M + H]^+$, ^{37}Cl), which gave fragment ions at *m/z* 282, 265, 196, 150 and 123. These results strongly suggest that the oxidation had occurred at the benzisothiazole moiety. Based on addition of 16 Da to the benzisothiazole moiety, three structures were originally considered for **M9**: oxidation at the nitrogen of the benzisothiazole ring to form an *N*-oxide, **1**; aromatic hydroxylation of the benzisothiazole moiety, **2** and reductive cleavage of the benzisothiazole followed by methylation of the resulting thiophenol to form **3** (Fig. 18).

Treatment of **M9** with aqueous $TiCl_3$, *tert*-butyldimethylsilyl-*N*-trifluoroacetamide or diazomethane did not change the HPLC retention time or the molecular ion of metabolite **M9**. These results indicate that M9 was neither an *N*-oxide (**1**) nor

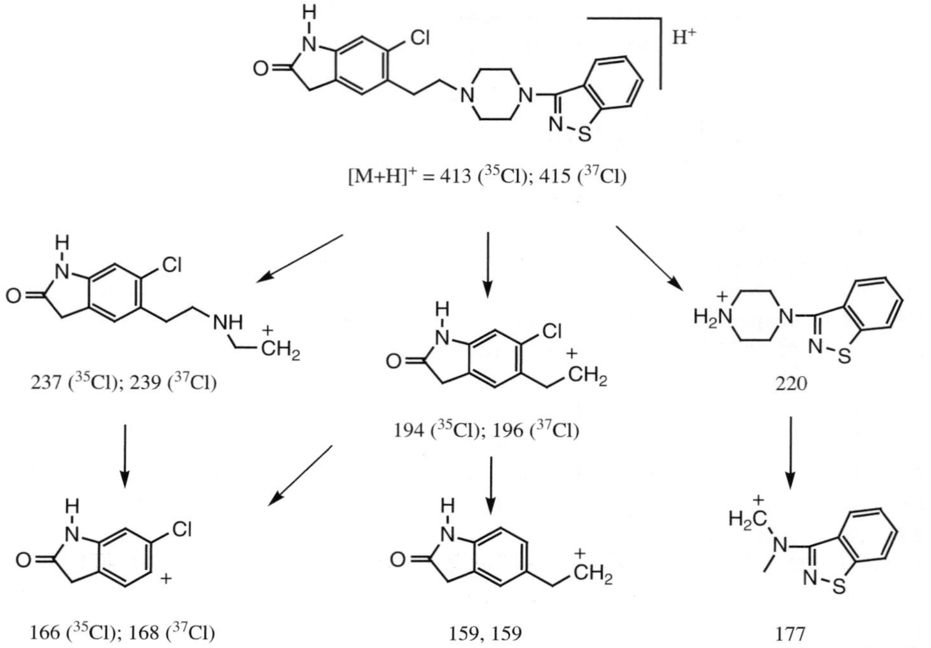

$[M+H]^+ = 413\ (^{35}Cl);\ 415\ (^{37}Cl)$

$237\ (^{35}Cl);\ 239\ (^{37}Cl)$

$194\ (^{35}Cl);\ 196\ (^{37}Cl)$

220

$166\ (^{35}Cl);\ 168\ (^{37}Cl)$

159, 159

177

Figure 15.
Tentative structural assignments for the most informative fragment ions of protonated ziprasidone

a hydroxy metabolite (**2**). Based on these data, **M9** was identified as *S*-methyl-dihydro-ziprasidone, **3**, formed by the reductive cleavage of the benzisothiazole moiety. Accurate mass measurements obtained with a Q-TOF mass spectrometer provided additional information. Table 5 summarizes the molecular formula assignments for the protonated molecule of ziprasidone and its two isobaric metabolite **M9** and **M10**. The significant difference between the observed *m/z* value for protonated **M9** and **M10** indicates that **M9** and **M10** have different molecular formulas, which is in keeping with their structural assignments (see Fig. 17).

6.5.4. Biotransformation of an Anxiolytic Drug Candidate, CP-93,393

CP-93,393, a pyrimidinylpiperazine analog (Fig. 19), exhibits highly selective 5-hydroxytryptamine (5-HT) serotonin autoreceptor agonist activity (Schmidt *et al.*, 1995). It differs in mechanism from the benzodiazepines, which are generally believed to act by potentiating the neural inhibition mediated by GABA, but bears a mechanistic relationship to a new class of nonbenzodiazepins anxiolytics, azapirones (Rollema *et al.*, 1996). But unlike these agents, CP-93,393 itself exhibits

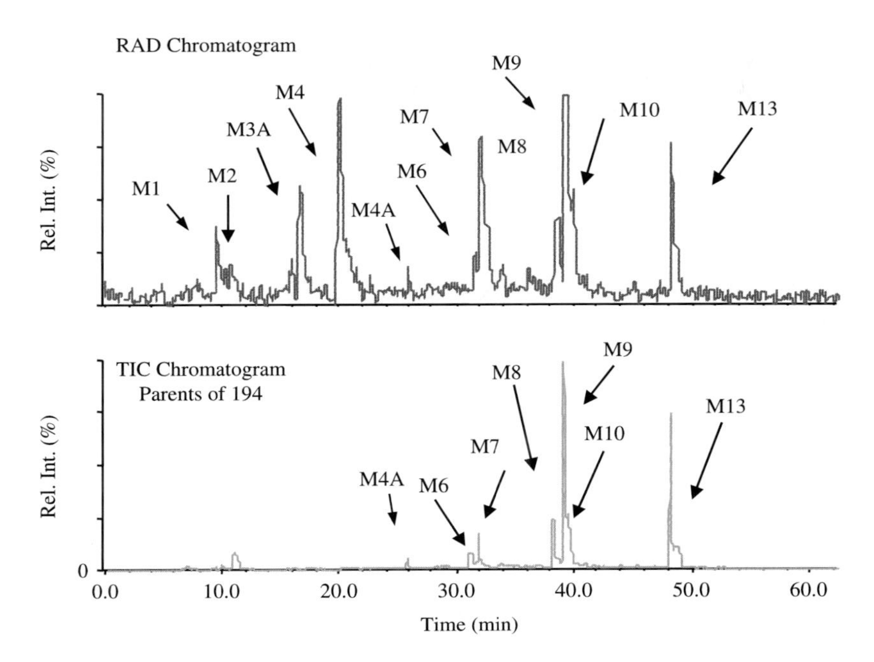

Figure 16.
HPLC-radiochromatogram and TIC chromatogram (precursor ion spectrum of *m/z* 194) of ziprasidone metabolites in rat urine

α2-adrenoceptor antagonist activity that may have therapeutic relevance for antidepressant activity. It is being evaluated for its effectiveness in the treatment of generalized anxiety disorder and major depressive disorder. Here, we describe the metabolism of CP-93,393, {1-(2-pyrimidin-2-yl)-octahydro-pyrido[1,2-a]pyrazin-7-ylmethyl)-pyrrolidine-2,5-dione, in rat urine as well as in monkey liver microsomes.

CP-93,393 and 15 metabolites were detected in the radiochromatogram from a urine sample of rats following oral administration of ^{14}C-labeled CP-93,393 (Fig. 20). The identification of metabolites was achieved by chemical derivatization, β-glucuronidase/sulfatase treatment and LC–MS/MS (Prakash and Soliman, 1997). Based on the structures of hydroxylated metabolites, four primary metabolic pathways of CP-93,393 were identified: hydroxylation at the pyrimidine ring **(M15)**, hydroxylation at the succinimide ring **(M16)**, hydroxylation α to the nitrogen of the piperazine ring (**M10**) and hydrolysis of imide bond of the succinimide ring (**M9**). The major oxidative metabolites were excreted as sulfate and/or glucuronide conjugates (**M7**, **M12** and **M13**) (Fig. 21).

CP-93,393 and metabolites displayed very intense protonated molecules, $[M + H]^+$. The product ion mass spectra of the $[M + H]^+$ provided very characteristic fragment ions; at least six structurally informative fragment ions were observed for CP-93,393 and its hydroxylated analogs (fragments a–f, Fig. 22A). For CP-93,393,

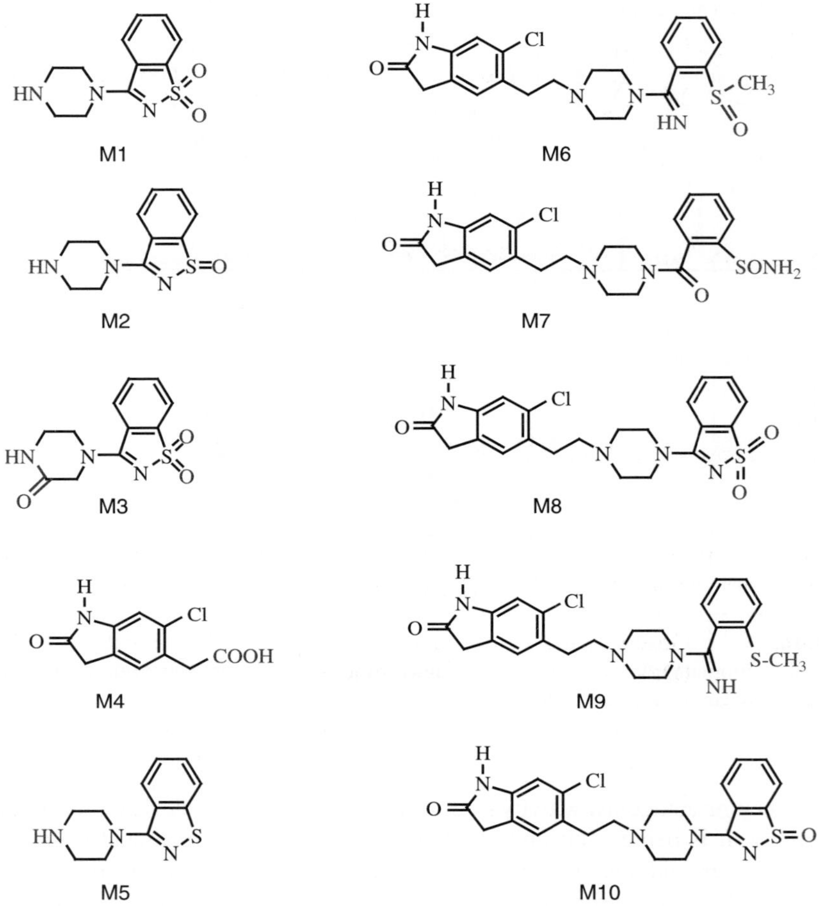

M1 M6 M2 M7 M3 M8 M4 M9 M5 M10

Figure 17.
Proposed structures of metabolites of ziprasidone in rats

fragmentation occurred across the piperazine ring to give ions at *m/z* 235 (fragment a), 209 (fragment b) and 195 (fragment c) with charge retention at the succinimidyl part of the molecule. Hydroxylation at the succinimide ring, such as **M16**, resulted in these ions appearing 16 Da higher, at *m/z* 251 (fragment a), 225 (fragment b) and 211 (fragment c). Absence of substitution at the succinimide moiety, such as **M10**, resulted in these ions being observed at the same *m/z* values as CP-93,393: *m/z* 235, 209 and 195 (fragments a, b and c, respectively). Two fragmentation reactions across the piperazine ring with charge retention at the pyrimidine portion of the CP-93,393 gave ions at *m/z* 136 (fragment d) and 122 (fragment e). Therefore, hydroxylation at the pyrimidine ring, such as **M15**, resulted in ions at *m/z* 152 and 138, 16 Da higher than CP-93,393. In addition, a very characteristic fragment ion due

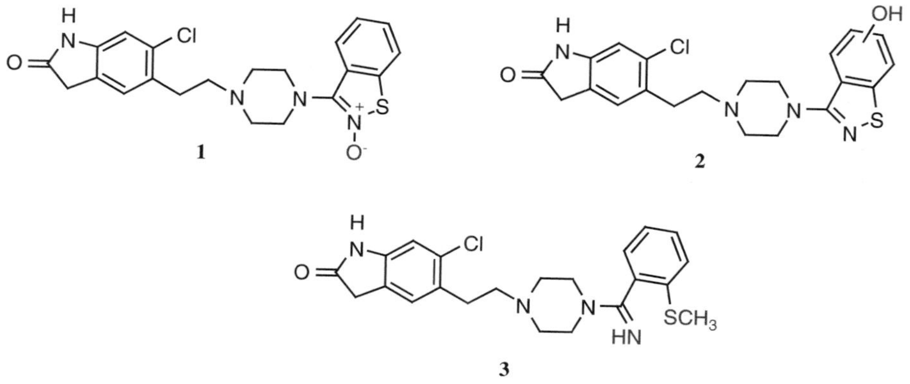

Figure 18
Possible structures of metabolite **M9**

Table 5.
Accurate mass measurements and molecular formula assignments of ziprasidone and its two isobaric metabolites **M9** and **M10** obtained with a micromass Q-TOF II

Compound	Observed mass (Da)	Formula assignment	Theoretical mass (Da)	Error (ppm)
Ziprasidone	413.1205	$C_{21}H_{22}N_4OSCl$	413.1203	0.4
M9	429.1520	$C_{22}H_{26}N_4OSCl$	429.1516	0.9
M10	429.1151	$C_{21}H_{22}N_4O_2SCl$	429.1152	−0.3

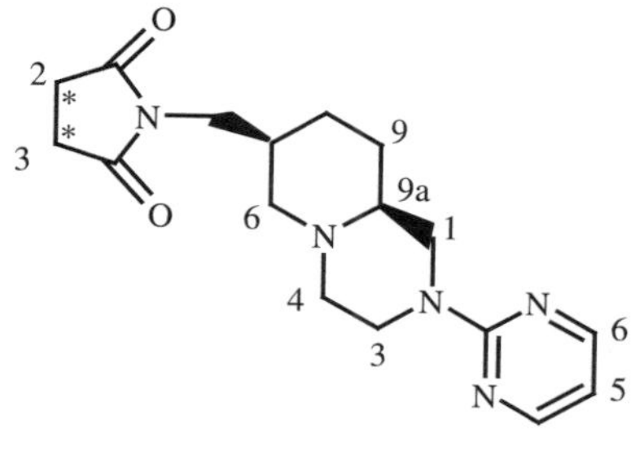

Figure 19.
Structure of CP-93,393 (* indicates the positions of the ^{14}C-labels)

to the loss of succinimide ring was observed for most of the compounds (fragment f, Fig. 22A).

For hydrolysis products, such as 1-(2-pyrimidin-2-yl-octahydro-pyrido[1,2-a]pyrazin-7-ylmethyl)-succinamic acid (NPMSA), fragmentation also occurred across

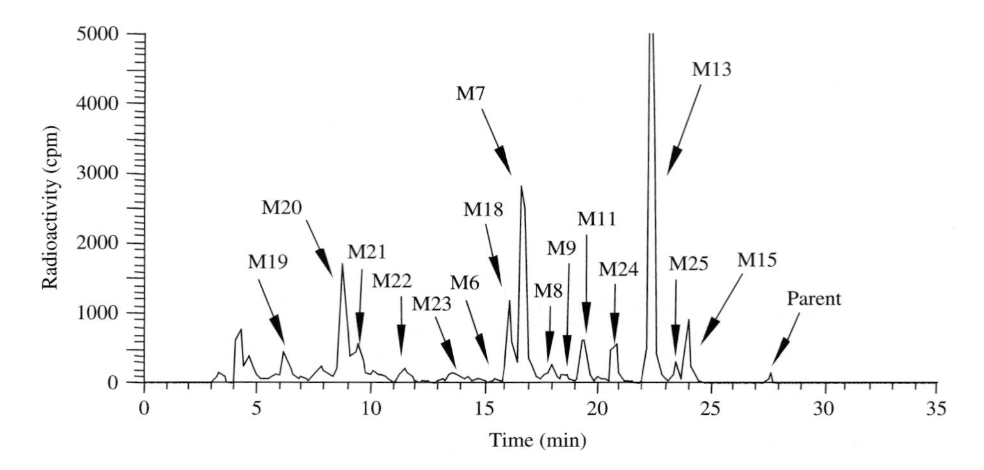

Figure 20.
HPLC-radiochromatogram of urinary metabolites of ^{14}C-labeled CP-93,393 following oral administration to rats

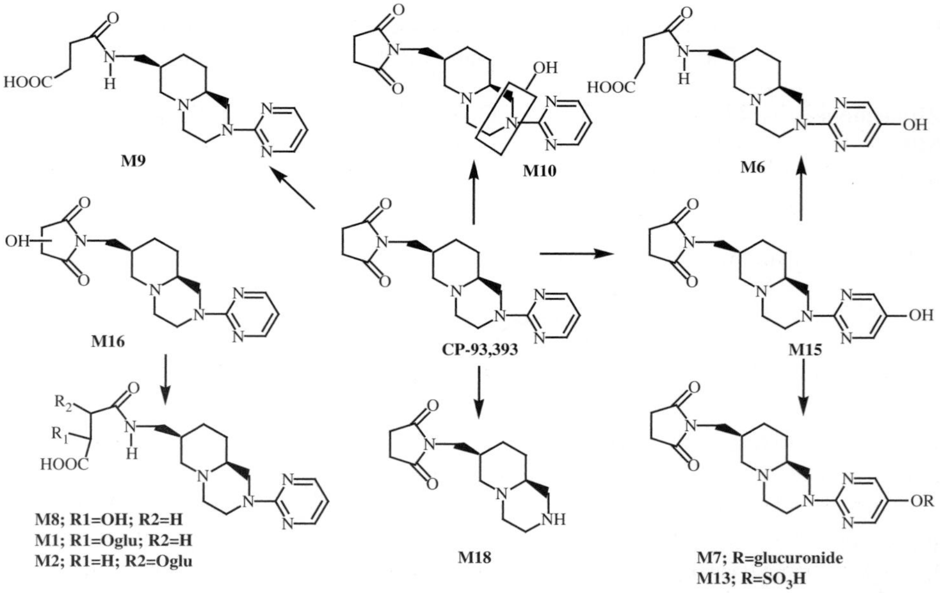

M9

M10

M6

M16

CP-93,393

M15

M8; R1=OH; R2=H
M1; R1=Oglu; R2=H
M2; R1=H; R2=Oglu

M18

M7; R=glucuronide
M13; R=SO_3H

Figure 21.
Proposed biotransformation pathways of CP-93,393 in rats

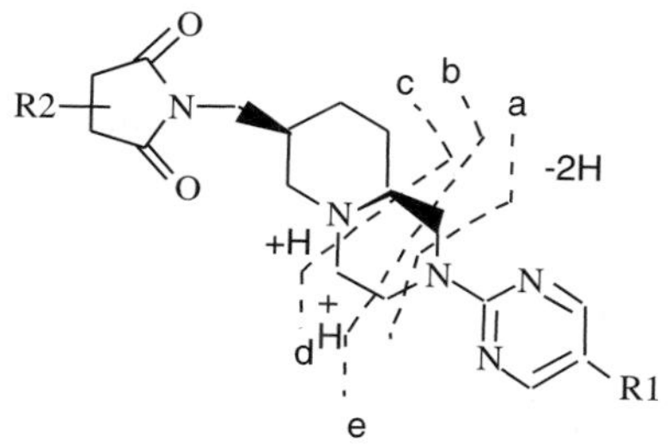

Compound	R1	R2	MH+	a	b	c	d	e	f
CP-93,393	H	H	330	235	209	195	136	122	231
5-OH-CP-93,393	OH	H	346	235	209	195	152	138	247
2 or 3-OH-CP-93,393	H	OH	346	251	225	211	136	122	231

(A)

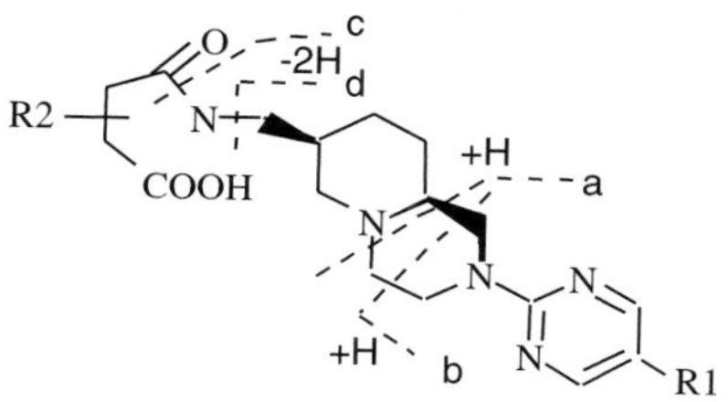

Compound	R1	R2	MH+	a	b	c	d
NPMSA	H	H	348	136	122	248	231
NHPMSA	OH	H	364	152	136	264	247
2 or 3-NPMHSA	H	OH	364	136	122	248	231

(B)

Figure 22.
Fragmentation pattern of protonated CP-93,393 and its putative metabolites

the piperazine ring with charge retention only at the pyrimidine portion of the molecule, generating characteristic ions at *m/z* 136 (fragment a) and at *m/z* 122 (fragment b). An additional fragment ion at *m/z* 248 resulted from the cleavage of the amide bond (fragment c). Hydroxylation at the pyrimidine ring, such as 1-[2-(5-hydroxy-pyrimidin-2-yl)-octahydro-pyrido[1,2-a]pyrazin-7-ylmethyl]-succinamic acid (5-NHPMSA), resulted in ions at *m/z* 152 (fragment a), 138 (fragment b) and 264 (fragment c), all 16 Da higher than NPMSA. For 1-(2-pyrimidin-2-yl-octahydro-pyrido[1,2-a]pyrazin-7-ylmethyl)-2-hydroxy-succinamic acid (2-NPMHSA) and-(2-pyrimidin-2-yl-octahydro-pyrido[1,2-a]pyrazin-7-ylmethyl)-3-hydroxy-succinamic acid (3-NPMHSA), these fragment ions were observed at the same *m/z* values as observed for NPMSA (Fig. 22B).

Two metabolites **M1** and **M2** eluting at retention times of ~10.8 and 12.9 min correspond to isomeric glucuronides of 2-NPMHSA and 3-NPMHSA, respectively. The structures of these isomeric glucuronides were elucidated by constant neutral loss scanning and LC–MS/MS/MS (MS^3) techniques. The glucuronide conjugates were dissociated at the orifice and the resulting aglycones were subjected to collision-induced dissociation. The product ion spectrum of **M1** (*m/z* 364; generated at the orifice) gave fragment ions at *m/z* 318, 290, 248, 231, 136 and 122. The characteristic fragment ion at *m/z* 290 (loss of –CH(OH)COOH) indicates the presence of a hydroxy group α to the carboxyl moiety. On the other hand, the MS/MS/MS spectrum of **M2** showed fragment ions at *m/z* 304, 248, 136 and 122. The characteristic fragment ion at *m/z* 304 (loss of CH_3COOH from the ion at *m/z* 364; generated at the orifice) suggests the presence of a hydroxy group β to the carboxyl moiety.

The radiochromatogram obtained from monkey liver microsomes showed the presence of a metabolite, **M18,** which was not identified in rat urine. **M18** displayed a protonated molecule at *m/z* 294, 36 Da lower than that of the parent molecule. The fragment ions at *m/z* 252, 235, 209 and 195 in the product ion spectrum of **M18** indicate the modification of the pyrimidine ring. Furthermore, the fragment ion at *m/z* 252, generated via loss of 42 Da from the molecular ion, suggest the presence of an acetyl or a carboxamidino group (Fig. 23). H/D exchange and derivatization techniques were very useful for differentiating these structures. When H_2O was replaced with D_2O in the mobile phase, the full-scan mass spectrum of **M18** revealed

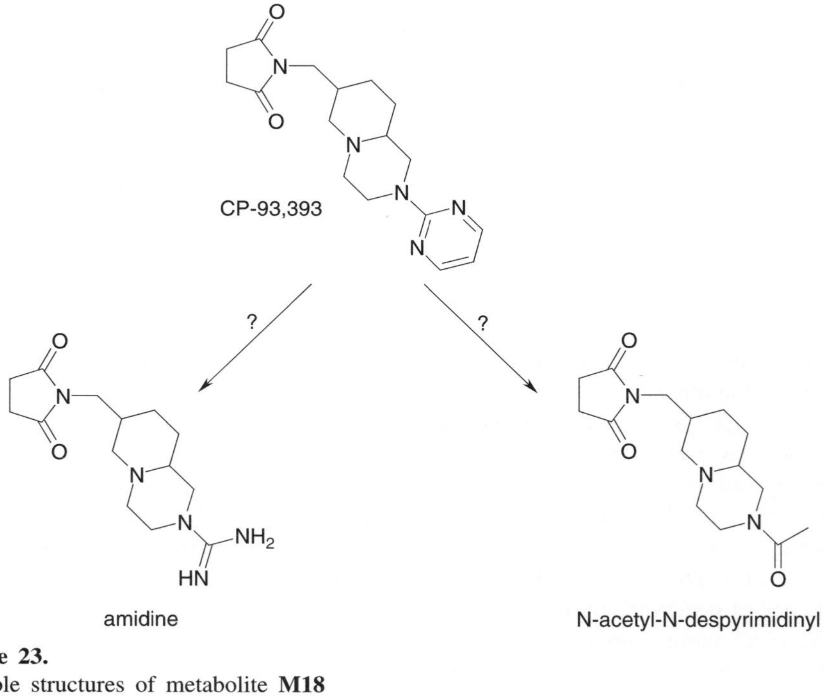

Figure 23.
Possible structures of metabolite **M18**

a molecular ion $[M_D + D]^+$ at m/z 294 (M_D represents the molecular weight following H/D exchange in D_2O), 4 Da higher than $[M_H + H]^+$. The increase of 4 Da is in agreement with the presence of three exchangeable hydrogen atoms, which is consistent with the carboxamidino analog structure for **M18**. The carboxamidine structure of **M18** was further confirmed by treatment with hexafluoroacetylacetone, which resulted in the formation of *bis*-trifluoromethyl-CP-93,393 (Fig. 24), a reaction characteristic for the detection of carboxamidines (Prakash and Cui, 1997). Comparison of the product ion spectrum of the derivatized metabolite with that of **M18** indicates that the product ion was compatible with the assigned structure.

6.6. Future Trends

Metabolite identification has proven to be of great value in drug discovery and deve-lopment and LC–MS is the preferred analytical tool. However, metabolite identification is a labor-intensive activity and depends on the availability of highly skilled scientists. Software has recently become available to automate certain steps in the metabolite identification process (Lim *et al.*, 1999; Yu *et al.*, 1999). For example, after acquiring full scan MS data, the software will interrogate the MS data for anticipated metabolites considering the mass changes associated with the biotransformations listed in Tables 1 and 2. Subsequently, product ion spectra will automatically be acquired for tentatively identified biotransformations. Although this approach can be successful for in vitro samples (liver microsomes, hepatocytes, etc.), it is of limited use for in vivo samples because of the large amount of endogenous material present, which can have the same molecular weight as the anticipated metabolites. Correlating the full scan MS data of the sample of interest with those of an appropriate control sample (t = 0 min, minus NADPH, inactivated enzymes, etc.) will prevent acquisition of product ion spectra of endogenous components with the same molecular weight as anticipated metabolites. However, proper controls may not always be available and, sometimes, the drug affects the abundance of certain endogenous components. Although the software

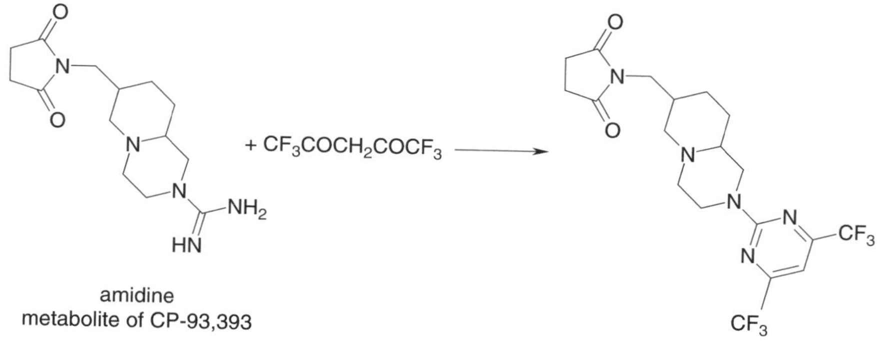

amidine metabolite of CP-93,393 + $CF_3COCH_2COCF_3$ →

Figure 24.
Derivatization of **M18** with hexafluoroacetylacetone

can facilitate the experimental component of the metabolite identification process, ultimately interpretation of the spectra will still be the rate limiting. Thus, tools to automate data interpretation are urgently needed. Reliable in silico tools, which predict metabolite formation, are valuable as well and combining these with the LC–MS data acquisition software should provide a powerful combination.

Acknowledgments

The authors thank numerous former colleagues at Merck Research Laboratories, in particular Drs. Yanfeng Wang, Arun Agrawal and Maria Silva Elipe, and Drs. Amin Kamel, Dan Cui, Wayne Anderson, Keith McCarthy, Kathleen Zandi and Ms. Kim Johnson at Pfizer Global Research & Development for intellectual input and experimental assistance.

References

Baillie, T.A., Pearson, P.G., 2000. The impact of drug metabolism in contemporary drug discovery: new opportunities and challenges for mass spectrometry. In: Burlingame, A.L., Carr, S.A., Baldwin, M.A. (Eds.), Mass Spectrometry in Biology & Medicine Humana Press, Totowa, NJ, pp. 481–496.

Bakhtiar, R., Ramos, L., Tse, F.L.S., 2002. High-throughput mass spectrometric analysis of xenobiotics in biological fluids. J. Liq. Chrom. Rel. Technol. 25, 507–540.

Berger, J., Moller, D.E., 2002. The mechanism of action of PPARs. Annu. Rev. Med. 53, 409–435.

Borchardt, R.T., Freidinger, R.M., Sawyer, T.K., Smith, P.L., 1998. Integration of Pharmaceutical Discovery and Development: Case Histories. Plenum Press, New York, NY.

Brewer, E., Henion, J., 1998. Atmospheric pressure ionization LC/MS/MS techniques for drug disposition studies. J. Pharm. Sci. 87, 395–402.

Burton, K.I., Everett, J.R., Newman, M.J., Pullen, F.S., Richards, D.S., Swanson, A.G., 1997. On-line liquid chromatography coupled with high field NMR and mass spectrometry: a new technique for drug metabolite structure elucidation. J. Pharm. Biomed. Anal. 15, 1903–1912.

Clayton, E., Taylor, S., Wright, B., Wilson, I.D., 1998. The application of high performance liquid chromatography, coupled to nuclear magnetic resonance spectroscopy and mass spectrometry (HPLC–NMR–MS), to the characterisation of ibuprofen metabolites from human urine. Chromatographia 47, 264–270.

Cox, K.A., White, R.E., Korfmacher, W.A., 2002. Rapid determination of pharmacokinetic properties of new chemical entities: in vivo approaches. Comb. Chem. High Throughput Screen 5, 29–37.

Dalvie, D.K., O'Donnell, J.P., 1998. Characterization of polar urinary metabolites by ionspray tandem mass spectrometry following dansylation. Rapid Commun. Mass Spectrom. 12, 419–422.

Dear, G.J., Ayrton, J., Plumb, R., Fraser, I.J., 1999. The rapid identification of drug metabolites using capillary liquid chromatography coupled to an ion trap mass spectrometer. Rapid Commun. Mass Spectrom. 13, 456–463.

Desai, R.C., Gratale, D.F., Han, W., Koyama, H., Metzger, E., Lombardo, V.K., MacNaul, K.L., Doebber, T.W., Berger, J.B., Leung, K., Franklin, R., Moller, D.E., Heck, J.V., Sahoo, S.P., 2003. Aryloxazolidinediones: identification of potent orally active PPAR dual α/γ agonists. Bioorg. Med. Chem. Lett. 13, 3541–3544.

Eddershaw, P.J., Dickins, M., 1999. Advances in in vitro drug metabolism screening. Pharm. Sci. Technol. Today 2(1), 13–19.

Hop, C.E.C.A., Yu, S., Xu, X., Singh, R., Wong, B., 2001. Elucidation of fragmentation mechanisms involving transfer of three hydrogen atoms using a quadrupole time-of-flight mass spectrometer. J. Mass Spectrom. 36, 575–579 .

Hop, C.E.C.A., Wang, Y., Kumar, S., Silva Elipe, M.V., Raab, C.E., Dean, D.C., Poon, G.K., Keohane, C.-A., Strauss, J., Chiu, S.-H.L., Curtis, N., Elliott, J., Gerhard, U., Locker, K., Morrison, D., Mortishire-Smith, R., Thomas, S., Watt, A.P., Evans, D.C., 2002. Identification of metabolites of a substance P (neurokinin 1 receptor) antagonist in rat hepatocytes and rat plasma. Drug Metab. Dispos. 30, 937–943.

Hopfgartner, G., Chernushevich, I.V., Covey, T., Plomley, J.B., Bonner, R., 1999. Exact mass measurement of product ions for the structural elucidation of drug metabolites with a tandem quadrupole orthogonal-acceleration time-of-flight mass spectrometer. J. Am. Soc. Mass Spectrom. 10, 1305–1314.

Hopfgartner, G., Bourgogne, E., 2003. Quantititative high-throughput analysis of drugs in biological matrices by mass spectrometry. Mass Spec. Rev. 22, 195–214.

Howard, H.R., Seeger, T.F., 1993. Novel antipsychotics. In: Bristol, J.A. (Ed.), Annual Reports in Medicinal Chemistry, Vol. 28. Academic Press, New York, NY, pp. 39–47.

Howard, H.R., Prakash, C., Seeger, T.F., 1994. Ziprasidone hydrochloride. Drugs of the Future 19, 560–563.

Johnson, K., Shah, A., Jaw, S., Baxter, J., Prakash, C., 2003. Metabolism, pharmacokinetics, and excretion of a highly selective NMDA receptor antagonist, traxoprodil, in human cytochrome P450 2D6 extensive and poor metabolizers. Drug Metab. Dispos. 31, 76–87.

Knapp, D.R., 1979. Handbook of Analytical Derivatization Reactions. Wiley, New York, NY.

Korfmacher, W.A., Cox, K.A., Bryant, M.S., Veals, J., Ng, K., Watkins, R., Lin, C.-C., 1997. HPLC–API/MS/MS: a powerful tool for integrating drug metabolism into the drug discovery process. Drug Disc. Today 2, 532–537.

Kramer, M.S., Cutler, N., Feighner, J., Shrivastava, R., Carman, J., Sramak, J.J., Reines, S.A., Liu, G., Snavely, D., Wyatt-Knowles, E., Hale, J.J., Mills, S.G., MacCoss, M., Swain, C.J., Harrison, T., Hill, R.G., Hefti, F., Scolnick, E.M., Cascieri, M.A., Chicchi, G.G., Sadowski, S., Williams, A.R., Hewson, L., Smith, D., Carlson, E.J., Hargreaves, R.J., Rupniak, N.M.J., 1998. Distinct mechanism for antidepressant activity by blockade of central substance P receptors. Science 281, 1640–1645.

Lee, M.S., Kerns, E.H., 1999. LC/MS applications in drug development. Mass Spectrom. Rev. 18, 187–279.

Lee, M.S., 2002. LC/MS Applications in Drug Development. Wiley Interscience,New York, NY.

Lim, H.K., Stellingweif, S., Sisenwine, S., Chan, K.W., 1999. Rapid drug metabolite profiling using fast liquid chromatography, automated multiple-stage mass spectrometry and receptor binding. J. Chrom. A 831, 227–241.

Lin, J.H., Lu, A.Y.H., 1997. Role of pharmacokinetics and metabolism in drug discovery and development. Pharmacol. Rev. 49, 403–449.

Liu, D.Q., Hop, C.E.C.A., Beconi, M.G., Mao, A., Chiu, S.-H.L., 2001. Use of on-line hydrogen/deuterium exchange to facilitate metabolite identification. Rapid Commun. Mass Spectrom. 15, 1832–1839.

Louden, D., Handley, A., Taylor, S., Lenz, E., Miller, S., Wilson, I.D., Sage, A., 2000. Reversed-phase high-performance liquid chromatography combined with on-line UV diode array, FT infrared, and ^{1}H nuclear magnetic resonance spectroscopy and time-of-flight mass spectrometry: application to a mixture of nonsteroidal antiinflammatory drugs. Anal. Chem. 72, 3922–3926.

Mayol, R.F., Jajoo, H.K., Klunk, J., Blair, I.A., 1991. Metabolism of the antipsychotic drug tiospirone in humans. Drug Metab. Dispos. 19, 394–399.

Meltzer, H.Y., 1995. Role of serotonin in the action of atypical antipsychotic drugs. Clin. Neurosci. 3, 64–75.

Moller, D.E., 2001. New drug targets for type 2 diabetes and the metabolic syndrome. Nature 414, 821–827.

Navari, R.M., Reinhardt, R.R., Gralla, R.J., Kris, M.G., Hesketh, P.J., Khojasteh, A., Kindler, H., Grote, T.H., Pendergrass, K., Grunberg, S.M., Carides, A.D., Gertz, B.J., 1999. Reduction of cisplatin-induced emesis by a selective neurokinin-1-receptor antagonist. N. Engl. J. Med. 340, 190–195.

Niessen, W.M.A., 1998. Advances in instrumentation in liquid-chromatography-mass spectrometry and related liquid-introduction techniques. J. Chrom. A 794, 407–435.

Olsen, M.A., Cummings, P.G., Kennedy-Gabb, S., Wagner, B.M., Nicol, G.R., Munson, B., 2000. The use of deuterium oxide as a mobile phase for structural elucidation by HPLC/UV/ESI/MS. Anal. Chem. 72, 5070–5078.

Ortiz, A., Gershon, S., 1986. The future of neuroleptic psychopharmacology. J. Clin Psychiatry 47 (Suppl.), 3–11.

Poon, G.K., 1997. Drug metabolism and pharmacokinetics. In: Cole, R.B. (Ed.), Electrospray Ionization Mass Spectrometry: Fundamentals, Instrumentation & Applications. Wiley, New York, NY, pp. 499–525.

Prakash, C., Kamel, A., Anderson, W., Howard, H., 1997. Metabolism and excretion of the antipsychotic drug ziprasidone in rat following oral administration of a mixture of ^{14}C- and ^{3}H-labeled ziprasidone. Drug Metab. Dispos. 25, 206–218.

Prakash, C., Soliman, V., 1997. Metabolism and excretion of a novel antianxiety drug candidate, CP-93,393 in Long Evans rats: differentiation of regioisomeric glucuronides by LC/MS/MS and LC/MS/MS/MS. Drug Metab. Dispos. 25, 1288–1297.

Prakash, C., Cui, D., 1997. Metabolism and excretion of a new antianxiety drug candidate, CP-93,393, in cynomolgus monkeys. Drug Metab. Dispos. 25, 1395–1406.

Qin, X., Frech, P., 2001. Liquid chromatography/mass spectrometry (LC/MS) identification of photooxidative degradates of crystalline and amorphous MK-912. J. Pharm. Sci. 90, 833–844.

Riley, R.J., Martin, I.J., Cooper, A.E., 2002. The influence of DMPK as an integrated partner in modern drug discovery. Curr. Drug Metab. 3, 527–550.

Rodriguez, R.J., Proteau, J.P., Marquez, B.L., Hetherington, C.L., Buckholz, C.J., O'Connell, K.L., 1999. Flavin-containing monooxygenase-mediated metabolism of *N*-deacetyl ketoconazole by rat hepatic microsomes. Drug Metab. Dispos. 27, 880–886.

Rollema, H., Clarke, T., Lu, Y., Schmidt, A.W., Sprouse, J.S., 1996. Comparison of the effects of CP-93,393 and buspirone on 5HT and NE release: microdialysis studies in the hippocampus of freely moving rat and guinea pig. J. Neurochem. 66 (Suppl. 2), S37.

Rossi, D.T., Sinz, M.W., 2002. Mass Spectrometry in Drug Discovery. Marcel Dekker, Inc., New York, NY.

Rushmore, T.H., Reider, P.J., Slaughter, D., Assang, C., Shou, M., 2000. Bioreactor systems in drug metabolism: synthesis of cytochrome P450-generated metabolites. Metab. Eng. 2, 115–125.

Schmidt, A.W., Fox, C.B., Lazzaro, J., McLean, S., Ganong, A., Schulz, D.W., Desai, K., Bright, G.M., Heym, J., 1995. CP-93,393, a novel anxiolytic-antidepressant agent with both 5-HT1A agonist and alpha-2 adrenergic properties: in vitro studies. Neurosci. Abstr. 21, 2106.

Shockcor, J.P., Unger, S., Wilson, I.D., Foxall, P.J.D., Nicholson, J.K., Lindon, J.C., 1996. Combined HPLC, NMR spectroscopy, and ion-trap mass spectrometry with application to the detection and characterization of xenobiotic and endogenous metabolites in human urine. Anal. Chem. 68, 4431–4435.

Smith, D., Schmid, E., Jones, B., 2002. Do drug metabolism and pharmacokinetic departments make any contribution to drug discovery. Clin. Pharmacokinet. 41, 1005–1019.

Tarsey, D., 1983. Neuroleptic-induced extrapyramidal reactions: classification, description, and diagnosis. Clin. Neuropharmacol. 6 (Suppl. 1), S9–S26.

Tattersall, F.D., Rycroft, W., Cumberbatch, M., Mason, G., Tye, S., Williamson, D.J., Hale, J.J., Mills, S.G., Finke, P.E., MacCoss, M., Sadowski, S., Ber, E., Cascieri, M., Hill, R.G., MacIntyre, D.E., Hargreaves, R.J., 2000. The novel NK1 receptor antagonist MK-0869 (L-754,030) and its water soluble prodrug, L-758,298, inhibit acute and delayed cisplatin-induced emesis in ferrets. Neuropharmacology 39, 652–663.

Tiller, P.R., Land, A.D., Jardine, I., Murphy, D.M., Sozio, R., Ayrton, A., Schaefer, W.H., 1998. Application of liquid chromatography–mass spectrometry^n analyses to the characterization of novel glyburide metabolites formed in vitro. J. Chrom. A 794, 15–25.

Tiller, P.R., Raab, C., Hop, C.E.C.A., 2001. An unusual fragmentation mechanism involving the transfer of a methyl group. J. Mass Spectrom. 36, 344–345.

White, R.E., 2000. High-throughput screening in drug metabolism and pharmacokinetic support of drug discovery. Annu. Rev. Pharmacol. Toxicol. 40, 133–157.

Wienkers, L.C., Steenwyk, R.C., Mizsak, S.A., Pearson, P.G., 1995. In vitro metabolism of tirilazad mesylate in male and female rats. Drug Metab. Dispos. 23, 383–392.

Wienkers, L.C., Steenwyk, R.C., Sanders, P.E., Pearson, P.G., 1996. Biotransformation of tirilazad in human: 1. cytochrome P450 3A-mediated hydroxylation of tirilazad mesylate in human liver microsomes. J. Pharmacol. Exp. Ther. 277, 982–990.

Willoughby, R., Sheehan, E., Mitrovich, S., 1998. A Global View of LC/MS: How to Solve Your Most Challenging Analytical Problems. Global View Publishing, Pittsburgh, PA.

Woolf, T.F., 1999. Handbook of Drug Metabolism. Marcel Dekker, Inc., New York, NY.

Yu, X., Cui, D., Davis, M.R., 1999. Identification of in vitro metabolites of indinavir by "Intelligent Automated LC–MS/MS" (INTAMS) utilizing triple quadrupole tandem mass spectrometry. J. Am. Soc. Mass Spectrom. 10, 175–183 and references therein.

Zimmet, P, Alberti, K.G.M.M., Shaw, J., 2001. Global and societal implications of the diabetes epidemic. Nature 414, 782–787.

Identification and Quantification of Drugs, Metabolites and Metabolizing Enzymes by LC–MS
Swapan K. Chowdhury, editor.

Chapter 7

THE UTILITY OF IN VITRO SCREENING FOR THE ASSESSMENT OF ELECTROPHILIC METABOLITE FORMATION EARLY IN DRUG DISCOVERY USING HPLC–MS/MS

Diane E. Grotz, Nigel A. Clarke, and Kathleen A. Cox

7.1. Introduction

"Between the time research begins to develop a new prescription medicine until it receives approval from the Food and Drug Administration (FDA) to market the drug in the United States, a drug company typically spends $802 million over the course of 10 to 15 years."[1]

Generation of successful drugs is expensive and time-consuming. Ever-rising costs for the clinical trials critical to drug development and the increasing breadth and complexity of pre-clinical safety and efficacy assessment have increased the burden. Since the cost and time needed to develop a successful drug continue to rise, the pressure on scientists to screen and identify drug candidates also grows rapidly. Effective screening, however, is not simple. The current challenge in Drug Discovery is to obtain as much information as possible about a new chemical entity (NCE) prior to its advancement into the more expensive stages of Drug Development such as long-term laboratory animal studies and, later, clinical trials. According to the Pharmaceutical Research and Manufacturers of America, only five in 5000 compounds that enter pre-clinical testing make it to human testing. Only one of those five is approved. The expense and failure rate make it critical to have better screening earlier in the drug development process. The more known about an NCE's likely biological reactivity and its metabolic fate in humans, the more effective the decisions will be about assessments of its therapeutic potential and risk/benefit ratio.

The right experiments, such as high-throughput receptor binding studies, enzyme inhibition studies, rapid pharmacokinetic studies (Cox *et al.*, 1999), metabolic stability studies (Boyle *et al.*, 2002, Lau *et al.*, 2002), and preliminary metabolite

[1]Tufts Center for the Study of Drug Development, Tufts University, November 30, 2001.

characterization, can all reduce the attrition rate of NCEs due to easily identifiable problems such as possible drug–drug interactions, low bioavailability, and potential metabolic liabilities.

7.1.1. Metabolite Characterization

Metabolite characterization is critical to the NCE screening program. Drugs that fail after clinical trials usually fail due to so-called idiosyncratic reactions. By definition these reactions are impossible to predict, and they occur so sporadically that they are not observed until the drug has been administered to large populations. Often, the reactions are related to metabolism: metabolism in some individuals can be slightly different than it is in others. Some metabolites have biologic activity significantly different from the parent compound and may have increased target-receptor binding. Some metabolites have significant ability to bind proteins or interact with other drugs. Others may have extended half-lives, which can have a positive or negative effect, or be retained in body fat stores due to increased lipid solubility. Techniques that can help predict human metabolism can help characterize the risks (and potential benefits) of NCEs.

Often the metabolic liabilities of a drug are not discovered until the compound is well into clinical trials. However, even large clinical trials may not access the few patients that metabolize the drug differently from the "normal" heterogeneous population. While the clinical implications of an idiosyncratic drug reaction are far beyond the scope of Drug Discovery, the knowledge that an animal species can form a potentially liable metabolite might change the future of an NCE. Depending on the therapeutic area, the decision may be made not to progress the compound out of Discovery, or to progress the compound at risk, but with a specific outlook to the toxicology and safety studies as well as to the marketing strategies. There have been many examples of drugs shown to be hepatotoxic in man that were approved because their therapeutic value outweighs their potential toxicity, e.g. tacrine, a drug approved for use in treating Alzheimer's disease (Ballet, 1997).

To get an early determination of potential metabolic pathways for an NCE, we have chosen to characterize metabolites of lead drug candidates using in vivo experiments. For example, our laboratory has utilized bile-duct-cannulated animals (rats and monkeys) to get an early look at metabolites in both bile and urine. Hepatocyte in vitro incubations are also conducted to compare metabolism across species; this provides the first look at potential human metabolites. A complete metabolic profile of every NCE would be ideal; unfortunately this is impractical in a Discovery setting, so the effective use of information obtained from the resources available must be maximized. Early critical information includes the identification of major metabolic pathways, potential metabolic liabilities, and interspecies variations.

The consequences of not addressing a metabolic liability can be severe. Various non-steroidal anti-inflammatory drugs (NSAIDs) have been withdrawn from the market because of severe adverse events (e.g. Zomepirac (Wang and Dickinson, 2000) and benoxaprofen (Qiu *et al.*, 1998)). The successful marketing of a drug can also be derailed by the issuance of a "Black Box" warning to physicians, prominently displayed in the Physicians Desk Reference and in the package insert labeling to alert physicians to serious risks associated with use of the drug. If patients taking the drug develop potentially life-threatening idiosyncratic reactions that were not observed in the clinical trials the FDA will require a notice to warn physicians ("Special problems, particularly those that lead to death or serious injury, may be required to be placed in a prominently displayed box" – 21 CFR 201.57 [e]). Approximately 8% of the drugs approved between 1975 and 1999 were issued at least one black box warning; approximately 3% were withdrawn from the market. Of the drugs that were withdrawn from the market, half of them were withdrawn within two years of their approval (Lasser *et al.*, 2002). Clearly, a pharmaceutical company needs to know all the issues about a new drug, including potential contra-indications, in order to be able to promote it accurately and with confidence.

Zomepirac, benoxaprofen, and ibuprofen all show covalent protein-adduct formation. Zomepirac and benoxaprofen have been withdrawn from the market due to adverse drug reactions. Unfortunately, there is no consistent link between organ toxicity and the covalent binding of a drug or its metabolites to proteins. Because the potential for a causal relationship exists, Drug Discovery programs try to reduce or eliminate the ability of an NCE to bind to proteins.

7.1.2. In Vitro Metabolic Screening Methodology

The pharmaceutical company's goal is to manufacture safe and effective drugs and to avoid potentially problematic drugs from being used incorrectly. The risk-to-benefit profile acceptable to the therapeutic team may vary depending on the expected prescribed use of the drug. While the contributions of Drug Discovery may be forgotten for a drug that is in the midst of clinical trials, the ability to identify metabolic liabilities presents itself first in the discovery setting. In many cases, these metabolic pathways can be accessed and observed in well-designed in vitro metabolism studies in a fairly rapid manner.

Early in Drug Discovery, high-throughput screens are an invaluable tool used to select compounds with desirable properties and also to *deselect* compounds with undesirable properties. In early Drug Discovery, typically only a small amount of compound (mgs) is available. A maximum amount of information needs to be obtained from this limited quantity of compound. This constraint, along with the interest in decreasing the number of animals used in pre-clinical testing, leads directly to an interest in in vitro screening.

The availability of radiolabeled compounds in early Drug Discovery is limited. Often an NCE is not radiolabeled until it is considered a potential lead, and some pharmaceutical companies do not generate a radiolabeled drug until after the compound progresses out of Drug Discovery. A screen using non-radiolabeled compounds will give a qualitative answer to the question of formation of reactive metabolites; while this does not give a rigorous answer to the concern about covalently bound proteins, it does give a first look at compounds that may have a structural propensity for forming reactive metabolites. In some cases, this information is enough for the chemists to consider a structural modification that can then also be tested in the screen.

In vitro metabolism experiments have mostly centered on the liver, as the liver is the major site of metabolism for most drugs. Purified enzymes, sub-cellular fractions (e.g. S9, microsomes), whole cells and precision-cut liver slices are in widespread use. Liver slices and isolated hepatocytes are used because they are a more holistic system, containing all of the hepatic metabolic systems, and have the ability to give both Phase I and Phase II metabolism. The preparation and incubation times needed for liver slices and hepatocytes to generate useful metabolic profiles of many NCEs can be on the order of several hours, which preclude their general use in a high-throughput screen. Some studies have shown significant metabolism in hepatocytes and liver slices (Axelsson *et al.*, 2003) in comparatively short time frames, which appear to be exceptions to the longer incubation times more normally observed in the literature. Hepatic microsomes however, are more robust and typically require only short incubation times (10 min to 2 h), lending themselves fairly well to use in higher-throughput screening assays. Whether a study is classified as high-throughput or not depends not only on the amount of samples that can be run in one day but also on the time needed to prepare the samples (e.g. incubation, dosing, protein precipitation), and the time needed to analyze the data set that comes from those samples.

In vitro screening makes it possible to focus on specific metabolic pathways, depending on the materials used. Screening for electrophilic metabolites is a deselection screen; the intent of such a screen is to find compounds with the lowest risk for toxic liabilities. Compounds put into this type of screen are compared to others in the same therapeutic area. This comparison may influence the decision to progress one compound over another, or may simply provide information that allows a large compound set to be assessed, giving higher priority to low-risk compounds for further, more resource-demanding studies. Setting up this kind of in vitro screen requires decisions such as which hepatic system to use; this choice is governed by a number of factors. Screens that are set up to "trap" the reactive electrophilic metabolites are most often performed in hepatic microsomes with trapping agents added. Glutathione (GSH), potassium cyanide, methoxylamine, and semicarbazide can all be used as trapping agents (Evans *et al.*, 2004).

To set up a useful, higher-throughput screen analysts need to consider factors such as cost, ease of use, availability, and reproducibility. In all of those aspects, microsomes and S9 have an advantage over hepatocytes and liver slices. Sub-cellular fractions cost less initially and also cost less to maintain. Hepatocytes and liver slices are living systems; the cells can die due to stress or mishandling, rendering them useless. They require special handling, have a fairly short useful lifetime, and if not used fresh, need to be specially preserved and stored in liquid nitrogen. Microsomes and S9 are much more rugged; they can almost be treated as an off-the-shelf reagent, and can be kept upward of 10 years at −80°C. Commercial S9 and microsomes are generally pooled preparations representing at least five animals. Fresh hepatocytes and liver slices reflect one animal, although cryopreserved hepatocytes can be pooled. A pooled system should more closely represent an "average" metabolism, and should be more consistent than cells from one animal. A large enough supply of a particular pooled fraction allows for direct comparison of results between compounds across a structural class, without the worries inherent in using a different supply for each experiment.

Another important fact to consider is that human hepatic sub-cellular fractions are accessible for studies in Drug Discovery, while human in vivo data may not be available for many years after a compound leaves discovery. Setting up a parallel screen that uses human as well as an animal species may increase the amount of useful information obtained.

7.1.3. The Significance of Electrophilic Metabolites

Many obstacles arise between the synthesis of an NCE and the introduction of a new drug to market. Sometimes, a new structural series has problems that were not and could not have been foreseen by the chemists. Electrophilic metabolites often fall into that category; they may indicate potential serious metabolic consequences for the NCE. Electrophilic reactive intermediates have the potential to covalently modify proteins, eliciting impaired cellular function, genotoxicity or immune-mediated toxic responses in vivo. Reactive metabolites can covalently bind to proteins, DNA, RNA, and enzymes (Hinson, 1992). This binding may cause adverse and sometimes idiosyncratic reactions, including haptenation, immune responses, and inactivation of particular proteins, all of which have been implicated in organ damage (Boelsterli, 2002). For example, the toxicity of acetaminophen is due to an electrophilic reactive intermediate, *N*-acetyl-*p*-benzoquinoneimine, that covalently binds to cellular macromolecules (Manautou *et al.*, 1994).

As mentioned previously, benoxaprofen has been found to be associated with covalent protein binding and has a very high incidence of hepatotoxicity. Idiosyncratic cholestatic reactions were observed after the introduction of benoxaprofen. Previous to this, cholestatic reactions, an interruption of bile flow were normally considered benign. However, these cholestatic reactions led to the death of 11 of 14 patients in the U.S. and more than 70 deaths worldwide (Tolman, 1998). While predicting toxicity within a chemical class can be exceedingly difficult, an in vitro screen can highlight the propensity of compounds to form electrophilic metabolites. It will not, however, determine whether these electrophilic metabolites will be a liability to the drug once it is progressed. If this screen highlights a particular metabolic liability, the therapeutic team may have enough flexibility to pursue other structural series. If not, and a compound goes forward, the therapeutic team has been alerted to the possible risks.

Electrophilic compounds or metabolites can undergo Phase II metabolism to form conjugates. While conjugation is typically a detoxification mechanism, some conjugates are more reactive than the precursor electrophile. Two metabolites formed from electrophilic reactive metabolites, and which can be considered a warning sign of possible toxicity are acylglucuronides and glutathiones. Acylglucuronide and glutathione conjugations are regularly targeted in metabolite profiling studies. Along with acylglucuronides, other acylglycosides are known to covalently bind to proteins. This report considers four types of electrophilic reactive metabolites that form the following conjugates: glutathiones, acylglucuronides, acylglucosides, and *N*-acetylglucosaminides.

7.1.3.1. Glutathiones

Glutathione conjugation is generally considered a process for the detoxification of electrophilic compounds to less toxic and more hydrophilic thioether compounds. As such, it is a major mechanism for protection against oxidative stress, and therefore glutathione conjugates are regularly screened for in metabolite profiling studies. It is important to note that there are examples where the conjugation product is still reactive and in some cases more reactive than the parent compound. An example of this is 1, 2-dichloroethane, which is mutagenic in the Ames test for bacterial mutagenicity. This mutagenicity is dependant on the presence of GSH and glutathione-*S*-transferases (GSTs), due possibly to the formation of an episulfonium ion (Monks and Lau, 1988).

GSH and GSTs are found in different parts of the body, including the kidneys and the liver. GSH traps electrophilic substrates either by direct nucleophilic attack, or by a GST catalyzed reaction. When formed in the liver, the conjugate is excreted through the bile or to the kidney where, through further modifications, the metabolized compound may be excreted in urine as a mercapturic acid derivative. GSTs are generally soluble enzymes, with the exception of a small group of microsomal GSTs. In humans, GSTs constitute 4% of the soluble protein in the liver (van Bladeren, 2001). Glutathione transferases far outweigh the amount of glutathione available for conjugation. Theoretically, with the correct amount of an appropriate substrate in vivo, GSTs are capable of completely depleting the glutathione pool in seconds. Severe toxicity follows the depletion of GSH and the resynthesis of GSH takes time (resynthesis in rat liver takes 3–4 h (Rinaldi *et al.*, 2002)). While depletion of the GSH pool is not a problem for potent drugs, it is a problem for lower potency drugs such as acetaminophen and in cases of acute overdose where the circulating levels of drug, and consequently the reactive electrophilic metabolite, are high. The toxicity of electrophilic reactive metabolites not trapped by glutathione is not limited to the liver; damage can also be caused in bone marrow, the pulmonary system, and the kidneys (Monks and Lau, 1988). Some of the substrates subject to glutathione conjugation are also known to lead to alkylation of proteins and DNA and are carcinogens, such as dichloromethane (Sherratt *et al.*, 2002); or are nephrotoxic, such as hexachlorobutadiene (Dekant, 2001).

Figure 1 shows three glutathione reactions observed in vivo. The first example is aflatoxin B_1 (Sherratt and Hayes, 2002). This compound is metabolized to a highly electrophilic species, aflatoxin B_1-8, 9-epoxide, which can covalently bind to DNA. In a detoxification process, the epoxide ring is opened by the nucleophilic attack of the glutathione. The second example is also a detoxification process. Here acetaminophen is metabolized to *N*-acetyl-*p*-benzoquinone imine, a metabolite associated with liver damage. Subsequent conjugation with glutathione detoxifies the reactive intermediate (James *et al.*, 2003). The third example is a bioactivation process; the glutathione conjugation of dibromoethane produces a reactive conjugate. This metabolic pathway for dibromoethane creates an episulfonium ion, which is highly electrophilic and can covalently bind to DNA (Boelsterli, 2003a).

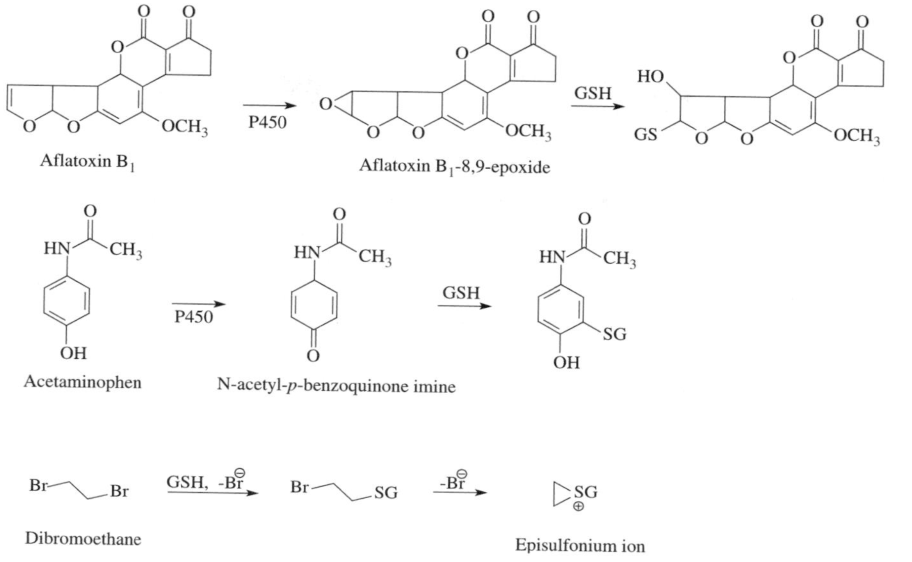

Aflatoxin B_1 → P450 → Aflatoxin B_1-8,9-epoxide → GSH →

Acetaminophen → P450 → N-acetyl-*p*-benzoquinone imine → GSH →

Dibromoethane → GSH, $-Br^{\ominus}$ → → $-Br^{\ominus}$ → Episulfonium ion

Figure 1.
Examples of metabolic glutathione conjugations

7.1.3.2. Acylglucuronides

Acylglucuronides are formed by esterification of glucuronic acid with a carboxylic acid containing compound or metabolite. Glucuronidation is catalyzed by a family of uridine 5′-diphospho-glucuronsyltransferases (UGTs). The UGTs are a membrane-bound superfamily of enzymes that are found in the highest concentration in the endoplasmic reticulum.

Conjugation with glucuronic acid is a fundamental metabolic process for detoxification and elimination of xenobiotics. In general, glucuronide conjugates are less biologically or chemically reactive than the corresponding aglycone[2]. However, because acylglucuronides are electrophilic they are an exception to this rule. The reactivity of acylglucuronides is due to the electrophilic nature of the carboxylic acid moiety, which among other things, can react with nucleophilic targets on macromolecules and form covalently bound protein adducts. Acylglucuronides can also covalently bind to UGTs (e.g. ketoprofen), which are then irreversibly inactivated (Terrier *et al.*, 1999), and can cause immune-mediated toxicity (e.g. diclofenac) (Boelsterli, 2003b).

Acylglucuronides are chemically unstable due to the susceptibility of the acyl group to nucleophilic attack. Their instability depends in part on pH (Hasegawa *et al.*, 1982), and matrix (Smith *et al.*, 1985). An acylglucuronide can spontaneously undergo hydrolysis (conversion back to the aglycone), which can be subject

[2]The non-sugar portion of a conjugated glycosidic metabolite such as a glucuronide conjugate is called an aglycone.

to enterohepatic circulation and reconjugation. Once formed, the conjugates can also undergo intramolecular rearrangement to form multiple anomers through acyl migration (Akira *et al.*, 1997), shown in Fig. 2. Consequently, one carboxylic acid moiety can give rise to a number of acylglucuronides. Acylglucuronides also undergo intermolecular reactions with proteins, which lead to covalently bound drug-protein adducts (Bailey and Dickinson, 2003). Covalently modified cellular proteins may lead to altered signal transduction, haptenation of proteins and subsequent immune responses, and direct inactivation of the protein function (Boelsterli, 2002).

Acylglucuronides are postulated to adduct to proteins in three ways: (1) transacylation, which is a direct displacement of the glucuronic acid by nucleophilic attack of the protein; (2) adduction after acylmigration of the drug moiety, which produces an imine bond between the opened glucuronic acid ring and the protein; and (3) adduction of the ring-opened glucuronic acid to the protein after loss of the aglycone (Georges *et al.*, 1999). The last process frees the aglycone to go through another cycle of conjugation and possible protein binding (Fig. 3).

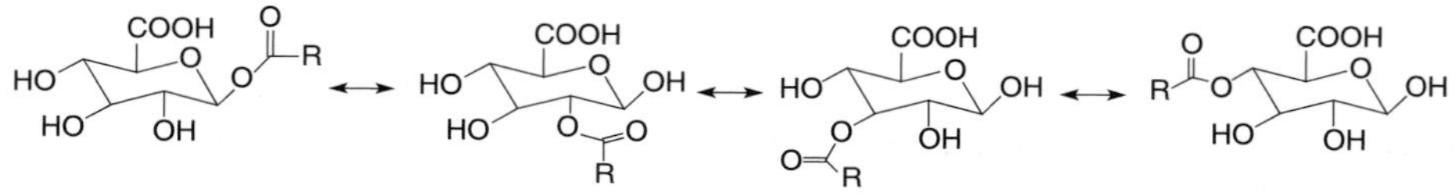

Figure 2.
Migration of the aglycone around the glucuronide structure

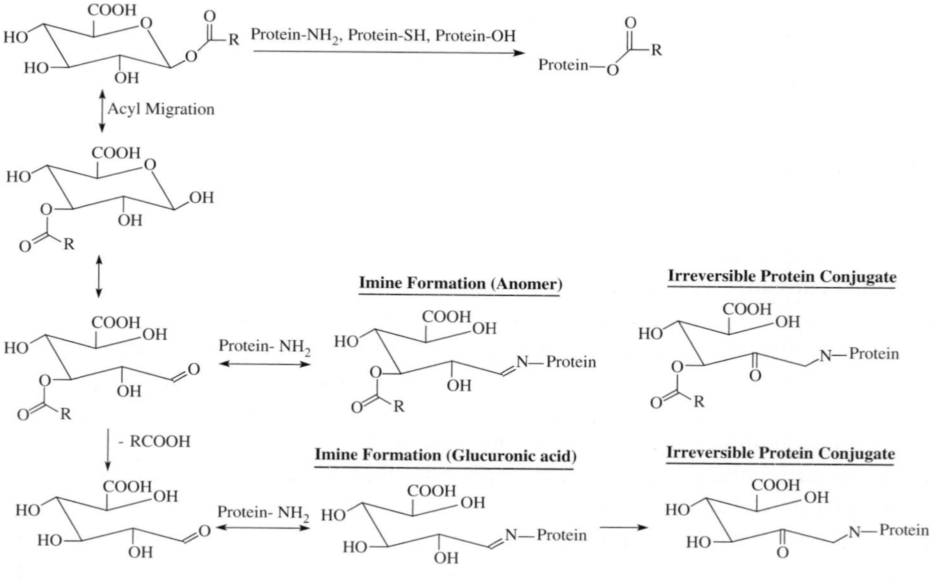

Figure 3.
Protein conjugates of acylglucuronides

Acylglucuronide metabolites of two NSAIDs, diclofenac (Boelsterli, 2003a) and Zomepirac (Wang *et al.*, 2001), have been observed adducted to proteins at the canalicular membrane of hepatocytes. Metabolites were observed adducted to the canalicular ectoenzyme, dipeptidyl peptidase IV (DPP IV). Acylglucuronide metabolites of diclofenac have also been observed adducted to plasma proteins and intestinal mucosal proteins. After exposure to diclofenac, rats were observed to have adducts on DPP IV, and the peptidase function of that enzyme was found to have decreased. Formation of ulcers was also observed in the small intestine. It is important to note that no causal links have been proven for the adduction to DPP IV and the functional decrease of the enzyme or for the formation of ulcers at the intestinal sites where protein adducts were observed (Boelsterli, 2003a).

7.1.3.3. Acylglucosides and *N*-Acetylglucosaminides

Acylglucosidation and *N*-acetylglucosaminidation are rare in mammals, although common in plants and invertebrates. Glucosidation may serve as an alternative pathway when the more common detoxification pathway via glucuronidation is not available, e.g. in the case of depletion of uridine 5′-diphosphoglucuronic acid (UDPGA) due to mechanism-based inhibition or, in some cases, due to disease-impaired livers.

Two examples of acylglucosidation in human cellular subfractions are known to Grotz. One example is mycophenolic acid glucoside generated by kidney microsomes (Shipkova *et al.*, 2001); mycophenolic acid was used in treating recalcitrant psoriasis. The other is Compound A [(+)-(5*S*,6*R*,7*R*)-2-isopropylamino-7-[4-methoxy-2-((2*R*)-3-methoxy-2-methylpropyl)-5-(3,4-methylenedioxyphenyl) cyclopenteno [1,2-*b*] pyridine 6-carboxylic acid], a selective endothelin ETA receptor antagonist, generated by human liver microsomes and human hepatocytes (Tang *et al.*, 2003).

7.2. Current Approaches

7.2.1. Background

Historically, precision-cut liver slices, isolated hepatocytes, and sub-cellular fractions (e.g. S9, microsomes) have been used to predict or confirm the occurrence of metabolites seen in vivo (De Graaf *et al.*, 2002). Hepatocytes contain all the enzymes and cofactors necessary to form the hepatic metabolites one would expect to see generated by a compound in vivo. For results that reflect a large population, experiments are carried out with pooled isolated hepatocytes. Pooling hepatocytes from a number of animals normalizes the metabolism so that it more closely reflects an "average" metabolism. Two types of hepatocytes are used, cryopreserved and fresh. In our experience, fresh hepatocytes are more likely to reflect the in vivo metabolism. Cryopreserved hepatocytes have some definite advantages over fresh hepatocytes, however. Cryopreserved hepatocytes can be stored and then used at the scientist's convenience, while the preparation of fresh hepatocytes or their arrival (if purchased from an outside vendor) dictates the timing of experiments. The most common hepatocyte incubations involve the following species: Sprague–Dawley rat, beagle dog, cynomolgus monkey, and human.

Since in early Drug Discovery there are no human in vivo experiments, we can only estimate human metabolism using human liver fractions: liver slices, hepatocytes, microsomes, and S9. Because fresh human hepatocytes come from one donor, the results from an in vitro experiment using these hepatocytes may not reflect a representative heterogeneous population. Since there is no realistic way of pooling fresh human hepatocytes, using a cryopreserved pooled liver fraction is the most representative method we have for screening human metabolism of NCEs.

Hepatocyte incubation is a well-established technology. However, some drawbacks to hepatocytes make them a less likely candidate for a higher throughput screen. Most prominent is the incubation time. For metabolite characterization studies with fresh hepatocytes, the incubation time in our lab is 24 h. Even with cryopreserved hepatocytes the incubations can take as long as 5 h (5 h is currently the standard in our lab). This kind of time frame requirement precludes a high-throughput screen. While the viability of hepatocytes diminishes dramatically over this period, metabolite characterization is not a kinetic assessment of metabolite formation, but an attempt to assess all potential metabolic pathways and to produce as much of the metabolites as possible to enhance detection.

For a more rapid turn-around, hepatic microsomes and S9 are often used in higher-throughput screening assays. For the purposes of our screen, liver microsomes and S9 are sufficient. As mentioned previously, while most of the GSTs are in the soluble fraction there are a small percentage that are membrane-bound. The Discovery setting does not allow the time necessary to elucidate which GSTs may be responsible for the conjugation, so our method uses a combination of hepatic S9 and microsomes. UGTs are membrane-bound and so are present in hepatic microsomes. For the acylglucuronide, acylglucoside and *N*-acetylglucosaminide conjugation incubations, hepatic microsomes are used. The cofactors glutathione, UDP-glucuronic acid, UDP-glucose, and UDP-*N*-acetylglucosamine are added at the time of the incubation.

Although it can often be that not all metabolites observed in an in vivo study will be observed in vitro, the reverse is also true, there are instances where in an in vitro environment the compound may exhibit metabolic pathways not observed in vivo. Although every effort is made to keep the in vitro incubations as close to the in vivo biological conditions as possible, it is impossible to mimic every detail. The rate of metabolism in vitro may also differ from that observed in vivo.

Since radiolabeled compounds are not always available early in Drug Discovery, this in vitro screen was set up to utilize non-radiolabeled compounds. An alternative to the use of radiolabeled substrates would be the use of radiolabeled cofactors. In either case, the compound conjugate would be detectable, and presumably quantifiable, by radiography. However, in previous experiments, in our lab to determine the utility of in vitro screening for electrophilic metabolites, it was determined that the use of commercially available radiolabeled cofactors was not helpful (data not shown). The rate of turnover was such that unless a prohibitively expensive amount of radiolabeled cofactor was used, the ability of detecting the radiolabeled conjugate, much less quantitatively ranking compounds in order of the metabolite formation obtained, was severely limited (Grotz *et al.*, 2001). In most cases the results obtained were, at best, semiquantitative.

To perform quantitative reactive metabolite protein binding experiments similar to those performed at Merck (Evans *et al.*, 2004), it is necessary to have radiolabeled compounds. These experiments can be used to rank the order of NCEs by formation of reactive intermediates. The Merck group's upper limit threshold target value is 50 pmol/mg protein, which is 1/20th the dose of a prototype hepatotoxin (e.g. acetaminophen) sufficient to cause hepatic necrosis. This target is not a fixed limit, as other factors such as therapeutic indication, projected dose, and projected duration of therapy all need to be considered. This screen gives Medicinal Chemistry the opportunity to modify the structures and to lessen the potential for covalent protein binding (Baillie, 2004), increasing the chance for developing a successful drug.

Recently a method was developed to quantitatively assess the formation of glutathione using hepatocytes incubated with ^{35}S-labeled methionine to intracellularly generate ^{35}S-labeled glutathione (Hartman *et al.*, 2002). This innovative technique eliminates some of the problems experienced in our lab. Unfortunately, this is a time-intensive and expensive process, and is not readily adaptable to a higher-throughput screen.

7.2.2. Strategy

7.2.2.1. Materials and Methods

Chemicals: Acetaminophen, alamethicin, diclofenac, furosemide, glucose-6-phosphate, glucose-6-phosphate dehydrogenase, glutathione, ibuprofen, magnesium chloride, NADPH, D-saccharic acid 1,4-lactone, uridine-5-diphosphate glucuronic acid, uridine-5-diphosphate glucose, and uridine-5-diphosphate *N*-acetylglucosamine were obtained from Sigma-Aldrich (St. Louis, MO). Mycophenolic acid was obtained from Alexis Corporation (San Diego, CA). Furosemide and Ursodeoxycholic acid were obtained from ICN Biochemicals (Aurora, OH). Acetonitrile, formic acid, potassium phosphate monobasic, and potassium phosphate dibasic were purchased from Fisher Scientific (Pittsburgh, PA).

Liver Fractions: Sprague–Dawley rat liver microsomes and S9, and cynomolgus monkey liver S9 and microsomes were purchased from In Vitro Technologies (Baltimore, MD); Instrumentation: Waters Alliance 2690 HPLC system (Milford, MA); Thermo-Finnigan LCQ Classic ion trap mass spectrometer (San Jose, CA).

7.2.2.2. Glutathione Conjugation Incubations

Alamethicin at a level of 50 μg/mg protein was added to a volume of S9 and microsomes containing 0.5 mg protein each. This was diluted to approximately half the total incubation volume with 100 mM potassium phosphate (pH 7.4) and was mixed and left on ice for 20 min. A volume of test compound was then added to give a final incubation concentration of 1 mM. An NADPH regenerating system (containing NADPH, glucose-6-phosphate, glucose-6-phosphate dehydrogenase, and magnesium chloride in 100 mM potassium phosphate buffer at pH 7.4) and a volume of glutathione to give a final concentration of 2 mM were added to the incubation mixture. The mixture was then lightly vortexed and placed in a shaking water bath at a temperature of 37°C and incubated for 2 h. A volume of cold acetonitrile, equal to the incubation volume, was added to the samples to quench the incubation. The samples were vortexed and then centrifuged for 20 min at 4000 rpm. The supernatants were then transferred to vials and injected on to the LC/MS system for analysis.

7.2.2.3. Acylglucuronide Conjugation Incubations

Alamethicin at a level of 50 μg/mg protein was added to a volume of microsomes containing 1.0 mg protein. This was diluted to approximately half the total incubation volume with 100 mM potassium phosphate (pH 7.0) and was mixed and left on ice for 20 min. A volume of test compound was then added, giving a final incubation concentration of 0.1–1 mM. An NADPH regenerating system (containing NADPH, glucose-6-phosphate, glucose-6-phosphate dehydrogenase, and magnesium chloride in 100 mM potassium phosphate buffer at pH 7.0), saccharic acid 1, 4-lactone giving a final incubation concentration of 5 mM, and a volume of UDP-glucuronic acid giving a final concentration of 2 mM were added to the incubation mixture. The mixture was then lightly vortexed and placed in a shaking water bath at a temperature of 37°C and incubated for 2 h. A volume of 9:1 cold acetonitrileformic acid, equal to the incubation volume, was added to the samples to quench the incubation and stabilize any acylglucuronides that had formed. The samples were vortexed and then centrifuged for 20 min at 4000 rpm. The supernatants were then transferred to vials and injected on to the LC/MS system for analysis.

7.2.2.4. Acylglucoside and *N*-Acetylglucosaminide Conjugation Incubations

Alamethicin at a level of 50 μg/mg protein was added to a volume of microsomes containing 1.0 mg protein. This was diluted to approximately half the total incubation volume with 100 mM potassium phosphate (pH 7.0) and was mixed and left on ice for 20 min. A volume of test compound was then added to give a final incubation concentration of 0.1–1 mM. An NADPH regenerating system (containing NADPH, glucose-6-phosphate, glucose-6-phosphate dehydrogenase, and magnesium chloride in 100 mM phosphate buffer at pH 7.0), and a volume of UDP-glucose or UDP-*N*-acetylglucosaminide giving a final concentration of 2 mM were added to the incubation mixture. The mixture was then lightly mixed and placed in a shaking water bath at a temperature of 37°C and incubated for 2 h. A volume of cold acetonitrile, equal to the incubation volume, was added to the samples to quench the incubation. The samples were vortexed and then centrifuged for 20 min at 4000 rpm. The supernatants were then transferred to vials and injected on to the LC/MS system for analysis.

7.2.2.5. Screening Method Details

Compounds were incubated using Sprague–Dawley rat and cynomolgus monkey liver microsomes and S9. The compounds used to validate the screen were chosen with prior knowledge (either from our lab or from literature) of their formation of the metabolites of interest, either in vivo or in vitro, or from an expectation based on the structure. Compounds that included a carboxylic acid in their structure were incubated with UDP-glucose and UDP-*N*-acetylglucosamine as well as with UDPGA, with the idea that they might form acylglucosides and *N*-acetylglucosaminides along with the previously observed acylglucuronides, whether we knew of previous citations in the literature or not.

7.2.2.6. High Performance Liquid Chromatographic (HPLC) Conditions

Chromatographic conditions were chosen to give a robust, reverse phase gradient run that would accommodate all tested compounds. The gradient was held at 100% A for 3 min to elute all unconjugated cofactors. The chromatography was a simple 15 min gradient with a flow rate of 0.8 mL/min on a Jones Chromatography Genesis AQ, 4 μm, 150 × 4.6 mm analytical column. The gradient ran from 100% A to 95% B. Mobile phase: A – 99:1 water: methanol, 10 mM ammonium acetate, 100 μl/l acetic acid (aqueous pH ~5.5); B – 1:99 water: methanol, 10 mM ammonium acetate, 100 μl/l acetic acid.

7.2.2.7. Mass Spectrometric Analysis

Samples were analyzed by LC/MS/MS using a Thermo-Finnigan LCQ Ion Trap. Acylglucuronide, acylglucoside, and *N*-acetylglucosaminide metabolites are thermally labile. To maintain structural integrity of the conjugated metabolites for identification, positive electrospray mode was used, as this “soft” ionization technique was the most suitable to maximize transmission and detection of the intact conjugated metabolites. Expected metabolites were targeted using MS/MS experiments, monitoring the fragmentation of specific masses expected for the conjugated metabolites. MS^3 was utilized to provide further structural confirmation on the glucuronide, glucoside and *N*-acetylglucosaminide conjugations. It has been our experience that further fragmentation of a glutathione conjugate on an ion trap mass spectrometer provides little additional compound specific structural information.

Results obtained from this approach were qualitative in nature. This type of screen gives a “Yes/No” answer for the presence of specific metabolites, and is of great utility to the Discovery team when the targeted data are taken in context. For an ongoing screen samples could be run on a triple-quadrupole mass spectrometer monitoring constant neutral losses that are characteristic of the conjugated metabolites (i.e. 129, 176, 162, and 203 for glutathione, glucuronide, glucoside, and *N*-acetylglucosaminide, respectively). The use of neutral loss scans permits detection of non-predicted metabolites. Using the neutral loss scan technique, the scientist needs no prior knowledge of the metabolite structure. In these screens assessment of non-radio-labeled compounds is qualitative only.

7.3. Results/Discussion

Compounds known to form the types of conjugated metabolites discussed above were used as positive controls to establish and optimize the incubation system. Compounds used as positive controls for the glutathione screen were acetaminophen, diclofenac (Poon *et al.*, 2001), and one NCE (SCH 1). Compounds used as positive controls for the acylglucuronide, acylglucoside, and *N*-acetylglucosaminide screens were diclofenac (Seitz and Boelsterli, 1998), furosemide (Bolze *et al.*, 2002), mycophenolic acid (Shipkova *et al.*, 2001), ursodeoxycholic acid (Momose *et al.*, 1998), and one NCE (SCH X). In a cynomolgus monkey in vivo study SCH X was found to form three acylglucuronides, one acylglycoside and one *N*-acetylglycosaminide in bile, while in rat only the acylglucuronide was observed in bile. SCH 1 formed 5 glutathione conjugates in rat in vivo and was not dosed in monkeys.

UGTs are a superfamily of membrane-bound enzymes. The active site for the UGTs is localized inside the endoplasmic reticulum. While most GSTs are soluble, a small percentage is membrane bound (Rinaldi *et al.*, 2002). Alamethicin is a peptide that forms pores in microsomes and creates access to membrane-bound enzymes for the catalyzation of conjugations (Fisher *et al.*, 2000). In our experience, alamethicin has been extremely useful in increasing the metabolic turnover resulting from the conjugation reactions. Saccharic acid 1,4-lactone is a beta-glucuronidase inhibitor and therefore, stabilizes the formation of acylglucuronides by preventing enzymatic hydrolysis to the aglycone. Saccharic acid 1,4-lactone was used for the acylglucuronide incubations only.

Following the validation of the screen as a routine procedure, several NCEs along with commercially available compounds as controls were incubated to determine the in vitro formation of the expected metabolites under the optimized conditions. The results from our in vitro screen were qualitatively similar to those obtained in vivo, supporting the predictive utility of this screen. A comparison of the screen results for the NCEs and data previously obtained from in vivo or from fresh hepatocyte in vitro studies was made, and the results for all compounds analyzed either matched or exceeded the results (i.e. more conjugates formed) from the previously analyzed studies. Tables 1 and 2, illustrating rat and monkey in vitro screens, respectively, indicate the presence (or absence) of the conjugated metabolites from the screening incubations.

The targeted metabolic profile of the commercially available, non-Schering compounds agreed with literature results (published), and the results for the seven Schering NCEs qualitatively agreed with either in vivo data and/or fresh hepatocytes in vitro data. SCH 1 formed eight glutathione conjugates, exceeding the five seen in vivo.

SCH X not only formed the metabolites observed in cynomolgus monkey and rat in vivo studies, but also formed a rat acylglucoside that had not been observed in vivo. Overall, more glutathiones were observed for the NCEs tested in our in vitro

Table 1.
Results of rat in vitro screens

Compound	Glutathione	Acylglucuronide	Acylglucoside	*N*-Acetyl-glucosaminide
Acetaminophen	Yes	—	—	—
Diclofenac	Yes	Yes	No	No
Furosemide	—	Yes	No	No
Mycophenolic acid	—	Yes	No	No
Ursodeoxycholic acid	—	Yes	No	No
SCH X	—	Yes	Yes	No
SCH Y	—	No	No	No
SCH 1	Yes	—	—	—
SCH 2	Yes	—	—	—
SCH 3	Yes	—	—	—
SCH 4	Yes	—	—	—
SCH 5	Yes	—	—	—

Note: Yes: metabolite observed in screen; No: metabolite not observed; —: no experiment performed.

Table 2.
Results of cynomolgus monkey in vitro screens

Compound	Glutathione	Acylglucuronide	Acylglucoside	*N*-Acetyl-glucosaminide
Acetaminophen	Yes	—	—	—
Diclofenac	Yes	Yes	No	No
Furosemide	—	Yes	Yes	Yes
Mycophenolic acid	—	Yes	Yes	Yes
Ursodeoxycholic acid	—	Yes	Yes*	Yes
SCH X	—	Yes	Yes	Yes
SCH Y	—	Yes	No	No
SCH 1	Yes	—	—	—
SCH 2	Yes	—	—	—
SCH 3	Yes	—	—	—
SCH 4	Yes	—	—	—
SCH 5	Yes	—	—	—

Note: Yes: metabolite observed in screen; No: metabolite not observed; —: no experiment run; *: site of attachment not determined, possibly *O*-glucoside.

screen than had been previously observed either in vivo or in fresh hepatocytes in vitro (data not shown).

In monkey in vivo experiments performed in our lab, acylglucoside and *N*-acetylglucosaminide metabolites have been observed on rare occasions (data not shown). Possibly due to the limited database generated in our lab to date, the formation of these metabolites appears to be highly compound-specific.

Figures 4–8 show typical results from the screen, where the mass spectra show the transitions that are characteristic of the different conjugations (i.e. loss of 129 Da for glutathiones, loss of 176 Da for glucuronides, loss of 162 Da for glucosides, and loss of 203 Da for *N*-acetylglucosaminides), along with characteristic spectra of the parent compounds in MS/MS mode.

Figure 4 shows the MS/MS spectra of acetaminophen and a glutathione conjugate metabolite from the in vitro monkey glutathione screen. Acetaminophen was used as a positive control for the glutathione in vitro screen and has an MH+ *m/z* value of 152 Da. The glutathione conjugate has an MH+ *m/z* value of 457 Da. The MS/MS spectrum of the glutathione conjugate shows the characteristic transition of MH+ to (MH+ −129) that is considered diagnostic for the detection of glutathiones. This spectrum also shows a transition corresponding to the loss of 75 Da that is characteristic of glutathiones, and is also used as a diagnostic ion to confirm the presence of a glutathione conjugate.

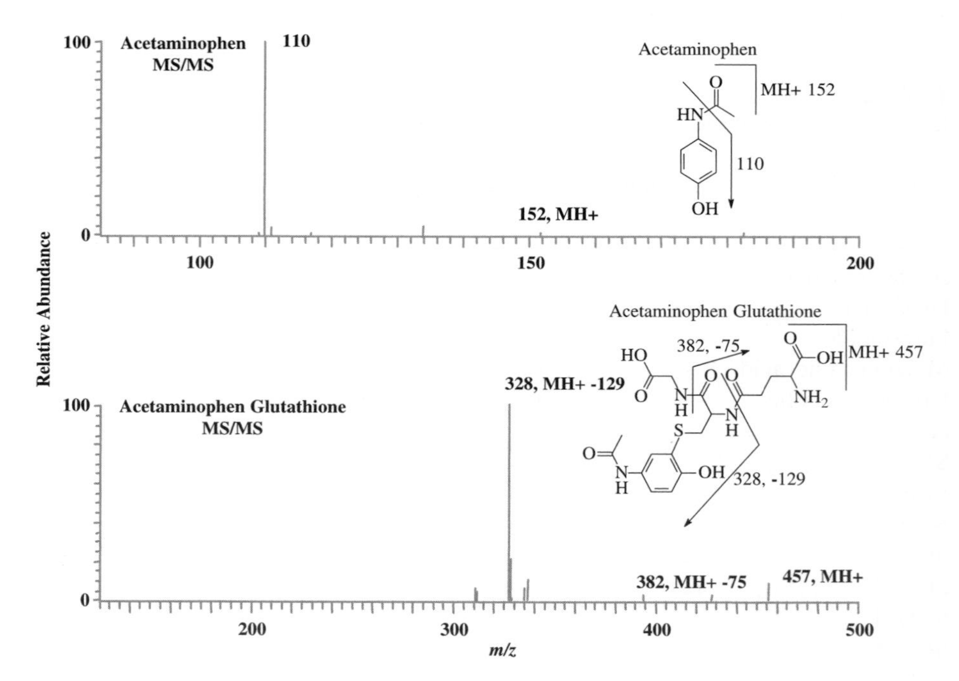

Figure 4.
MS/MS of acetaminophen and acetaminophen glutathione

Figure 5 shows the extracted ion chromatogram (XIC) of the observed glutathione conjugates of SCH 1 from the in vitro monkey glutathione screen. SCH 1 formed 8 chromatographically distinct glutathione conjugates in this screen. Two types of glutathiones were observed previously in vivo and were targeted in this assay. The first type observed was a conjugation of glutathione to the parent with an observed mass increase of 305 Da (Type 1 in Fig. 5). The second type involved a demethylation and subsequent conjugation of glutathione with an observed mass increase of 291 Da (Type 2 in Fig. 5).

The use of a triple quadrupole mass spectrometer is ideal for this kind of screening. Whereas on an ion trap mass spectrometer, the analyst would have to meticulously interrogate the full scan data to detect possible glutathione conjugates, the use of a triple quadrupole instrument operating in the CNL mode, scanning for the characteristic neutral loss of 129 Da, should indicate the presence of any glutathione conjugate metabolites in one experiment, without requiring any prior knowledge of the compound's structure.

Furosemide was used as a positive control for the acylglucuronide, acylglucoside and *N*-acetylglucosaminide in vitro screens. Figures 6–8 show the mass spectral data from the monkey positive control incubations.

The MS/MS spectra of furosemide and furosemide acylglucuronide and the MS^3 spectrum of the furosemide acylglucuronide from the positive control from the monkey acylglucuronide screen are shown in Fig. 6. Furosemide has an MH+ *m/z* value of 329 Da. The acylglucuronide conjugate has an MH+ *m/z* value of 505 Da. The

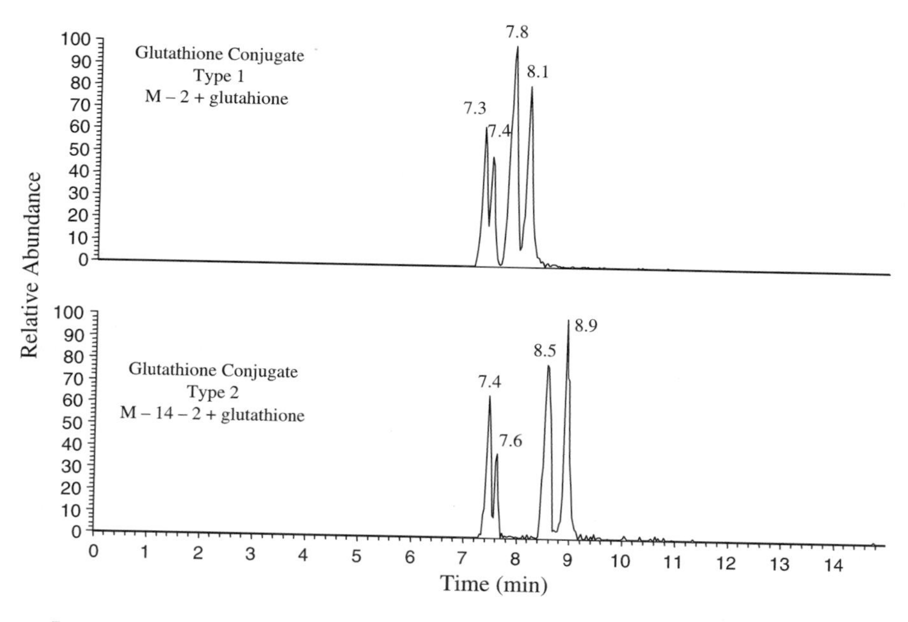

Figure 5.
XIC of two types of glutathione conjugates formed by SCH 1

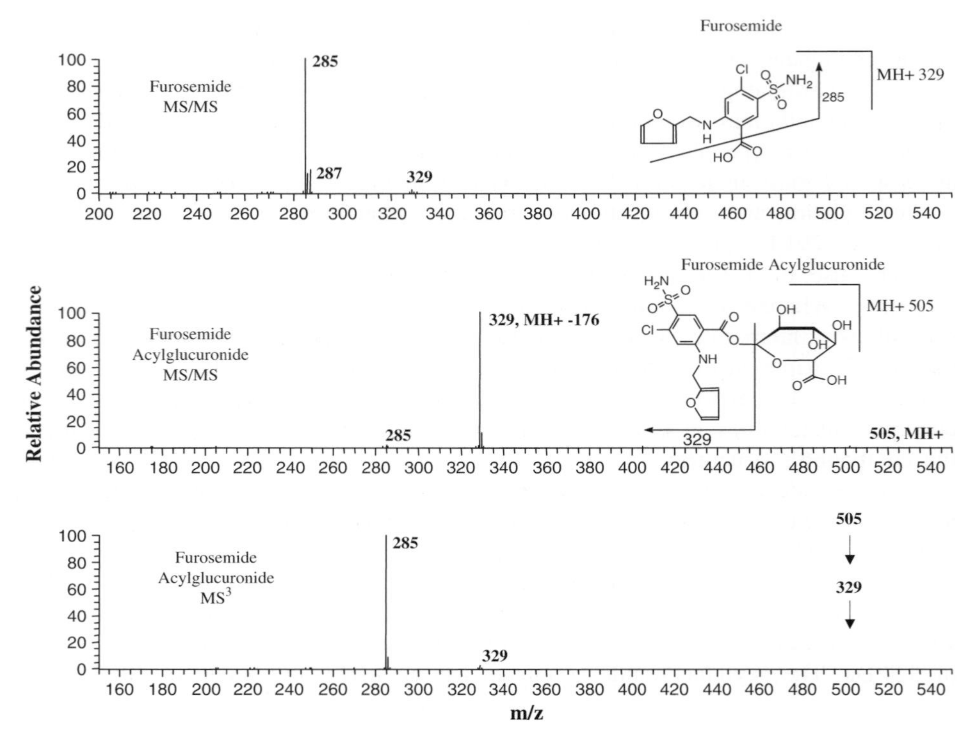

Figure 6.
MS/MS of furosemide and MS/MS and MS^3 of furosemide acylglucuronide

MS/MS spectrum of the acylglucuronide conjugate (middle panel) shows the characteristic transition of MH+ to (MH+ −176) that is considered diagnostic for the detection of glucuronides. Further fragmentation of the ion corresponding to the aglycone (MH+ −176) results in a product ion characteristic of the starting material (bottom panel), further confirming the identity of the metabolite as an acylglucuronide conjugate.

The MS/MS spectra of furosemide and furosemide acylglucoside and the MS^3 spectrum of the furosemide acylglucoside from the positive control from the monkey acylglucoside incubation are shown in Fig. 7. The acylglucoside conjugate has an MH+ *m/z* value of 491 Da. The MS/MS spectrum of the acylglucoside conjugate (middle panel) shows the characteristic transition of MH+ to (MH+ − 162) that is considered diagnostic for the detection of glucosides. The MS^3 spectrum (bottom panel) shows the same fragmentation of the aglycone as the parent, further confirming the identity of the acylglucoside conjugate.

Figure 8 shows the MS/MS spectrometric data obtained from the furosemide *N*-acetylglucosaminide positive monkey control for the monkey *N*-acetylglucosaminide screen. The top panel displays the MS/MS spectrum of the parent compound and the middle panel displays the MS/MS spectrum of the *N*-acetylglucosaminide conjugate.

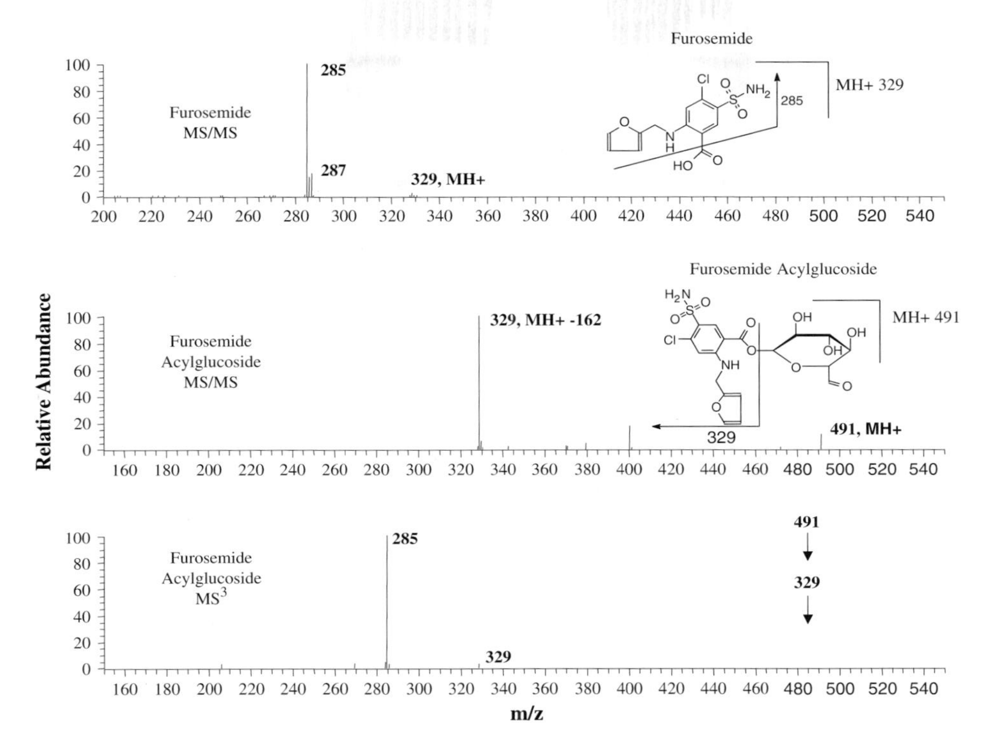

Figure 7.
MS/MS of furosemide and MS/MS and MS^3 of furosemide acylglucoside

The conjugate has an MH+ *m/z* value of 532 Da and fragments in a characteristic manner by loss of 203 Da. Additional fragmentation of the (MH+ −203) ion (bottom panel) results in a product ion characteristic of the starting material.

7.4. Conclusions

We have successfully developed an in vitro screening methodology that allows for the early identification of potential compound metabolic liabilities, for chosen NCEs or for structural series, in the form of specific electrophilic metabolite generation. Such metabolites were trapped as glutathione, acylglucuronide, acylglucoside, and *N*-acetylglucosaminide conjugates through the use of hepatic microsomal and S9 incubations. Multi-stage mass spectrometric detection was employed, focusing on characteristic fragmentation patterns for the individual types of conjugates. The incubation conditions were optimized, and the results were compared to known literature results for commercial compounds and to in-house in vivo and/or in vitro data for the Schering NCEs.

The goal of this screen is to give the chemists and therapeutic teams early information about potential metabolic liabilities and to show probable outcomes. Since

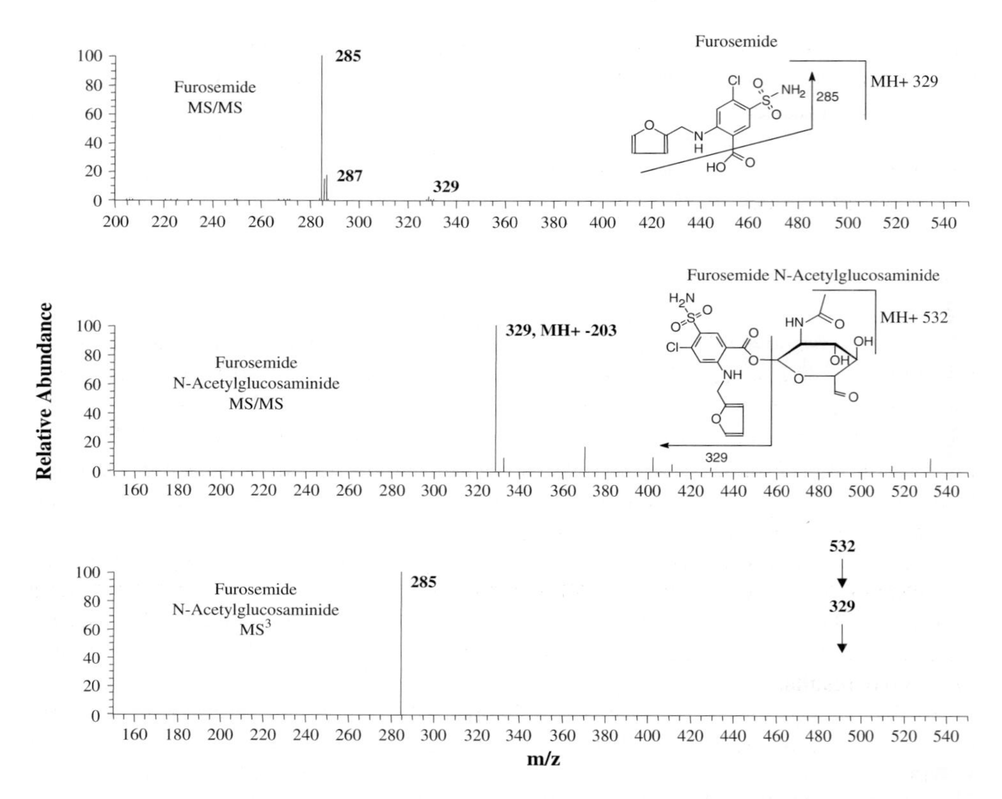

Figure 8.
MS/MS spectrum of furosemide and MS/MS and MS^3 spectra of furosemide *N*-acetylglucosaminide

this is an early screen, false positives are to be expected. What this screen should not do is produce false negatives – the failure to raise red flags when compounds exhibit these types of possible toxic liabilities.

To make this screen effective at assessing the possible formation of electrophilic metabolites, the system was optimized to find conditions that mimic in vivo conditions and/or give the greatest chance of successful metabolism. This optimization could potentially give results not seen in vivo, such as the acylglucoside observed for SCH X in the rat screen, but not in the rat in vivo study. The in vitro result is classed as a false positive. No false negatives have been observed in this screen to date.

As it stands, this system allows for the use of non-radiolabeled compounds, which is important in early Drug Discovery when radiolabeled compounds may not be available. These screens are strictly qualitative and give a "Yes/No" response to the potential of NCEs to form electrophilic metabolites.

These screens are now being incorporated into the metabolite identification strategy utilized within our department. This screen was developed using a Thermo-Finnigan

LCQ ion trap mass spectrometer, but for future screening samples will be analyzed by a triple quadrupole mass spectrometer using constant neutral loss experiments. Since the conjugates of interest give characteristic, well-documented neutral losses by MS/MS, this approach should detect expected conjugated metabolites as well as any unexpected conjugates, without previous in vivo metabolite characterization studies of the parent compounds.

7.5. Future

An extension of this work would be the ability to provide quantitative information regarding the levels of conjugates formed. The use of radiolabeled compounds should provide the information desired. This would be useful for rank ordering compounds within a structural class or therapeutic program. Since a screen is an early test, and in our department generally, only compounds that are considered possible lead candidates are synthesized with a radiolabel, a rank ordering of all compounds in a therapeutic area is not feasible at this point.

Analysis of a radiolabeled lead compound using this methodology, providing quantitative information, would provide historical data to compare with future lead candidates. The success of this methodology as a quantitative screen is dependent on degree and rate of conjugation in vitro, and the correlation of the in vitro results to in vivo results.

This methodology is designed to be a qualitative screen used in early Drug Discovery. This in vitro screen highlights compounds with respect to their propensity to form electrophilic metabolites. It can also highlight a structural series as having a potential liability. It will not, however, determine whether these electrophilic metabolites will be a liability to a drug once it is progressed. The information provided to the Discovery team from this screen adds to the knowledge base created for a therapeutic program. This knowledge base, along with all the other data obtained for a compound within the Discovery phase, will serve as the basis for the decision to progress the compound forward to development for investigation in clinical trials, and to a possible future as a successful drug.

Acknowledgements

Diane Grotz would like to acknowledge the following people at Schering-Plough Research Institute for their willingness to discuss the science and techniques, and for their general support: Len Favreau, Diane Rindgen, and Philip Sherratt.

References

Akira, K., Taira, T., Shinohara, Y., 1997. Direct detection of the internal acyl migration reactions of benzoic acid 1-*O*-acylglucuronide by ^{13}C-labeling nuclear magnetic resonance spectroscopy. J. Pharmacol. Toxicol. 37, 237–243.

Axelsson, H., Granhall, C., Floby, E., Jaksch, Y., Svedling, M., Sohlenius-Sternbeck, A.-K., 2003. Rates of metabolism of chlorzoxazone, dextromethorphan, 7-etoxy-coumarin, imipramine, quinidine, testosterone and verapamil by fresh and cryo-preserved rat liver slices, and some comparisons with microsomes. Toxicol. In Vitro 17, 481–488.

Bailey, M., Dickinson, R., 2003. Acyl glucuronide reactivity in perspective: biological consequences. Chem.-Biol. Interact. 145, 117–137.

Baillie, T.A., 2004. Chemically reactive metabolites in drug discovery and development. In: Lee, J.S., Obach, R. S., Fisher, M.B. (Eds.), Drug Metabolizing Enzymes: Cytochrome P450 and Other Enzymes in Drug Discovery and Development, Marcel Dekker: New York, pp. 147–154.

Ballet, F., 1997. Hepatotoxicity in drug development: detection, significance and solutions. J. Hepatol 26 (Suppl. 2), 26–36.

van Bladeren, P., 2001. Glutathione conjugation as a bioactivation reaction. Chem.-Biol. Interact. 129, 61–76.

Boelsterli, U., 2002. Xenobiotic acyl glucuronides and acyl-CoA thioesters as protein-reactive metabolites with the potential to cause idiosyncratic drug reactions. Curr. Drug. Metab. 3, 439–450.

Boelsterli, U., 2003a. Mechanistic toxicology, the molecular basis of how chemicals disrupt biological targets. New York: Taylor & Francis.

Boelsterli, U., 2003b. Diclofenac-induced liver injury: a paradigm of idiosyncratic drug toxicity. Toxicol. Appl. Pharm. 192, 307–322.

Bolze, S., Bromet, N., Gay-Feutry, C., Massiere, F., Boulieu, R., Hulot, T., 2002. Development of an in vitro screening model for the biosynthesis of acyl glucuronide metabolites and the assessment of their reactivity toward human serum albumin. Drug Metab. Dispos. 30, 404–413.

Boyle, C.D., Vice, S., Campion, J., Chackalamannil, S., Lankin, C., McCombie, S., Billard, W., Binch, H., III, Crosby, G., Cohen-Williams, M., Coffin, V., Cox, K., Grotz, D., Duffy, R.A., Ruperto, V., Lachowicz, J., 2002. Enhancement of pharmacokinetic properties and in vivo efficacy of benzylidene ketal M2 muscarinic receptor antagonists via benzamide modification. Bioorg. Med. Chem. Lett. 12, 3479–3482.

Cox, K.A., Dunn-Meynell, K., Korfmacher, W.A., Broske, L., Nomeir, A.A., Lin, C.-C., Cayen, M.N., Barr, W.H., 1999. Novel *in vivo* procedures for rapid pharmacokinetic screening of discovery compounds in rats. Drug Discov. Today 4, 232–237.

De Graaf, I.A.M., van Meijeren, C.E., Pektas, F., Koster, H.J., 2002. Comparison of in vitro preparations for semi-quantitative prediction of in vivo drug metabolism. Drug Metab. Dispos. 30, 1129–1136.

Dekant, W., 2001. Chemical induced nephrotoxicity mediated by glutathione *S*-conjugate formation. Toxicol. Lett. 124, 21–36.

Evans, D.C., Watt, A.P., Nicoll-Griffith, D.A., Baillie, T.A., 2004. Drug-protein adducts: An industry perspective on minimizing the potential for drug bioactivation in drug discovery and development. Chem. Res. Toxicol. 17, 3–16.

Fisher, M.B., Campanale, K., Ackerman, B.L., Vandenbranden, M., Wrighton, S.A., 2000. In vitro glucuronidation using human liver microsomes and the pore-forming peptide alamethicin. Drug Metab. Dispos. 28, 560–566.

Georges, H., Jarecki, I., Netter, P., Magdalou, J., Lapicque, F., 1999. Glycation of human serum albumin by acylglucuronides of nonsteroidal anti-inflammatory drugs of the series of phenylpropionates. Life Sci. 65, (12), 151–156.

Grotz, D.E., Clarke, N.J., Cox, K.A., Korfmacher, W., 2001. An in vitro method for the assessment of acyl glucuronide metabolite formation early in drug discovery using HPLC–MS/MS. ASMS Conference on Mass Spectrometry and Allied Topics, Poster WPQ 322.

Hartman, N.R., Cysyk, R.L., Bruneau-Wack, C., Thenot, J.-P., Parker, R.J., Strong, J.M., 2002. Production of intracellular 35 *S*-glutathione by rat and human hepatocytes for the quantification of xenobiotic reactive intermediates. Chem.-Biol. Interact. 142, 43–55.

Hasegawa, J., Smith, P.C., Benet, L.Z., 1982. Apparent intramolecular acyl migration of zomepirac glucuronide. Drug Metab. Dispos. 10, 469–473.

Hinson, J.A. 1992. Role of covalent and noncovalent interactions in cell toxicity: effects on proteins. Annu. Rev. Pharmacol. Toxicol. 32, 471–510.

James, L.P., Mayeux, P.R., Hinson, J.A., 2003. Acetaminophen-induced hepatotoxicity. Drug Metab. Dispos. 31, 1499–1506.

Lasser, K.E., Allen, P.D., Woolhandler, S.J., Himmelstein, D.U., Wolfe, S.M., Bor, D.H., 2002. Timing of new black box warnings and withdrawals for prescription medications. J. Am. Med. Assoc. 287, 2215–2220.

Lau, Y.Y., Krishna, G., Yumibe, N.P., Grotz, D.E., Sapidou, E., Norton, L., Chu, I., Chen, C., Soares, A.D., Lin, C.-C., 2002. The use of *in vitro* metabolic stability for rapid selection of compounds in early drug based on their expected hepatic extraction ratios. Pharmaceut. Res. 19, 1606–1610.

Manautou, J.E., Hoivik, D.J., Tveit, A., Emaigh Hart, S.G., Khairallah, E.A., Cohen, S.D., 1994. Clofibrate pretreatment diminishes acetaminophen's selective covalent binding and hepatotoxicity. Toxicol. Appl. Pharmacol. 129, 252–263.

Momose, T., Hirata, H., Iida, T., Goto, J., Nambara, T., 1998. Simultaneous analysis and retention behavior of the glucuronide, glucoside, and *N*-acetylglucosaminide conjugates of bile acids in conventional and inclusion high-performance liquid chromatographic methods. J. Chromatogr. A 803, 121–129.

Monks, T.J., Lau, S.S., 1988. Reactive intermediates and their toxicological significance. Toxicology 52, 1–53.

Pharmaceutical Research and Manufacturers of America, 2002. The drug development and approval process, New Medicines in Development for Biotechnology: 2002 Survey. (http://www.phrma.org/newmedicines/resources/2002-10-21.93.pdf)

Poon, G.K., Chen, Q., Teffera, Y., Ngui, J.S., Griffin, P.R., Braun, M.P., Doss, G.A., Freeden, C., Stearns, A., Evans, D.C., Baillie, T.A., Tang, W., 2001. Bioactivation of diclofenac via benzoquinone imine intermediates – identification of urinary mercapturic acid derivatives in rats humans. Drug Metab. Dispos. 29, 1608–1613.

Qiu, Y., Burlingame, A.L., Benet, L.Z., 1998. Mechanisms for covalent binding of benoxaprofen glucuronide to human serum albumin: studies by mass spectrometry. Drug Metab. Dispos. 26, 246–256.

Rinaldi, R., Eliasson, E., Swedmark, S., Morgenstern, R., 2002. Reactive intermediates and the dynamics of glutathione transferases. Drug Metab. Dispos. 30, 1053–1058.

Seitz, S., Boelsterli, U., 1998. Diclofenac acyl glucuronide, a major biliary metabolite, is directly involved in small intestinal injury in rats. Gastroenterology 115, 1476–1482.

Sherratt, P.J., Hayes, J.D., 2002. Glutathione *S*-transferases. In: Ioannides, C. (Ed.), Enzyme Systems That Metabolise Drugs and Other Xenobiotics. John Wiley & Sons, UK, 319–352.

Sherratt, P.J., Williams, S., Foster, J., Kernohan, N., Green, T., Hayes, J.D., 2002. Direct comparison of the nature of mouse and human GST T1-1 and the implications on dichloromethane carcinogenicity. Toxicol. Appl. Pharmacol. 179, 89–97.

Shipkova, M., Strassburg, C.P., Braun, F., Streit, F., Grone, H.J., Armstrong, V.W., Tukey, R.H., Oellerich, M., Wieland, E., 2001. Glucuronide and glucoside conjugation of mycophenolic acid by human liver, kidney and intestinal microsomes. Br. J. Pharmacol. 132, 1027–1034.

Smith, P. C., Hasegawa, J., Langendijk, P.M.N., Benet, L.Z., 1985. Stability of acyl glucuronides in blood, plasma, and urine: studies with zomepirac. Drug Metab. Dispos. 13, 110–112.

Tang, C., Hochman, J.H., Ma, B., Subramanian, R., Vyas, K.P., 2003. Acylglucuronidation and glucosidation of a new and selective endothelin ETA receptor antagonist in human liver microsomes. Drug Metab. Dispos. 31, 37–45.

Terrier, N., Benoit, E., Senay, C., Lapicque, F., Radominska-Pandya, A., Magdalou, J., Fournel-Gigleux, S., 1999. Human and rat liver UDP-glucuronosyltransferases are targets of ketoprofen acylglucuronide. Mol. Pharmacol. 56, 226–234.

Tolman, K. G., 1998. Hepatotoxicity of non-narcotic analgesics. The Am. J. Med. 105, 13S–19S.

Wang, M., Dickinson, R.G., 2000. Bile duct ligation promotes covalent drug-protein adduct formation in plasma but not in liver of rats given zomepirac. Life Sci. 68, 525–537.

Wang, M., Gorrell, M.D., McCaughan, G.W., Dickinson, R.B., 2001. Dipeptidyl peptidase IV is a target for covalent adduct formation with the acyl glucuronide metabolite of the anti-inflammatory drug zomepirac. Life Sci. 68, 785–797.

Identification and Quantification of Drugs, Metabolites and Metabolizing
Enzymes by LC–MS
Swapan K. Chowdhury, editor.

Chapter 8

LIQUID CHROMATOGRAPHY TANDEM MASS SPECTROMETRY-BASED METABOLITE IDENTIFICATION STRATEGIES IN PHARMACEUTICAL RESEARCH

M. Reza Anari, Philip R. Tiller, and Thomas A. Baillie

8.1. Introduction

The assessment of the metabolic fate of drug candidates, knowledge of the routes of metabolism in animals and humans, and evaluation of the biological properties of metabolites represent important objectives in contemporary pharmaceutical research (Baillie *et al.*, 2002). Metabolite identification investigations are imperative to support various phases of pharmaceutical research, i.e., drug discovery, preclinical development, and clinical development (Fig. 1). At each phase of drug evolution, the goal of a particular metabolite identification study has to be recognized and addressed using suitable strategies and appropriate resources.

In the drug discovery stage, the major goal is to locate substructural moieties that are vulnerable to high rate of intrinsic metabolic clearance, resulting in short, effective half-life and low oral bioavailability. In the later stage of drug development, the goal is to compare metabolic profiles of human in vitro systems with those of the animal species used in safety assessment. In the late stages of preclinical and clinical drug developments, the goal of metabolite identification is to unambiguously identify all metabolites in animal species and humans, and communicate these detailed studies to regulatory agencies. Thus, the objectives of metabolite identification in early preclinical studies require higher throughput analytical approaches to support medicinal chemistry efforts in designing superior drug-like molecules. On the other hand, the objectives in the late developmental stage require more detailed comprehensive studies with particular emphasis on the quality of the data that has to be submitted to the regulatory agencies. Although it is possible to use the latter detailed characterization approach in early drug discovery, the urgent need for drug metabolism data at the discovery phase does not warrant the resources required for such detailed characterization. By the same token, high throughput approaches will not be suitable to provide a complete understanding of metabolic pathways and their potential liabilities. Such data may raise serious safety concerns and could delay drug registration with regulatory agencies.

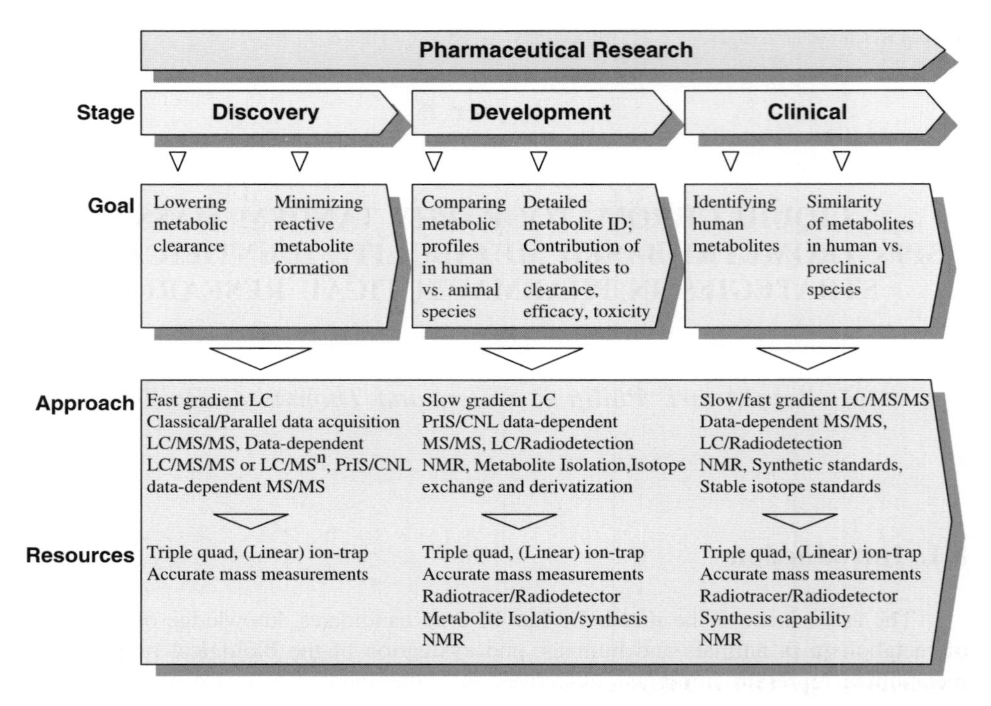

Figure 1.
Roadmap of the metabolite identification strategic focus in pharmaceutical research

It is the authors' intention in this chapter to highlight the desired goal of metabolite identification studies at each stage of the evolution of a new chemical entity (NCE), and discuss the appropriate strategy and resources essential to effectively address each goal. The advent of new technologies in the area of mass spectrometry and bioinformatics has created significant opportunities to improve both throughput and quality of information obtained in support of development of novel therapeutic agents. The importance of balancing throughput and the quality of data is discussed in this chapter for each paradigm, with an emphasis on the application of the best strategy and appropriate resources.

8.2. Metabolite Identification in Drug Discovery

An early assessment of metabolic pathways provides valuable information in the context of drug discovery efforts, in which metabolic "soft spots" that lead to high clearance may be recognized promptly (Sinz and Podoll, 2002; Watt *et al.*, 2003). The knowledge of biotransformation pathways is essential to the design of metabolically stable leads with acceptable pharmacokinetic characteristics. Several literature reviews have highlighted the successful design of NCEs with improved pharmacokinetic properties based on available drug metabolism information (Lin, 2002; Lin *et al.*, 1999; Smith, 1994; Smith *et al.*, 1996). Generating such information for an NCE and communicating the data promptly to the discovery team is vital in designing new leads and/or backup drugs with optimum exposure, biological half-life, and bioavailability.

In addition, early information on the structures of metabolites can guide efforts aimed at reducing a compound's propensity for metabolic activation. Thus, concerns over the exposure of biological systems to chemically reactive intermediates, which may react with cellular macromolecules leading to cytotoxic, apoptotic, or idiosyncratic responses (Baillie, 2003; Baillie and Kassahun, 2001; Evans *et al.*, 2004; Gillette *et al.*, 1974; Uetrecht, 1997; Williams *et al.*, 2002), have led to efforts to screen out such species early in the discovery phase (Evans *et al.*, 2004; Kassahun *et al.*, 2001). To successfully eliminate those drug candidates that lead to appreciable levels of irreversibly bound protein adducts, one has to understand the nature of the reactive metabolites as well as the mechanism of the biotransformation reactions involved in their formation (Anari *et al.*, 1997; Chen *et al.*, 2002; Hecht, 1996; Maggs *et al.*, 2000; Samuel *et al.*, 2003; Stevens *et al.*, 1997; Tang *et al.*, 1999; Tingle *et al.*, 1995).

8.3. Classical Mass Spectrometry-Based Metabolite Identification Approach in Drug Discovery

The objective of early drug discovery mandates a quick metabolite identification approach in order to find the major metabolic pathways involved in clearance or bioactivation of the NCE. The analyst has to rely mainly on the appearance of pseudo-molecular ions, ($[M+H]^+$ or $[M-H]^-$), for the detection of drug-related materials in biological samples since radiolabelled compounds (^{14}C, ^{3}H) usually are not available to facilitate the detection of metabolites at this early stage. The classical metabolite identification approach emerged sometime ago when the first commercial triple quadrupole mass spectrometers became available. A common procedure includes analysis of the test and control samples in parallel with screening over an appropriate *m/z* range followed by comparison of the reconstructed ion chromatograms (RIC) of potential metabolites in the test samples versus control (Fig. 2) (Anari and Baillie, 2005; Mutlib and Shockcor, 2003; Sinz and Podoll, 2002). The objective at this step is to determine the *m/z* value of all drug-related materials that are present only in the test sample. The next step includes product ion scan (PIS) or multistage PIS (MS^n) experiments on the *m/z* of the suspected metabolite. The PIS experiment in tandem mass spectrometry plays a critical role in metabolite identification, since comparison of the spectrum of product ions of metabolites with that of the parent drug is essential for the identification of metabolite structure (Clarke *et al.*, 2001). Since the product ion spectra are obtained in the final step following full-scan MS analysis (Fig. 2), this approach suffers from iterative analysis and low throughput, which often requires large amounts of samples and significant user intervention (Anari and Baillie, 2005; Clarke *et al.*, 2001).

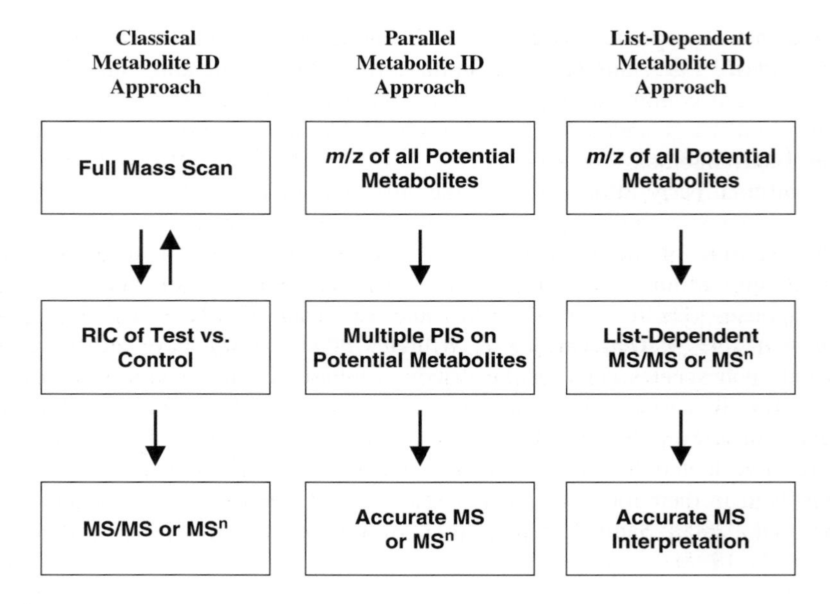

Figure 2.
Comparison of various tandem mass spectrometry-based metabolite ID approaches used in drug discovery

8.4. Parallel Mass Spectrometry-Based Metabolite Identification Approach in Drug Discovery

The low sample throughput and problems associated with interference from biological matrices have compromised the widespread utility of classical metabolite identification strategy in early discovery (Anari and Baillie, 2005). Direct PIS experiments on parent ions of potential metabolites have been suggested to expedite the process of obtaining useful structural information that could guide medicinal chemists in the early stage of discovery (Anari and Baillie, 2005; Clarke *et al.*, 2001). To enhance throughput further, multiple MS/MS experiments may be carried out within a single LC/MS/MS cycle (Clarke *et al.*, 2001), which affords fragment ion data for more than one expected metabolite at a time (Fig. 2).

In this so-called "parallel data acquisition strategy", knowledge of xenobiotic biotransformation is essential to determine the mass-to-charge ratio of potential metabolite parent ions (Anari and Baillie, 2005). Since many NCEs under investigation in drug discovery are structurally similar, the biotransformation pathways determined for previous lead compounds can help guide the analysis of the metabolites formed via similar metabolic routes (Clarke *et al.*, 2001). We have presented a comprehensive list of biotransformation-associated mass shifts on various functional groups, which could be used to facilitate the determination of the *m/z* of potential metabolite MH^+ ions resulting from a single reaction or multistep metabolic pathways (Anari and Baillie, 2005; Anari *et al.*, 2004). Potential metabolic reactions also may be predicted via knowledge-based cheminformatics approaches, which were discussed in greater detail elsewhere (Anari and Baillie, 2005).

8.5. Integrated List-Dependent Fast LC/MS^n in Identifying Metabolic Soft Spots

Historically, intelligent data-dependent scripts have been utilized to enhance the throughput of metabolite identification investigations in drug discovery allowing rapid acquisition of multiple complementary mass scan data for each metabolite "on the fly" (Fernandez-Metzler *et al.*, 1999; Gu and Lim, 2001; Yu *et al.*, 1999). Such data-dependent functionalities have now been incorporated into most data acquisition software packages, e.g., Xcalibur (Thermo Finnigan), Analyst QS (Applied Biosystems), and MetaboLynx (Waters Corporation). This is of significant interest in discovery absorption, distribution, metabolism, and excretion (ADME) studies as both survey and PISs could be obtained in a single analysis for the major metabolites, often associated with metabolic clearance. Thus, both throughput and the amount of qualitative mass spectral information could be maximized for each LC/MS analysis (Anari *et al.*, 2004; Lim *et al.*, 1999; Lopez *et al.*, 1998; Nassar and Adams, 2003; Tiller *et al.*, 1998; Xia *et al.*, 2003).

In applying such data-dependent technology in metabolism studies of biological samples in preclinical research, it has been the author's experience that a directed search for metabolites that have been predicted on the basis of the structure of the parent compound offers the greatest probability of success in detecting drug-related materials (Fig. 2) (Anari and Baillie, 2005; Anari *et al.*, 2004; Fernandez-Metzler *et al.*, 1999). The main reason for the lower chances of success in the identification of metabolites by a full data-dependent approach is the presence of highly intense matrix ion signals, which inadvertently trigger MS/MS or MS^n scan functions. We often obtained the product ion spectra of non-drug-related matrix components as well as unchanged parent drug, when full, automatic data-dependent acquisition was applied in a search for metabolites of a given NCE in biological samples (Anari *et al.*, 2004). This is particularly the case for samples with a high level of endogenous components such as bile, plasma, faeces, and crude post-mitochondrial hepatic preparations (Anari and Baillie, 2005; Anari *et al.*, 2004). To address this issue, use of list-dependent data acquisition (i.e., a targeted analysis based on the list of parent ions of potential metabolites) has been shown to provide product ion spectra of many drug-related materials within a single LC run (Fig. 2), without any interference from overwhelming matrix signals (Anari and Baillie, 2005; Anari *et al.*, 2004; Clarke *et al.*, 2001; Lopez *et al.*, 1998). For example, an approach based on the integration of knowledge-based metabolic prediction with list-dependent tandem mass spectrometry has been demonstrated to improve quality, throughput, and detailed qualitative information obtained on the metabolism of the human immunodeficiency virus (HIV) protease inhibitor indinavir (Anari and Baillie, 2005; Anari *et al.*, 2004; Lopez *et al.*, 1998).

If the compound of interest contains Cl or Br, it is possible to utilize the characteristic isotope clusters to reduce the impact of matrix interference in a way similar to that observed by using list-dependent data acquisitions. This is accomplished by using the ratio of the two halogen isotope signals (M:M+2 of 3:1 for Cl and M: M+2 of 1:1 for Br) as a filter in the data-dependent method. This approach has been demonstrated to dramatically reduce the MS/MS data derived, in error, from matrix signals (Baillie *et al.*, 1989; Drexler *et al.*, 1998).

8.6. Metabolite Identification in Preclinical Development

In the early stage of preclinical drug development, comparative metabolic profiling is performed using in vitro systems from preclinical species and humans to support the selection of the animal species employed in safety assessment studies (Baillie *et al.*, 2002). The qualitative similarity between in vitro metabolic "profiles", typically obtained from incubations with liver preparations from animal species versus humans, represents an important criterion in the choice of non-rodent species for preclinical safety testing (Baillie *et al.*, 2002; F.D.A., 2002). Such information provides assurance that metabolites likely to be formed in humans also are being formed in preclinical species (Baillie *et al.*, 2002).

Following the selection of safety species, the detailed assessment of the metabolic fate of a new drug candidate and knowledge of the routes of metabolism in preclinical species represent the major goals of ADME studies at the preclinical stage (Baillie *et al.*, 2002). This is often carried out by in-depth investigation of the biotransformation of the lead compound in rats and another non-rodent species. The quality of data and supporting documentation is of significant importance as the results of such investigations are audited and submitted to regulatory agencies. The use of a radiotracer mixed with cold drug helps to quantitatively determine the level and route of excretion of each metabolite in animals. At the same time, comparison of the radiochromatogram with mass chromatograms helps to identify metabolite structures that are poorly ionized or not detected by MS-based techniques. For major metabolites, isolation and detailed structural elucidation by NMR is often carried out, followed by synthesis of the metabolites and confirmation of the structures using LC/MS techniques.

Evaluation of the biological activity of a synthesized metabolite may provide additional information that is used to design backup compounds with enhanced potency, and improved metabolic stability and pharmacokinetic profile (Watt *et al.*, 2003).

8.7. Identification of Unanticipated Metabolites Using Data-Dependent Precursor Ion and Constant Neutral Loss Scans

Based on the authors' experiences with a number of preclinical drug candidates, the integrated metabolite identification strategy discussed above is able to reliably and rapidly identify metabolites of more than 90% of NCEs within a single LC/MS analysis (Anari and Baillie, 2005; Anari *et al.*, 2004; Tiller *et al.*, 1998). However, when using radiochromatographic profiles, it is not uncommon to find unexpected metabolites whose mass-to-charge ratios have not been anticipated (Anari and Baillie, 2005). These metabolites often are formed by extensive metabolism of the parent drug via multiple enzymatic reactions in addition to non-enzymatic processes. The tandem mass spectrometry approach to revealing the molecular masses of unexpected metabolites has focused on the use of precursor ion scan (PrIS) or constant neutral loss (CNL) experiments (Clarke *et al.*, 2001; Jemal *et al.*, 2003; Liu *et al.*, 2004; Xia *et al.*, 2003; Zhang *et al.*, 2000). As discussed earlier for the classical metabolite identification strategy, these additional experiments performed on triple quadrupole mass spectrometers often require iterative analysis in order to identify the pseudo-molecular ions and confirm the presence of drug-related materials by separate MS/MS experiments. However, with the availability of advanced linear ion trap and quadrupole tandem mass spectrometers, a new systematic approach may be envisaged in the search for the presence of both expected and unexpected metabolites based on data-dependent PrIS or data-dependent CNL (Jemal *et al.*, 2003). The fragment ions and neutral losses that are complementary in a molecule are identified in advance in order to conduct an effective data-dependent PrIS or CNL experiment. The ions detected from such survey scans are used to automatically trigger the product ion MS/MS experiments.

8.8. Detailed Characterization of Metabolites Using Radiochromatographic Techniques Coupled to LC/MS

Identification of metabolites using "cold" material always leaves the possibility of unexpected or unusual metabolites being formed that are not detectable using the mass spectrometry-based approaches discussed previously. Use of radiolabelled drugs provides greater confidence that all significant metabolites have been identified, though there is the risk that the radiolabel itself may be lost due to the biotransformation process. The loss of the label can be minimized by using the metabolite profiling data obtained, in the early drug discovery phase, on the "cold" compound to guide the choice as to the location of the radioactive nuclide within the structure of the drug. The LC–MS experiments utilizing radiolabelled material affords both radiochromatic and mass spectral data that aid in identifying and quantifying the metabolites and their contribution to elimination pathways (Martin *et al.*, 2003; van sen Bongard *et al.*, 2002).

As discussed earlier, the radiochromatograms occasionally contain a peak(s) at a retention time that does not correspond to any previously identified metabolite using LC–MS detection and unlabelled drug. This is usually due to unexpected metabolites being formed that are often more polar than the parent drug, and therefore elute in the early part of the chromatogram where matrix interferences can mask their presence in the LC–MS experiments. Isolation and identification of these metabolites using other complementary techniques, such as NMR spectroscopy are commonly carried out at a later stage of drug development (Mutlib and Shockcor, 2003; Mutlib *et al.*, 1995).

The LC–MS data always are influenced by differences in the ionization efficiency of parent drug and metabolites, and are not inherently quantitative. In contrast, the use of radiochromatograms to quantify metabolites is straightforward as the radio-detector response is linear with the amount of radiolabel present. Thus, unless authentic standards are available, the radiochromatographic data are frequently used to quantify the levels of particular metabolites (Nassar and Adams, 2003), which necessitates the use of appropriate chromatography to unambiguously separate these species.

8.9. Metabolite Identification in Clinical Drug Development

At the clinical stage of drug development, human metabolism studies provide similar information on routes of elimination, mass balance, and metabolic profiles in plasma and urine. Taken together, the results of preclinical and clinical ADME studies allow direct comparisons to be made of (a) the pathways of biotransformation followed by the drug candidate in the safety species and humans and (b) the qualitative similarity of the metabolic profiles in plasma (and/or other relevant matrices) of animals versus humans (Baillie *et al.*, 2002). The latter is the key point, in that it addresses directly the question raised above of whether human plasma contains any unique metabolites to which the safety species are not being exposed systemically (Baillie *et al.*, 2002; F.D.A., 2002; Hastings *et al.*, 2003).

8.10. Detailed Characterization of Metabolites in Humans

The approach taken to characterize drug metabolism in detail in human subjects is not appreciably different from that taken in the later stages of preclinical drug development (Martin *et al.*, 2003; Mutlib and Shockcor, 2003; van sen Bongard *et al.*, 2002). Thus, the use of radio-tracers, together with in-depth LC–MS/MS and LC–NMR analyses, serve to establish the presence or absence of "unique" human metabolites, which may have implications for safety assessment (Baillie *et al.*, 2002).

Because much would already be known about the metabolism of the drug in question at this point in its development, the masses of the previously observed metabolites can be used to guide list-dependent analyses, while knowledge of the MS/MS behaviour of these metabolites allows for very specific application of precursor ion-scanning approaches. These mass spectrometric techniques, in combination with the use of radio-tracers, will enable the researcher to determine the presence or absence of unique human metabolites.

8.11. Quantitative Similarity of the Circulating Metabolites in Human versus Preclinical Species

Having determined the metabolite profile in humans, the next question to be addressed is how similar that profile is to those obtained from the preclinical species. It is not merely sufficient to determine that the same metabolites are present, but it is necessary to demonstrate that the metabolites circulating in blood in humans are present in the preclinical species. There are two main reasons for this comparison (Baille *et al.*, 2002). The first reason is to provide validation to the toxicological studies that have taken place or are going to take place, providing assurance that any

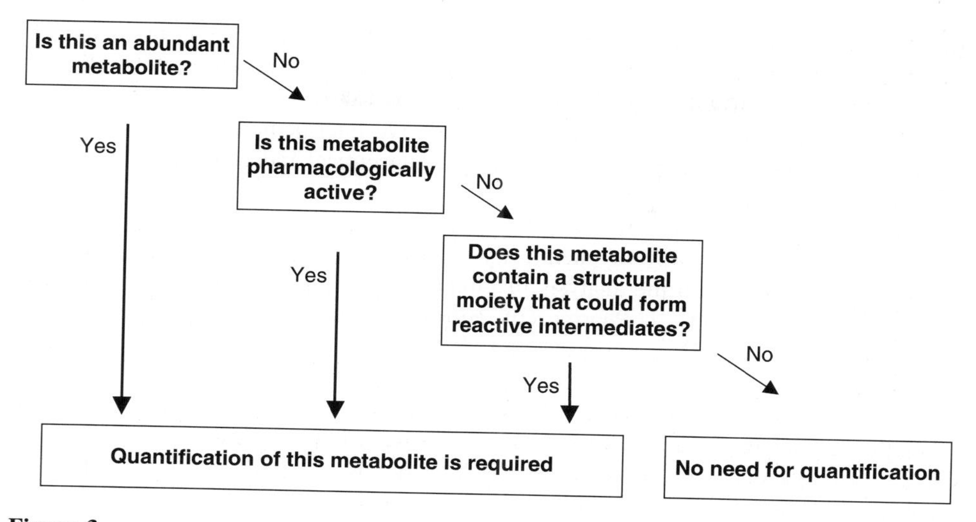

Figure 3.
Decision tree to determine the need for quantification of a metabolite observed in vivo in human

metabolites present in humans have been included in the toxicological assessment of the drug. The second reason is to determine if a metabolite is present in the systemic circulation at a significant concentration, in which case the metabolite itself may need to be characterized in terms of its pharmacological activity and pharmacokinetic profile. The decision tree to determine whether quantification of a human metabolite is necessary is shown in Fig. 3. A discussion on the need to quantify such circulating metabolites has been presented elsewhere (Baillie *et al.*, 2002).

References

Anari, M.R., Baillie, T.A., 2005. Bridging cheminformatic metabolite prediction and tandem mass spectrometry. Drug Discov. Today 15(10), 711–717.

Anari, M.R., Josephy, P.D., Henry, T., O'Brien, P.J., 1997. Hydrogen peroxide supports human and rat cytochrome P450 1A2-catalyzed 2-amino-3-methylimidazo[4,5-f] quinoline bioactivation to mutagenic metabolites: significance of cytochrome P450 peroxygenase. Chem. Res. Toxicol. 10(5), 582–588.

Anari, M.R., Sanchez, R.I., Bakhtiar, R., Franklin, R.B., Baillie, T.A., 2004. Integration of knowledge-based metabolic predictions with liquid chromatography data-dependent tandem mass spectrometry for drug metabolism studies: application to studies on the biotransformation of indinavir. Anal. Chem. 76, 823–832.

Baillie, T.A., 2003. Chemically reactive metabolites in drug discovery and development. In: Lee, J.S., Obach, R.S., Fisher, M.B. (Eds.), Drug Metabolizing Enzymes: Cytochrome P450 and Other Enzymes in Drug Discovery and Development, 1st ed. Marcel Dekker Inc, New York, pp. 147–154.

Baillie, T.A., Cayen, M.N., Fouda, H., Gerson, R.J., Green, J.D., Grossman, S.J., Klunk, L.J., LeBlanc, B., Perkins, D.G., Shipley, L.A., 2002. Drug metabolites in safety testing. Toxicol. Appl. Pharmacol. 182(3), 188–196.

Baillie, T.A., Kassahun, K., 2001. Biological reactive intermediates in drug discovery and development: a perspective from the pharmaceutical industry. Adv. Exp. Med. Biol. 500, 45–51.

Baillie, T.A., Pearson, P.G., Rashed, M.S., Howald, W.N., 1989. The use of mass spectrometry in the study of chemically-reactive drug metabolites. Application of MS/MS and LC/MS to the analysis of glutathione- and related *S*-linked conjugates of *N*-methylformamide. J. Pharmaceut. Biomed. Anal. 7(12), 1351–1360.

Chen, Q., Ngui, J.S., Doss, G.A., Wang, R.W., Cai, X., DiNinno, F.P., Blizzard, T.A., Hammond, M.L., Stearns, R.A., Evans, D.C., Baillie, T.A., Tang, W., 2002. Cytochrome P450 3A4-mediated bioactivation of raloxifene: irreversible enzyme inhibition and thiol adduct formation. Chem. Res. Toxicol. 15(7), 907–914.

Clarke, N.J., Rindgen, D., Korfmacher, W.A., Cox, K.A., 2001. Systematic LC/MS metabolite identification in drug discovery. Anal. Chem. 73(15), 430A–439A.

Drexler, D.M., Tiller, P.R., Wilbert, S.M., Bramble, F.Q., Schwartz, J.C., 1998. Automated identification of isotopically labeled pesticides and metabolites by intelligent real time liquid chromatography tandem mass spectrometry using a bench-top ion trap mass spectrometer. Rapid Commun. Mass Spectrom. 12(13), 1501–1507.

Evans, D.C., Watt, A.P., Nicoll-Griffith, D.A., Baillie, T.A., 2004. Drug-protein adducts: an industry perspective on minimizing the potential for drug bioactivation in drug discovery and development. Chem. Res. Toxicol. 17, 3–6.

F.D.A., 2002. U.S. Food and Drug Administration. Guideline for metabolism studies and for selection of residues for toxicological testing. U.S. Food and Drug Administration, Center for Veterinary Medicine.

Fernandez-Metzler, C.L., Owens, K.G., Baillie, T.A., King, R.C., 1999. Rapid liquid chromatography with tandem mass spectrometry-based screening procedures for studies on the biotransformation of drug candidates. Drug Metab. Dispos. 27(1), 32–40.

Gillette, R., Mitchell, J.R., Brodie, B.B., 1974. Biochemical mechanisms of drug toxicity. Annu. Rev. Pharmacol. 14, 271–288.

Gu, M., Lim, H.K., 2001. An intelligent data acquisition system for simultaneous screening of microsomal stability and metabolite profiling by liquid chromatography/mass spectrometry. J. Mass Spectrom. 36(9), 1053–1061.

Hastings, K.L., El-Hage, J., Jacobs, A., Leighton, J., Morse, D., Osterberg, R.E., 2003. Drug metabolites in safety testing. Toxicol. Appl. Pharmacol. 190(1), 91–92.

Hecht, S.S., 1996. Recent studies on mechanisms of bioactivation and detoxification of 4-(methylnitrosamino)-1-(3-pyridyl)-1-butanone (NNK), a tobacco-specific lung carcinogen. Crit. Rev. Toxicol. 26(2), 163–181.

Jemal, M., Ouyang, Z., Zhao, W., Zhu, M., Wu, W.W., 2003. A strategy for metabolite identification using triple-quadrupole mass spectrometry with enhanced resolution and accurate mass capability. Rapid Commun. Mass Spectrom. 17(24), 2732–2740.

Kassahun, K., Pearson, P.G., Tang, W., McIntosh, I., Leung, K., Elmore, C., Dean, D., Wang, R., Doss, G., Baillie, T.A., 2001. Studies on the metabolism of troglitazone to reactive intermediates in vitro and in vivo. Evidence for novel biotransformation pathways involving quinone methide formation and thiazolidinedione ring scission. Chem. Res. Toxicol. 14(1), 62–70.

Lim, H.K., Stellingweif, S., Sisenwine, S., Chan, K.W., 1999. Rapid drug metabolite profiling using fast liquid chromatography, automated multiple-stage mass spectrometry and receptor-binding. J. Chromatogr. A. 831(2), 227–241.

Lin, J.H., 2002. The role of pharmacokinetics in drug discovery: finding drug candidates with the greatest potential for success. Ernst. Schering. Res. Found. Workshop 37, 33–47.

Lin, J.H., Chiba, M., Baillie, T.A., 1999. Is the role of the small intestine in first-pass metabolism overemphasized? Pharmacol. Rev. 51(2), 135–158.

Liu, D.Q., Karanam, B.V., Doss, G.A., Sidler, R.R., Vincent, S.H., Hop, C.E., 2004. In vitro metabolism of MK-0767 [(+/−)-5-[(2,4-dioxothiazolidin-5-yl)methyl]-2-methoxy-N-[[(4-trifluoromethyl)-phenyl]methyl]benzamide], a peroxisome proliferator-activated receptor alpha/gamma agonist. II. Identification of metabolites by liquid chromatography–tandem mass spectrometry. Drug Metab. Dispos. 32(9), 1023–1031 [Erratum in: Drug Metab. Dispos. 2004 Nov; 32(11), 1331].

Lopez, L.L., Yu, X., Cui, D., Davis, M.R., 1998. Identification of drug metabolites in biological matrices by intelligent automated liquid chromatography/tandem mass spectrometry. Rapid Commun. Mass Spectrom. 12(22), 1756–1760.

Maggs, J.L., Naisbitt, D.J., Tettey, J.N., Pirmohamed, M., Park, B.K., 2000. Metabolism of lamotrigine to a reactive arene oxide intermediate. Chem. Res. Toxicol. 13(11), 1075–1081.

Martin, P.D., Warwick, M.J., Dane, A.L., Hill, S.J., Giles, P.B., Phillips, P.J., Lenz, E., 2003. Metabolism, excretion, and pharmacokinetics of rosuvastatin in healthy adult male volunteers. Clin. Ther. 25(11), 2822–2835.

Mutlib, A.E., Shockcor, J.P., 2003. Application of LC/MS, LC/NMR, NMR and stable isotopes in identifying and characterizing metabolites. In: Lee, J.S., Obach, R.S., Fisher, M.B. (Eds.), Drug Metabolizing Enzymes: Cytochrome P450 and Other Enzymes in Drug Discovery and Development, 1st ed. Marcel Dekker Inc, New York, pp. 33–86.

Mutlib, A.E., Strupczewski, J.T., Chesson, S.M., 1995. Application of hyphenated LC/NMR and LC/MS techniques in rapid identification of in vitro and in vivo metabolites of iloperidone. Drug Metab. Dispos. 23(9), 951–964.

Nassar, A.E., Adams, P.E., 2003. Metabolite characterization in drug discovery utilizing robotic liquid-handling, quadruple time-of-flight mass spectrometry and in-silico prediction. Curr. Drug Metab. 4(4), 259–271.

Samuel, K., Yin, W., Stearns, R.A., Tang, Y.S., Chaudhary, A.G., Jewell, J.P., Lanza, T. Jr., Lin, L.S., Hagmann, W.K., Evans, D.C., Kumar, S., 2003. Addressing the metabolic activation potential of new leads in drug discovery: a case study using ion trap mass spectrometry and tritium labeling techniques. J. Mass Spectrom. 38(2), 211–221.

Sinz, M.W., Podoll, T., 2002. The mass spectrometer in drug metabolism. In: Rossi, D.T., Sinz, M.W. (Eds.), Mass Spectrometry in Drug Discovery, 1st ed., Marcel Dekker Inc, New York, pp. 271–336.

Smith, D.A., 1994. Design of drugs through a consideration of drug metabolism and pharmacokinetics. Eur. J. Drug Metab. Pharmacokinet. 19(3), 193–199.

Smith, D.A., Jones, B.C., Walker, D.K., 1996. Design of drugs involving the concepts and theories of drug metabolism and pharmacokinetics. Med. Res. Rev. 16(3), 243–266.

Stevens, G.J., Hitchcock, K., Wang, Y.K., Coppola, G.M., Versace, R.W., Chin, J.A., Shapiro, M., Suwanrumpha, S., Mangold, B.L., 1997. In vitro metabolism of *N*-(5-chloro-2-methylphenyl)-*N*′-(2-methylpropyl)thiourea: species comparison and identification of a novel thiocarbamide-glutathione adduct. Chem. Res. Toxicol. 10(7), 733–741.

Tang, W., Stearns, R.A., Bandiera, S.M., Zhang, Y., Raab, C., Braun, M.P., Dean, D.C., Pang, J., Leung, K.H., Doss, G.A., Strauss, J.R., Kwei, G.Y., Rushmore, T.H., Chiu, S.H., Baillie, T.A., 1999. Studies on cytochrome P-450-mediated bioactivation of diclofenac in rats and in human hepatocytes: identification of glutathione conjugated metabolites. Drug Metab. Dispos. 27(3), 365–372.

Tiller, P.R., Land, A.P., Jardine, I., Murphy, D.M., Sozio, R., Ayrton, A., Schaefer, W.H., 1998. Application of liquid chromatography–mass spectrometry analyses to the characterization of novel glyburide metabolites formed in vitro. J. Chromatogr. A. 794(1–2), 15–25.

Tingle, M.D., Jewell, H., Maggs, J.L., O'Neill, P.M., Park, B.K., 1995. The bioactivation of amodiaquine by human polymorphonuclear leucocytes in vitro: chemical mechanisms and the effects of fluorine substitution. Biochem. Pharmacol. 50(7), 1113–1119.

Uetrecht, J.P., 1997. Current trends in drug-induced autoimmunity. Toxicology 119(1), 37–43.

van sen Bongard, H.J., Pluim, D., Rosing, H., Nan-Offeringa, L., Schot, M., Ravic, M., Schellens, J.H., Beijnen, J.H., 2002. An excretion balance and pharmacokinetic study of the novel anticancer agent E7070 in cancer patients. Anticancer Drugs 13(8), 807–814.

Watt, A.P., Mortishire-Smith, R.J., Gerhard, U., Thomas, S.R., 2003. Metabolite identification in drug discovery. Curr. Opin. Drug Discov. Devel. 6(1), 57–65.

Williams, D.P., Kitteringham, N.R., Naisbitt, D.J., Pirmohamed, M., Smith, D.A., Park, B.K., 2002. Are chemically reactive metabolites responsible for adverse reactions to drugs? Curr. Drug Metab. 3(4), 351–366.

Xia, Y.Q., Miller, J.D., Bakhtiar, R., Franklin, R.B., Liu, D.Q., 2003. Use of a quadrupole linear ion trap mass spectrometer in metabolite identification and bioanalysis. Rapid Commun. Mass Spectrom. 17(11), 1137–1145.

Yu, X., Cui, D., Davis, M.R., 1999. Identification of in vitro metabolites of Indinavir by "intelligent automated LC-MS/MS" (INTAMS) utilizing triple quadrupole tandem mass spectrometry. J. Am. Soc. Mass Spectrom. 10(2), 175–183.

Zhang, J.Y., Wang, Y., Dudkowski, C., Yang, D., Chang, M., Yuan, J., Paulson, S.K., Breau, A.P., 2000. Characterization of metabolites of Celecoxib in rabbits by liquid chromatography/tandem mass spectrometry. J. Mass Spectrom. 35(11), 1259–1270.

Identification and Quantification of Drugs, Metabolites and Metabolizing Enzymes by LC–MS
Swapan K. Chowdhury, editor.

Chapter 9

QUANTIFICATION AND STRUCTURAL ELUCIDATION OF LOW QUANTITIES OF RADIOLABELED METABOLITES USING MICROPLATE SCINTILLATION COUNTING (MSC) TECHNIQUES IN CONJUNCTION WITH LC–MS

Mingshe Zhu, Donglu Zhang, and Gary L. Skiles

9.1. Introduction

Radioactive isotopes, mainly ^{14}C, are routinely used to trace the fate of drugs in disposition and metabolism studies conducted during their development. These studies include: (1) tissue distribution and mass balance studies wherein the total amounts or concentrations of all drug-related components in biological fluids or tissues are determined; (2) metabolite profiling studies, to determine the number and relative concentrations of individual metabolites in various biological matrices; (3) enzymatic studies to identify the enzymes involved in the metabolism of the drug, and the kinetics of the reactions; and (4) structural elucidation of metabolites. Information derived from these studies is a critical component of developing safe and effective new drugs [1] and is also a regulatory requirement for registration approval [2,3]. For example, proper toxicological evaluation of prospective drugs is assured by demonstrating that all drug-related components present in human plasma are also present in the plasma of preclinical species during toxicity testing. Indeed, comparison of these metabolite profiles is a key element in the selection of animal species for toxicological evaluations [4]. Knowledge of the relative drug and metabolite concentrations in urine, feces, and bile, along with mass balance data, is essential for understanding the role of metabolism in drug clearance and the major metabolic pathways in humans and animals [1,3]. Radioisotopically labeled drugs not only facilitate the detection, isolation, and identification of metabolites, they also provide a facile means for quantification of these metabolites based on their radioactivity without the need for synthetic metabolite standards [4,5]. Due to these significant advantages, radiolabels have recently been used increasingly earlier in the drug discovery process for biotransformation studies. For these early phase studies, tritium ^{3}H-labeled drugs are often used because their preparation is often much easier than synthesis of ^{14}C-labeled compounds.

Planar radiochromatography was the first widely used technique for separation and detection of radiolabeled metabolites [6,7]. The technique is very sensitive, fast, and has relatively low equipment and operating costs; however, due to low separation resolution and poor precision in quantification, it has largely been replaced by HPLC-based radiochromatographic techniques in the pharmaceutical industry. On-line radio flow detection (RFD) and off-line liquid scintillation counting (LSC) have been extensively used for analysis of radiolabeled metabolites. HPLC–RFD provides quick results with high resolution, and by splitting of the HPLC effluent it can be coupled with a mass spectrometric detection for simultaneous metabolite characterization [8–14]. HPLC–LSC is at least 25-fold more sensitive than HPLC–RFD because of the longer time individually collected effluent fractions can be counted. LSC has, therefore, been employed for quantification of low levels of radiolabeled metabolites, such as plasma metabolites [15]. Radioactivity analysis by off-line LSC consists of four steps: HPLC separation, fraction collection into individual vessels, mixing with scintillation cocktail, and counting radioactivity one fraction at a time. The entire process is time-consuming and relatively labor-intensive.

Recently, microplate scintillation counting (MSC) has been introduced as a new HPLC radioactivity detection technique for sensitive metabolite detection and profiling in drug metabolism studies [16–22]. In addition, HPLC–MSC is often used for the determination of metabolite concentrations in enzyme kinetic and reaction phenotyping studies [23,24]. Compared to LSC, MSC not only significantly increases sensitivity and analytical throughput, but also reduces radioactive waste and the need for manual handling of individual samples. In combination with various mass spectrometric techniques for metabolite identification, MSC is utilized for determination of formation pathways and structures of secondary metabolites [25], selective detection of molecular ions of unknown metabolites in a complex biological matrix using a high-resolution LC–MS-based mass defect filter (MDF) approach [26,27], and sensitive characterization of plasma [28] and tissue [29] metabolites using micro flow LC–MS/MS. Because of these attributes, HPLC–MSC has become the method of choice for high sensitivity analysis of radiolabeled metabolites in some metabolism laboratories in the pharmaceutical industry.

9.2. The HPLC–MSC Technique

The first commercially available microplate scintillation counter (TopCount® Packard, Meriden, CT) was originally designed for high-throughput bioassays in 96- and 384-microplate formats, such as cell proliferation and receptor binding assays. A TopCount can count 6–12 wells in a microplate at a time and has a capacity of up 40 microplates. The TopCount utilizes newly developed single photomultiplier tube technology that enhances signal-to-noise, reduces optical crosstalk to negligible levels, and enables counting of samples using liquid or solid scintillators. In the late 1990s, an opaque 96 deep-well microplate (Deep-Well LumaPlate™), in which a solid scintillator (yttrium silicate) is coated onto the bottom of each well, was introduced specifically for HPLC radiochromatographic analysis [18]. Later, the MicroBeta® plate counter (Wallac Instruments), together with Scintiplates® from the same manufacturer, were marketed for radiolabeled metabolite profiling [21]. In Scintiplates, solid scintillators are incorporated into the plate material itself rather than coated onto the bottom of each well. Up to 350 μl HPLC effluent per well can be collected in both the LumaPlate and Scintiplate microplates. HPLC–MSC, like its cousin HPLC–LSC, is an off-line technique and a system to incorporate this technique must be designed accordingly. A general set-up for an HPLC–MSC system is shown in Fig. 1.

Depending on HPLC flow rates, HPLC–MSC can be categorized into conventional, low, and micro flow methods (Fig. 1). The conventional HPLC–MSC method (Fig. 1, method A) was developed for conventional HPLC flow system using a 3.9–4.6 mm ID HPLC column and flow rates of 0.8–1.2 ml/min [16,17,21]. After sample injection, HPLC effluent fractions are collected into microplates, and solvent is removed using a centrifugal vacuum system (e.g. SpeedVac®, Savant Instruments, Holbrook, NY). The radioactivity in residues remaining in each well is determined by counting up to 12 wells at a time using a plate counter. A radiochromatogram of the sample being analyzed can then be prepared by plotting the CPM values of the HPLC fractions against their elution times (Fig. 2A). Concomitantly, the relative concentration or percent distribution of a radioactive component in the sample can be simply calculated by dividing the radioactivity of the individual peak by the total radioactivity of all radioactive components determined in the entire HPLC analysis. Conventional flow HPLC–MSC is the most commonly used method for metabolite profiling. It provides consistent retention times, high resolution of sample

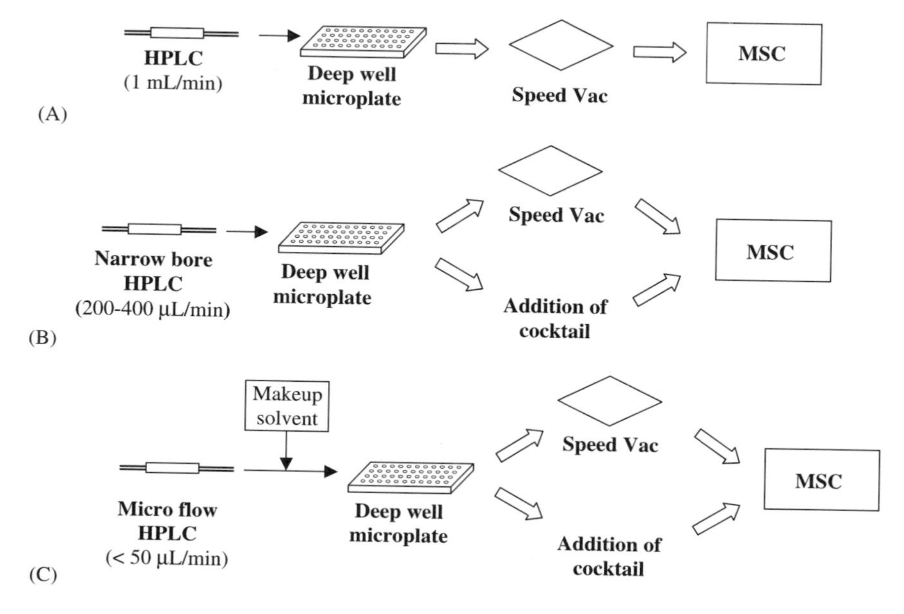

Figure 1.
General set-ups of HPLC–MSC systems

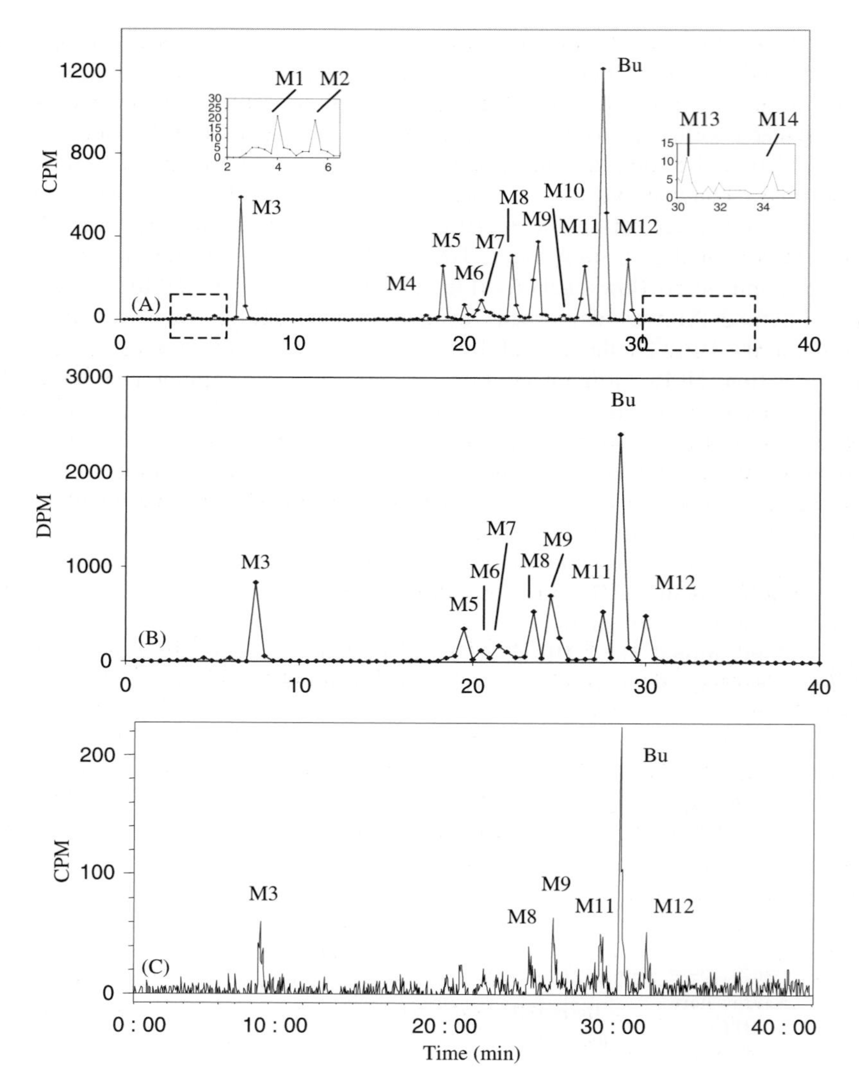

Figure 2.
Metabolite profiles of buspirone determined by HPLC with MSC, LSC, and RFD [30]. Buspirone metabolites from human liver microsome incubations were injected onto HPLC (1 ml/min). Radioactivity in HPLC eluate was determined by (A) TopCount: 8000 DPM injected, four fractions per min and 10 min counting time; (B) LSC: 8000 DPM injected, two fractions per min, 10 min counting time; and (C) RFD: 32000 DPM injected

components, and minimal matrix effects from complex biological samples such as bile and feces. In addition, this method provides much better radiodetection sensitivity than low- and micro-flow HPLC methods since conventional HPLC has a much larger sample loading capacity.

The low flow HPLC–MSC method [19] is suitable for a narrow-bore HPLC column (2 mm ID and flow rates of 200–400 μl/min, Fig. 1, method B). The HPLC effluent can be collected into Deep-Well LumaPlates or Scintiplates in a process similar to method A, or fractions can be collected into regular 96-well plates (no solid scintillators) followed by the direct addition of liquid scintillation cocktail. Direct addition of the cocktail not only shortens the total analytical time by eliminating the drying process that can take at least 5 h, but also prevents the possible loss of volatile metabolites during the drying process. This method is suitable for in vitro samples that contain limited biological matrices. It may not, however, be the best technique for samples obtained from in vivo sources due to the low loading capacity of narrow-bore HPLC columns. For a micro flow HPLC–MSC system (Fig. 1, method C), a makeup solvent is needed to ensure uniformity of the fractions [18]. Without a makeup solvent flow, a difference of a single drop between two fractions results in excessive variability because one drop is such a substantial proportion of each fraction. HPLC–MSC can also be used in conjunction with micro-separation techniques (flow rate less than ≤50 μl/min), such as capillary LC, capillary electrophoresis, and microdialysis.

9.3. Analytical Characteristics of HPLC–MSC

Unlike LSC, MSC uses external standardization for calculation of DPM values from the measured CPM values. This necessitates at least 1000 DPM of radioactivity in each fraction to be analyzed [31]. Because of this limitation, CPM values must be used to calculate relative concentrations of radioactive peaks for profiling low-level radiolabeled metabolites by HPLC–MSC. If the counting efficiency of each fraction is consistent or stays within an acceptable range, then CPM values can be reliably used to determine the relative concentrations in each fraction from an entire HPLC run. Counting efficiencies can, however, be greatly affected by color quenching and other phenomena when analyzing biological samples [22]. Consequently, the accuracy and precision of the MSC method can be significantly affected with certain sample types that contain colored matrix components which co-elute with radiolabeled analytes. Thus, the understanding of analytical characteristics of the MSC technique and the use of a validated HPLC–MSC method are essential for ensuring the quality of HPLC–MSC analysis. In this section, sensitivity, precision, and accuracy of radioactivity profiling by HPLC-TopCount are described. In addition, a LSC-based method for determining radioactivity recovery in HPLC–MSC analysis and matrix effects on HPLC–MSC performance are discussed.

9.3.1. Sensitivity

A primary factor in the superior sensitivity in off-line scintillation counting techniques is the availability of the greater counting time. The limit of detection in the quantitation of radioactivity can be calculated using Eq. (1) [22,32–34].

$$\mathrm{LD} = 2.7 + 4.65\sqrt{C} \tag{1}$$

where LD is the limit of detection expressed as the total counts, C is the total counts of background in the counting time interval, and 2.71 and 4.65 are empirically determined constants.

In Eq. (1), C is equal to the product of B (background radioactivity expressed in DPM), T (counting time) and E (counting efficiency). When Eq. (1) is divided by T and E, it becomes

$$\text{LD} = 2.71/TE + 4.65\sqrt{B/TE} \qquad (2)$$

where LD is expressed in DPM.

As is apparent from Eq. (2), radiodetection sensitivity can be improved by increasing counting time. In HPLC–RFD, an eluted radioactive component is typically counted for only for 5–15 s. Longer counting times in larger flow cells can increase sensitivity, but radiochromatographic resolution is diminished when multiple components are present simultaneously in the cell. The consequence of these limitations is relatively poor sensitivity with limits of detection ranging from 250 to 500 DPM for ^{14}C.

Equation (2) also shows that decreasing background counts and improving counting efficiency can achieve improved limits of detection. The limit of detection for ^{14}C on a TopCount instrument with 10 min counting was approximately 5 DPM (Table 1), which was approximately two-fold better than LSC [30], and 50–100-fold better than RFD. The limits of detection for a MicroBeta counter were 7 DPM for ^{14}C and 16 DPM for ^{3}H [22]. Extremely lower background radioactivity (1–2 DPM) of TopCount is the main factor leading to the superior sensitivity [19,20]. A comparison of the sensitivity using MSC, LSC, and RFD is illustrated in Fig. 2. Minor metabolite peaks M1, M2, M4, M10, M13, and M14 (Table 2) were detected in all repeat analyses with a TopCount (Fig. 1A), but not detected by RFD even when four times as much sample was injected (Fig. 1C). Most minor metabolites were detected by LSC, but several minor peaks detected by MSC were not observed with LSC analysis. For example, M14 (15 CPM) was not observed in the LSC radiochromatogram shown in Fig. 2B.

Table 1.
Analytical sensitivity of TopCount and LSC for metabolite profiling

Instrument	Background (CPM)	Counting efficiency (%)	Counting time (min)	Limit of detection[a] (DPM)
LSC	25	90	10	10
MSC (TopCount)	2	70	10	5

[a]Limit of detection was calculated based on Eq. (1)

Table 2.
Precision of TopCount for quantitative analysis of low levels of radioactive metabolites

Metabolite[a]	M1	M2	M4	M10	M13	M14
Mean radioactivity[b] (CPM)	32	30	32	42	23	15
Relative SD[c] (%)	14.6	10.8	14.6	5.3	9.0	27.8

[a]Buspirone metabolites in HLM were profiled by HPLC with TopCount (see Fig. 2A)
[b]Mean of five injections
[c]Standard deviation

9.3.2. Precision and Accuracy

Table 2 lists the precision obtained on a TopCount instrument for quantitative analysis of low levels of radioactive metabolites after an in vitro sample was repeatedly analyzed ($N = 5$) [30]. With a 10 min counting time, the TopCount provided good precision (RSD $< 15\%$) for minor metabolites present in quantities of 23–42 CPM). The RSD for a very minor peak (M14, 15 CPM) was 28%. The accuracy obtained on a TopCount for an HPLC–MSC determination of a metabolite profile (% distribution of radioactivity) were ±1–12% of nominal values obtained by LSC and were similar to RFD (Table 3). Excellent accuracy for metabolite profiling was also demonstrated when a plasma sample was analyzed by HPLC with MicroBeta counter [22].

9.3.3. Matrix Effect of Biological Samples

TopCount uses solid scintillators to convert radioactivity decay energy into photons. Biological samples, such as plasma, urine, and feces, contain significant amounts of proteins, salts, and small organic chemicals which can co-elute with radiolabeled analytes and precipitate on the bottom of 96 wells after removing solvent. These chemicals may interfere with the performance of MSC by color quenching or other unknown mechanisms [22]. The matrix effect of biological samples on the performance of a TopCount was recently evaluated [30]. Extracted or concentrated samples from human biological matrices, including plasma, urine, and feces as well as human liver microsome (HLM) incubations, were tested. Figure 3 shows radiochromatograms of an infused radiolabeled compound with and without injection of the biological sample extracts. For both the TopCount (Fig. 3A) and LSC (Fig. 3B), the measured radioactivity in each fraction remained within ±15% of the mean when a biological sample was not injected (controls).

Extracts or concentrates of plasma (equivalent to 1 ml of original sample, Fig. 3D), urine (equivalent to 2 ml original sample, Fig. 3E), and HLM incubation (equivalent to 1 ml incubation, Fig. 3C) samples did not significantly affect measurement of

Table 3.
Accuracy of metabolite profiling by HPLC–MSC and HPLC–RFD[a]

Radioactivity peak	Accuracy of radioactivity distribution (%)[b]	
	TopCount	RFD
M3	+11.7	+4.3
M5	+0.7	−9.4
Sum of M6 and M7	+3.2	+0.9
M8	−6.6	−2.1
M9	−2.1	+1.7
M11	+1.0	−1.1
M12	−0.6	+6.6
Buspirone	+9.1	−5.8

[a]Profiles of buspirone metabolites were determined by HPLC with three radio-detection techniques (Fig. 2). The mean values ($N = 5$) of radioactivity distribution in percentage of each metabolite were calculated by dividing the radioactivity of the metabolite peak by the total radioactivity determined in each respective HPLC analysis

[b]Accuracy of radioactivity distribution in percentage was calculated by using the distribution determined by HPLC–LSC ($N = 5$) as "true values"

radioactivity in any fractions except for corresponding to the retention time of the HPLC dead volume in the chromatograms of HLM and urine. Unlike the plasma sample, the urine and HLM samples were not treated with solid phase extraction and, therefore, proteins, salts, and other polar components were not removed from these samples. Elution of these components in the dead volume and deposition of them in the 96-well plates likely acted as a physical barrier between the radiolabeled compound and the solid scintillators, resulting in lower counting efficiency. Lower CPM values were also observed in radiochromatograms of the fecal extract (fractions 84–88, Fig. 3F). In a manner similar to that caused by salts eluting at the solvent front, this might have been caused by a shielding effect by endogenous components that eluted at this time rather than color quenching because these fractions did not display an intense color.

The matrix effect of the biological samples is proportional to the amount of the endogenous material injected, which is dependent on the injection volume and concentration of the extract. The effect can, therefore, be minimized by limiting the amount of the sample injected. Table 4 lists the recommended maximum amounts of human samples for metabolite profiling of a [^{14}C]-labeled drug using HPLC with MSC on a TopCount. These amounts should meet most needs for metabolite profiling in these matrices. Quantitative analysis of very polar metabolites that elute with the HPLC dead volume would not be accurate and so care must be taken to properly resolve analytes from the HPLC solvent front. Feces contain food residues and other

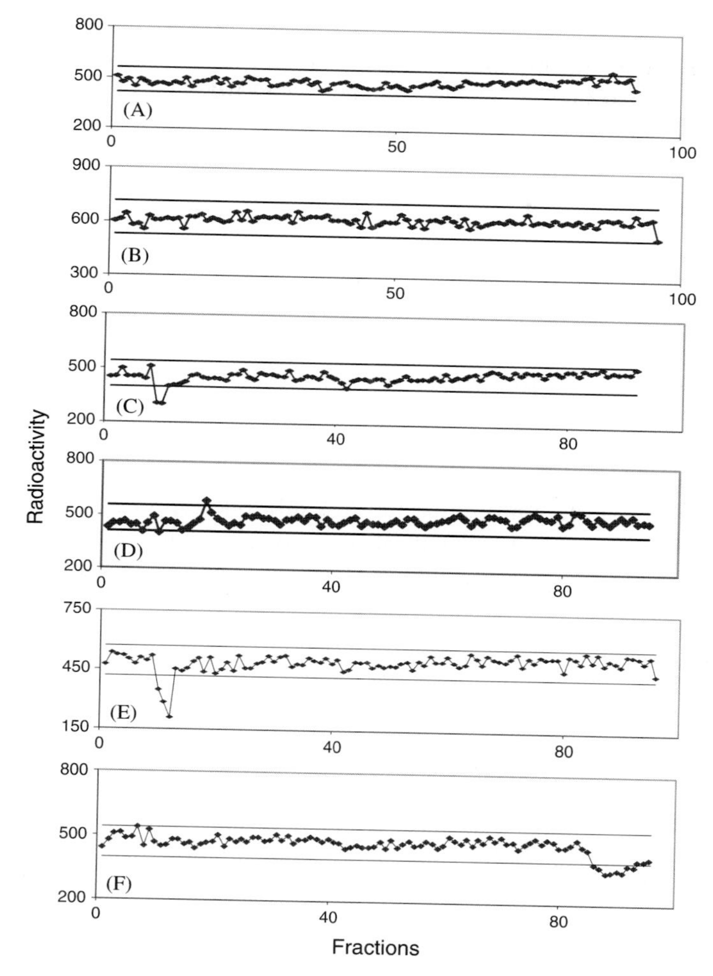

Figure 3.
Matrix effect of human plasma, urine, feces, and liver microsomal incubations [30]. Extracted or concentrated biological samples were injected onto HPLC and eluted with solvents A (water) and B (acetonitrile) up to 80% solvent B. The samples were pretreated with solid phase extraction (plasma), liquid extraction (feces), protein precipitation by methanol (HLM). Extraction fractions from plasma, supernatant from liver microsome sample and urine were dried with nitrogen gas, and then were reconstituted with the HPLC solvents before injection. A [^{14}C] labeled drug was infused into HPLC eluent using a post column Tee. The mixed eluate was collected into 96-well plates for TopCount analyses (Panels A, C–F) or test tubes for LSC analysis (Panel B) at four fractions per min for 10 min counting. (A) TopCount control, no injection of a biological sample (B) LSC control, no injection of a biological sample; (C) HLM (equivalent to 1 ml HLM incubation); (D) human plasma (equivalent to 1 ml plasma); (E) human urine (equivalent to 2 ml urine); and (F) human feces (equivalent to 100 mg feces). The up and down lines in each figure represent ±15% of the mean values

Table 4.
Recommended maximum injection of extracts or concentrates equivalent to volumes or masses of original human samples for metabolite profiling of a ^{14}C labeled drug by HPLC with TopCount[a]

Human samples[b]	HLM (ml)	Plasma (ml)	Urine (ml)	Feces (mg)	Bile[c] (μl)
Maximum injection equivalent to volumes or masses of original sample[b]	1	1	2	50	50

[a]The recommendations for human liver microsomes (HLM), urine, plasma are based on the data from the experiments described in Fig. 3. The recommendation for human bile is proposed based on comparison of metabolite profiles obtained by TopCount and LSC

[b]Human plasma (up to 1 ml), urine (up to 2 ml), bile (up to 50 μl), and HLM (up to 1 ml) were tested, and did not display significant matrix effects. Human feces (up to 500 mg) was tested. Treatment of HLM, plasma, urine, and feces samples is described in Fig. 3

[c]Organic solvent was added into bile (1:1) prior to analysis

biological components that are dependent on dietary and other factors that are difficult to control or predict. The result shown for a fecal extract in Fig. 3 is, therefore, likely to vary considerably with samples obtained from other sources. Because of this variability, it is recommended to inject as low a quantity of fecal extract as possible ($\leq$50 mg feces equivalents). In addition, when quantitating radioisotopes other than ^{14}C or when analyzing biological matrices different from those listed in Table 4, potential matrix effects on MSC performance should be considered.

9.3.4. Radioactivity Recovery Determination

Determination of radioactivity recovery from an HPLC analysis is necessary to ensure accurate determination of the distribution of all components in the original sample. In addition, because metabolite profiling by MSC requires the additional step of in vacuo removal of HPLC solvents, it is also important to determine whether volatile metabolites are lost in the process (plate recovery). We have developed a simple method to determine both HPLC column recovery and plate recovery for an HPLC–MSC analysis [30]. In general, aliquots of a sample are injected onto an HPLC with and without an HPLC column. The entire eluates from each HPLC are separately collected in graduated cylinders and then aliquots are analyzed for total radioactivity by LSC with and without solvent evaporation. The DPM values obtained are used to calculate the recoveries. To test the method a dog bile sample, which contained multiple non-volatile metabolites, was profiled by both HPLC–LSC and HPLC–MSC methods. HPLC–LSC analysis indicates that the column recovery of the HPLC method is excellent. Radioactivity recoveries of the HPLC–MSC methods were 96–97% (Table 5). The results demonstrate that HPLC–MSC radioactivity recovery can be rapidly and reliably determined using this method.

9.4. Application of MSC to Quantitative Analysis of Drugs and Metabolites

Radiolabeling of drugs provides a means for quantitative determination of their in vitro and in vivo metabolites without the use of synthetic standards. Radiolabeled drugs are used for most biotransformation studies in drug development and many in drug discovery when quantitative measurement of metabolites is required. These studies include: (1) inter-species comparison of in vitro metabolism in support of the selection of appropriate species for safety evaluation; (2) ADME studies in animals and humans to determine the major circulating metabolites and major metabolic pathways [2,4,15]; (3) in vitro enzymology studies to identify the enzyme(s) involved in specific metabolic pathways; and (4) characterization of metabolism kinetics [5,23,24]. In addition, radiolabels are sometimes used for the determination of plasma metabolite concentrations [29] in early preclinical and toxicokinetic studies when LC–MS methods for metabolite quantification are not available. All of these uses of radiolabeled drugs are amenable to and benefit from the advantages afforded by MSC. The improved sensitivity, shortened analytical times, and reduction of radioactive waste provided by MSC have led to its replacing LSC as the main tool for quantitative analysis of low levels of metabolites in some biotransformation laboratories.

9.4.1. Detection and Profiling of Low Level Metabolites

One of the major applications of MSC is to detect and profile in vivo metabolites, especially plasma metabolites [16,17,19,21]. Usually, plasma sample volumes available for analysis are limited, and metabolite concentrations in plasma are relatively low compared to their concentrations in urine and bile. Figure 4 is a plasma radiochromatogram obtained using an HPLC coupled with a TopCount (10 min counting time) of a plasma sample containing 493 DPM of radioactivity [30]. A minor metabolite (7 CPM) that was approximately 2% of the total radioactivity injected was clearly detected. Additionally, the percent distribution of all drug-related components in the plasma sample was determined in the same analysis. Based on the sensitivity provided by TopCount analysis (Table 1), samples containing more than 500 DPM of radioactivity are required for detection of minor metabolites (10 DPM, 10 min counting time) corresponding to 2% of the total radioactivity. For quantitative analysis of the same minor components, at least 1500 DPM of radioactivity may be required for LC–MSC analysis [30]. Alternatively, increasing the MSC counting time to 20–30 min can significantly improve the sensitivity and the same analytical accuracy and precision can be achieved with less sample.

Table 5.
Radioactivity recovery determination in metabolite profiling by the HPLC–MSC analysis[a]

Injection number	DPM without column before drying (A)	DPM with column before drying (B)	DPM with column after drying (C)	DPM without column after drying (D)	Column recovery (%)[b]	Plate recovery (%)[c]	Total recovery (%)[d]
1	71	71	80	81			
2	85	73	73	72			
3	78	79	76	78			
4	80	77	75	74			
Average	78	75	76	76	96	97	97

[a]HPLC was performed on a Shimadzu LC-10AT system. An Xterra MS C18, 3.5 μM, 4.6×150 mm column, was used. The mobile phase flow rate was 1 ml/min. The HPLC solvent system was a 60 min gradient and consisted of two solvents: (A) 100% 0.01 M ammonium bicarbonate, pH 9, and (B) 100% acetonitrile. Aliquots of a human sample are injected onto an HPLC with and without the HPLC column. The entire eluates from each HPLC are separately collected in graduated cylinders and then aliquots (2 ml) are analyzed for total radioactivity (30 min counting time) by LSC with and without solvent evaporation

[b]Column recovery is the average DPM in (B) over the average DPM in (A)

[c]Plate recovery is the average DPM in (D) over the average DPM in (A)

[d]Total recovery is the average DPM in (C) over the average DPM in (A)

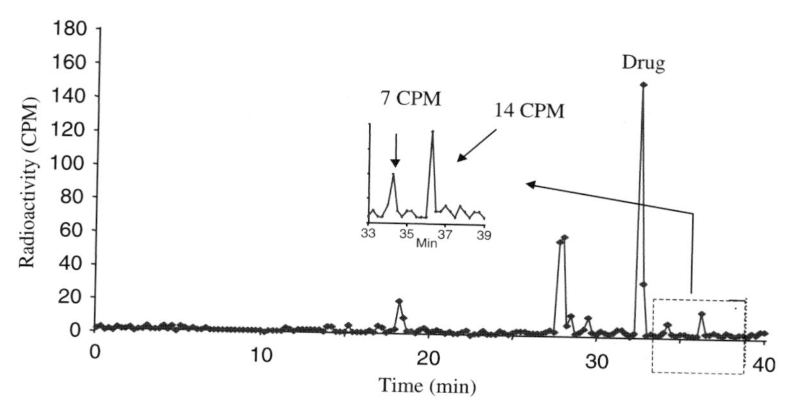

Figure 4.
HPLC–MSC profile of low-level radioactive metabolites in rat plasma [30]. A plasma sample from rats after oral administration of a ^{14}C-labeled drug was analyzed by HPLC (1 ml/min) with TopCount (4 wells/min, 10 min counting time). The sample was pretreated with solid phase extraction. Approximately 493 DPM radioactivity was injected

MSC is often used for detection and profiling of in vitro metabolites when analytical sensitivity is critical. For instance, comparative metabolism studies in liver microsomes and hepatocytes are usually carried out at substrate concentrations in the range of 1–10 μM. If the specific activity of the radiolabeled drug is low, metabolic turnover is low, or quantification of minor metabolites is required, the LC–MSC method could provide the necessary sensitivity to achieve the experimental objectives. Another application of HPLC–MSC in metabolite profiling is the analysis of phase II metabolites when in vitro incubations are carried out with non-radiolabeled drugs in the presence of radiolabeled co-factors such as glutathione (GSH). Figure 5 shows radioactivity profiles of a non-radiolabeled drug incubated with HLMs in the presence of 1 mM [^{3}H]GSH, with and without the addition of NADPH. Reactive metabolites M1 and M2 trapped as GSH conjugates were detected and quantified by HPLC–MSC (Fig. 5A). These same conjugates were not detected in the incubation in which NADPH was omitted (Fig. 5B). Formation of reactive metabolites typically represents minor metabolic pathways and the detection of low quantities of labeled conjugates is further complicated by the overwhelming quantity of unconjugated GSH and oxidized glutathione (GSSG) present after incubation. The sensitivity of MSC allows detection of low quantities of GSH adducts even when the specific activity and total concentration of [^{3}H]GSH in the incubations are kept low. If the same experiment is attempted using a radio-flow detector, 50–100-fold more radiolabeled GSH would be required to achieve the same level of detection. In the example shown here, MSC substantially reduced the amounts of [^{3}H]GSH needed to achieve the experimental objective and, in doing so, simplified the analysis and reduced the amount of radioactive material that had to be purchased and ultimately disposed.

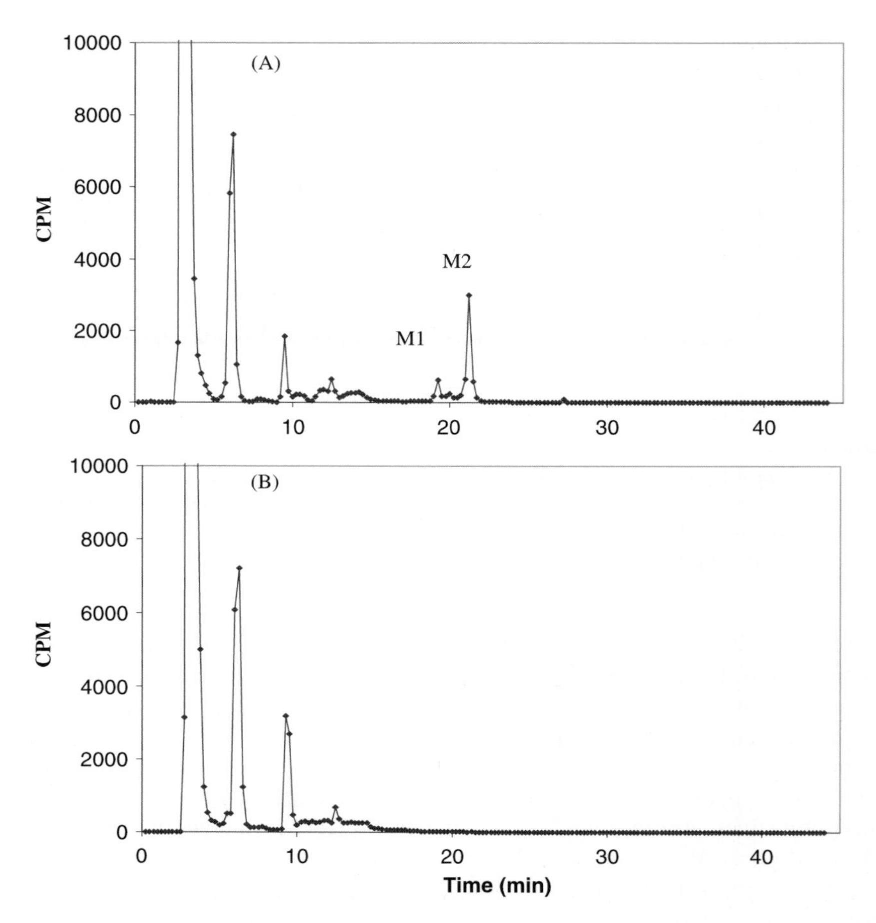

Figure 5.
Analysis of [^{3}H]GSH trapped reactive metabolites by HPLC with TopCount [30]. A non-labeled drug (50 mM) was incubated with a mixture of GSH (1 mM) and trace [^{3}H]GSH (1–2 μCi/ml) in human liver microsomes. After precipitating proteins, the samples were analyzed by HPLC (1 ml/min) with TopCount (4 wells/min, 10 min counting time). (A) Radioactivity profile of the incubation. M1 and M2 were GSH-trapped reactive metabolites; (B) Radioactivity profile of a control incubation sample (without NADPH)

9.4.2. Analysis of Drug and Metabolite Concentrations

Determination of metabolite concentrations in biological fluids and tissues obtained during preclinical pharmacokinetics (PK), toxicokinetic (TK), and clinical PK studies is often needed to support drug development. For example, non-linear pharmacokinetic/pharmacodynamic (PK/PD) correlations may be due to the presence of an active metabolite or metabolites in the circulation. In addition, determination of metabolite concentrations in tissues could provide useful information for understanding the chemical mechanism of tissue-specific toxicity in animals, which in some cases may be due to accumulation of metabolites in the tissue.

Quantification of radiolabeled metabolites in these samples can be readily achieved utilizing LC–MSC [29]. A general procedure for the LS–MSC approach is: (1) determination of total radioactivity in plasma using LSC and in tissues using quantitative whole-body autoradiography (QWBA); (2) determination of the percent distribution of metabolites in these samples using LC–MSC; and (3) calculation of the concentrations of each metabolite according to the following equation:

$$\text{Concentration of metabolite} = \%\text{Distribution of metabolite} \times \text{Concentration of total radioactivity}$$

An approximate 15–20 CPM limit of quantification for ^{14}C (10 min counting times 30) should meet most needs for quantitative analysis of plasma metabolites in PK or TK studies.

9.4.3. Enzymology Studies

Enzymology studies in drug metabolism include the determination of kinetic parameters such as K_m and V_{max}, and reaction phenotyping of enzymes that catalyze formation of metabolites [2]. These studies require the quantification of metabolites formed. Two analytical approaches are employed for metabolite quantification: (1) LC–MS/MS using synthetic standards of metabolites; and (2) liquid radiochromatography using radiolabeled test drugs. The traditional radiochromatographic technique used in enzyme kinetic studies is the on-line RFD method. Many of these studies require quantification of low levels of radioactivity either because metabolites are present at low concentrations or because the experiment is conducted with drug at a low specific activity. These conditions may result from low turnover of substrate, low substrate concentrations necessitated because of low K_m values, or scarcity of radiolabeled drug. The off-line HPLC–LSC approach is often used in these circumstances because of its greater sensitivity than the on-line RFD method. The HPLC–MSC approach is, however, 50–100-fold more sensitive than RFD and 5–10-fold faster than HPLC–LSC when analyzing a large number of samples; therefore, this technique has been used as a tool for enzyme kinetic studies in some laboratories [23,24].

An enzyme kinetic study of the *N*-glucuronidation of MaxiPost is a good example of the use of HPLC–MSC for the determination of K_m and V_{max} [23]. MaxiPost was incubated in HLM and recombinant UDPGT 2B7 enzyme with UDPGA. A representative radioactivity profile of these incubations determined by HPLC–MSC is shown in Fig. 6. The relative distribution of *N*-glucuronide in the sample was determined and the concentrations of the *N*-glucuronide in the incubations were calculated based on the following equation:

$$\text{Concentration of metabolite} = \%\text{Distribution of metabolite} \times \text{Initial concentration of drug in the incubation}$$

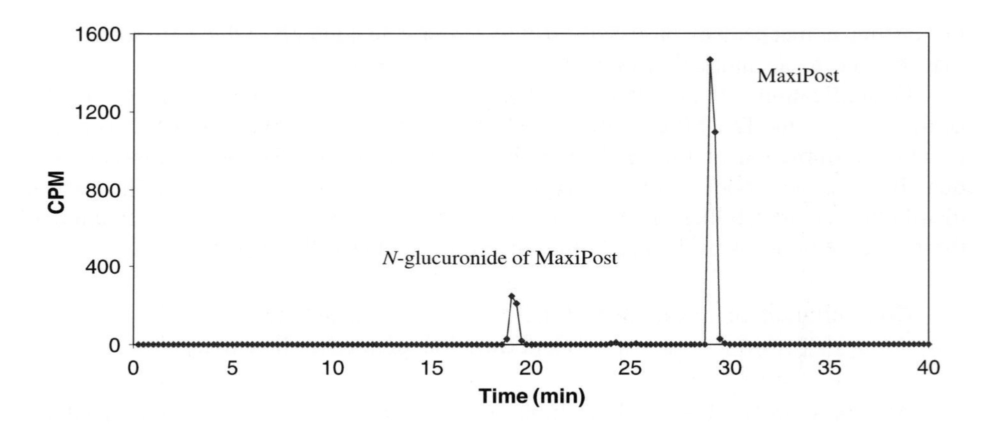

Figure 6.
HPLC-radiochromatogaphic profile of incubation of ^{14}C MaxiPost with human liver microsomes in the presence of UDPGA

Accordingly, the relationship of *N*-glucuronidation rates with substrate concentrations was determined (Fig. 7) and utilized for the calculation of K_m and V_{max}.

9.5. Application of a Combined of MSC and LC–MS to Metabolite Identification

The use of radiolabeled drugs is not only crucial for metabolite quantification, but has also long played a critical role in metabolite identification studies. The utility of radiolabeled substrates includes: (1) their use as tracers of drug-related components during sample cleanup, selective metabolite detection, and large-scale isolation from complex biological metrics; (2) facilitation of LC–MS/MS detection of molecular ions of unknown metabolites based on their HPLC retention times and in some cases isotopic ratios; and (3) enhancement of structural elucidation based on isotopic patterns or ratios in metabolite fragments formed by collision-induced dissociation (CID). For ^{14}C-radiolabeled drug candidates with specific activities ranging from 5 to 50 μCi/mg, MSC provides a level of detection comparable to LC–MS/MS and can be achieved without the need for synthesis of every known or suspect metabolite for characterization of their mass spectral properties.

9.5.1. A General LC–MSC–MS Approach

A general configuration of a combined HPLC–MSC and mass spectrometer (LC–MSC–MS) for structural elucidation of low-level radioactive metabolites is shown in Fig. 8A. The HPLC eluate is split so that a small portion flows to the mass

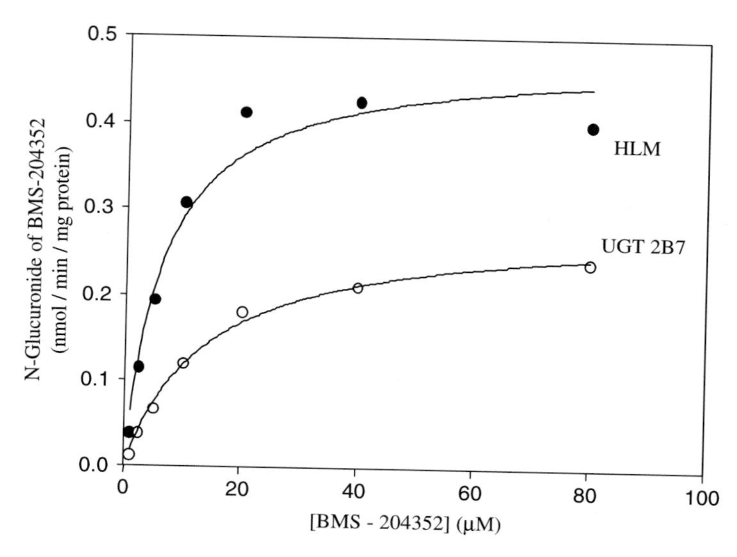

Figure 7.
Concentration-dependent formation of *N*-glucuronide of MaxiPost BMS-204352 in incubations with HLM and cDNA-expressed UGT 2B7 [23]. HPLC was performed on a Shimadzu Class VP® system. A Zorbax® RX C-18 column (4.6 × 250 mm, 5 μm) equipped with a guard column was used. The mobile phase flow rate was 1 ml/min. HPLC effluent was collected at 15-s intervals after sample injection into 96-Deep-Well LumaPlates® with a Gilson Model 202 fraction collector (Gilson Medical Electronics). The effluent in the plates was dried with a SpeedVac® (Savant) and the plates were counted for 10 min per well with a TopCount® scintillation analyzer (Packard Instrument Co.). Radiochromatographic profiles were prepared by plotting the CPM values in HPLC fractions against time-after-injection

spectrometer and the remaining portion flows to 96-well microplates. Radioactivity in the 96-well plates is counted by MSC after either removing solvent under vacuum or direct addition of scintillation cocktail. Because HPLC eluate flows directly to the mass spectrometer, this technique requires HPLC solvent systems suitable for atmospheric pressure ionization. Conventional HPLC columns (3.9–4.6 mm ID) are typically employed in LC–MSC–MS analyses because the large column provides consistent retention times and separation resolution of complex biological samples that is superior to shorter, narrow-bore columns. It also provides better MSC and mass spectrometric sensitivity because larger sample volumes can be injected without scarifying chromatographic performance.

Identification of metabolites by LC–MSC–MS usually involves two steps. The first step is to determine the molecular ion of unknown components represented by each radioactive peak. For in vitro incubation samples, full scan spectra corresponding to radiolabeled metabolite chromatographic peaks often reveal the molecular ions of the unknown component either as the predominate ion (*m/z* 402, Fig. 9A) or as one of a few major ions. Assignment of the molecular ion of an in vitro metabolite is, therefore, relatively straightforward. Molecular ion determination of in vivo metabolites is

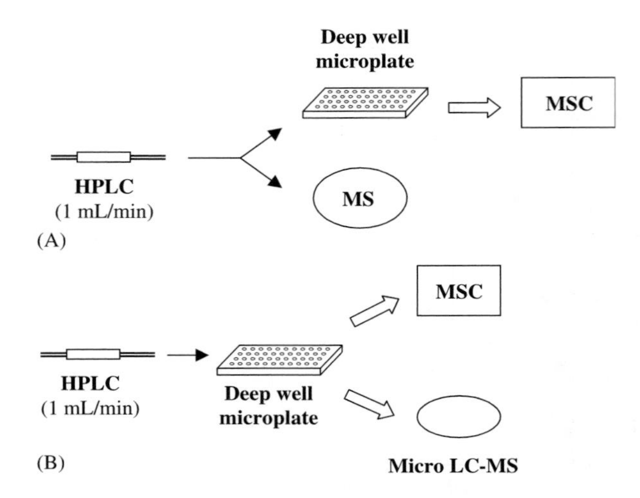

Figure 8.
Set-ups of the combination HPLC–MSC with mass spectrometry. System A is the set-up for simultaneous radioactivity profiling and metabolite identification. The HPLC eluent is spilt into both MSC and MS. System B is the set-up for tandem analysis of metabolite profile by HPLC–MSC and metabolite characterization by micro LC–MS

more challenging due to interference by significant amounts of co-eluting endogenous components (*m/z* 402, Fig. 9B) that often occurs. Usually, metabolite peaks are hardly visible in total ion chromatograms (TIC) of in vivo samples (Fig. 10A). Molecular ions of metabolites formed by common biotransformation reactions can be easily calculated and used to ascertain their presence in the sample. For instance, the *m/z* of a protonated molecule ($[M+H]^+$) of the glucuronide conjugate of a drug is the sum of 176 mass units and that of the $[M+H]^+$ of the parent drug. When metabolites are formed by uncommon or unpredicted biotransformation reactions, additional techniques may be necessary to determine the structure of the metabolite in an HPLC fraction containing radiolabeled material. These techniques can include conventional mass spectrometric techniques that are not dependent on a radiolabeled analyte. These techniques, such as neutral loss and parent ion scan analyses, usually require triple quadrupole mass spectrometry. Other techniques such as isotope ratio scanning can take advantage of known $^{12}C/^{14}C$ ratios, or can be based on chlorine and bromine atoms present in the parent molecule. Confirmation of an expected or determined molecular ion can be aided by comparing its retention time to that of the corresponding radioactive peak. An example of this approach is shown in Fig. 10. The protonated molecular ion of M6 was confirmed to be the ion at *m/z* 402 (a monohydroxylated metabolite), because it exhibited the same retention time in both the selected ion chromatogram (Fig. 10B) and the radiochromatogram (Fig. 10C). The peak shape and width in both chromatograms were also similar, adding confidence that the peak was properly assigned.

The second step of radioactive metabolite identification using LC–MSC–MS is to acquire product-ion spectra of the molecular ions, followed by interpretation of the spectra. Triple quadrupole mass spectrometry provides abundant data on

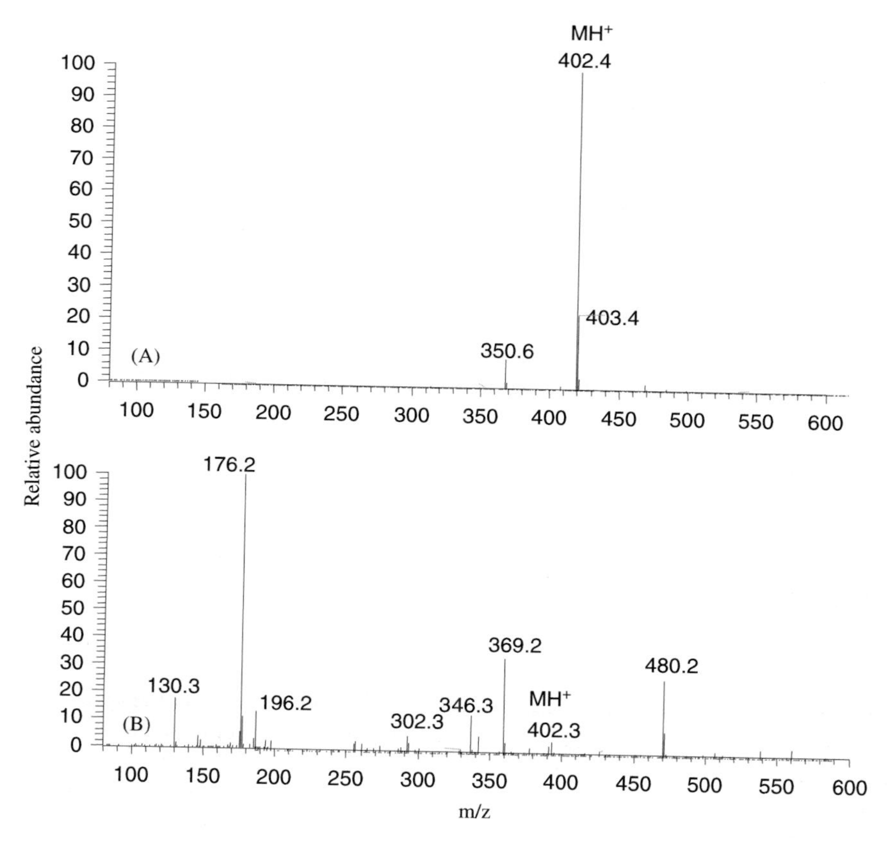

Figure 9.
Full scan mass spectra of buspirone metabolite M6 in human liver microsomes and human urine. [^{14}C] buspirone was incubated with HLM. After precipitating proteins the sample was analyzed by ion trap LC/MS (electrospray, positive mode). (A) Full scan spectrum of M6 of the HLM sample. The ion at *m/z* 402 corresponding to the molecular ion is 16 amu greater than the parent drug; (B) Full scan spectrum of M6 in human urine, in which the HLM sample was spiked into a control human urine (see Fig. 10)

small fragment ions or ions formed from multiple fragmentation steps in a single spectrum. Ion trap MS provides the ability to generate mass spectra through multiple stages of CID. This enables determination of parent–product-ion relationships that are not readily apparent in a triple quadrupole mass spectrum but are very helpful for understanding fragmentation pathways. Accurate-mass MS analysis by time of flight (TOF) MS or Fourier transform MS provides the ability to infer molecular formulas of both parent and fragment ions. Knowledge of the molecular formula of every ion in the spectrum greatly facilitates structural elucidation. These mass spectrometric techniques are highly complementary and their use in combination has been recommended to facilitate metabolite identification [35].

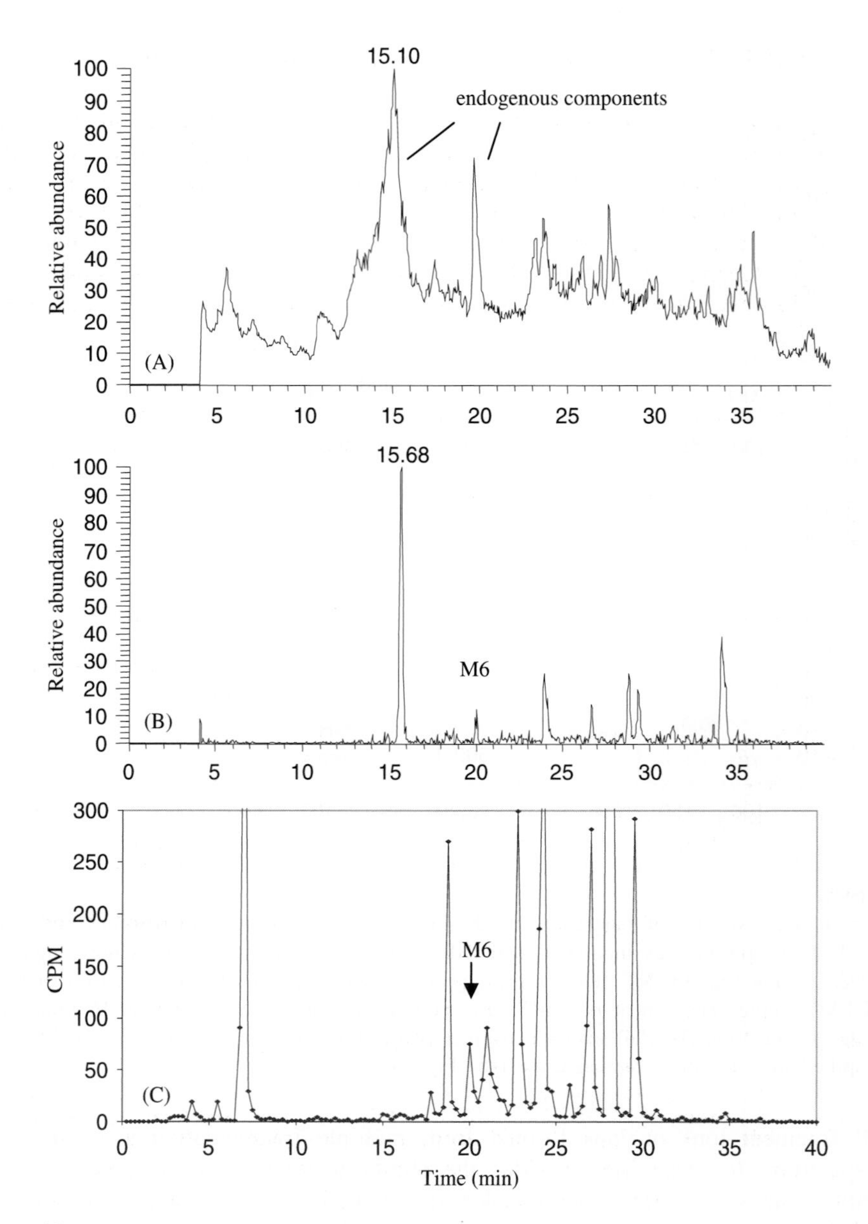

Figure 10.
Comparison of ion chromatograms and radiochromatogram of buspirone metabolites spiked into a control human urine sample. Buspirone metabolites in human urine were separated by HPLC and detected by ion trap MS (electrospray, positive mode) and TopCount (4 wells/min, 10 min counting time). (A) Total ion chromatogram; (B) Reconstructed ion chromatogram corresponding to *m/z* 402; (C) Radiochromatogram determined by TopCount. Full scan spectrum of M6 in this sample is presented in Fig. 9B

9.5.2. Use of High Resolution LC–MS to Enhance Analytical Selectivity

Mass spectrometric identification of low abundance drug metabolites in complex biological fluids such as plasma and urine is challenging because endogenous species can suppress metabolite ionization and overshadow metabolite ions. Typically, the TIC obtained from the LC–MS analysis of complex biological samples reveals no metabolite peaks or a few metabolite peaks with low intensities (Fig. 11A). Recently, a high resolution LC–MS-based mass defect filter (MDF) technique has been developed for selective detection of metabolite ions in complex

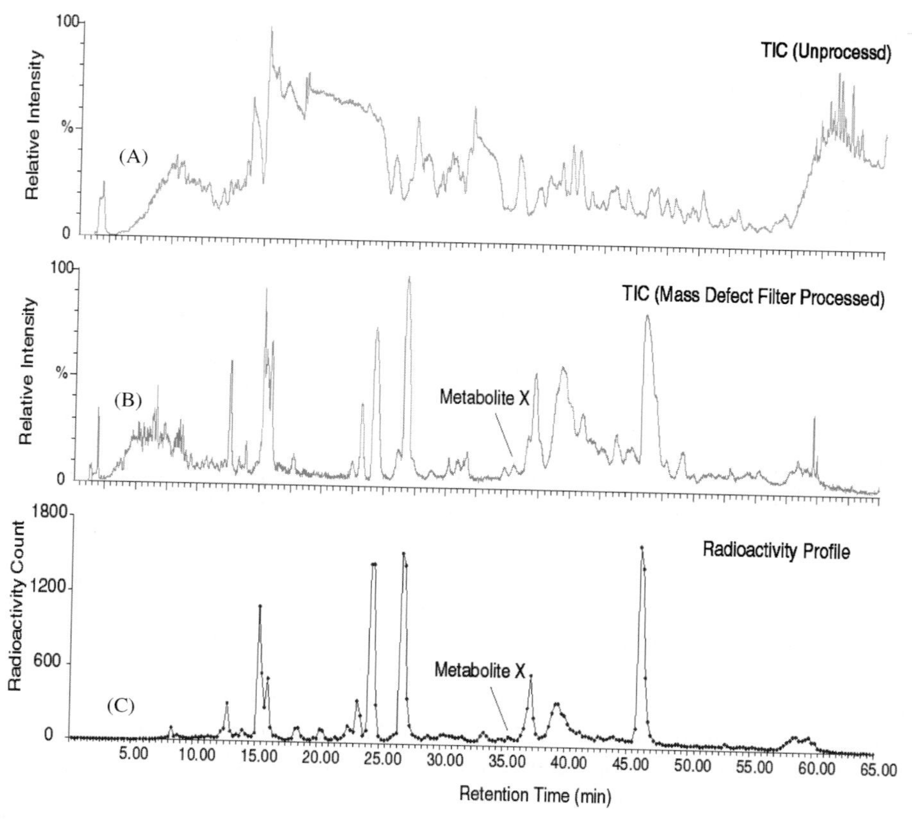

Figure 11.
Metabolite profile of compound A in dog bile [26]. (A) Total ion chromatogram of unprocessed LC/MS data. (B) Total ion chromatogram of processed LC/MS data. (C) Corresponding radioactivity chromatogram. Bile was collected from dogs and orally administered with ^{14}C-labeled compound A (20 mg/kg). HPLC solvents were water and acetonitrile containing 10 mM NH_4HCO_3 (pH 9.0). A portion of the HPLC effluent was collected in 15-s fractions for the radioactivity chromatogram. Another portion of the HPLC effluent was directed to a QToF Ultima mass spectrometer in positive ion electrospray mode

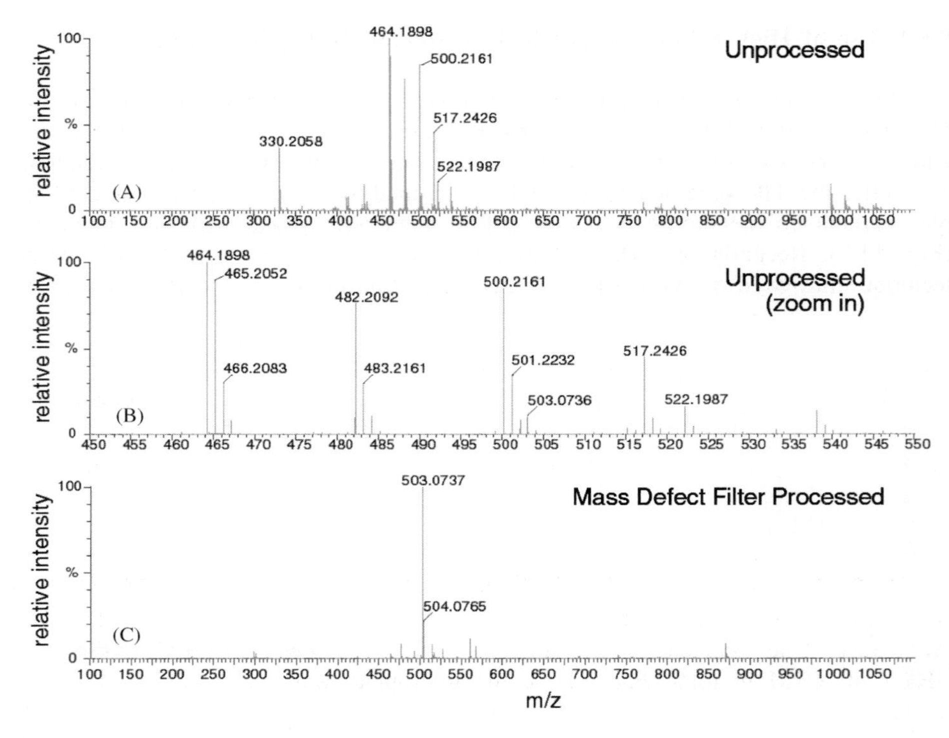

Figure 12.
Mass spectra of metabolite X at retention time 35.5 min [26]. (A) Full scan spectrum of metabolite X form the unprocessed total ion chromatogram (Fig. 11A). (B) Detail of (A) in mass range 450–550 Da. (C) Full scan spectrum of metabolite X (the molecular ion was at *m/z* 503.0737 from the MDF-processed total ion chromatogram (Fig. 11B)

matrices [26]. MDF can distinguish metabolite ions from interfering ions because the mass defects of drug-related components are often within a narrow range. As shown in Fig. 11B, the MDF-processed TIC is similar to the corresponding radiochromatogram (Fig. 11C). Most metabolites, including some minor metabolites such as metabolite X, were evident in the filtered TIC. Knowledge of the presence of metabolite X at the specific retention time provided the ability to determine its molecular ion (*m/z* 503) in the full scan spectrum (Fig. 12C) after the interfering ions were removed from the spectrum by MDF processing. In contrast, it was very difficult to recognize the molecular ion of metabolite X without MDF due to the presence of many ions originating from endogenous components (Fig. 12A and B). Although additional mass spectrometric analysis is required to determine the structures of metabolites detected by the MDF approach, several examples [26,27] have demonstrated that MDF is a very useful alternative to neutral loss and parent ion scanning for selective detection of metabolites formed via uncommon biotransformation pathways.

9.5.3. Use of Micro Flow LC–MS to Enhance Analytical Sensitivity

The identification of unknown metabolites requires molecular ions and CID fragment product ions of sufficient intensity for proper assignment of their mass spectra. Deficiencies in these requirements often occur in the LC–MS/MS analysis of in vivo metabolites because of low abundance of metabolites or suppression of metabolite ionization by co-eluting endogenous components and dosing formulation components. Solid phase extraction is often employed for sample cleanup and concentration, but can typically remove only polar components such as proteins and salts, and not the co-eluting less-polar components. To enhance MS sensitivity for analysis of radioactive metabolites, a new strategy has been developed as shown in Fig. 8B [28,29]. First, a radioactivity profile is determined by MSC without splitting a part of the HPLC eluate into the mass spectrometer. Based on the resultant metabolite profile, radioactive metabolites of interest are then recovered from the microplates and structurally characterized by nano and capillary LC–MS.

The detailed procedures and utility of this method are illustrated in the following example [28]. A rat plasma sample was treated by solid phase extraction, and then injected into a conventional analytical HPLC coupled with off-line MSC (MicroBeta). An HPLC profile of radiolabeled metabolites was obtained (Fig. 13) after solvent evaporation. The radioactivity residues corresponding to a minor metabolite peak, GH7 (240 CPM in total, Fig. 13) in the microplate was redissolved in methanol (200 μl methanol/well) and transferred into a test tube. After removing methanol, the

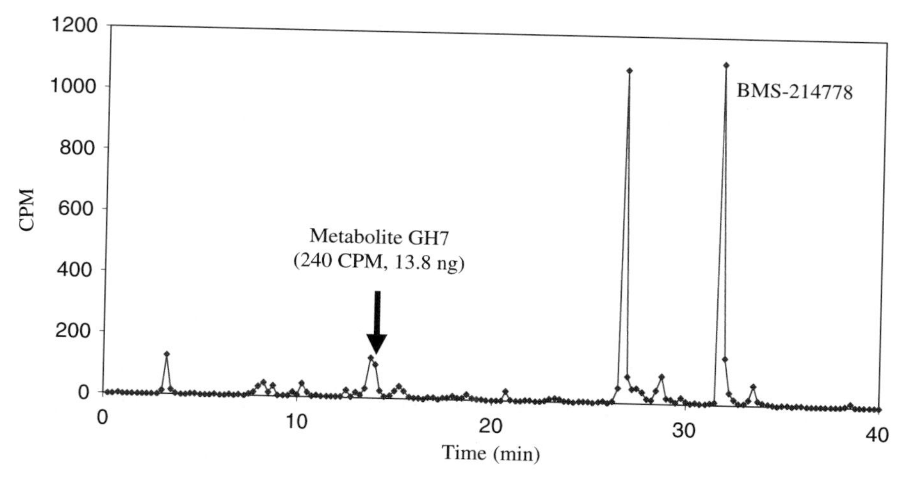

Figure 13.
Metabolite profile of [^{14}C] BMS-214778 in rat plasma [28]. A mixture of BMS-214778 and its stable isotope analog at 1:1 ratio was administrated into rats. Rat plasma extract (2856 DPM) was injected into HPLC (1 ml/min), followed by radioactivity determination by MicroBeta (4 wells/min, 10 min counting time) after removing solvent from microplates using SpeedVac. The metabolite peak had the total radioactivity 240 CPM

recovered materials were reconstituted in the HPLC solvents and analyzed by nano LC–MS/MS (200 nl/min, positive electrospray). Significant amounts of non-drug-related components, which co-eluted with GH7 in the first chromatographic step,

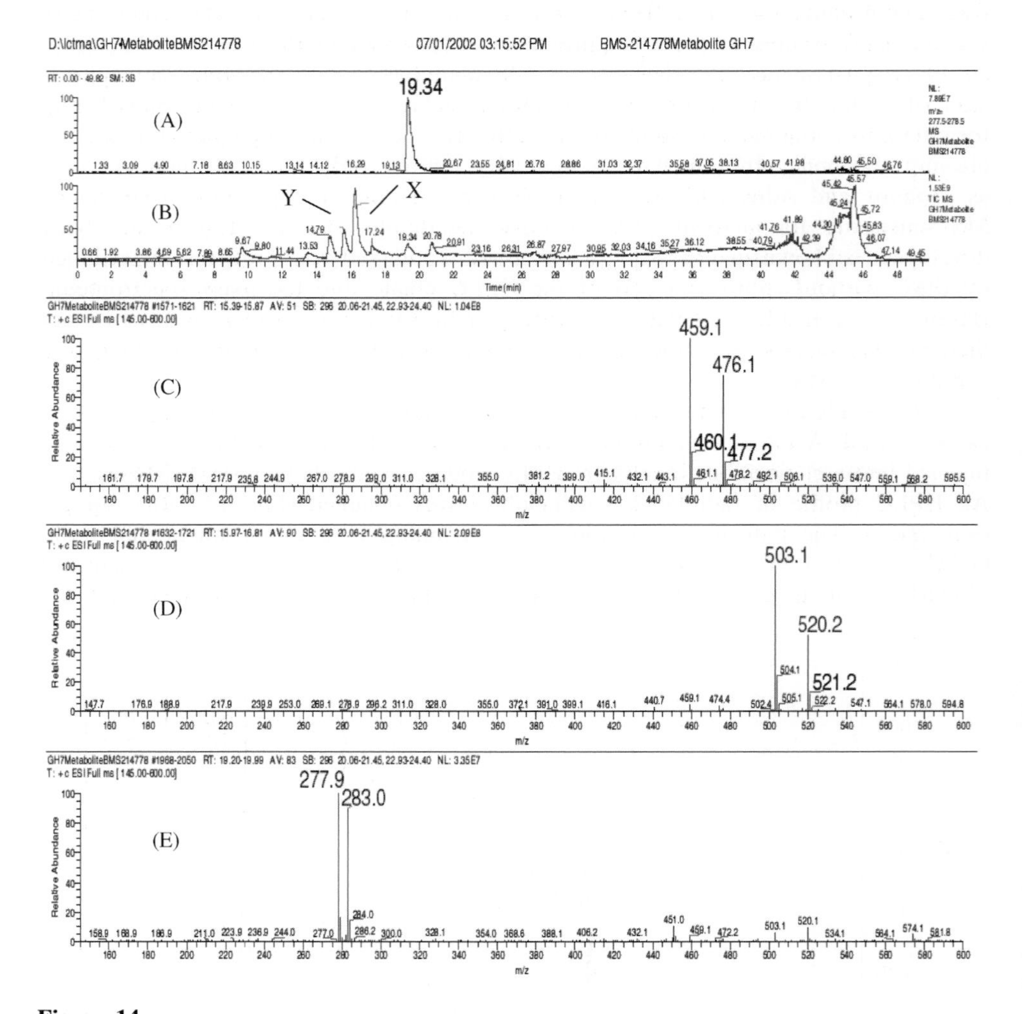

Figure 14.
Nano LC–MS analysis of metabolite GH7 recovered from 96-well plate [28]. Metabolite GH7 was recovered from the MicroBeta microplate (Fig. 13) with methanol (200 µl). A portion of the recovered radioactivity was analyzed by nano LC (200 nl/min) coupled with an ion trap mass spectrometer (electrospray, positive mode). (A) Reconstructed full scan ion chromatogram corresponding to *m/z* 278; (B) Total ion chromatogram; (C) Full scan spectrum of peak X; (D) Full scan spectrum of peak Y. Both X and Y may be associated with formulation components; (E) Full scan spectrum of metabolite GH7. The ion at *m/z* 278 is the molecular ion of GH7 and the ion at *m/z* 283 is the molecular ion of its stable isotope analog

were separated from the metabolite in the second, nano-flow chromatographic step (Fig. 14A and B). As a result, an intense molecular ion (*m/z* 278) of GH7 was observed in a full scan spectrum (Fig. 14E). Further product-ion analysis of GH7 indicated that it was a carboxylic acid metabolite (Fig. 15). Full scan spectra (Fig. 14C and D) of the co-eluting components suggested that those were components originating from the dosing formulation. In addition, in this example, detection of the molecular ion and interpretation of the mass spectrum was facilitated (Fig. 14E) by the use of a stable isotope labeled analog that was mixed with the drug in a 1:1 ratio in the dosing formulation.

There are several advantages to the combination of conventional HPLC for radioactivity profiling and micro- or nano-flow LC/MS for structural characterization: (1) Radiodetection retains superior sensitivity due to the high loading capacity of the large HPLC column; (2) Suppression of metabolite ionization by co-eluting components can be significantly reduced or eliminated by the dual stages of chromatographic separation; (3) Exceptionally high sensitivity is achieved by the use of

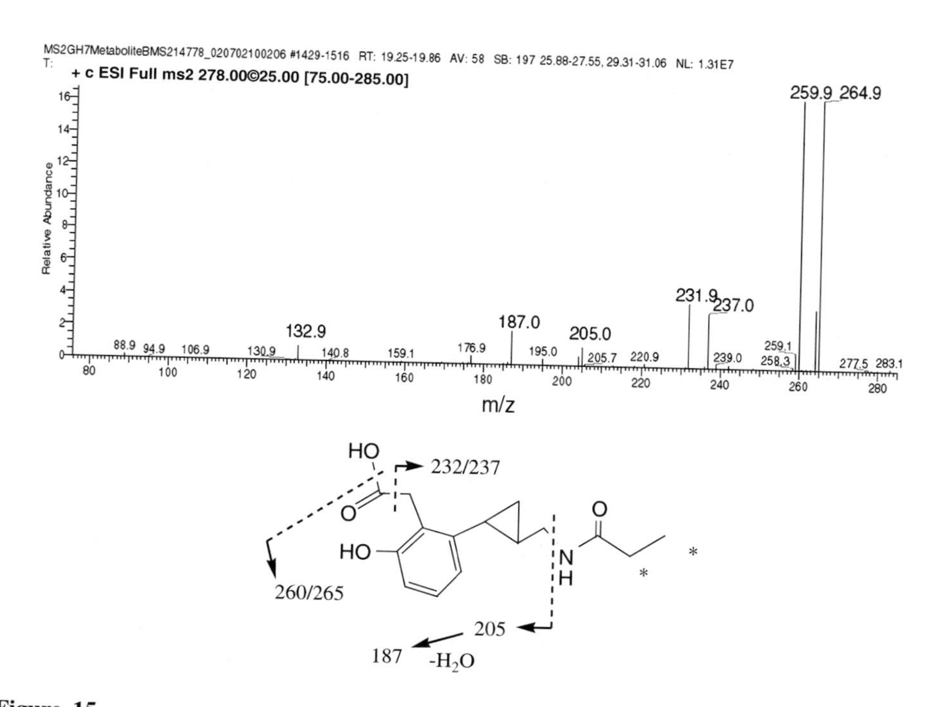

Figure 15.

Product-ion spectrum and assigned structure of metabolite GH7 [28]. The product-ion scan analysis was carried out on the molecular ion (*m/z* 278–283) of GH7 metabolite recovered from a microplate (Fig. 13). The * sign indicates location of five ^{2}H labels. The structure of the GH7 metabolite is proposed based on its molecular- and product-ion spectrum

very low flow rates; and (4) Use of the second HPLC allows LC–MS conditions to be optimized for the analysis of each metabolite.

9.6. Conclusions

MSC is an off-line liquid chromatographic detector that has been available for only the past few years; however, in that short time it has shown great promise as a powerful and versatile tool for detection and quantitative analysis of low levels of radiolabeled metabolites. Compared to conventional HPLC–LSC methods, HPLC–MSC not only has comparable precision, better sensitivity and increased analytical throughput, it also significantly reduces radioactive waste and requires minimal manual operation. HPLC–MSC is useful for enzyme kinetic studies where low concentrations of metabolites are formed by drugs that are slowly metabolized or have low K_m values, or where multiple metabolites that are each present in low concentrations are formed. HPLC–MSC coupled with an electrospray ionization mass spectrometer is an ideal choice for structural elucidation of low abundance radiolabeled metabolites. Additionally, LC–MSC has been used in conjunction with high mass-resolution MS-based mass defect filtering technique for selective detection of metabolite ions in complex biological matrices. Finally, micro flow LC–MS/MS analysis of radioactive metabolites recovered from microplates has shown excellent sensitivity for identification/characterization of very small quantities of metabolites.

References

1. Lin, J.H., Lu, A.Y.H., 1997. Role of pharmacokinetics and metabolism in drug discovery and development. Pharm. Rev. 49, 403–449.
2. Tucker, G.T., Houston, J.B., Huang, S-M., 2001. Optimizing drug development: strategies to assess drug metabolism/transporter interaction potential-toward a consensus. Clin. Pharm. Ther. 70, 103–114.
3. The European Agency for the Evaluation of Medicinal Products: Human Medicines Evaluation Unit, 1997. Note for Guidance on the Investigation of Drug Interactions. (CPMP/EW/560/95).
4. Baillie, T.A., Cayen, M.N., Fouda, H., Gerson, R.J., Green, J.D., Grossman, S.J., Klunk, L.J., LeBlance, B., Perkins, D.G., Shipley, L.A., 2002. Contemporary issues in toxicology: drug metabolites in safety testing. Toxicol. Appl. Pharm. 182, 188–196.
5. Lautala, P., Ulmanen, I., Taskine, J., 1999. Radiochemical high-performance liquid chromatographic assay for the determination of catechol *O*-methyltransferase activity towards various substrates. J. Chromatogr. B 736, 143–151.
6. Veltkamp, A.C., 1990. Radiochromatography in pharmaceutical and biomedical analysis. J. Chromatogr. 531, 101–129.

7. Klebovich, I., Morovjan, G., Hazai, I., Mincsovice, E., 2002. Separation and assay of ^{14}C-labeled glyceryl trinitrate and its metabolites by OPLC coupled with on-line or off-line radioactivity detection. J. Planar. Chromatogr. 15, 404–409.
8. Egnash, L.A., Ramanathan, R., 2002. Comparison of heterogeneous and homogeneous radioactivity flow detectors for simultaneous profiling and LC–MS/MS characterization of metabolites. J. Pharm. Biomed. Anal. 27, 271–284.
9. Heath, T.G., Mooney, J.P., Broersma, R., 1997. Narrow-bore liquid chromatography–tandem mass spectrometry with simultaneous radioactivity monitoring for partially characterizing the biliary metabolites of an arginine fluoroalkyl ketone analog of D-MePhe-Pro-Arg, a potent thrombin inhibitor. J. Chromatogr. B 688, 281–289.
10. Vlasakova, V., Brezinova, A., Holik, J., 1998. Study of cytokinin metabolism using HPLC with radioisotope detection. J. Pharm. Biomed. Anal. 17, 39–44.
11. Onisko, B.C., 2001. Radiochemical detection for packed capillary liquid chromatography–mass spectrometry. J. Am. Soc. Mass Spectrom. 13, 82–84.
12. Mullen, W., Hartley, R.C., Crozier, A., 2003. Detection and identification of ^{14}C-labelled flavonol metabolites by high-performance liquid chromatography–radiocounting and tandem mass spectrometry. J. Chromatogr. A 1007, 21–29.
13. Morvovjan, G., Dalmadi-Kiss, B., Klebovich, I., Mincsovice, E., 2002. Metabolite analysis, isolation and purity assessment using various liquid chromatographic techniques combined with radioactivity detection. J. Chromatogr. Sci. 40, 603–608.
14. Iavarone, L., Scandola, M., Pugnaghi, F., Gross, P., 1995. Qualitative analysis of potential metabolites and degradation products of a new antiinfective drug in rat urine, using HPLC with radiochemical detection and HPLC–mass spectrometry. J. Pharm. Biomed. Anal. 13, 607–614.
15. Chando, F.J., Everett, D.W., Kahle, A.D., Starrett, A.M., Vachharajani, N., Shyu, W.C., Kripalani, K.J., Barbhaiya, R.H., 1998. Biotransformation of irbestartan in man. Drug Metab. Dispos. 26, 408–417.
16. Zhu, M., Bering, N., Mitroka, J.G., 2000. Use of HPLC with 96-well microplate scintillation counter for rapid profiling of low levels of radioactive metabolites in biological fluids. Drug Metab. Rev. 32 (Suppl. 2), 177.
17. Kelly, P.J., Harrell, A.W., North, S.E., Boyle, G.W., 2000. A rapid and efficient method for off-line radiodetection using HPLC with TopCount and 96-well plate technology. Drug Metab. Rev. 32 (Suppl. 2), 178.
18. Boernsen K.O., Floeckher, J.M., Bruin, G.J.M., 2000. Use of microplate scintillation counter as a radioactivity detector for miniaturized separation techniques in drug metabolism. Anal. Chem. 72, 3956–3959.
19. Yin, H., Greenberg, G.E., Fischer, V., 2000. Application of Wallac MicroBeta radioactivity plate counter and Wallac Scintiplate in metabolite profiling and identification studies. Application Note. PerkinElmer Life Sciences.

20. Kiffe, M., Jehle, A., Ruembeli, R., 2003. Combination of high performance liquid chromatography and microplate scintillation counting for crop and animal metabolism studies: a comparison with classical on-line and thin-layer chromatography radioactivity detection. Anal. Chem. 75, 723–730.
21. Zhang, D., Organ, M., Gedamke, R., Roongta, V., Dai, R., Zhu, M., Rinehart, J., Klunk, L., Mitroka, J., 2003. Protein covalent binding of MaxiPost through a cytochrome P450-mediated ortho-quinone methoide intermediate in rats. Drug Metab. Dispos. 31, 837–845.
22. Nedderman, A.N.R., Savage, M.E., White, K.L., Walder, D.K., 2004. The use of 96-well scintiplates to facilitate definitive metabolism studies for drug candidates. J. Pharm. Biomed. Anal. 34, 607–617.
23. Zhang, D., Zhao, W., Roongta, V.A., Mitroka, J.G., Klunk, L.J., Zhu, M., 2004. Amide *N*-glucuronidation of MaxiPost catalyzed by UDP-glucuronosyltransferase 2B7 in humans. Drug Metab. Dispos. 32, 545–551.
24. Zhu, M., Zhao, W., Jimenez, H., Zhang, D., Yeola, S., Dai, R., Vachharajani, N., Mitroka, J., 2005. Cytochrome P450 3A-mediated metabolism of buspirone in human liver microsomes. Drug Metab. Dispos. 33, 500–507.
25. Zhu, M., Zhang, H., Zhao, W., Mitroka, J., Klunk, L., 2002. A novel in vitro approach, using a microplate scintillation counter and LC/MS, for determining formation pathways and structures of sequential metabolites: buspirone as an example. Drug Metab. Rev. 34, 167.
26. Zhang, H., Zhang, D., Ray, K., 2004. A software filter to remove interference ions from drug metabolites in accurate mass liquid chromatography/mass spectrometric analyses. J. Mass Spectrom. 38, 1110–1112.
27. Zhang, H., Ray, K., Zhao, W., Zhang, D., Gozo, S., Zhu, M., 2004. A new approach utilizing mass defect filter on high resolution LC/MS data to identify metabolite ions in biological matrices: omeprazole as an example. ASMS Archive Abstract.
28. Gedamke, R., Zhao, W., Gozo, S., Mitroka, J., Zhu, M., 2003. A sensitive method for plasma metabolite identification using nano LC/ion trap MS in conjunction with a microplate scintillation counter. ASMS Archive Abstract.
29. Zhu, M., Wang, L., Gedamke, R., Zhao, W., Zhang, D., Gozo, S., 2003. Determination of radioactivity distribution and metabolite profiles in tissues using a combination of microplate scintillation counting, capillary LC/MS and whole-body autoradiography. Metab. Rev. 35, 76.
30. Zhu, M., Zhao, W., Vazquez, N., Mitroka, J.G., 2005. Analysis of low level radioactive metabolites in biological fluids using high-performance liquid chromatography with microplate scintillation counting: method validation and application. J. Pharm. Biomed. Anal. (in press).
31. Packard, 1999. TopCount NXT™ microplate scintillation & luminescence counters. Reference Manual. Packard Instrument Company, Meriden, CT, USA.
32. Currie, L.A., 1968. Limits for qualitative detection and quantitative determination: application to radiochemistry. Anal. Chem. 40, 586–593.

33. Nassar, A.-E.F., Bjorge, S.M., Lee, D.Y., 2003. On-line liquid chromatography—accurate radioisotope counting coupled with a radioactivity detector and mass spectrometer for metabolite identification in drug discovery and development. Anal. Chem. 75, 785–790.
34. Packard, 1996. LSC Handbook of Environmental Liquid Scintillation Spectrometry. Packard Instrument Company, Meriden, CT, USA.
35. Clarke, N.J., Rindgen, D., Korfmacher, W.A., Cox, K., 2001. Systematic LC/MS metabolite identification in drug discovery. Anal. Chem. 73, 431A–439A.

Identification and Quantification of Drugs, Metabolites and Metabolizing
Enzymes by LC–MS
Swapan K. Chowdhury, editor.

Chapter 10

OXIDATIVE METABOLITES OF DRUGS AND XENOBIOTICS: LC–MS METHODS TO IDENTIFY AND CHARACTERIZE IN BIOLOGICAL MATRICES

Ragulan Ramanathan, Swapan K. Chowdhury, and Kevin B. Alton

10.1. Introduction

Discovery and development of a new chemical entity (NCE) as a human therapeutic agent has become a tremendously complex process, takes over 10 years and costs 1–2 billion dollars (Dickson and Gagnon, 2004; Rawlins, 2004). Early understanding of the metabolic fate of NCEs is critical for reducing preclinical and clinical phase attrition of drug candidates. In late stage development, characterization of metabolites in human and non-clinical species is required to assure that the non-clinical species undergoing safety evaluations are adequately exposed to human metabolites. Therefore, studies on drug metabolism have become an increasingly important component of the drug development process.

Drugs and xenobiotics, when exposed to living systems undergo metabolism or biotransformation as a detoxification process to form more polar derivatives that can be readily eliminated from the body. Although metabolism can occur in different parts of the body, most of the metabolizing enzymes are localized in the gut and liver. These enzymes and their role in metabolism have been discussed in detail in Chapter 12 (Ghosal *et al.*) of this book.

Briefly, the two principal pathways involved in the detoxification are phase I and phase II metabolic reactions. The phase I reaction includes oxidation, reduction, hydrolysis, etc. and usually involves small changes in hydrophilicity of drugs or xenobiotics, while phase II processes in general involve modification of a functional group (-OH, -NH_2, -SH, or -COOH) by glucuronidation, sulfation, and amino acid or glutathione conjugation. Conjugation often leads to greater aqueous solubility such that these metabolic species can be excreted more readily in urine and bile. Only a handful of phase II metabolites have been found to be pharmacologically active, for example, the phenolic glucuronide conjugate of ezetimibe (Patrick *et al.*, 2002) and morphine-6-glucuronide (Ishii *et al.*, 1997).

The primary reason for identification of metabolites is to assure human safety of drugs under clinical investigation. Initially, metabolites of a drug are characterized with in vitro systems (microsomes, hepatocytes, S9 fractions, etc.) and in vivo using the mouse, rat, rabbit, dog, and monkey, species commonly used for safety assessment studies. Subsequently, metabolites in humans are identified following drug administration to assure that the non-clinical species undergoing safety assessment are adequately exposed to human metabolites of the drug. Additionally, metabolite(s) can be pharmacologically active drug and/or may have a better safety profile than the parent drug (Van Heek *et al.*, 1997), thus redirecting the search for the next generation drug.

Although drugs can be metabolized to a plethora of biotransformation products, this chapter focuses, using specific drug and xenobiotic examples, on the oxidative biotransformation (the most common phase I reactions) and the liquid chromatography–mass spectrometry (LC–MS) methods to detect and characterize them. For the sake of simplicity, oxidative biotransformation processes can be divided into two groups: (1) microsomal oxidation catalyzed mainly by Cytochrome P450s (CYP450) and (2) non-microsomal oxidations involving (but not limited to) monoamine oxidase (MAO), peroxidases and flavin-containing monooxygenase (FMO). Although both groups of drug metabolizing enzymes have been the focus of many studies, the CYP superfamily has been extensively studied because they account for more than 90% of oxidative biotransformation of drugs and xenobiotics (Guengerich *et al.*, 2000). About 750 CYP450s or polymorphs have been sequenced to date and 55 of them have been characterized as human isoforms (Venkatakrishnan *et al.*, 2001). Microsomal oxidations include aromatic and side-chain hydroxylation, *N*-oxidation, *S*-oxidation (sulfoxidation and sulfonation), *N*-hydroxylation, *N*-, *O*-, *S*-dealkylation, deamination, dehalogenation, and desulfation. Although MAOs, FMOs, and peroxidases have been associated with several of the above-mentioned biotransformation processes, their involvement is of lesser importance. This chapter is organized according to oxidative biotransformation processes with specific examples for detecting and characterizing the metabolites in biological matrices by LC–MS and MS/MS techniques.

10.2. Dealkylation

Dealkylation is a process that results in removal of alkyl group(s) from a heteroatom (nitrogen, oxygen, and sulfur) following oxidation of the α-carbon. The simplest form of dealkylation involves oxidative removal of a single or multiple methyl, ethyl, or larger alkyl moieties of a drug or xenobiotics to expose nitrogen, oxygen, or sulfur. These dealkylated biotransformation products are more hydrophilic and may be eliminated more extensively in the bile and urine compared to the administered drug. Although it is not the intent of this Chapter to list all the drugs and xenobiotics that undergo dealkylation, a few of the most commonly discussed drugs that extensively undergo dealkylation metabolism include Verapamil (VER), Sildenafil, Omeprazole, Imipramine, Selegiline (SG), Diazepam, Codeine, Erythromycin, Morphine, Tamoxifen (TX), Dextromethorphan (DXM), and Fluoxetine.

Dealkylation always results in a reduction of molecular weight relative to the administered drug. Compared to *N*- and *O*-dealkylation, only a handful of drugs on the market have been shown to undergo biotransformation through *S*-dealkylation (e.g., 6-methylmercaptopurine). In the following section, selected drugs that undergo *N*- and/or *O*-delkylation are discussed by categorizing their biotransformation pathways into (a) mono-demethylation, (b) didemethylation or deethylation, (c) depropyl- or deisopropylation, and (d) dealkylation of larger moieties. When available, LC–MS spectra and/or MS/MS spectra will be presented and discussed.

10.2.1. Mono-Demethylation

A single *N*- or *O*-demethylation results in a negative mass shift of the metabolite by 14 Da compared to the parent compound. LC–MS spectra, will therefore show a mass-to-charge (*m/z*) shift of 14 Thompson (Th) for the $[M+H]^+$ ion of the metabolite. Structure elucidation using tandem mass spectrometry or LC–MS/MS can be complex or simple depending on the chemical structure of the compound and possible site(s) of *N*- or *O*-demethylation.

10.2.1.1. Verapamil (Calan/Verelan)

VER, a calcium channel blocker, widely used for arrhythmias, antianginal therapy, and myocardial ischemia, undergoes biotransformation to form *N*- and/or *O*-demethylated as well as *N*-dealkylated metabolites. The possible sites of dealkylations are shown below.

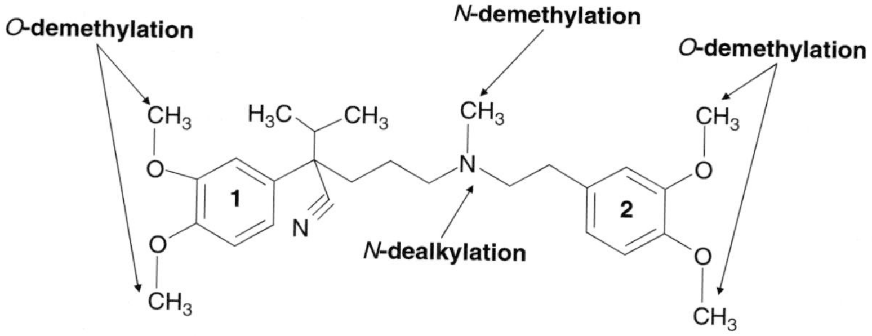

In the absence of reference standards and chromatographic separation, using LC–MS spectra alone it would not be possible to distinguish an *O*-demethylated metabolite of VER from that following *N*-demethylation. Both biotransformation pathways lead to metabolites with identical elemental composition and protonated molecular ions ($[M+H]^+$) with the same mass-to-charge ratio, in this case 441 Th, which is 14 Th lower than that of the molecular ion of VER (*m/z* 455).

The proposed fragmentation scheme for VER is shown below (Fig. 1). Careful interpretation of the LC–MS/MS spectrum (Fig. 2, modified from Walles *et al.*) suggests that M2 results from *N*-demethylation and not from *O*-demethylation.

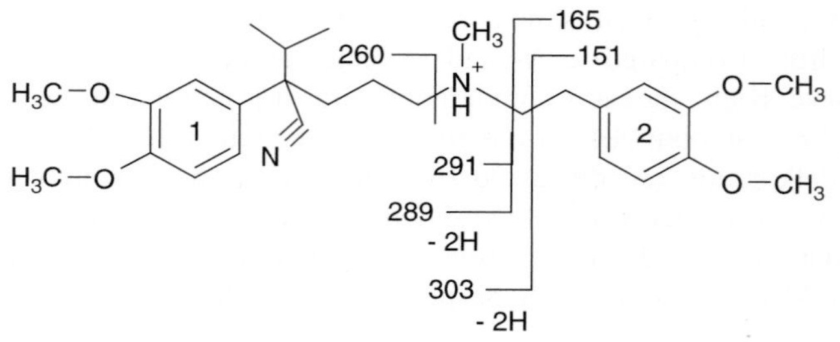

Figure 1.
Proposed fragmentation scheme for protonated ions of VER

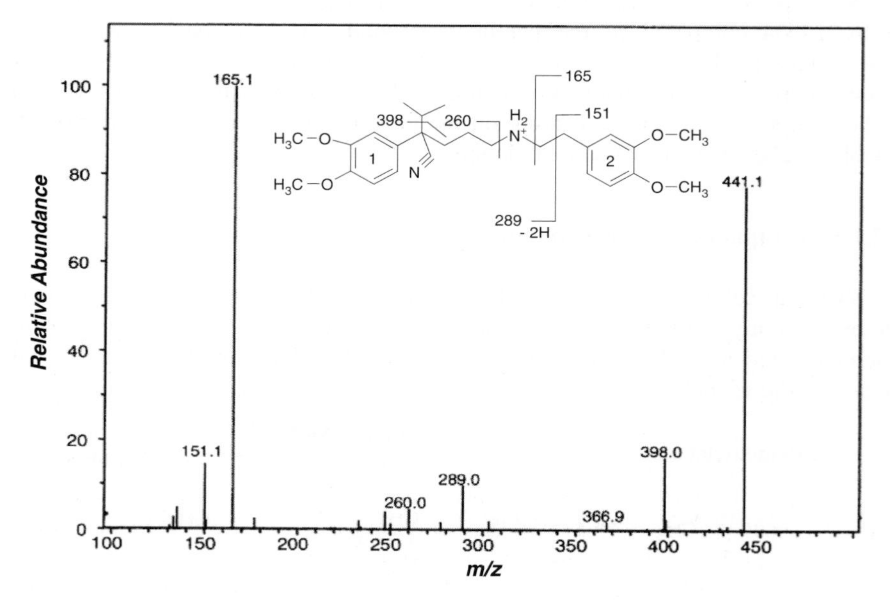

Figure 2.
LC–MS/MS spectrum and the proposed fragmentation scheme for protonated ions of M2 (norverapamil)

Comparison of the fragmentation schemes shown in Figs. 1 and 2 and the observation of the fragment ions at *m/z* 165 and 260 in the LC–MS/MS spectrum of M2 show that both dimethoxyphenyl moieties are intact, whereas the methyl amino moiety has undergone demethylation. In addition to M2, M3, and M4 are also demethylated metabolites of VER; the LC–MS/MS spectrum of M3 is shown in Fig. 3. Observation of a fragment ion at *m/z* 165 Th suggested that the dimethoxyphenyl group (2) is intact. The absence of a fragment ion at *m/z* 260 Th observed with VER and the appearance of the fragment ion at *m/z* 246 in the MS/MS spectrum

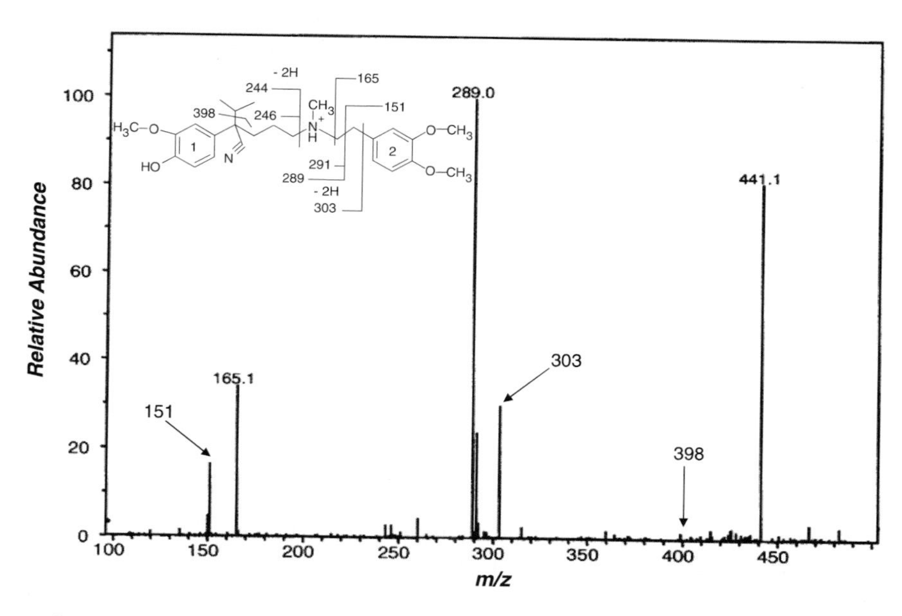

Figure 3.
LC–MS/MS spectrum and the proposed fragmentation scheme for protonated ions of M3

of M3 suggest that demethylation occurred on the dimethoxyphenyl group (1) adjacent to the isopropylvaleronitrile moiety and M3 indeed is a *O*-desmethylverapamil metabolite. Similar MS/MS fragmentation was used to distinguish M4, from M2 and M3. M4 was characterized as another *O*-desmethylverapamil metabolite with demethylation taken place from the dimethoxyphenyl group.

Although the nature of demethylation (*N*- or *O*-) could be easily characterized by LC–MS/MS, the precise location of *O*-demethylation (within the same dimethoxyphenyl group) cannot be identified by LC–MS/MS alone. Nuclear magnetic resonance (NMR) or LC–NMR analysis and LC–MS/MS data on synthetic reference standards may provide the ultimate identification of any metabolite where structure cannot easily be deduced from MS/MS data.

10.2.1.2. Diazepam (Valium)

Diazepam is widely used as muscle relaxant, sedative, and anticonvulsant agent. Under LC–MS conditions, protonated ions are observed at *m/z* 285. Diazepam (structure follows) contains a chlorine atom that serves as a distinct marker for detection of all drug-derived components in the LC–MS spectra. Diazepam undergoes CYP2C19- and CYP3A4-mediated *N*-demethylation to form nordiazepam, which gives protonated ions at *m/z* 271 (Jung *et al.*, 1997). Since only one metabolic demethylation is possible for diazepam, simple *m/z* determination by LC–MS indicates not only occurrence of demethylation, but also possible site of demethylation. This is an example of the simplest form of demethylation that can be readily identified by LC–MS.

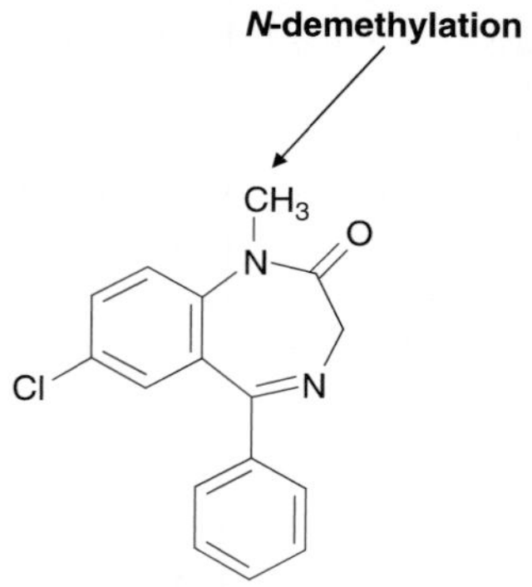

N-demethylation

CH_3 N O Cl N

In general, when a drug contains element(s) (e.g., Cl and Br) with a distinctive natural abundance isotope pattern, the mass spectral detection and structural characterization of drug-derived material becomes easier. In the absence of an inherent diagnostic isotope pattern, a similar benefit can be derived by using test article which has been isotopically enriched (Gopaul *et al.*, 2001).

10.2.1.3. Omeprazole (Prilosec®)

Omeprazole is a proton pump inhibitor, which is used for controlling various gastrointestinal disorders by regulating acid production in the stomach.

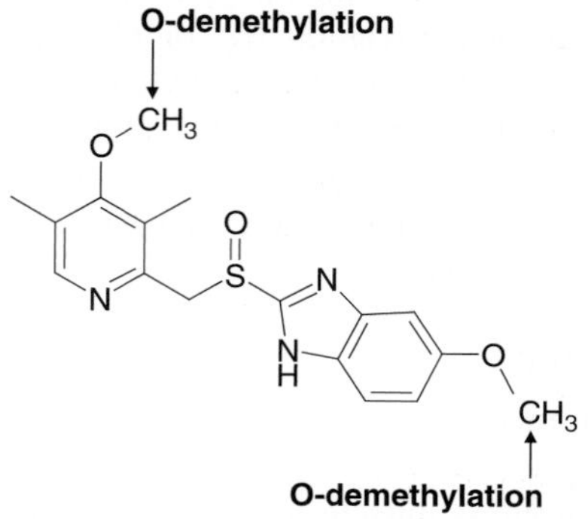

O-demethylation

O CH_3 O S N N N H O CH_3

O-demethylation

Omeprazole can undergo *O*-demethylation from either the 4-methoxy-3,5-dimethyl-2-pyridinyl or methoxybenzimidazole moiety (Äbelö *et al.*, 2000). Under LC–MS conditions, a negative mass shift of 14 Da was observed between the parent drug (*m/z* 346) and the demethylated metabolites (*m/z* 332). Fragmentation under LC–MS/MS conditions can be used to identify the site of *O*-demethylation quite easily due to the different chemical environment of each methyl group.

10.2.2. Didemethylation and Deethylation

Didemethylation, regardless of whether the process involves the same heteroatom (*O*- or *N*-) or heteroatoms from two different sites, results in a 28 Da negative shift of the LC–MS measured mass of the metabolite relative to the parent. Similarly, LC–MS-measured mass following a single deethylation, irrespective of the heteroatom involved would shift lower by 28 Da. However, ring cleavage deethylation involving two heteroatoms (most often involves two nitrogens but possible from combinations of *N*-, *O*- and *S*-) results in a net loss of 26 Da from the LC–MS-measured mass of the parent drug. Selected examples of drugs that undergo multiple demethylation or deethylation biotransformations are as follows.

10.2.2.1. Dextromethorphan

DXM is a synthetic analog of codeine, which is widely used as a cough suppressant. In the LC–MS spectrum, protonated molecular ions for DXM are observed at *m*/*z* 272. DXM undergoes biotransformation via *N*- and *O*-demethylation to form 3-methoxymorpinan and dextrorphan (DXP), respectively. The *O*-demethylation of DXM has been demonstrated as a suitable in vivo and in vitro probe for assessing P450 2D deficiency (Marshall *et al.*, 1992). Either *N*- or *O*-demethylation produces molecular ions at *m*/*z* 258 (Yu *et al.*, 2001). 3-Methoxymorpinan and DXP then undergo a second demethylation process to yield didemethylated metabolite of DXM (3-hydroxymorphinan). Under LC–MS conditions, 3-hydroxymorphinan, which is an *N*-, *O*-didesmethyl metabolite of DXM, shows protonated ions at *m*/*z* 244. Since there are only two demethylation sites and there is no other reasonable way to lose 28 Da, the molecular weight determination by LC–MS alone is sufficient for characterization of 3-hydroxymorphinan. Formation of the 3-hydroxy metabolite allows phase II conjugation and rapid elimination of DXM from the body.

***N*-demethylation**

N– CH_3

H_3C–O

***O*-demethylation**

10.2.2.2. Lidocaine

Lidocaine (LIDO) is a commonly used local anesthetic, which has been on the market for over 50 years. LIDO is also used in patients with acute myocardial infraction as an agent to prevent ventricular fibrillation. The principal biotransformation pathway of LIDO in humans involves CYP3A4- and CYP1A2-catalyzed oxidative *N*-deethylation to *N*-(*N*-ethylglycyl)-2,6-xylidine (MEGX). Following *N*-deethylation, protonated molecular ions of the metabolite is observed at *m/z* 207, which is 28 Da lower than those observed for protonated ions of LIDO (Wang *et al.*, 2000). This example further illustrates that in many cases when there is no other reasonable way to lose 28 Da, simple molecular weight determination can provide general clues on the nature of biotransformation, and LC–MS/MS experiments may only be necessary to confirm the proposed metabolic modification.

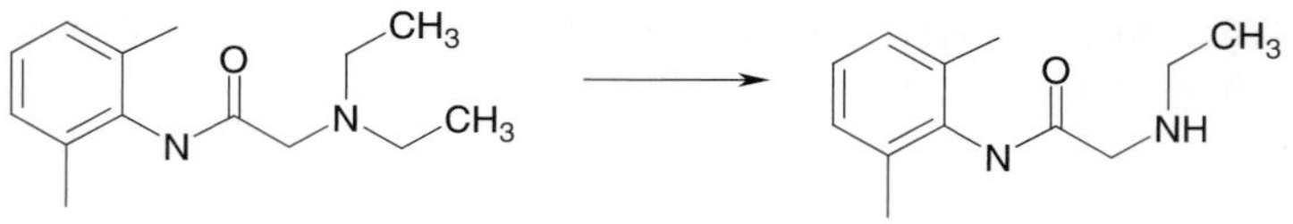

Lidocaine
$[M + H]^+$ @ m/z 235

Desethyllidocaine (MEGX)
$[M + H]^+$ @ m/z 207

10.2.2.3. Phenacetin

An analgesic and antipyretic drug withdrawn from the market in 1996 due to potential toxicity. *O*-deethylation of phenacetin is catalyzed by CYP1A2; phenacetin-*O*-deethylation is often used as a probe for CYP1A2 enzyme activity. Protonated ions of phenacetin and the *O*-deethylated metabolite are detected at *m/z* 180 and 152 (Δ = 28 Th), respectively. In this case, a simple molecular weight determination by LC–MS provide evidence for deethylation and the most likely site of modification.

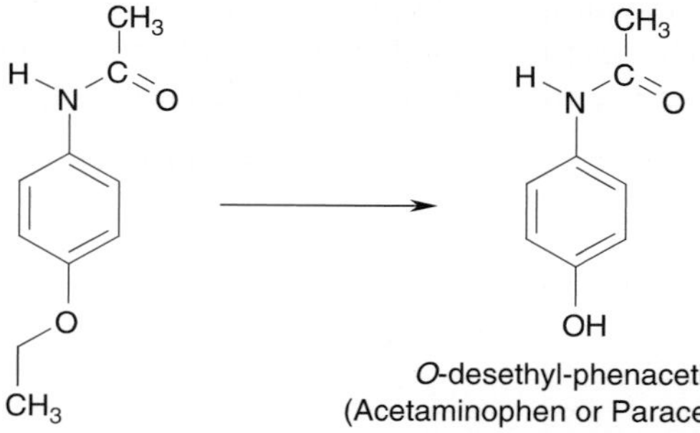

Phenacetin
$[M + H]^+$ @ m/z 180

O-desethyl-phenacetin
(Acetaminophen or Paracetamol)
$[M + H]^+$ @ m/z 152

10.2.2.4. Reboxetine (Edronax®)

Reboxetine is an antidepressant with a molecular weight of 313 Da. Following *O*-deethylation, protonated molecular ions for the metabolite are observed at *m/z* 286 (Wienkers *et al.*, 1999). The *O*-deethylation of reboxetine is primarily mediated by CYP3A4 and one of the primary routes of metabolism involved in the elimination of this drug.

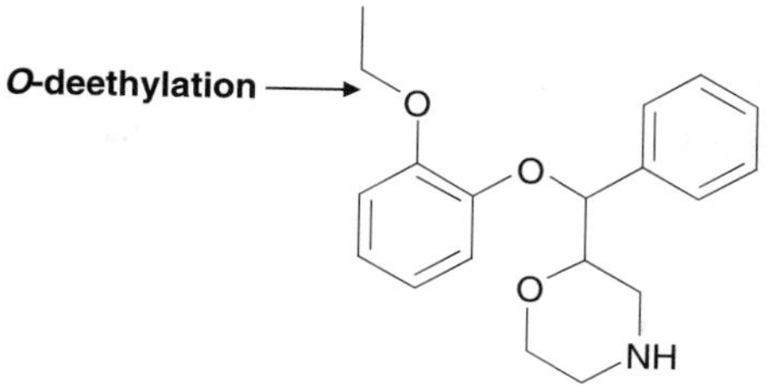

O-deethylation →

10.2.2.5. Sildenafil (Viagra®)

An inhibitor of phosphodiesterase type 5 enzyme (PDE5), which is used for treatment of erectile dysfunction (Walker *et al.*, 1999). In male subjects, following oral (PO) and intravenous (IV) dosing, the circulating levels of piperazine-*N*-desmethyl-sildenafil and piperazine-*N, N*-desethyl-sildenafil metabolites collectively accounted for 32 and 16% of the profiled radioactivity, respectively. In the excreta, piperazine-*N*-desmethyl-sildenafil and piperazine-*N, N*-desethyl-sildenafil and piperazine-*N*-desmethyl-*N, N*-desethyl-sildenafil each accounted for 3, 22, and 3% of the administered dose, respectively.

The metabolism of sildenafil represents an example, where a single or a combination of different metabolic processes (e.g., dealkylation) can lead to the formation of a number of metabolites with identical molecular weight. Therefore, molecular weight determination of metabolites by LC–MS alone is not sufficient to identify site(s) of dealkylation of sildenafil. However, as shown in Figs. 4–6, careful interpretation of LC–MS/MS and LC–MS^3 spectra can lead to the identification of each metabolite.

In the LC–MS spectrum, the protonated molecular ions of Sildenafil appeared at *m/z* 475. Under MS/MS conditions, sildenafil ($[M+H]^+$) yielded diagnostic fragment ions at *m/z* 99, 283, 311, and 377 (Fig. 5).

In addition to the fragment ions observed at *m/z* 283, 311, and 377 (Figs. 4 and 5), the fragmentation of the -NH–SO_2- bond is critical to understanding the structural changes that have occurred in the piperazine moiety. For example, in the MS/MS spectrum of piperazine-*N*-desmethylsildenafil (Fig. 6), the corresponding fragment ion is observed at *m/z* 85 rather than at *m/z* 99. This value is 14 Th lower than that observed for sildenafil (Fig. 5) thus indicating that demethylation has occurred. The MS/MS spectra provided in Figs. 4 and 5 underscore the importance that low-mass fragment ions can have in the structural elucidation of metabolites.

LC–MS/MS and LC–MS^3 mass spectra of Sildenafil obtained using an ion trap mass spectrometer are shown in Fig. 7.

In comparison to the MS/MS spectra obtained using a triple quadrupole mass spectrometer (Figs. 5–6), the MS/MS spectrum obtained using an ion trap mass spectrometer gave fragment ions only up to *m/z* 163 (Fig. 7) and the MS^3 spectra did not yield any additional fragment ions, particularly lower *m/z* fragment ions (*m/z* 99 and 85).

The inability of ion trap mass spectrometers to efficiently trap low *m/z* ions in the MS/MS experiments is often a drawback for this class of MS systems in structural elucidation of metabolites. Multiple step MS/MS experiments (MS^3 and MS^4) help to understand the origin and mechanism of the fragmentation, specifically the ability to distinguish

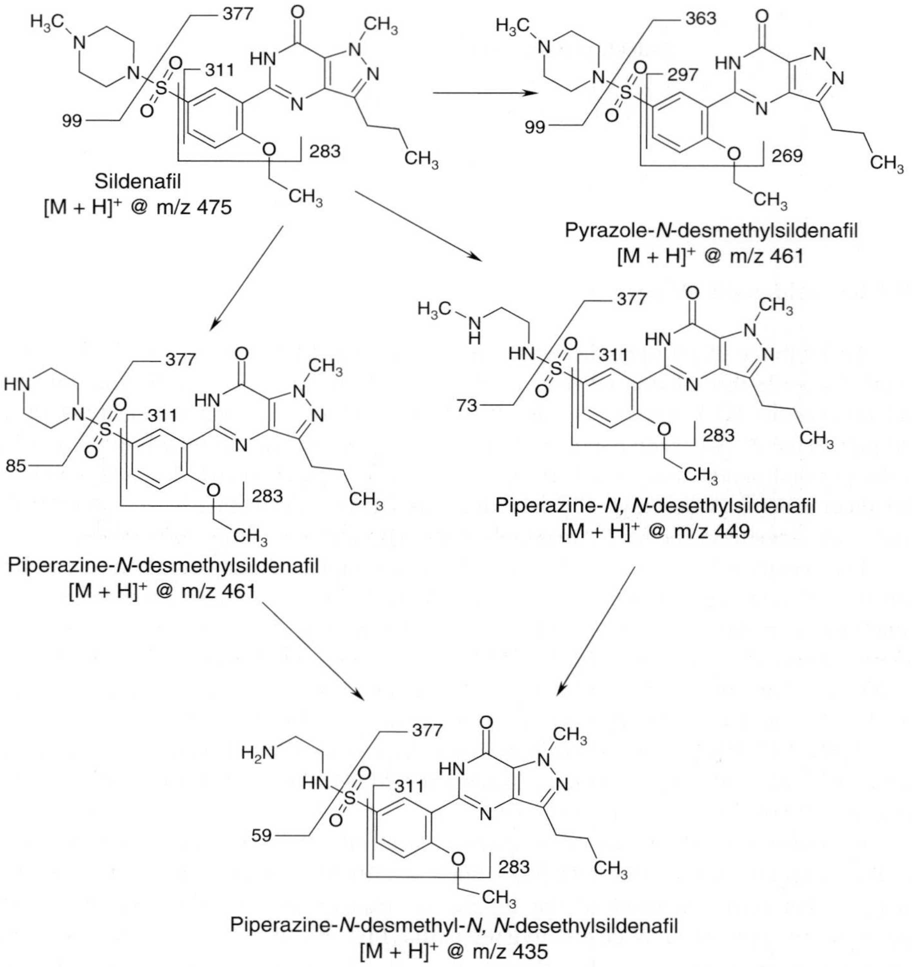
377
311
99
283
Sildenafil
[M + H]+ @ m/z 475
363
297
99
269
Pyrazole-N-desmethylsildenafil
[M + H]+ @ m/z 461
377
311
85
283
Piperazine-N-desmethylsildenafil
[M + H]+ @ m/z 461
377
311
73
283
Piperazine-N, N-desethylsildenafil
[M + H]+ @ m/z 449
377
311
59
283
Piperazine-N-desmethyl-N, N-desethylsildenafil
[M + H]+ @ m/z 435

Figure 4.
Proposed fragmentation schemes for sildenafil and its human and animal metabolites (created based on Walker *et al.*, 1999 and Qin *et al.*, 2002)

the primary and secondary fragment ions under low energy multiple collision conditions. For example, MS^3 spectrum of *m/z* 377 shows that fragment ions of *m/z* 311, 331, and 349 are secondary fragments of *m/z* 377. This example clearly shows the need for mass spectrometers with different MS/MS capabilities for metabolite structure elucidation.

10.2.2.6. Linezolid (Zyvox®)

The first of a new class of antibiotics used for the treatment of Gram-positive bacterial infections by selectively inhibiting the initiation phase of bacterial protein

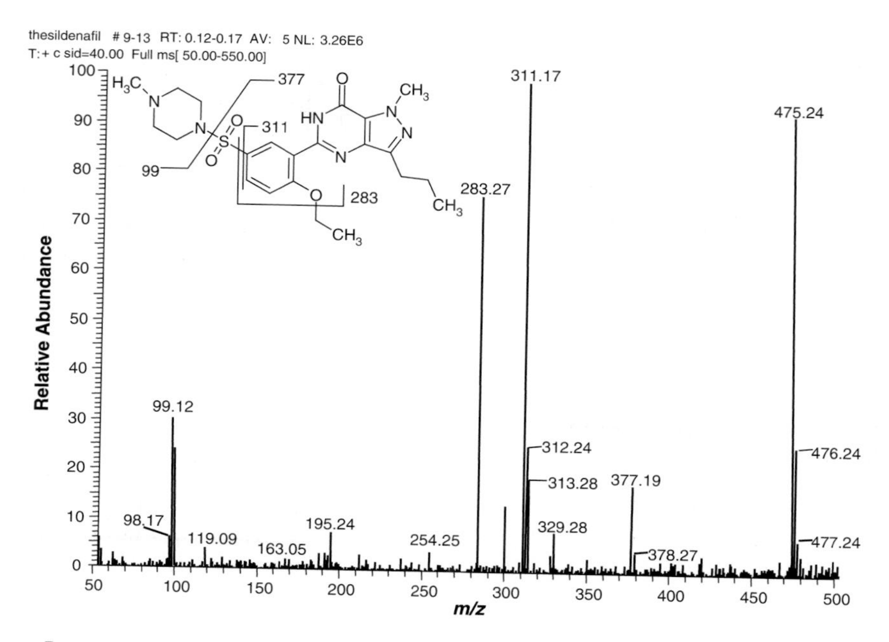

Figure 5.
Source collision induced dissociation (MS/MS) spectrum and the proposed fragmentation schemes for sildenafil (viagra) (modified from Qin *et al.*, 2002)

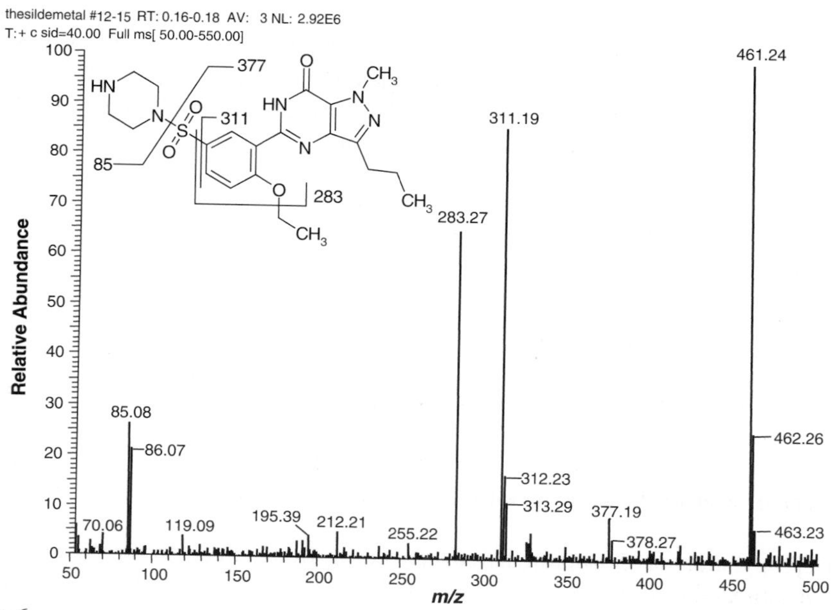

Figure 6.
Source collision induced dissociation (MS/MS) spectrum and the proposed fragmentation scheme for piperazine-*N*-desmethylsildenafil (modified from Qin *et al.*, 2002)

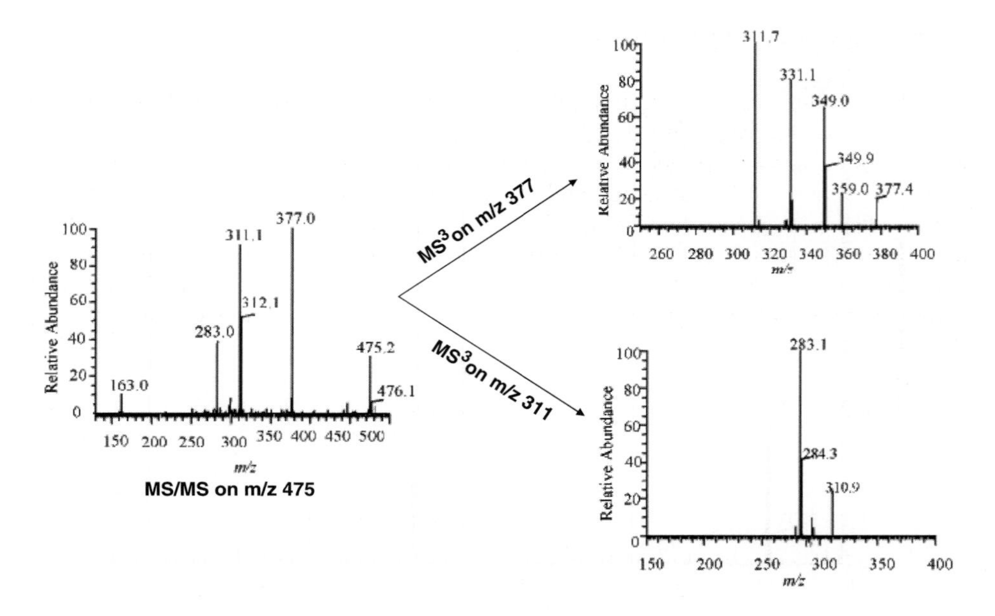

Figure 7.
LC–MS/MS and LC–MS^3 spectra of protonated ions of sildenafil and primary and secondary fragment ions (modified from Zhong *et al.*, 2002)

synthesis. Under LC–MS conditions, the parent drug gives a molecular ion peak at *m/z* 338. As shown in the following scheme, linezolid undergoes *N*, *O*- deethylation via formation of the morpholine ring opened metabolite (Slatter *et al.*, 2001). LC–MS-detected molecular ion peak for *N*, *O*-dealkylated metabolite is observed at *m/z* 312, which is 26 Da lower than that of the parent drug.

Linezolid
$[M + H]^+$ @ m/z 338

Desethyllinezolid
$[M + H]^+$ @ m/z 312

10.2.3. Depropyl or Deisopropylation

Depropylation or deisopropylation can occur from either *O*- or *N*- and results in the negative shift of the LC–MS-measured mass of the metabolite by 42 Da from that of the parent. SG, Buprenorphine (BN), and Fluvastatin (FV) are reported to undergo depropylation or de-isopropylation.

10.2.3.1. Selegiline (Sd Deprenyl®/Eldepryl®)

SG is a selective irreversible inhibitor of monoamine oxidase B (MAO-B), which is co-administered with Levodopa for treating Parkinson's disease. In Parkinson's patients, SG is used to decrease dopamine metabolism at the target location to prolong the effect of Levodopa. Following oral administration to humans, SG is rapidly absorbed and metabolized via *N*-demethylation to form *N*-desmethylselegiline ($[M+H]^+$ at *m/z* 174) and *N*-despropynylselegiline or methamphetamine (MA) ($[M+H]^+$ at *m/z* 150) metabolites, which are further metabolized by *N*-dealkylation to form desmethyldespropynylselegiline or amphetamine (AP) ($[M+H]^+$ at *m/z* 136) (Shin *et al.*, 1997).

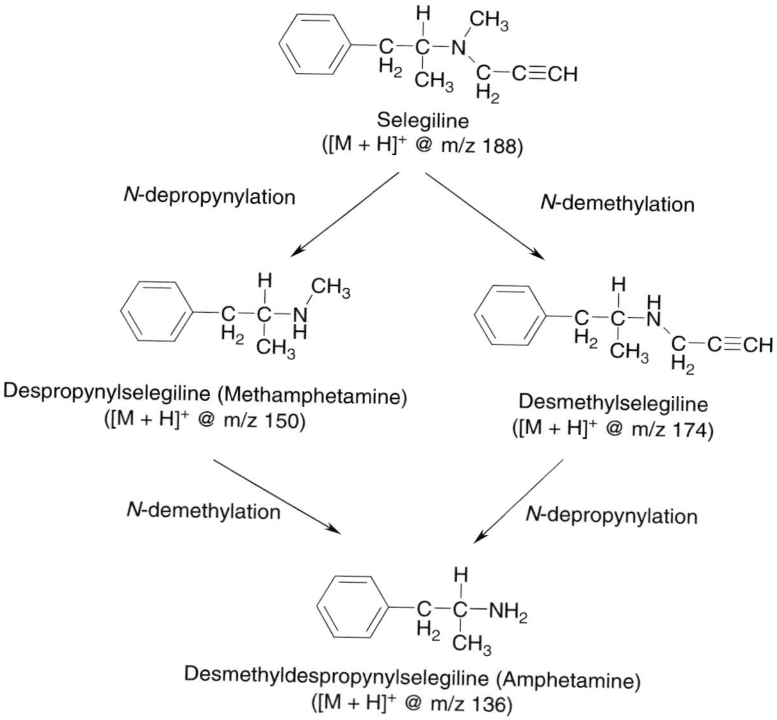

Selegiline
($[M + H]^+$ @ m/z 188)

N-depropynylation

N-demethylation

Despropynylselegiline (Methamphetamine)
($[M + H]^+$ @ m/z 150)

Desmethylselegiline
($[M + H]^+$ @ m/z 174)

N-demethylation

N-depropynylation

Desmethyldespropynylselegiline (Amphetamine)
($[M + H]^+$ @ m/z 136)

Detecting and characterizing SG and its metabolites were challenging for several reasons: (1) plasma concentrations of SG and metabolites were in the low ng/ml range, (2) detection of lower-mass range ions are obscured by the presence of matrix ions, (3) enantiomeric separation of SG metabolites was required to differentiate between medical and illicit use of SG and (4) SG metabolites were also common for its analogues such as dimethylamphetamine (DMA) and benzphetamine (BZP). Until most recently, SG and its metabolites were detected by fluorescence assay (Mahmood *et al.*, 1994) and GC–MS following chemical derivatization. However, these techniques failed to give online enantioselective detection of MA and AP.

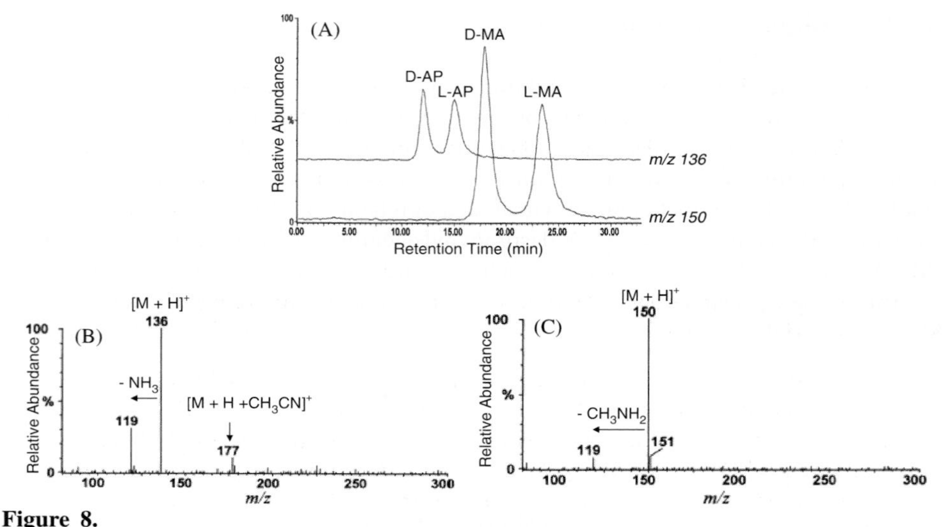

Figure 8.
Extracted ion chromatograms for *m/z* 150 and 136 (Panel A) and LC–MS spectra of L-AP (panel B) and L-MA (panel C)

Enantioselective detection of SG metabolites is important because MA and AP formed from biotransformation of SG are in L-form whereas most illicitly used MA is in the D-form. In 1996, for the first time, online enantioselectivite separation and MS characterization of MA and AP were achieved by using a β-cyclodextrin phenylcarbamate-bonded column and thermospray mass spectrometry. Later in 2003, the same column in combination with electrospray ionization (ESI)-mass spectrometry was used to develop an automated method sensitive enough to detect and characterize SG metabolites and distinguish them from those originating from illicit drugs. Figure 8 shows extracted ion chromatograms and LC–MS spectra of L-AP and L-MA from ESI using a β-cyclodextrin phenylcarbamate-bonded column.

10.2.3.2. Buprenorphine (Subutex®)

Buprenorphine (BN) is a synthetic derivative of morphine, which exhibits mixed agonist/antagonist properties. In comparison to other opiates, BN exhibits limited physical dependency and high potency (25–50 times more potent than morphine) (Lagrange *et al.*, 1998). In humans, BN functions as a potent opioid antagonist at the dose range of 2–32 mg and as an agonist at a dose range of 0.3–0.6 mg. In humans, one of the major in vivo and in vitro biotransformation pathways of BN involves CYP 3A4 mediated *N*-demethylcyclo-propylation to form norbuprenorphine (NBN) (Kobayashi *et al.*, 1998). Since analgesistic effects in humans are achieved at very low doses, detecting and structurally characterizing BN and NBN metabolites in biological matrices is very challenging. However, the introduction of atmospheric pressure ionization (API) based LC–MS methods (discussed in chapter 6 by Hop *et al.*) provided the sensitivity necessary to detect BN and NBN at concentrations as low as 0.01 and 0.05 ng/ml, respectively.

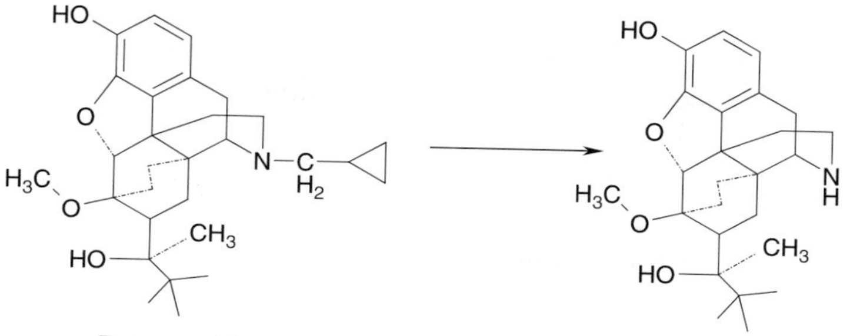

Buprenorphine
([M + H]$^+$ @ m/z 468)

Norbuprenorphine
([M + H]$^+$ @ m/z 414)

N-dealkylation of BN via loss of the methylcyclopropyl moiety results in a shift of the molecular ion peak by 54 Da; the protonated molecules for BN and NBN are detected at *m/z* 468 and 414, respectively. There is no other structural change which could add up to 54 Da. Demethylcyclopropylation is the only structural change of this size is possible for BN; simple *m/z* measurement by LC–MS may provide sufficient information for the characterization of this metabolite.

10.2.3.3. Fluvastatin (Lescol®)

Fluvastatin (FV) is used in the treatment and control of elevated low density lipoprotein cholesterol levels. Similar to other statins on the market (Simvastatin, Lovastatin, Atorvastatin, etc.), FV is a member of the 3-hydroxy-3-methylglutaryl coenzyme A (HMG-CoA) reductase inhibitor family of drugs that functions by blocking the body's production of cholesterol. Following a single oral administration to humans (2–40 mg/daily), more than 60% of the dose was excreted as metabolites in the feces. In addition to the *N*-deisopropylated metabolite of FV (shown in the following scheme) metabolites resulting from hydroxylation of the indole ring at the 5 and 6 positions collectively accounted for about 48% of the dose (Dain *et al.*, 1993)

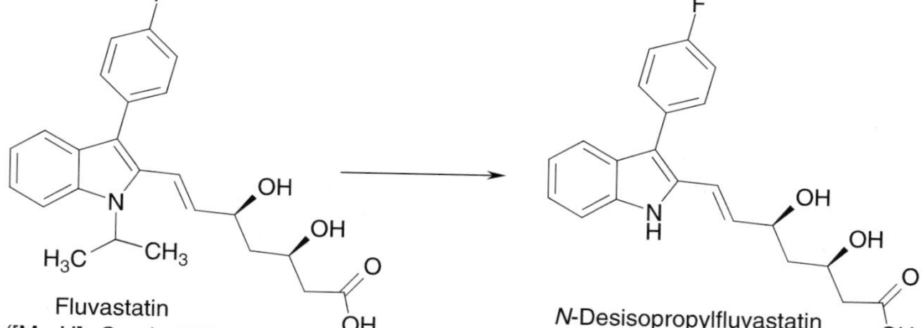

Fluvastatin
([M - H]$^-$ @ m/z 410)

N-Desisopropylfluvastatin
([M - H]$^-$ @ m/z 368)

The carboxylic acid moiety makes FV a good candidate for analyzing in negative ionization mode, where under LC–MS conditions, deprotonated molecules are observed at 410 Th. The *N*-deisopropylated metabolite of FV also ionizes well in the negative ionization mode to give deprotonated ions at 368 Th. Biotransformation of FV to *N*-desisopropyl-FV is another example, where the molecular weight information from LC–MS spectra alone can provide sufficient information for proposing the structure of the metabolite. Unambiguous confirmation of the *N*-despropyl-FV metabolite is then achieved by additional MS/MS experiments and comparison of the HPLC retention time with that of reference standards.

10.2.4. Dealkylation of Larger Moieties

Similarly, dealkylation of a larger structural moiety can involve either a single heteroatom (*O*- or *N*-) or two heteroatoms and results in a net loss of the LC–MS-measured *m*/*z* compared to that from the parent drug. The following examples are representative of this reaction.

10.2.4.1. Ritonavir (Norvir®)

A potent orally active HIV-1 protease inhibitor. Following oral administration to rats (20 mg/kg), dogs (20 mg/kg), and healthy male volunteers (600 mg), on average 96, 93, and 86% of the dose, respectively was excreted in the feces and the rest of the dose was recovered in the urine (Denissen *et al.*, 1997). In all species, *N*-dealkylation of ritonavir ($[M+H]^+$, 721 Th) via loss of either the thiazolyl carbamate moiety (M1) or loss of the isopropylthiazolylmethyl moiety (M11) was observed. As shown in the following scheme, M1 and M11 exhibited molecular ion peaks at *m*/*z* 580 and 582, respectively.

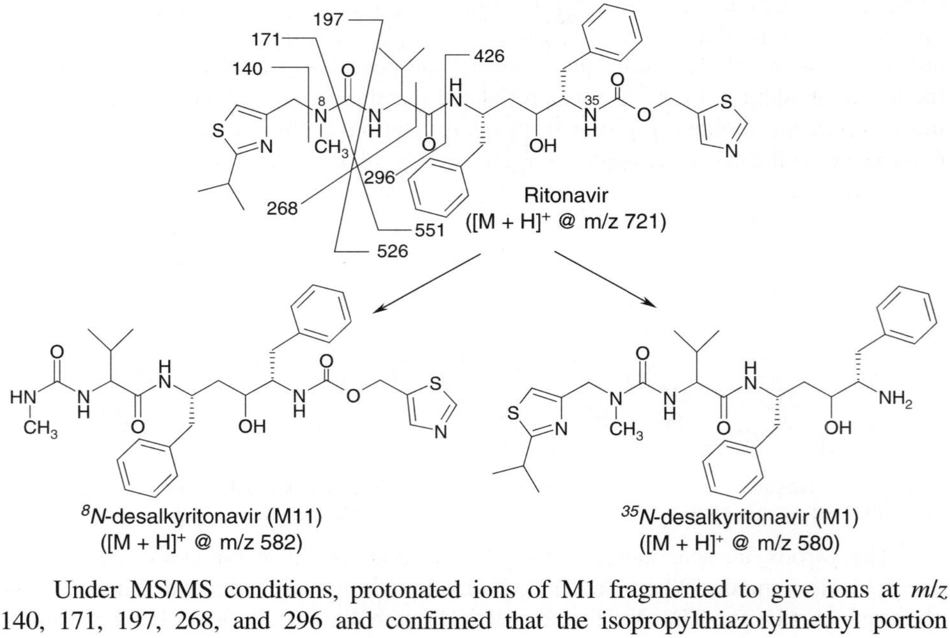

Under MS/MS conditions, protonated ions of M1 fragmented to give ions at *m*/*z* 140, 171, 197, 268, and 296 and confirmed that the isopropylthiazolylmethyl portion of the molecule was intact. Shifting of the fragment ions at *m/z* 551, 526, and 426 by 141 Da confirmed loss of the thiazolyl carbamate moiety. Similarly, the absence of fragment ions of *m/z* 140, 171, 197, 268, and 296 and the presence of ions at *m/z* 426, 526, and 551 in the MS/MS spectrum of M11 confirmed the identity of this metabolite.

10.2.4.2. Donepezil (Aricept®)

A potent acetylcholinesterase inhibitor used in the treatment of Alzheimer's disease. More than 15 metabolites have been identified in microsomal incubates and in plasma and excreta from rats, dogs and humans (Matsui *et al.*, 1999). One of the major routes of donepezil elimination involves *N*-dealkylation to form *N*-desbenzyldonepezil as shown below.

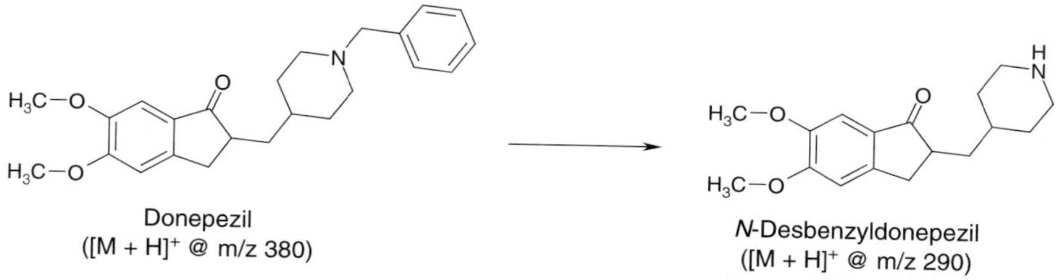

H_3C—O, H_3C—O, O, N

Donepezil
([M + H]$^+$ @ m/z 380)

H_3C—O, H_3C—O, O, H N

N-Desbenzyldonepezil
([M + H]$^+$ @ m/z 290)

Under LC–MS conditions, protonated ions for donepezil and *N*-desbenzyldonepezil were detected at *m/z* 380 and 290, respectively.

10.2.4.3. Tamoxifen (Nolvadex®)

Since 1971, TX has been used as an antiestogenic drug in the prevention and treatment of breast cancer. In humans, TX is metabolized via dealkylation, hydroxylation, and *N*-oxide formation by Cytochrome P450s and FMOs. Among the many TX metabolites that have been detected and characterized, 4-hydroxytamoxifen (discussed later) and *O*-desalkyltamoxifen are of particular importance. These two metabolites have been suggested as possible precursors to the formation of quinone methide metabolites of TX, which are potential cytotoxic/genotoxicic agents (Fan and Bolton, 2001).

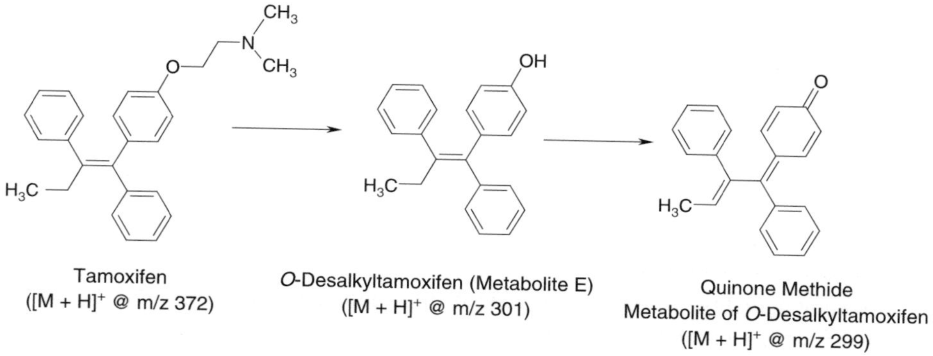

CH_3, N, CH_3, O, H_3C

Tamoxifen
([M + H]$^+$ @ m/z 372)

OH, H_3C

O-Desalkyltamoxifen (Metabolite E)
([M + H]$^+$ @ m/z 301)

O, H_3C

Quinone Methide
Metabolite of *O*-Desalkyltamoxifen
([M + H]$^+$ @ m/z 299)

10.2.4.4. Terfenadine (Seldane®)

A nonsedating H1 receptor antagonist used as an antihistamine until it was withdrawn from the market. Following administration to humans, terfenadine undergoes biotransformation via *N*-dealkylation and hydroxylation (Jones *et al.*, 1998). The LC–MS detected mass shift (Δ = 204 Da) alone is sufficient for proposing a structure for the azacyclonol metabolite (Chen *et al.*, 1991).

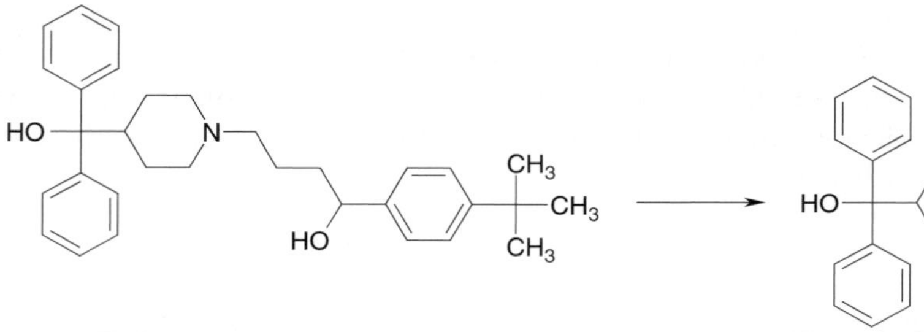

Terfenadine
([M + H]$^+$ @ m/z 472)

Azacyclonol (Desbutyrophenone-terfenadine)
([M + H]$^+$ @ m/z 268)

10.3. Oxidative Deamination

In addition to Cytochrome P450 enzymes, MAOs located in liver, also catalyze the oxidative deamination of many drugs and xenobiotics. Although similar to oxidative *N*-dealkylation, oxidative deamination involves the cleavage of a C–N bond. As can be seen in the following discussion, oxidative deamination can be as simple as that observed with AP or more complex as is the case with benzphetamine and terbinafine.

10.3.1. Amphetamine

AP is a chiral drug that displays stereoselective differences in its biological action. As shown below, oxidative deamination of AP results in phenylacetone (PA), which is detected in the LC–MS spectrum at *m/z* 134 (Shiiyama *et al.*, 1997). PA is subsequently oxidized to benzoic acid and excreted as a glucuronide or glycine (hippuric acid) conjugate. As shown below, the simplest oxidative deamination process results in a 2 Da molecular weight shift.

Amphetamine
([M + H]$^+$ @ m/z 136)

Phenylacetone
([M + H]$^+$ @ m/z 134)

As shown in Fig. 8, molecular weight information from the LC–MS spectrum, in-source induced fragmentation, and/or LC–MS/MS experiment would be sufficient to distinguish AP from PA. Under fragmentation conditions (in-source or LC–MS/MS), only AP would be expected to dissociate to form fragment ions corresponding to loss of an NH_3 molecule whereas PA would not. Other additional confirmation could be achieved by accurate mass determination and H/D exchange experiments, which are discussed in Chapter 6 by Hop *et al.*

10.3.2. Benzphetamine (Didrex)

Benzphetamine (BZP) is a derivative of AP, which is used to treat obesity. Oxidative deamination products and their associated LC–MS-detected *m/z* are shown in the following scheme.

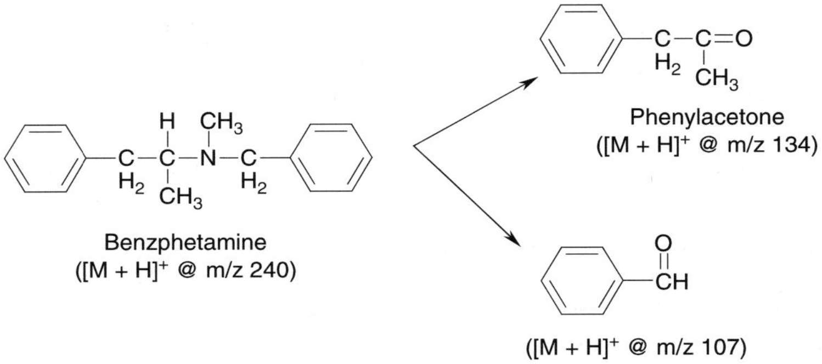

Benzphetamine
([M + H]$^+$ @ m/z 240)

Phenylacetone
([M + H]$^+$ @ m/z 134)

([M + H]$^+$ @ m/z 107)

10.3.3. Terbinafine (Lamisil®)

An orally active allylamine derivative that has demonstrated excellent fungicidal activity against dermatophytes. Following oral administration, terbinafine is rapidly absorbed and widely distributed to body tissues including toenails. Since terbinafine persists in toenails for at least 30 weeks after oral administration, this drug is efficient in the treatment of onychomycosis.

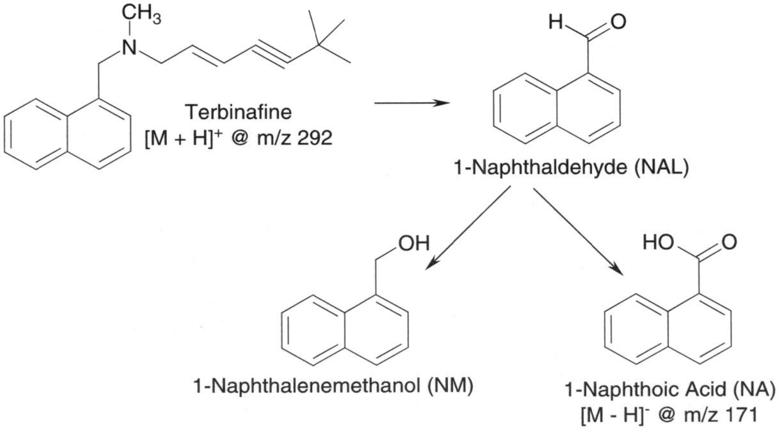

Terbinafine
[M + H]$^+$ @ m/z 292

1-Naphthaldehyde (NAL)

1-Naphthalenemethanol (NM)

1-Naphthoic Acid (NA)
[M - H]$^-$ @ m/z 171

Under LC–MS conditions (positive ion electrospray ionization (ESI)), the molecular ion for terbinafine is observed at *m/z* 292. Following CYP3A4-mediated oxidative deamination, terbinafine is metabolized to 1-naphthaldehyde (NAL) and subsequently reduced to 1-Naphthalenemethanol (NM) or 1-Naphtholic acid (NA). Both NAL and NM did not ionize in positive or negative mode ESI–MS conditions, however, NA ionized in the negative mode ESI conditions and exhibited a molecular ion ($[M-H]^-$) at *m/z* 171. The presence and identity of NAL and NM was confirmed using UV absorption of the corresponding reference standards (Vickers *et al.*, 1999).

10.4. Oxidation of Carbon

Oxidations at a single carbon atom resulting in the addition of an oxygen atom are commonly termed as hydroxylations while oxidation involving a single oxygen and two carbon atoms is referred to as epoxidation.

10.4.1. Hydroxylation

Hydroxylation can occur on an aromatic or aliphatic carbon. The fragmentation observed in the LC–MS or LC–MS/MS spectra can often be used to distinguish between these two types. Hydroxylations are mainly catalyzed by the Cytochrome P450 (CYP450) systems. The lack of substrate specificity of these systems can result in hydroxylation at multiple sites yielding the same molecular weight.

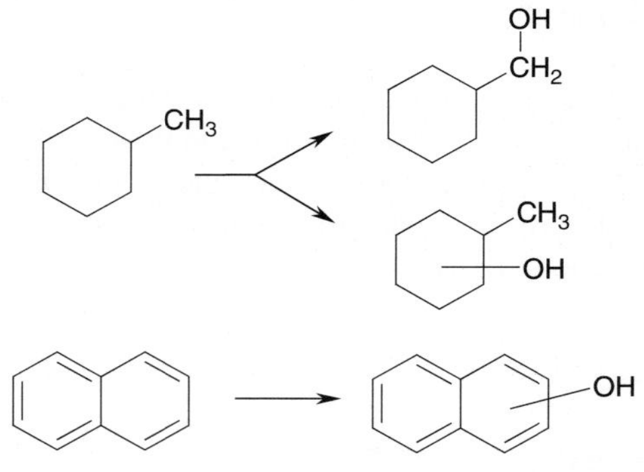

OH

CH_2

CH_3

CH_3

OH

OH

As shown above, the major products of aromatic hydroxylation are phenols and those of aliphatic hydroxylation involving alicyclic or aliphatic carbon atoms are alcohols. The subsequent biotransformation of ring and side-chain hydroxylated metabolites differ from each other. Phenolic and alcoholic derivative(s) on alicyclic carbons, once formed, can undergo phase II conjugation reactions. Aliphatic alcohols also can undergo phase II conjugation reaction or may undergo further oxidation to form carboxylic acid metabolites before undergoing phase II conjugation reactions.

Aromatic hydroxylation is generally considered to form via an epoxide intermediate, which is usually unstable. The evidence for epoxidation hinges on intramolecular hydrogen migration and the detection of NIH shift (NIH stands for US National

Institutes of Health where this phenomenon was first observed) in metabolites as shown below.

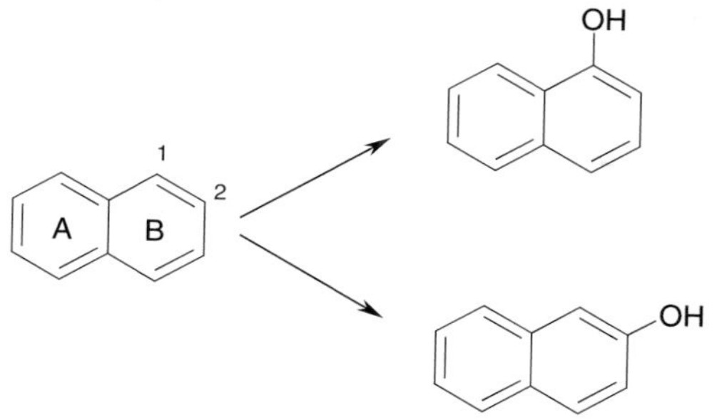

All monohydroxylations result in a shift of a LC–MS detected mass of the metabolite by 16 Da from that of the parent drug. Using mass spectrometry techniques alone, distinguishing between the 1-OH and 2-OH compounds shown above would be very difficult and would require NMR and/or comparison with reference standards. Some specific examples on the use of LC–MS and LC–MS/MS techniques for the characterization of hydroxylated metabolites are discussed using the following examples.

10.4.1.1. Loratadine (Claritin®)

In 1993, Loratadine (LOR) was approved as one of the new generation potent and long-acting non-sedating antihistamines for treating seasonal allergy symptoms. In 2002, FDA approved Claritin® to be sold over-the-counter as an allergy medicine. Following a 10 mg oral administration of [^{14}C]LOR to healthy male volunteers, LOR is rapidly metabolized to desloratadine (DL, Clarinex®) via decarboethoxylation and subsequently undergoes aromatic hydroxylation to form 3-OH-DL (M40). While 3-OH-DL-glucuronide (M13) is the major circulating as well as urinary metabolite, 3-OH-DL is the major metabolite in the feces (Ramanathan *et al.*, 2004). Metabolites resulting from the hydroxylation at the bridgehead, 5-OH-DL (M33) and 6-OH-DL (M31), were minor metabolites in human, but they were major metabolites in mouse, rat, and monkey (Ramanathan *et al.*, 2005).

Under LC–MS conditions, molecular ions of LOR and its descarboethoxy metabolite (DL) were observed at *m/z* 383 and 311, respectively. Hydroxylation of DL results in LC–MS-detected peak at *m/z* 327 Th. The glucuronide metabolite of 3-OH-DL was detected at *m/z* 503 (Fig. 9).

As mentioned earlier, LOR and its metabolite DL undergo both aromatic and aliphatic hydroxylation, and characterization of these two types of hydroxylations involved careful interpretation of both LC–MS and LC–MS/MS spectra and confirmation using reference standards. In a previous publication, we have discussed the use of LC–MS and LC–MS/MS methods to characterize hydroxylated metabolites and to distinguish between aromatic and aliphatic hydroxylations (Ramanathan *et al.*, 2000). Briefly, under LC–ESI/MS and LC–ESI/MS/MS conditions, dehydration (loss

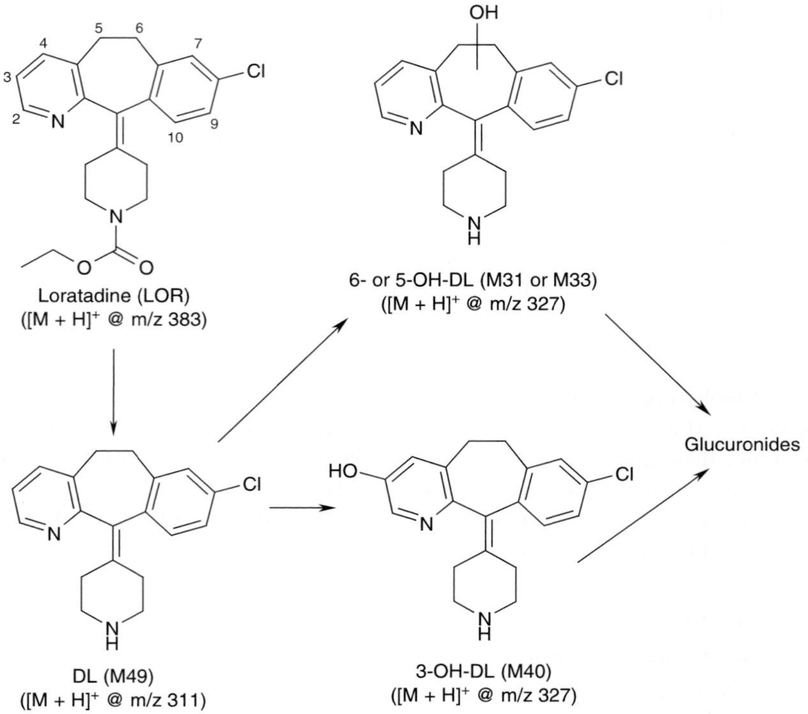

OH

Cl

N

Loratadine (LOR)
([M + H]$^+$ @ m/z 383)

6- or 5-OH-DL (M31 or M33)
([M + H]$^+$ @ m/z 327)

Glucuronides

DL (M49)
([M + H]$^+$ @ m/z 311)

3-OH-DL (M40)
([M + H]$^+$ @ m/z 327)

Figure 9.
Biotransformation of LOR in healthy male volunteers

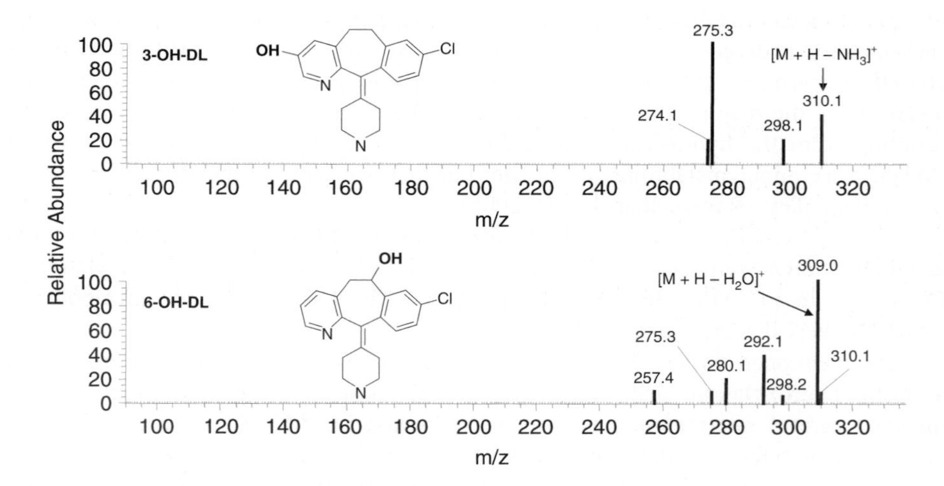

Figure 10.
Comparison of LC–MS/MS spectra of 3-OH-DL (aromatic hydroxylation) and 6-OH-DL (aliphatic hydroxylation)

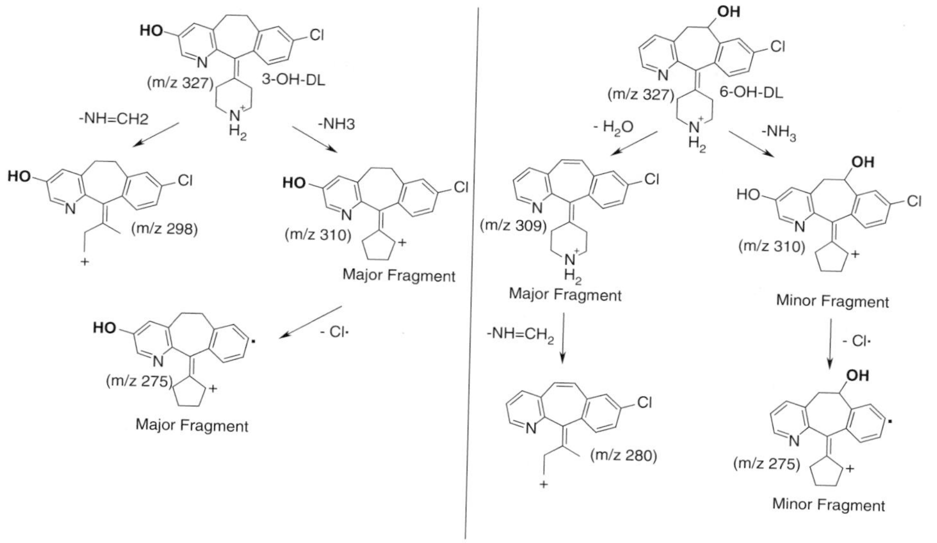

Figure 11.
A proposed fragmentation pathway for *m/z* 327 Th from 6-OH-DL and 3-OH-DL

of a water molecule from $[M+H]^+$ ions) was observed as a favored pathway for the aliphatic hydroxylated metabolites; these decompositions were minor to trace from aromatic hydroxylated metabolites such as 3-OH-DL.

As shown in Figs. 10 and 11, the LC–MS/MS spectrum of 3-OH-DL could be readily distinguished from that of 6-OH-DL (or 5-OH-DL) because the formation of *m/z* 309 (loss of water) was not as favored when hydroxylation was on the aromatic ring (3-OH-DL).

10.4.1.2. Diclofenac (Cataflam/Voltaren)

Diclofenac (D) is a widely used non-steroidal anti-inflammatory agent, which also has been shown to be effective in the treatment of rheumatoid arthritis and osteoarthritis. Following administration to humans, D is metabolized mainly via hydroxylation and glucuronidation. Among the hydroxylated metabolites, 4′-hydroxydiclofenac (4′-OH-D) (shown below) was the major hydroxylated metabolite.

Diclofenac (D) → 4'-Hydroxydiclofenac (4'-OH-D)

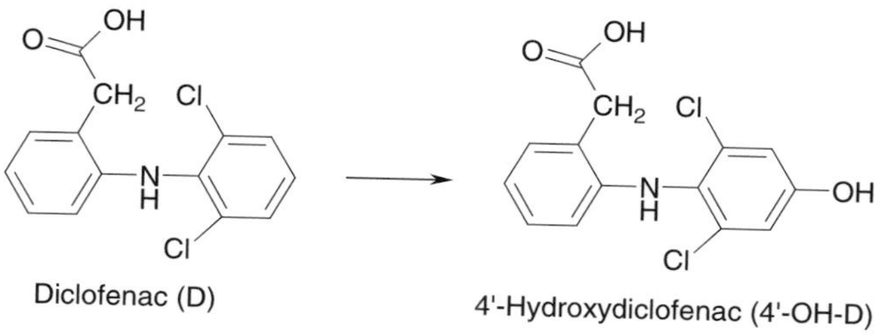

Under LC–MS conditions (negative mode ESI), deprotonated ions of D and 4′-OH-D are observed, respectively, at *m/z* 294 and 310. As shown in Fig. 12, similar to the LC–MS/MS fragmentation behavior of the aromatic hydroxylated metabolite of LOR and DL (3-OH-DL), 4′-OH-D does not undergo fragmentation to lose a water molecule from the precursor ions of *m/z* 310. The product ion from loss of a CO_2 molecule from the precursor ion is the most favored product for both D and 4′-OH-D. In addition to product ions due to $[M-H-CO_2]^-$, product ions from losses of HCl were also observed in the MS/MS spectra of D and 4′-OH-D. However, water loss from [M−H]- of 4′-OH-D was not observed.

Although the presence of two chlorine atoms in diclofenac facilitates detection of the drug-derived components in biological matrices, poor LC–MS/MS fragmentation limits full structural characterization of these compounds from the MS/MS spectra. For example, comparison of the MS/MS spectra shown in Fig. 12 can only provide information to support hydroxylation of either ring A or B of diclofenac. Identifying the exact hydroxylation position (either ring A or B) would require NMR analysis without reference standard.

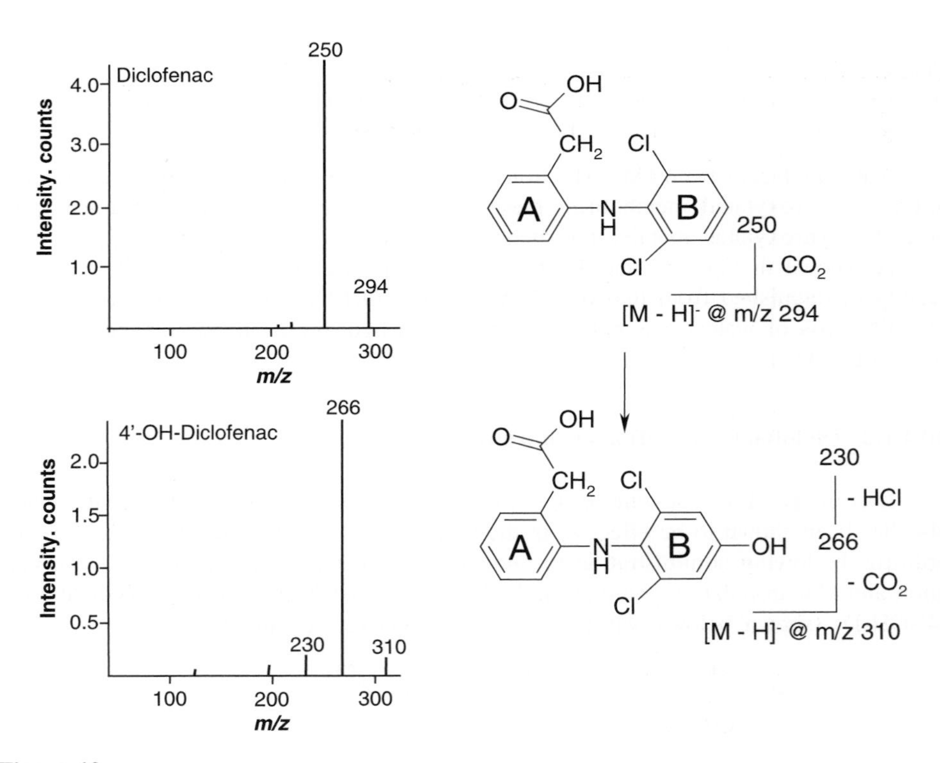

Figure 12.
MS/MS Spectra and Associated Fragmentation Schemes for Diclofenac and 4′-OH-diclofenac (modified from Scott *et al.*, 1999)

10.4.1.3. Debrisoquine

Debrisoquine is an antihypertensive drug whose molecular weight is 175 Da. Although hydroxylation of every available debrisoquine carbon atom has been described, several studies have shown debrisoquine 4-hydroxylation as a major in vivo CYP2D6-mediated metabolite. Until recently, debrisoquine had been used as a determinant of in vivo CYP2D6 activity. Experimental evidence from a recent study showed the involvement of CYP1A1 in the metabolism of debrisoquine to 4-hydroxy-debrisoquine (Granvil *et al.*, 2002), which limits the use of debrisoquine as a selective substrate for in vivo CYP2D6 activity.

Under LC–MS conditions, protonated molecular ions for debrisoquine and hydroxylated debrisoquine are observed at 176 and 192 Th, respectively. As shown in Fig. 13, under MS/MS conditions debrisoquine fragments to give ions at 159, 134, 117, 105, and 72 Th. The fragment ions of *m/z* 159 and 134 Th are due to losses of NH_3 and NH=C=NH moieties from the protonated debrisoquine at *m/z* 176.

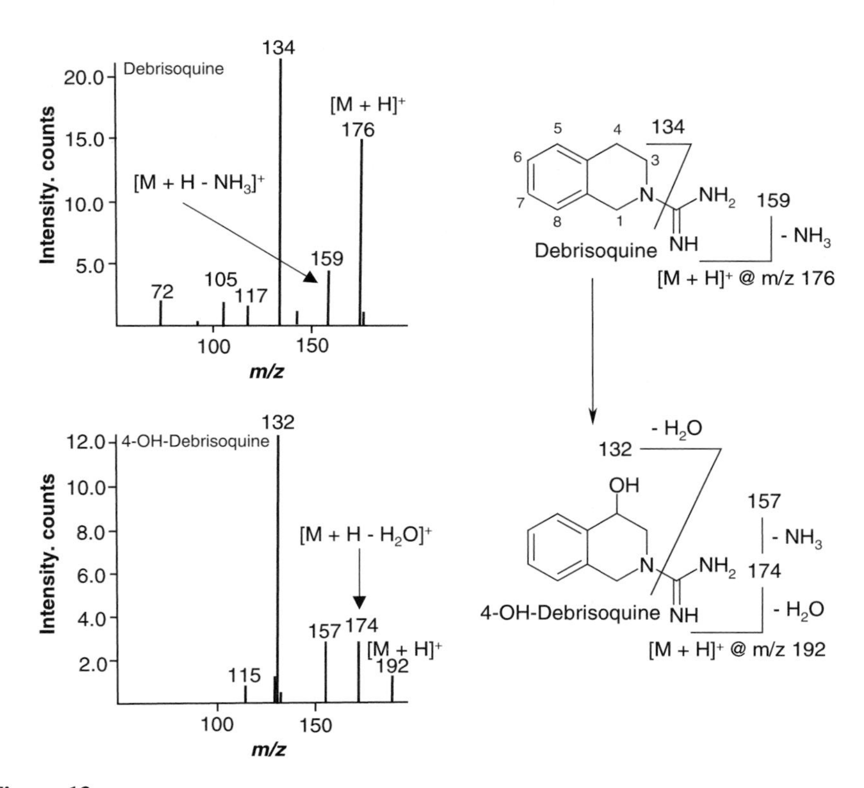

Figure 13.
MS/MS Spectra and Associated Fragmentation Schemes for Protonated debrisoquine and 4-OH-debrisoquine (an aliphatic hydroxylation) (modified from Scott *et al.*, 1999)

Protonated 4-hydroxydebrisoquine is detected at *m/z* 192. Unlike the MS/MS spectra of the parent drug where a loss of a NH_3 molecule from the precursor ions was observed, MS/MS spectra of the hydroxylated metabolites showed loses of a water molecule from the precursor ions (174 Th), which subsequently fragmented by eliminating NH_3 to form the ion at 157 Th or by eliminating NH=C=NH to form the ions of 132 Th.

The MS/MS spectra of all hydroxylated metabolites of debrisoquine exhibited product ions resulting from loss of a water molecule. The extent of water loss, however, depended on the proximity of available α-hydrogens (i.e., 4-hydroxy exhibited abundant ions corresponding to loss of a water compared to 5-, 6-, 7- or 8-hydroxy debrisoquine). This finding is consistent with other observations discussed in this chapter regarding the protonated ions of aliphatic hydroxylated metabolites undergoing more facile loss of H_2O in MS/MS than those with aromatic hydroxylation (e.g., see Ramanathan *et al.*, 2000).

10.4.1.4. Midazolam

Midazolam (MDZ) is used for sedation prior to minor medical procedures or surgery. As shown below, in humans, MDZ undergoes metabolism to form primarily 1-OH-MDZ and to a smaller extent 4-OH-MDZ. The formation of 1-OH-MDZ is rapid and almost exclusively mediated by CYP3A4. Therefore, MDZ is widely used as a probe to assess the activity of CYP3A4.

Under LC–MS conditions, protonated molecules for MDZ, 1-OH-MDZ, and 4-OH-MDZ were detected at *m/z* 326, 342, and 342 Th, respectively. In-source decomposition product ions corresponding to loss of a water molecule were observed for 1-OH-MDZ and not for 4-OH-MDZ (Fig. 14).

In 1-OH-MDZ, a hydrogen atom from the chlorine-containing ring is available for water elimination to form a stable five-membered ring as shown below, whereas such a stable product cannot be formed from 4-OH-MDZ.

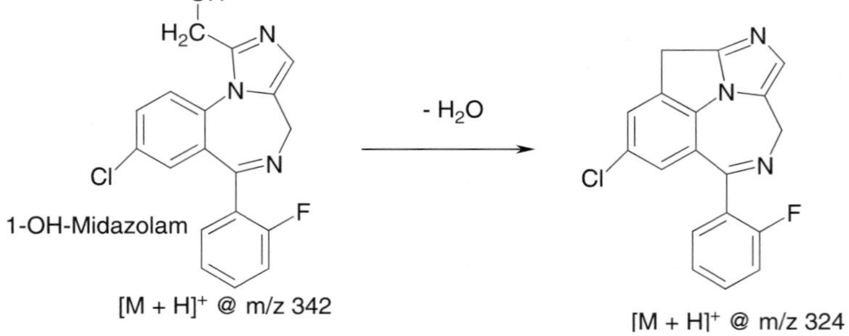

1-OH-Midazolam

$[M + H]^+$ @ m/z 342

$- H_2O$

$[M + H]^+$ @ m/z 324

Based on the LC–MS observation of $[M + H - H_2O]^+$ ions at *m/z* 324 (Fig. 14) and comparison of the MS/MS spectra of MDZ and 1-OH-MDZ (Fig. 15), one can distinguish the MS/MS spectrum of a side-chain-hydroxylated metabolite from the ring-hydroxylated metabolite. The water loss from the latter was not possible due to lack of α-hydrogens.

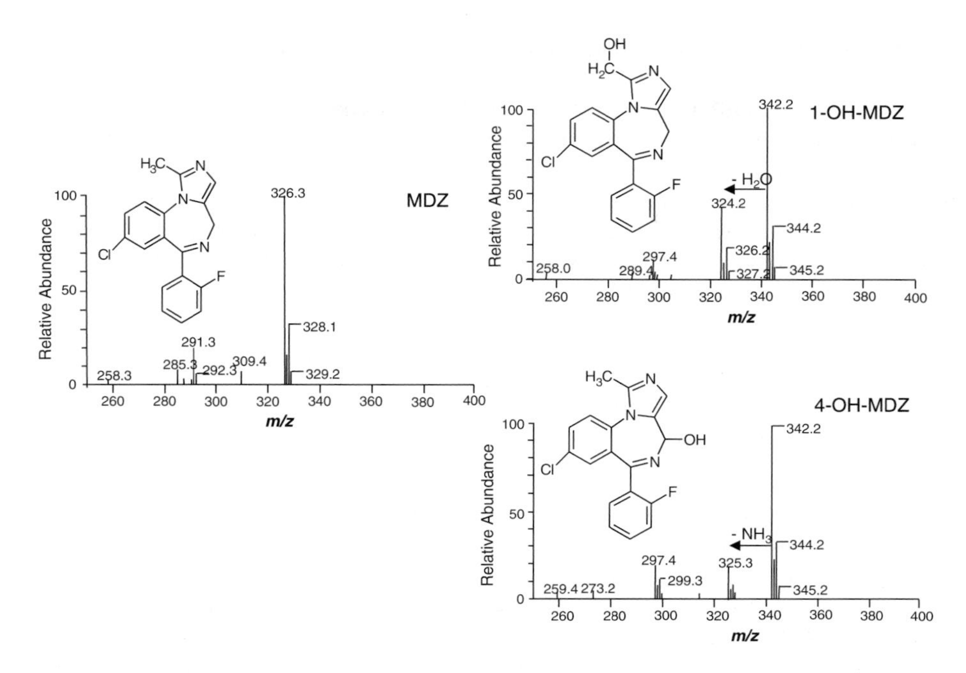

Figure 14.
LC–MS spectra of MDZ, 1-OH-MDZ, and 4-OH-MDZ (modified from Toyo'oka *et al.*, 2003)

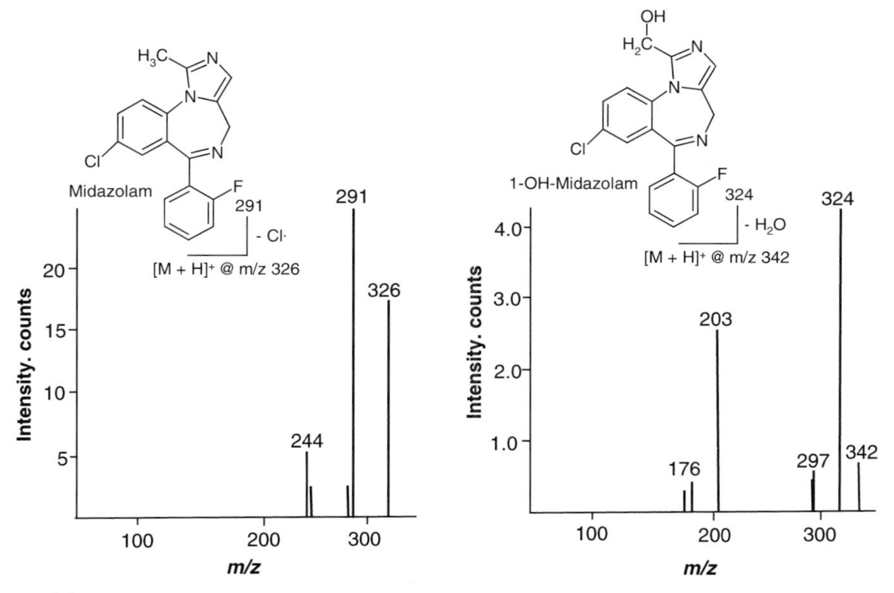

Figure 15.
MS/MS spectra and associated fragmentation schemes of MDZ and 1-OH-MDZ (modified from Scott *et al.*, 1999)

10.4.1.5. Bupropion (Wellbutrin®/Zyban®)

Bupropion (BUP) is prescribed as an antidepressant agent (Wellbutrin) as well as in the management of tobacco smoking (Zyban) (Holm *et al.*, 2000). Following a single oral administration of 200 mg BUP to healthy male volunteers about 88 and 10% of the administered radioactivity was excreted within 48 h in urine and feces, respectively. As shown below, a primary biotransformation pathway of BUP includes methyl hydroxylation of its tertiary butyl moiety to form hydroxyl-BUP (OH-BUP).

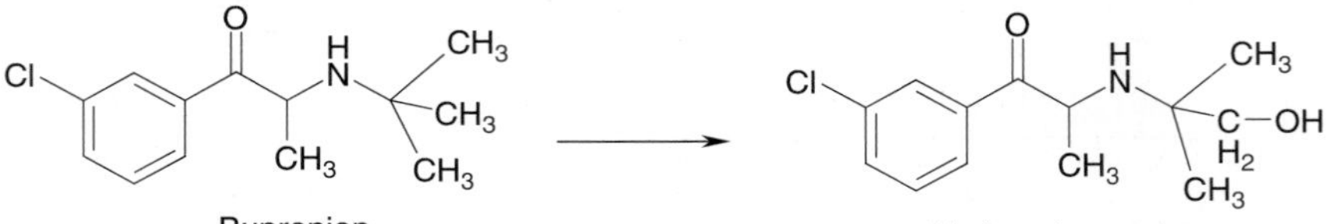

Bupropion
($[M + H]^+$ @ m/z 240)

Hydroxybupropion
($[M + H]^+$ @ m/z 256)

Under LC–MS conditions, BUP and OH-BUP showed protonated molecules at *m/z* 240 and 256 Th, respectively. In humans, circulating levels of OH-BUP were 16 times greater than BUP. Identification and characterization of OH-BUP was important because it showed more potent antidepressant activity and possessed greater toxicity potential compared with BUP and its other metabolites.

10.4.1.6. Terfenadine (Seldane®)

As mentioned previously, in humans, terfenadine undergoes biotransformation via *N*-dealkylation and hydroxylation. Hydroxylation of the tertiary butyl group, similar to any other hydroxylation, results in a positive mass shift of 16 Da from that of the parent drug. As shown in the following scheme, the hydroxyterfenadine metabolite undergoes further oxidation to form the carboxylic acid metabolite, which exhibited a molecular ion peak at *m/z* 502.

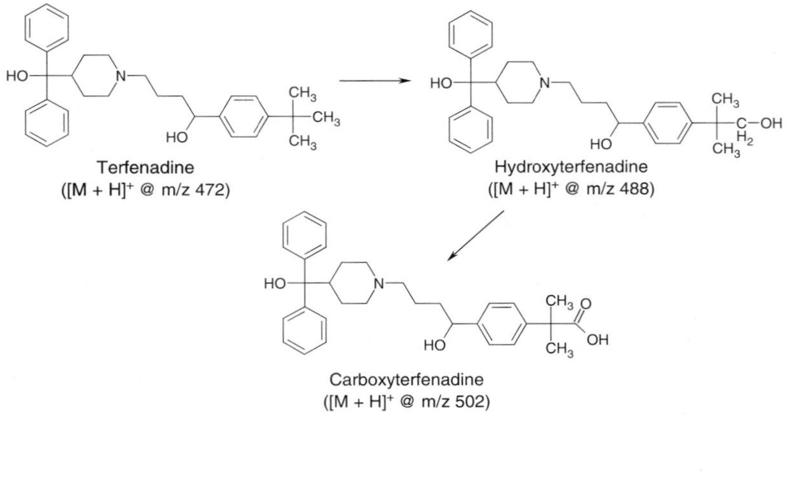

Terfenadine
($[M + H]^+$ @ m/z 472)

Hydroxyterfenadine
($[M + H]^+$ @ m/z 488)

Carboxyterfenadine
($[M + H]^+$ @ m/z 502)

Detection and characterization of the carboxylic acid metabolite led to the discovery of another biologically active antihistamine that had a much lower potential for human cardiotoxicity.

10.4.2. Epoxidation

Epoxidation results in addition of an oxygen atom but involves two carbon atoms. Similar to oxidation of carbon, nitrogen, and sulfur, following epoxidation, the molecular weight of the metabolite shifts by +16 Da from that of the parent drug. Protryptiline (PRO) and carbamazepine (CBZ) are discussed here as examples of drugs that undergo biotransformation to form epoxide metabolites. Interestingly, both drugs contain tricyclic ring systems with a double bond on the bridgehead of the seven-member ring moiety. Both LOR (Claritin) and amitryptiline (Elavil) also contain tricyclic ring systems, but without the double bond on the seven-membered ring moiety. For the latter two drugs epoxide metabolites were not reported. Most often epoxide metabolites are unstable and can lead to formation of glutathione conjugates.

10.4.2.1. Protryptiline (Vivactil)

PRO is a tricyclic antidepressant. As shown below, biotransformation of PRO to PRO-epoxide, leads to an LC–MS-detected peak, which is shifted by 16 Da to *m/z* 280.

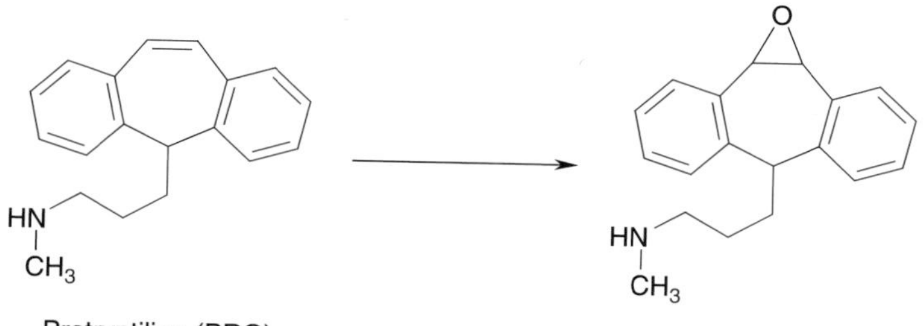

Protryptiline (PRO)
($[M + H]^+$ @ m/z 264)

Protryptiline-epoxide
($[M + H]^+$ @ m/z 280)

Based on the molecular weight information obtained from LC–MS experiments or accurate mass LC–MS measurements for elemental composition, one could not distinguish hydroxylated metabolites from the epoxide metabolite because both would have the same elemental composition and yield molecular ions with the same *m/z* value (isobaric). However, MS/MS experiments and H/D exchange-LC–MS experiments would provide sufficient information to distinguish a hydroxylated metabolite from the epoxide metabolite. The use of MS/MS and H/D exchange to differentiate isomeric epoxide and hydroxyl metabolites is further discussed and elaborated in the following example.

10.4.2.2. Carbamazepine (Atretol/Carbatrol/Epitol/Tegretol)

CBZ is used in the treatment for epilepsy. Although more than 30 in vivo metabolites have been identified in human and rat, the major biotransformation pathway involves oxidation to 10, 11-dihydro-10,11-epoxycarbamazepine (CBZ-epoxide) (Lertratanangkoon and Horning, 1982). Once formed, CBZ-epoxide undergoes hydration producing 10, 11-dihydro-10,11-dihydroxycarbamazepine, which in turn gets eliminated via glucuronidation. Monitoring levels of CBZ-epoxide in blood is important because CBZ-epoxide is active and its concentration has been shown to parallel clinical toxicities (Mesdjian *et al.*, 1999).

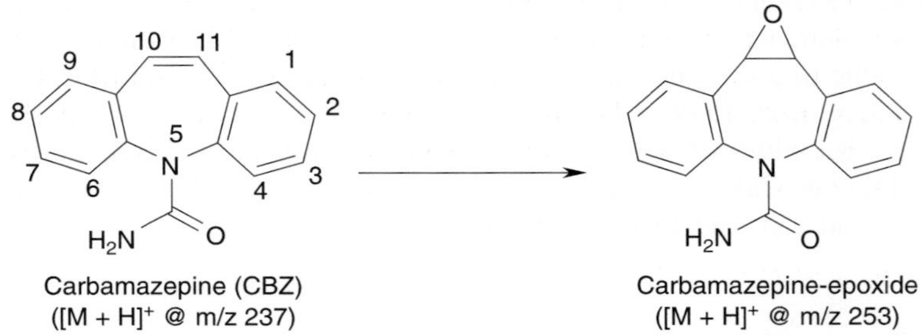

Carbamazepine (CBZ)
($[M + H]^+$ @ m/z 237)

Carbamazepine-epoxide
($[M + H]^+$ @ m/z 253)

Under LC–MS conditions, CBZ and CBZ-epoxide form protonated ions at *m/z* 237 and 253 Th, respectively. Direct monitoring of CBZ-epoxide levels using LC–MS

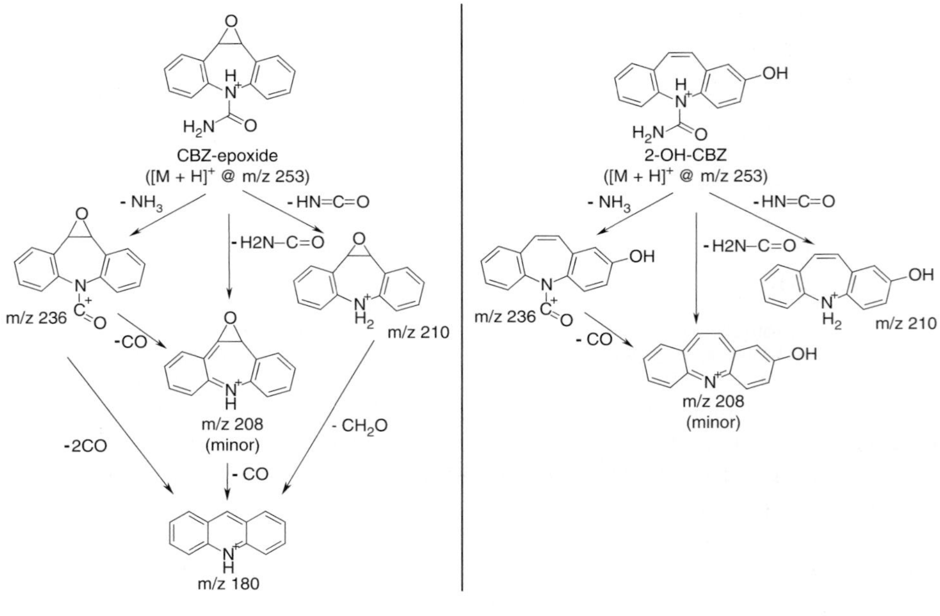

Figure 16.
Proposed LC–MS/MS fragmentation scheme for CBZ-epoxide and 2-OH-CBZ

could be misleading, since CBZ is also known to undergo hydroxylation at the 2 or 3 positions (2-OH-CBZ and 3-OH-CBZ) with identical molecular weights as CBZ-epoxide (m/z = 253 Th). However, under MS/MS conditions, 2-OH-CBZ and 3-OH-CBZ fragment differently from CBZ-epoxide. LC–MS/MS spectra of CBZ-epoxide and 2-OH-CBZ have been reported (Miao and Matcalfe, 2003). As shown in the following fragmentation scheme, under MS/MS conditions both CBZ-epoxide and 2-OH-CBZ yield common fragment ions: m/z 236, 210, and 208. Although the fragment ion at m/z 180 is formed from CBZ-epoxide but not from 2- or 3-OH CBZ, it can be formed from several other CBZ metabolites including from 10, 11-dihydroxy-CBZ (Miao *et al.*, 2003) (Fig. 16).

LC–MS following H/D exchange, however, can be very useful for distinguishing CBZ hydroxylated metabolites from a corresponding epoxide metabolite. The following table summarizes the H/D exchange approach that can be used to distinguish the above-discussed isomeric metabolites. Following H/D exchange, the net shift in m/z is 4 Th for the 2-OH-CBZ (or for any monohydroxy CBZ), whereas the net shift for CBZ-epoxide and CBZ is 3 Th.

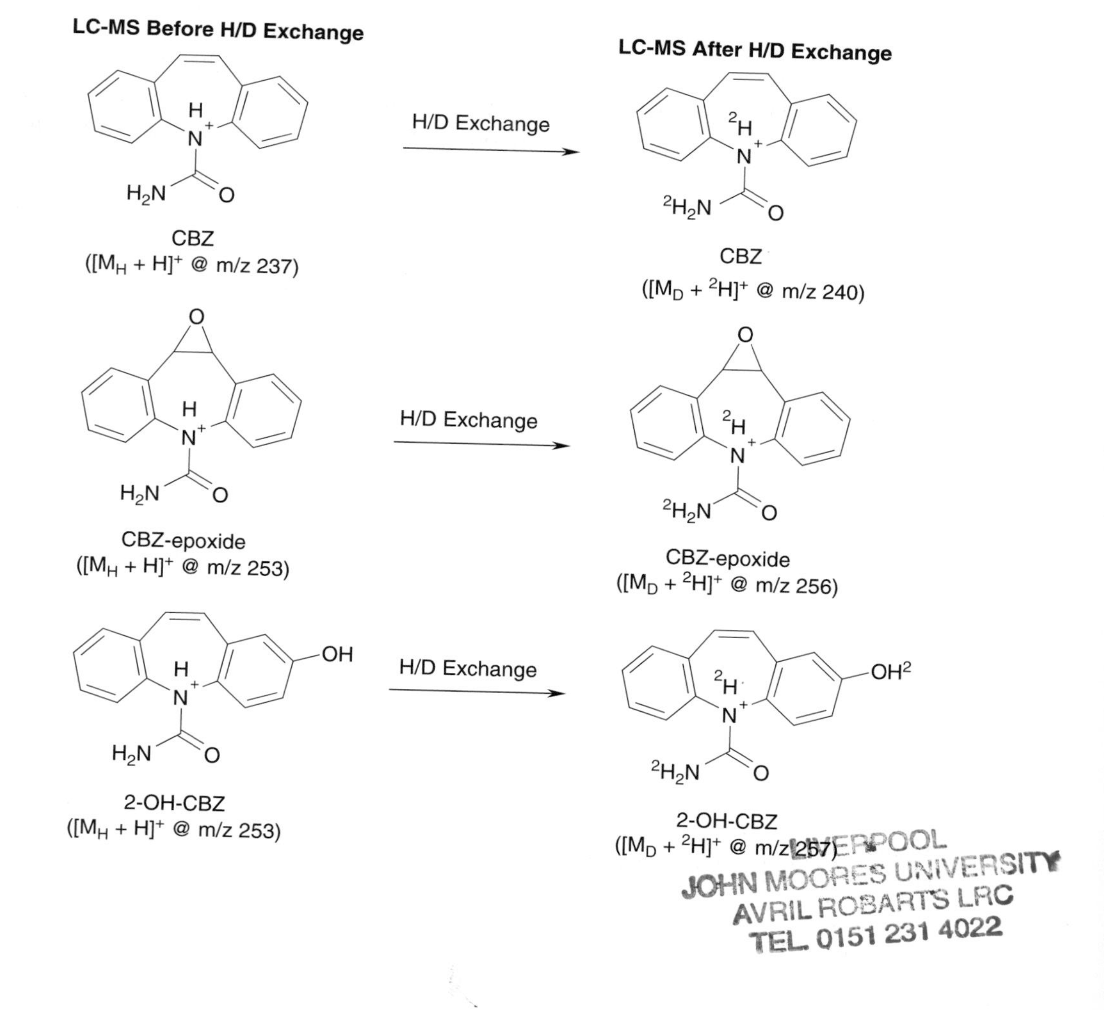

10.5. Oxidation of Nitrogen

Oxidation of nitrogen and other heteroatoms is primarily catalyzed by Cytochrome P450 enzymes. FMOs less frequently catalyze oxidations at nitrogen atoms. Tertiary amines can undergo oxidation to form *N*-oxide metabolites whereas primary and secondary amines form hydroxylamines (or *N*-oxides with dehydrogenation). Thus, from a standpoint of molecular weight, the *N*-oxides and hydroxylamine are indistinguishable from hydroxylated metabolites. Some of the common structural backbones that undergo biotransformation to form *N*-oxides and/or hydroxylamines are shown in Fig. 17.

10.5.1. *N*-Oxides

N-oxidation results in a 16 Da increase in molecular weight. As depicted in Fig. 17, when tertiary and aromatic amines undergo biotransformation to form *N*-oxides, LC–MS-detected mass of each metabolite increases by 16 Da from that of the parent. Secondary amines can also form *N*-oxides following oxidation (-2H) involving the α-carbon atom, but they lead to a net mass increase of 14 Da.

In two recent papers (Tong *et al.*, 2001; Ramanathan *et al.*, 2000), the use of API techniques for the characterization of *N*-oxide metabolites and distinguishing them

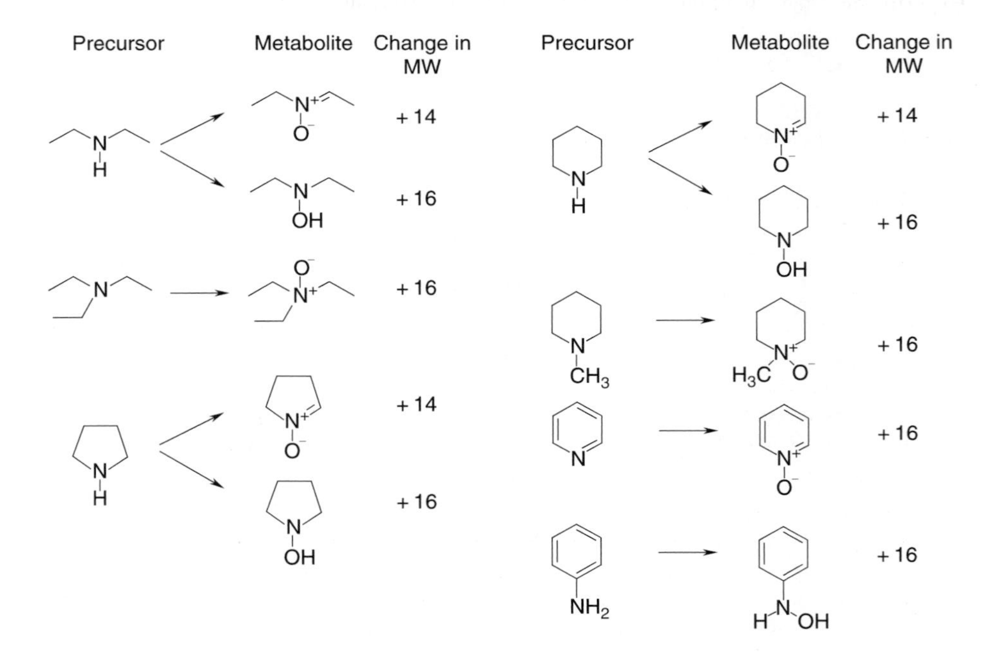

Figure 17.
Oxidation of primary, secondary, tertiary and aromatic amines and the corresponding *N*-oxide and hydroxylamine metabolites and associated mass shifts

from aliphatic- and aromatic-hydroxylated metabolites was described. These findings are summarized in the following scheme using common structural backbones.

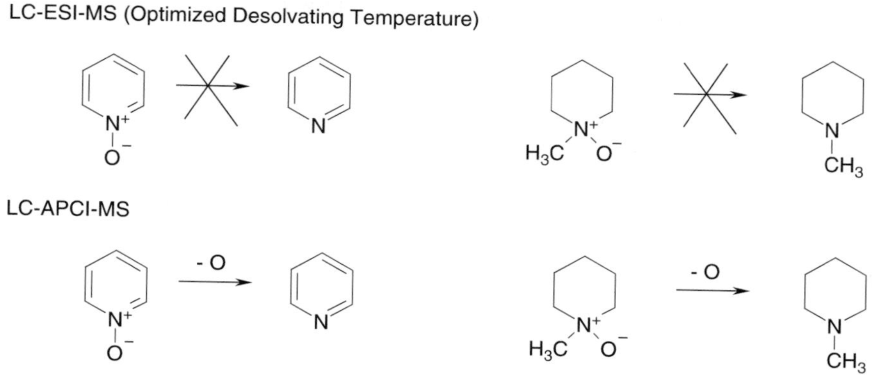

LC-ESI-MS (Optimized Desolvating Temperature)

LC-APCI-MS

- O

- O

Under LC–APCI–MS (APCI = atmospheric pressure chemical ionization) conditions but not under MS/MS conditions, *N*-oxides undergo in-source decomposition to form diagnostic $[M+H-O]^+$ ions, whereas the metabolites from oxidation of carbon (aliphatic and aromatic) do not form this diagnostic fragment ion. If significant thermal energy is imparted to the metabolite in the ion source before or after ionization, *N*-oxides can also undergo deoxygenation under LC–ESI–MS conditions (Tong *et al.*, 2002). Experiments that distinguish *N*-oxides from hydroxylated metabolites are discussed henceforth using LOR as an example. Although many different metabolites of LOR were observed, the focus here is the identification of M47, the *N*-oxide metabolite.

10.5.1.1. Loratadine

The disposition of LOR in male and female rats was studied following a single oral dose (8.0 mg/kg) of ^{14}C-LOR (Ramanathan *et al.*, 2005). LOR was extensively metabolized by rats via descarboethoxylation to form DL (M49) and hydroxylation at the 5- and/or 6-positions (M31 and M33). As shown in Fig. 18, the major circulating metabolite in male rats is a DL-derivative (M47) in which the piperidine ring was aromatized and oxidized to a pyridine-*N*-oxide. In contrast, DL was the major circulating metabolite in the female rats.

Under LC–ESI–MS conditions, M47 gave protonated molecular ions at 323 Th and $[M+H-OH]^+$ radical ions at 306 Th. The ion at 325 Th represents the ^{37}Cl component of the protonated M47. The net difference in the molecular weight between DL and M47 is 12 Da. Under LC–ESI–MS/MS conditions, protonated M47 fragmented to give ions at *m*/*z* 306, 305, 291, 270, and 228 (Fig. 19).

The fragment ion at *m*/*z* 228 provided evidence that the tricyclic portion of the DL is not modified and most likely the modification has taken place on the piperidine ring. However, the nature of the modification in M47, could not be deduced from the ESI–MS or ESI–MS/MS data alone.

As shown in Fig. 20, under LC–APCI–MS conditions, M47 underwent in-source decomposition to give the diagnostic fragment ion at *m*/*z* 307 corresponding to

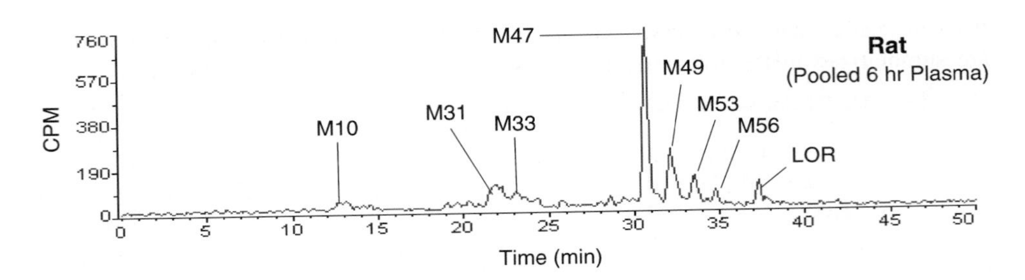

Figure 18.
Radiochromatographic profile from an extract from pooled male rat plasma collected 6 h following a single oral administration (8 mg/kg) of ^{14}C-SCH 29851

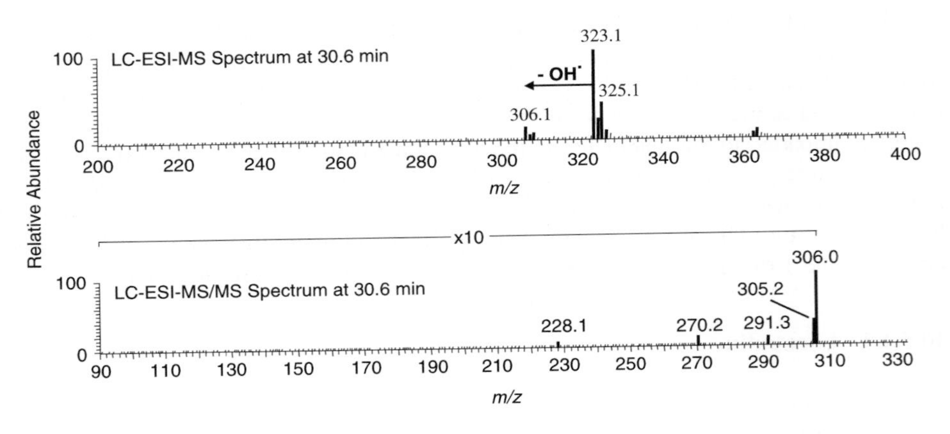

Figure 19.
LC–ESI–MS and LC–ESI–MS/MS spectra of metabolite M47 in rat plasma following administration of ^{14}C-LOR to rats

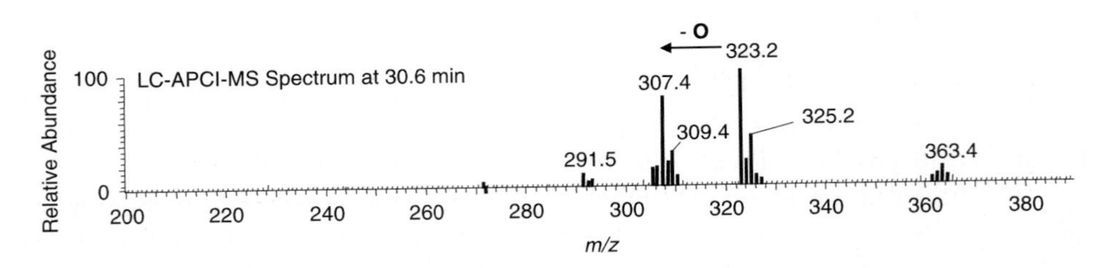

Figure 20.
LC–APCI–MS spectrum of metabolite M47 from rat plasma following administration of ^{14}C-LOR to rats

$[M+H-O]^+$ ions. Observation of the ion at *m/z* 307 indicated the possibility of *N*-oxidation. This information together with the molecular weight increase of 12 Da (addition of O and loss of four hydrogen) and the observation of *m/z* 228 in the MS/MS spectrum suggested the aromatization of the piperidine to pyridine and *N*-oxidation. Final structural elucidation was verified by NMR experiments. M47 is a novel biotransformation product because DL had undergone both ring aromatization and *N*-oxidation.

Once the structure was deduced from LC–APCI/MS and LC–NMR experiments, the LC–ESI/MS/MS spectrum can be explained as shown in the following scheme.

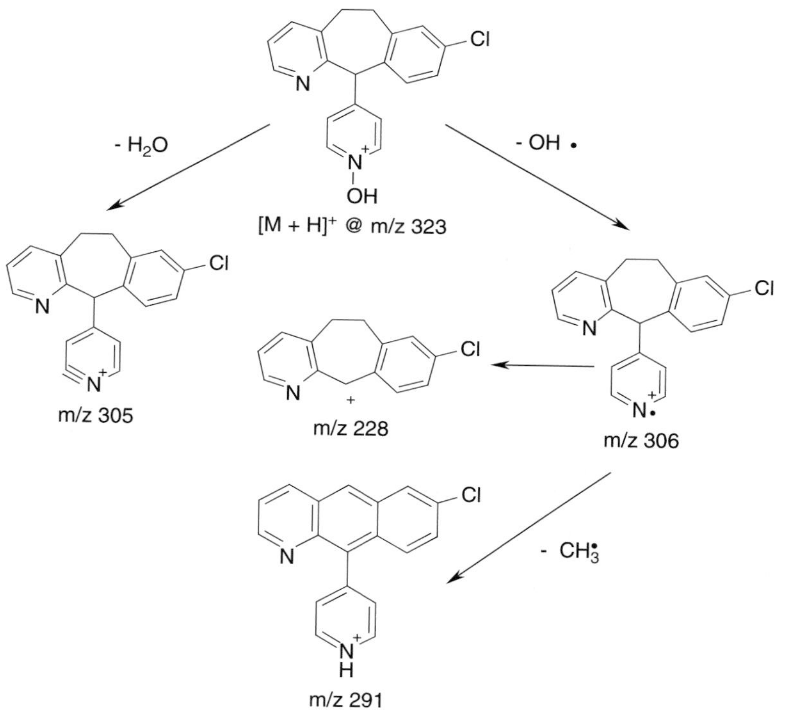

- H_2O - OH •

$[M + H]^+$ @ m/z 323

m/z 305   m/z 228   m/z 306

- $CH_3^•$

m/z 291

10.5.1.2. Clozapine (Clozaril®)

Clozapine (CLZ) is an atypical neuroleptic used in the treatment of schizophrenia. Following a single 50 mg oral dose of ^{14}C-CLZ to healthy male volunteers, about 30% and 50% of the dose was excreted over 6 days in the feces and urine, respectively. Major metabolites were formed via demethylation (desmethyl-CLZ), oxidation and glucuronidation (Dain *et al.*, 1997). Although clozapine-*N*-oxide (CLZ-*N*-oxide) was one of the major metabolites, it was also shown to undergo in vivo metabolic reduction to CLZ in human when CLZ-*N*-oxide was administered.

As illustrated below, CLZ can undergo *N*-oxidation at three possible sites, however, only the product from *N*-oxidation of the methyl-piperazinyl moiety has been reported as a metabolite following administration of CLZ to humans.

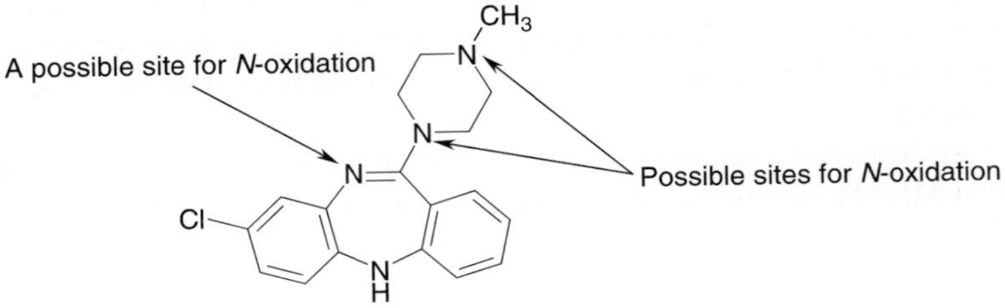

A possible site for *N*-oxidation

Possible sites for *N*-oxidation

Under LC–ESI–MS conditions, protonated molecular ions containing $^{35}Cl/^{37}Cl$ for CLZ and CLZ-*N*-oxide were detected at *m/z* 327/329 and 343/345, respectively. Under LC–ESI–MS/MS conditions, precursor ions at *m/z* 327 and 343 underwent fragmentation to give ions at *m/z* 99, 227 and 296 (Lin *et al.*, 1996).

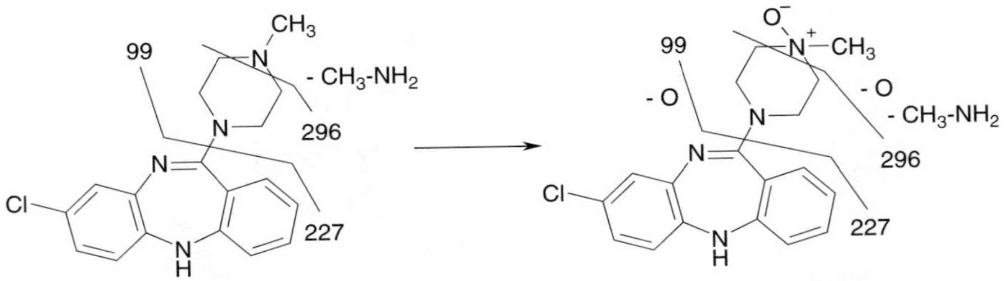

Clozapine (CLZ)
$[M + H]^+$ @ m/z 327

Clozapine-N-Oxide (CLZ-N-Oxide)
$[M + H]^+$ @ m/z 343

Observation of a fragment ion at 227, suggests most likely that the metabolite did not involve the tricyclic ring moiety. The fragment ion at *m/z* 296 simply indicates that this metabolite could be an *N*-oxide or an *N*-CH_3 moiety may have biotransformed to *N*-CH_2OH. Therefore, LC–ESI–MS and LC–ESI–MS/MS spectra could not have provided conclusive evidence about the site of *N*-oxidation. However, availability of reference standards allowed identification of the metabolites by co-chromatography.

In a recent paper, Kratzsch *et al.* (2003) reported LC–APCI–MS spectra of CLZ and CLZ-*N*-Oxide, where the diagnostic source-induced fragment ion corresponding $[M + H - O]^+$ ions was observed for CLZ-*N*-Oxide at *m/z* 327. Once again, this example highlights the importance of LC–APCI–MS for distinguishing hydroxylated metabolites from *N*-oxides.

10.5.2. Hydroxylamines

Similar to *N*-oxides, hydroxylamine metabolites result in an increase in molecular weight by 16 Da. Some of the common structural backbones that undergo biotransformation to form hydroxylamines are presented in Fig. 17.

Similar to distinguishing *N*-oxides from aliphatic and aromatic hydroxylations, distinguishing hydroxylamines from other monooxygenated metabolites also is not straightforward. Most recently, Pols *et al.* (2004) showed the applicability of H/D exchange and MS/MS for distinguishing isobaric but chromatographically separated hydroxylamines from *N*-oxides and hydroxylated metabolites using derivatives of DL. LC–MS following H/D exchange allowed the distinction of *N*-oxides and hydroxylamines from aliphatic and aromatic hydroxylation products because the two latter types of metabolites undergo an additional deuterium exchange. The utility of various experiments for the detection and identification of hydroxylamine metabolites is discussed in the following section using the metabolism of Dapsone (DDS) as an example.

10.5.2.1. Dapsone (Avlosulfon)

DDS is used for treating leprosy, various skin conditions, and pneumonia in people infected with the human immunodeficiency virus (HIV). One of the major routes of DDS elimination is via formation of DDS–hydroxylamine (DDS–NOH).

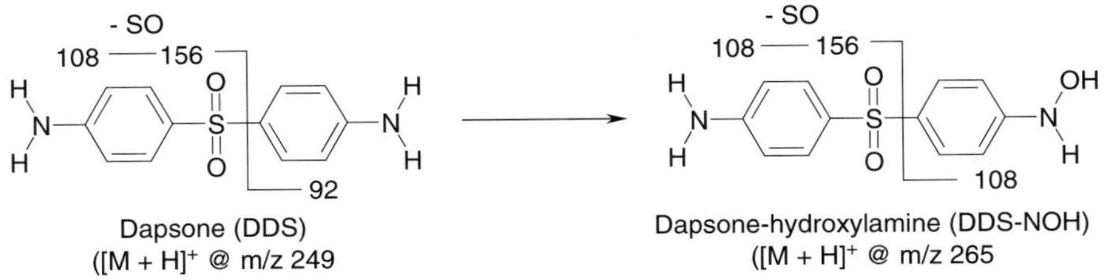

Dapsone (DDS)
($[M + H]^+$ @ m/z 249)

Dapsone-hydroxylamine (DDS-NOH)
($[M + H]^+$ @ m/z 265)

As shown above, protonated molecular ions for DDS and DDS–NOH are observed at *m/z* 249 and 265 Th, respectively. Similar to other oxidative processes, formation of hydroxylamine results in an *m/z* shift of 16 Th.

Under LC–MS/MS conditions, precursor ions of DDS fragmented to give product ions at *m/z* 92, 108, and 156 Th. Precursor ions of DDS–NOH fragmented to give ions at *m/z* 108 and 156 Th. However, the nature and the site of oxidation cannot be determined from the MS/MS data. If H/D exchange LC–ESI–MS experiments had been performed, valuable information regarding the nature of oxidation could have been obtained and to distinguish hydroxydapsone from dapsone hydroxylamine.

As shown in the following scheme, under H/D exchange conditions, DDS–NOH and DDS undergo the same number of exchanges and result in the shifting of the LC–MS-detected peak by 5 Th. In contrast, hydroxydapsone would undergo six exchanges. In general, formation of a hydroxylamine does not increase the number of exchangeable hydrogen atom(s) compared to the parent compound. This information together with molecular weight change and MS/MS data can be used to distinguish between different classes of oxidation and potentially the site(s) of oxidation.

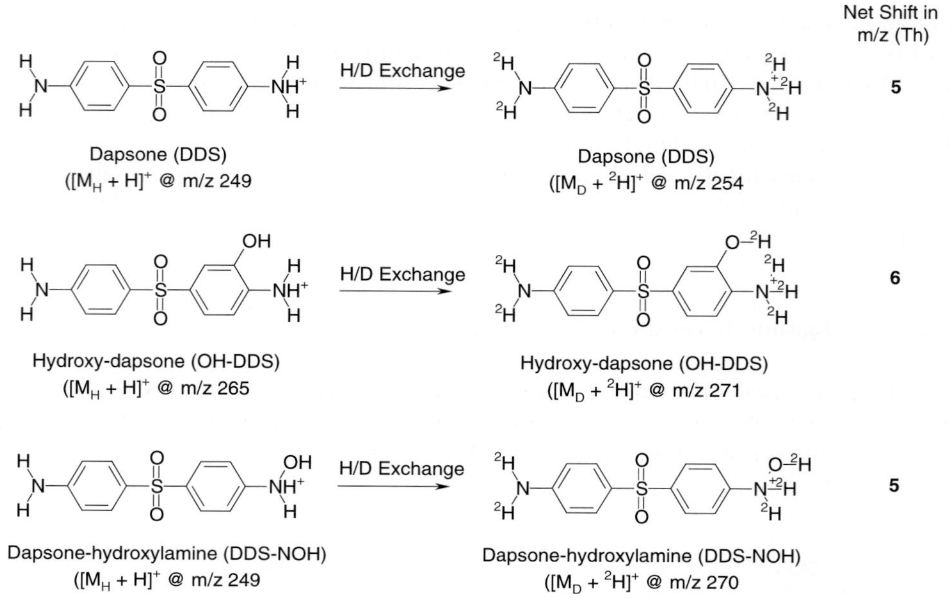

Net Shift in m/z (Th)

H/D Exchange

Dapsone (DDS) ($[M_H + H]^+$ @ m/z 249) → Dapsone (DDS) ($[M_D + ^2H]^+$ @ m/z 254) — 5

Hydroxy-dapsone (OH-DDS) ($[M_H + H]^+$ @ m/z 265) → Hydroxy-dapsone (OH-DDS) ($[M_D + ^2H]^+$ @ m/z 271) — 6

Dapsone-hydroxylamine (DDS-NOH) ($[M_H + H]^+$ @ m/z 249) → Dapsone-hydroxylamine (DDS-NOH) ($[M_D + ^2H]^+$ @ m/z 270) — 5

10.5.2.2. Debrisoquine

In addition to the formation of aromatic and aliphatic hydroxyl metabolites, debrisoquine also undergoes biotransformation to debrisoquine hydroxylamine (Granvil *et al.*, 2002). Under LC–MS conditions, protonated molecular ions for debrisoquine, and debrisoquine hydroxylamine were detected, respectively at *m/z* 176 and 192. In the absence of a reference standard, it should be possible to distinguish the hydroxylamine metabolite from aromatic hydroxylation (5-, or 6-, or 7-, or 8-hydroxy-debrisoquine) and aliphatic hydroxylation (1-, or 3-, or 4-hydroxydebrisoquine) by combination of LC–MS/MS, H/D exchange and APCI experiments described in the previous sections.

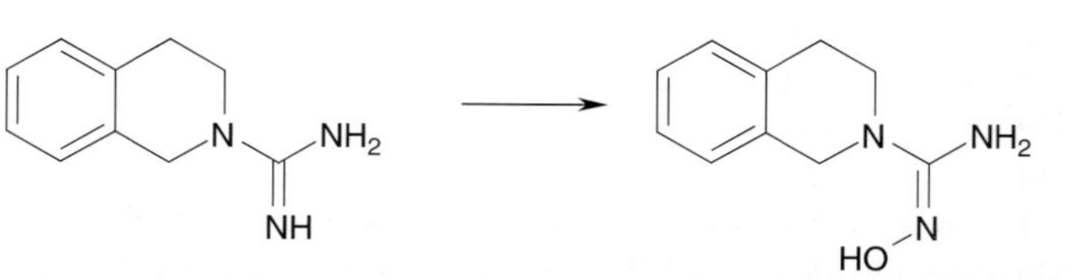

Debrisoquine ($[M + H]^+$ @ m/z 176) → Debrisoquine-hydroxylamine ($[M + H]^+$ @ m/z 192)

10.6. Oxidation of Sulfur

In addition to Cytochrome P450 enzymes, FMOs also catalyze *S*-oxidation (Lawton *et al.*, 1994). Metabolites produced from monooxidation of a sulfur atom are known as *S*-oxides or sulfoxides, whereas dioxidation of a sulfur atom results in a sulfone (Ziegler, 1988).

10.6.1. *S*-Oxidation/Sulfoxides

10.6.1.1. Cimetidine (Tagamet®)

Cimetidine is a histamine$_2$-receptor antagonist that is commonly used for treating peptic ulcers in the stomach or duodenum and acid reflux into the esophagus by reducing acid secretion in the stomach. Following oral administration to animals and humans, cimetidine is metabolized principally to cimetidine-*S*-oxide. Within 24 h following oral administration of cimetidine to humans, about 20% of the dose is excreted as the *S*-oxide (Taylor *et al.*, 1978).

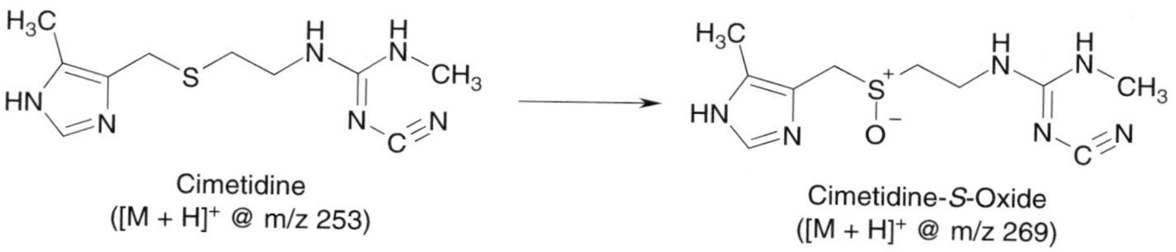

Cimetidine
($[M + H]^+$ @ m/z 253)

Cimetidine-*S*-Oxide
($[M + H]^+$ @ m/z 269)

Although LC–MS techniques were not available when cimetidine metabolite identification studies were initially carried out, chemical ionization–mass spectrometry (CI–MS) of chromatographically isolated cimetidine-*S*-oxide exhibited a protonated molecular ion at *m/z* 269 and a peak corresponding to loss of an oxygen atom at *m/z* 253 (Cashman *et al.*, 1993).

Today, on-line characterization of cimetidine-*S*-oxide could be achieved by using LC–APCI–MS, MS/MS and H/D exchange techniques. Similar to *N*-oxides, under LC–APCI–MS conditions, *S*-oxides would undergo decomposition in the APCI source to form $[M+H-O]^+$ ions.

10.6.1.2. Clindamycin (Cleocin®)

A Gram-positive bacterial antibiotic, clindamycin is excreted in urine and feces as clindamycin-sulfoxide and *N*-desmethyl-clindamycin following a single oral administration to humans. Recent studies have shown that Clindamycin is mainly catalyzed by CYP3A4 to clindamycin-sulfoxide (Wynalda *et al.*, 2003).

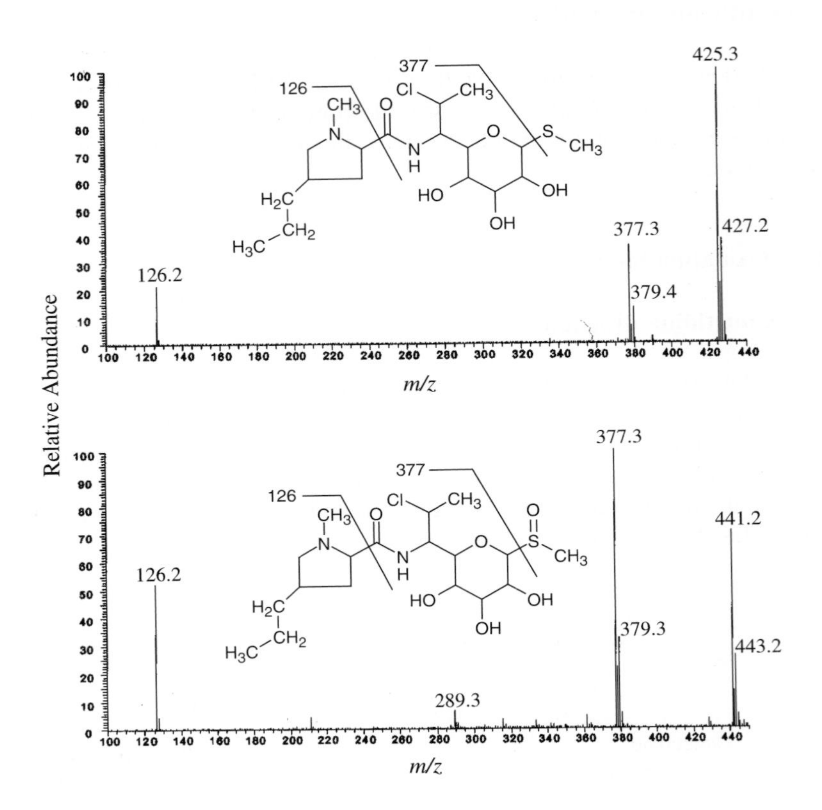

Figure 21.
LC–MS/MS spectra of clindamycin and clindamycin sulfoxide (modified from Wynalda *et al.*, 2003)

LC–MS spectra of clindamycin show protonated ions at *m/z* 425 (^{35}Cl) and 427 (^{37}Cl) Th. Molecular ions for the metabolite, clindamycin-sulfoxide, were detected at *m/z* 441/443, 16 Th higher than that for clindamycin. The presence of unshifted (Fig. 21) fragment ions at *m/z* 126 and 377 in the MS/MS spectra of clindamycin-sulfoxide facilitated the characterization of the metabolite. Although in principle, oxidation could have occurred on the terminal carbon atom as well, it can be readily distinguished from sulfoxidation using APCI–MS or H/D exchange LC–MS experiments.

10.6.1.3. Omapatrilat (Vanlev)

Omapatrilat is a drug recently developed for the treatment of hypertension and congestive heart failure. Following a 50 mg oral administration of ^{14}C-omapatrilat to healthy volunteers, about 64 and 8% of the dose were excreted, respectively, in urine and feces within 168 h. *S*-methyl-omapatrilat and its metabolites were prominent circulating as well as urinary metabolites (Iyer *et al.*, 2001). As shown in the following scheme, *S*-methyl-omapatrilat further undergoes *S*-oxidation at both sulfur atoms.

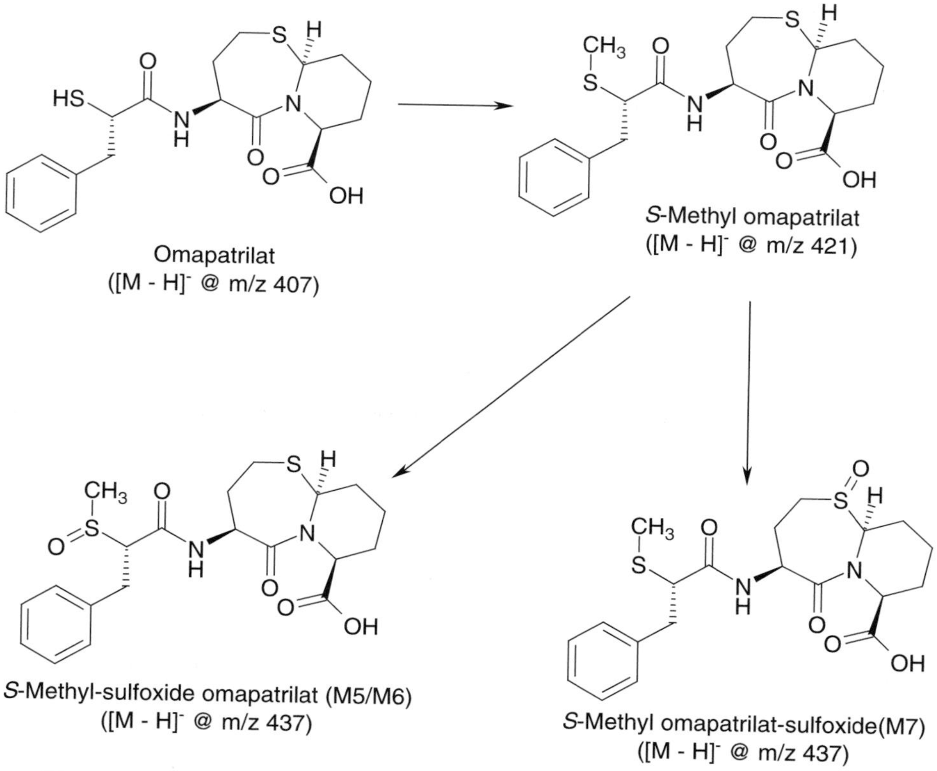

Omapatrilat
([M - H]$^-$ @ m/z 407)

S-Methyl omapatrilat
([M - H]$^-$ @ m/z 421)

S-Methyl-sulfoxide omapatrilat (M5/M6)
([M - H]$^-$ @ m/z 437)

S-Methyl omapatrilat-sulfoxide(M7)
([M - H]$^-$ @ m/z 437)

The fragmentation of *S*-methyl-omapatrilat and its sulfoxides in MS/MS experiments is complex and not easy to interpret (Fig. 22). However, there are enough differences in the MS/MS spectra of the isomeric *S*-methyl-omapatrilat that can be used to deduce each structure.

Deprotonated molecular ions for omapatrilat and *S*-methylomapatrilat were detected at *m/z* 407 and 421, respectively. Regioisomeric sulfoxides of *S*-methylomapatrilat were detected at *m/z* 437. Under LC–MS/MS conditions, the exocyclic sulfoxide of *S*-methylomapatrilat (M5/M6) dissociated to give major fragment ions at *m/z* 373, 310, and 246. The corresponding cyclic sulfoxide fragmented to yield ions at *m/z* 310 and 262. Observation of the fragment ion at *m/z* 373 due to loss of the methyl sulfoxide moiety (CH_3SO) provided evidence for oxidation of the exocyclic sulfur. By contrast, the fragment ion at *m/z* 373 was absent from the MS/MS spectrum of M7, whereas loss of the SO moiety (48 Da) from the fragment ion at *m/z* 310 is consistent with monooxidation of the cyclic sulfur.

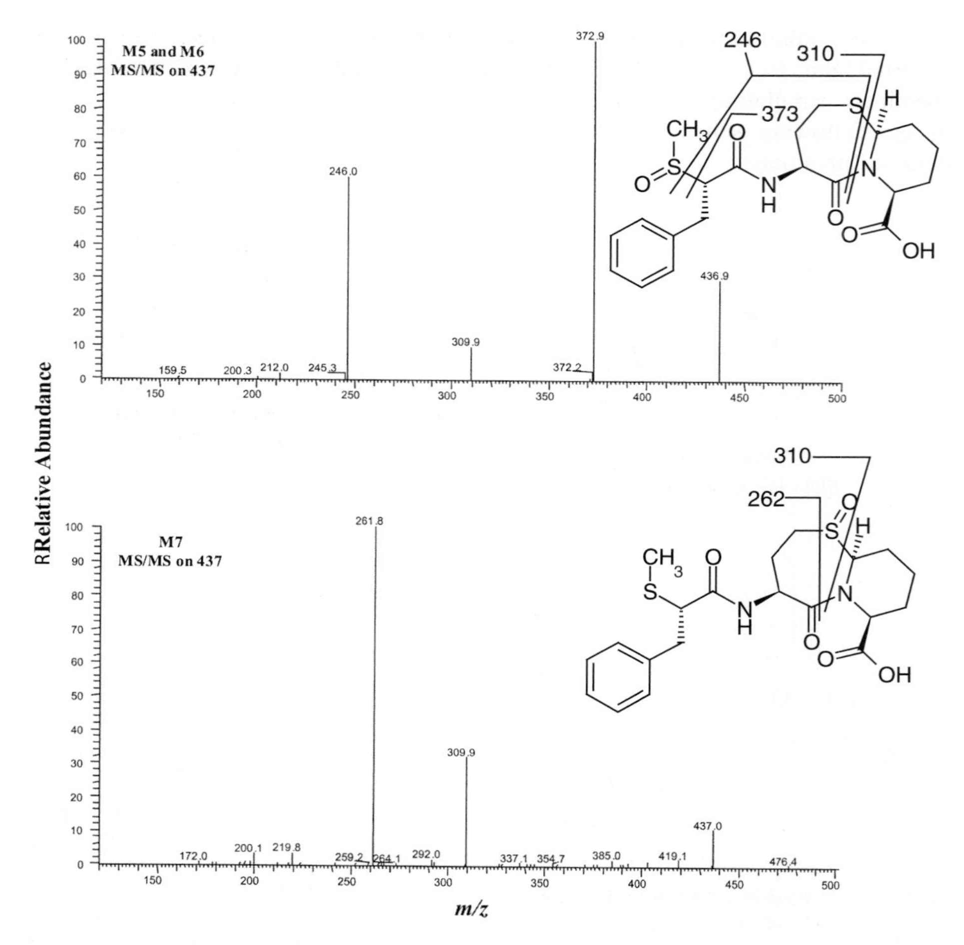

Figure 22.
LC–MS/MS spectra of regioisomeric sulfoxides of *S*-methyl omapatrilat (modified from Iyer *et al.*, 2001)

10.6.2. Sulfones

Sulfones result from incorporation of two oxygen atoms on a sulfur atom. Similar to *S*-oxide/sulfoxide metabolites, metabolism of sulfones is catalyzed by Cytochrome P450s and FMOs.

10.6.2.1. Chlorpromazine (Megaphen)

Since 1954, chlorpromazine (CPZ) has been used for the treatment of mental disorders such as schizophrenia. In mammals, minor to trace levels of *N*-didesmethyl-

CPZ-sulfone and *N*-didesmethyl-CPZ-sulfoxide metabolites are detected in the urine (Breyer-Pfaff *et al.*, 1978). Representative sulfoxide/sulfone metabolites of CPZ are illustrated as follows:

Chlorpromazine
([M + H]$^+$ @ m/z 319)

N-Didesmethyl-chlorpromazine-S-oxide
([M + H]$^+$ @ m/z 307)

N-Didesmethyl-chlorpromazine-sulfone
([M + H]$^+$ @ m/z 323)

Under LC–MS conditions (positive mode ESI), protonated molecular ions for CPZ, *N*-didesmethyl-CPZ-sulfone, and *N*-didesmethyl-CPZ-sulfoxide are detected at *m/z* 319, 323, and 307, respectively. LC–MS/MS characterization of CPZ has been discussed in detail by McClean *et al.* (2000).

10.6.2.2. Ziprasidone (Geodon®)

An effective antipsychotic agent with highly selective dopamine D2 and serotonin 5-HT2 receptor antagonistic properties (Howard *et al.*, 1994). Metabolic clearance of Ziprasidone (ZIP) involves: (1) *N*-dealkylation of the oxyindole ethyl moiety, (2) oxidation to sulfoxides and sulfones, (3) formation of methylated and benzisothizol ring opened products, and (4) dearylation of the benzisothiazole moiety (Prakash *et al.*, 1997). Use of dual radiolabels (^{14}C and ^{3}H), tracking the $^{35}Cl/^{37}Cl$ isotope pattern, and the unique fragmentation mechanism allowed the detection of several unique metabolites following administration of ZIP to animals and humans.

Within 11 days following administration of a 20 mg dose to healthy male volunteers, nearly 20 and 66% of the dose was excreted, respectively, in urine and feces. Those metabolites resulting from sulfoxidation and sulfonation are shown in the following scheme.

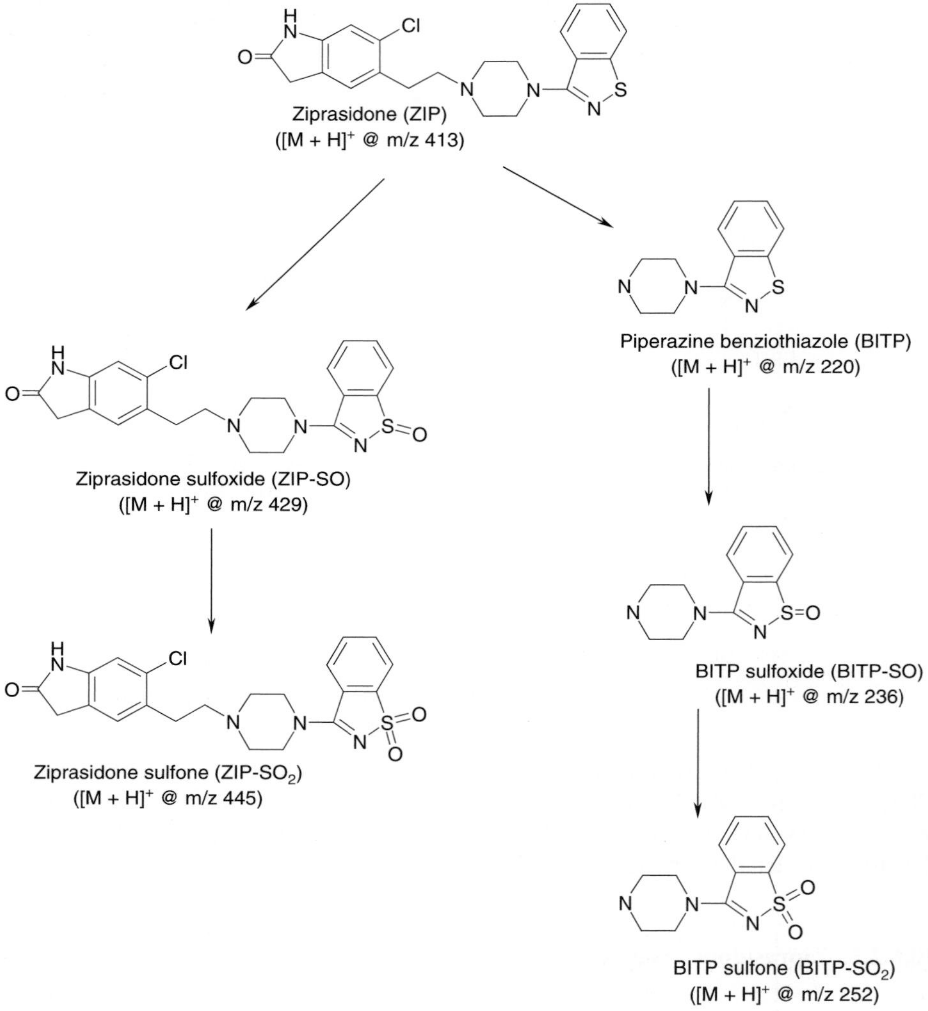

Ziprasidone (ZIP)
($[M + H]^+$ @ m/z 413)

Ziprasidone sulfoxide (ZIP-SO)
($[M + H]^+$ @ m/z 429)

Ziprasidone sulfone (ZIP-SO_2)
($[M + H]^+$ @ m/z 445)

Piperazine benziothiazole (BITP)
($[M + H]^+$ @ m/z 220)

BITP sulfoxide (BITP-SO)
($[M + H]^+$ @ m/z 236)

BITP sulfone (BITP-SO_2)
($[M + H]^+$ @ m/z 252)

10.7. β-Oxidation

Metabolites formed via β-oxidation are referred to as such because they resemble the stepwise oxidation of two-carbon chain observed in the β-oxidation of fatty acids. Drugs that undergo β-oxidation include valproic acid (discussed in detail by Bjorge and Baillie, 1991) and most of the statins. In the following section, β-oxidation of statins will be discussed using FV as an example.

Statins, inhibitors of HMG-CoA reductase involved in cholesterol biosynthesis, are used for lowering cholesterol. Most of the statins investigated to date are substrates for β-oxidation. In FV, the β-oxidation constitutes a minor biotransformation pathway. Following administration of a single oral dose (2–40 mg/day) of FV to healthy male volunteers, the major circulating metabolic products of FV were formed via β-oxidation, *N*-dealkylation (discussed previously) or a combination of both processes.

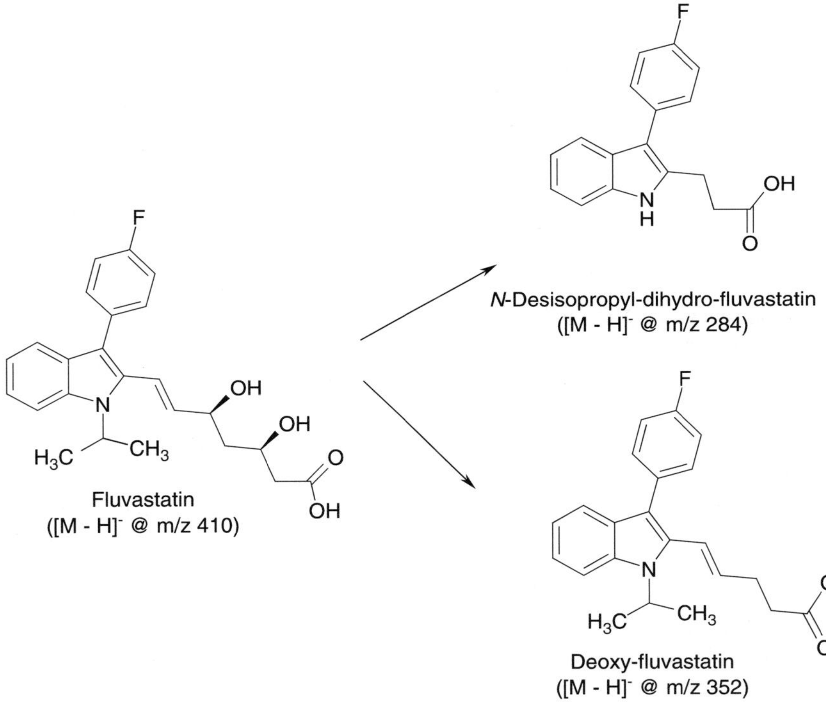

Fluvastatin
($[M - H]^-$ @ m/z 410)

N-Desisopropyl-dihydro-fluvastatin
($[M - H]^-$ @ m/z 284)

Deoxy-fluvastatin
($[M - H]^-$ @ m/z 352)

Under LC–MS conditions, FV, *N*-desisopropyl-dihydro-FV and deoxy-FV form deprotonated ions at *m/z* 410, 352, and 284 Th, respectively. Using reference standards, these metabolites were unambiguously characterized. However, identification could also have been achieved using accurate mass LC–MS for determination of elemental composition and LC–MS and LC–MS/MS experiments following H/D exchange.

10.8. Dehalogenation

Oxidative dehalogenation involves removal of halogen atoms such as fluorine, chlorine, and bromine, and together with simultaneous addition of an oxygen atom.

Oxidative dehalogenation involving a single oxygen atom and a single fluorine, chlorine, or bromine atom shifts molecular weight by 2, 18 and 62 Da, respectively. Under LC–MS conditions, oxidative dechlorination and oxidative debromination biotransformations would result in the disappearance of the cluster ion pattern observed in the mass spectra from $^{35}Cl/^{37}Cl$ and $^{79}Br/^{81}Br$, respectively. However, oxidative defluorination does not involve any isotope pattern change that can be discerned by mass spectrometry. In the following section, Halothane and Tolfenamic acid are discussed as examples of drug that undergo oxidative dehalogenation.

10.8.1. Halothane

Halothane, which has been used as an anesthetic in humans, undergoes biotransformation to form dehalogenated metabolites via oxidative elimination of chlorine and bromine atoms. These halothane/metabolites were reported to cause halothane hepatitis (Spracklin and Kharasch, 1998).

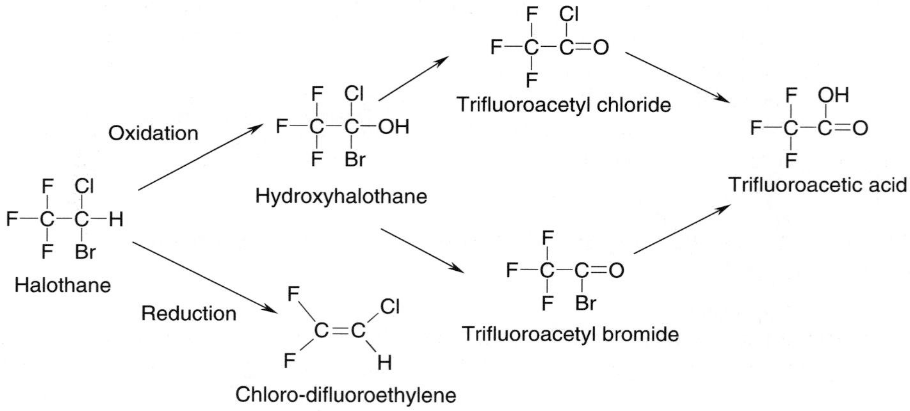

Halothane
Oxidation
Hydroxyhalothane
Trifluoroacetyl chloride
Trifluoroacetic acid
Trifluoroacetyl bromide
Reduction
Chloro-difluoroethylene

As shown in the scheme above, halothane is a unique substrate that can undergo both oxidative and reductive biotransformations. Under LC–MS conditions, molecular ions for trifluoroacetyl chloride, trifluoroacetyl bromide and trifluoroacetic acid would exhibit distinctly different isotope patterns from halothane due to losses of Br and/or Cl. Without dehalogenation, the isotope distribution pattern of hydroxyhalothane would be similar to that of halothane.

10.8.2. Tolfenamic Acid

Tolfenamic acid is an anti-inflammatory drug, which following administration to humans undergoes many oxidative biotransformations and is eliminated in urine as glucuronide conjugates. Monohydroxy-deschlorotolfenamic acid glucuronide is among the many metabolites formed following 200 mg oral administration of Tolfenamic acid to humans (sidelmann *et al.*, 1997). It is interesting to note that in this example, dehalogenation and oxidation took place on adjacent carbon atoms.

Tolfenamic acid
([M - H]$^-$ @ m/z 260)

Monohydroxy-deschlorotolfenamic acid glucuronide
([M - H]$^-$ @ m/z 418)

Under LC–MS conditions, [M − H]$^-$ ions for tolfenamic acid and monohydroxy-deschlorotolfenamic acid glucuronide are observed at *m/z* 260 and 418, respectively. Simulated mass spectra for tolfenamic acid and monohydroxy-deschlorotolfenamic acid glucuronide are shown in Fig. 23. The cluster ion pattern from ^{35}Cl and ^{37}Cl for tolfenamic acid is observed at *m/z* 260 and 262, respectively. Without Cl, a similar cluster ion pattern is not observed in the mass spectrum of the metabolite.

Although LC–MS data alone are sufficient to characterize the loss of a chlorine atom in the metabolite, MS/MS experiments would still be required to assure that the compound detected at *m/z* 418 is drug derived. Furthermore, NMR experiments would be necessary to locate the exact position of hydroxylation.

10.9. Future Directions of Metabolite Characterization/ Identification

With the advent of API techniques that allow for easy interface of a liquid chromatograph to a mass spectrometer (Whitehouse *et al.*, 1985; Covey *et al.*, 1986),

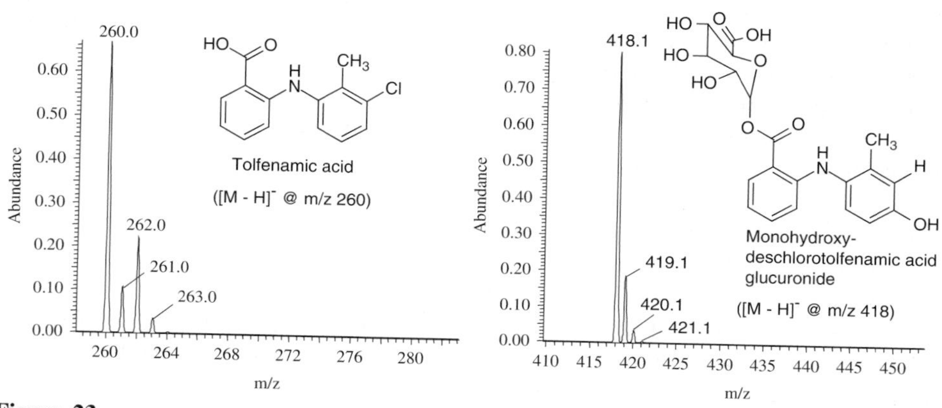

Figure 23.
Simulated mass spectra of tolfenamic acid and monohydroxy-deschlorotolfenamic acid glucuronide

metabolite characterization/identification from biological matrices is now routinely performed by LC–MS. Each advance in LC–MS technology offers the possibility that metabolite profiling and characterization studies can be conducted with ever increasing selectivity and sensitivity. With the recent introduction of linear ion trap mass spectrometers (LITs) as well as hybrid and pure-bred time-of-flight mass spectrometers (TOFMS), detection and identification of metabolites with very limited sample preparation are now possible. The TOFMS can also operate at high mass resolution (5000–20 000). As a result, these instruments provide accuracy ($m/z < 5$ ppm) sufficient to compute elemental composition of unknown metabolites as well as their fragment ions. Proposed structures derived from corresponding MS/MS experiments are greatly strengthened with evidence from accurate mass determination. It is thus obvious that LC-TOFMS systems will play a much wider role in metabolite identification in the future. As a result of the high sensitivity and fast scan speed, LITs and hybrid LITs offer great potential in rapid metabolite characterization through automation by data-dependent scans or simultaneously performed multiple scans in MS and MS/MS modes. These features will be of particular importance in drug discovery settings, where rapid metabolite characterization is required to speed up lead optimization process and candidate selection for development. Fourier Transform mass spectrometry (FTMS), coupled with its ultra-high resolution and high sensitivity, is another technology that is rapidly becoming an important analytical tool in the modern drug metabolism laboratory. Recent improvements in instrument ruggedness and the development of user-friendly software make FTMS a powerful tool for identification of unknown compounds. FTMS systems can routinely achieve sub-ppm accuracy in m/z measurements (Marshall *et al.*, 1998) leading to a very reliable determination of elemental composition thereby shortening the list of potential structures.

References

Äbelö, A., Andersson, T.B., Antonsson, M., Naudot, A.K., Skånberg, I., Weidolf, L., 2000. Stereoselective metabolism of omeprazole by human Cytochrome P450 enzymes. Drug Metab. Dispos. 28, 966–972.

Bjorge, S.M., Baillie, T.A., 1991. Studies on the beta-oxidation of valproic acid in rat liver mitochondrial preparations. Drug Metab. Dispos. 19, 823–829.

Breyer-Pfaff, U., Kreft, H., Rassner, H., Prox, A., 1978. Formation of sulfone metabolites from chlorpromazine and perazine in man. Drug Metab. Dispos. 6, 114–119.

Cashman, J.R., Park, S.B., Yang, Z.C., Washington, C.B., Gomez, D.Y., Giacomini, K.M., Brett, C.M., 1993. Chemical, enzymatic, and human enantioselective *S*-oxygenation of cimetidine. Drug Metab. Dispos. 21, 587–597.

Chen, T.M., Chan, K.Y., Coutant, J.E., Okerholm, R.A., 1991. Determination of the metabolites of terfenadine in human urine by thermospray liquid chromatography–mass spectrometry. J. Pharm. Biomed. Anal. 9, 929–933.

Covey, T.R., Lee, E.D., Bruins, A.P., Henion, J.D., 1986. Liquid chromatography/mass spectrometry. Anal. Chem. 58, 1451A–1465A.

Dain, J.G., Fu, E., Gorski, J., Nicoletti, J., Scallen, T.J., 1993. Biotransformation of fluvastatin sodium in humans. Drug Metab. Dispos. 21, 567–572.

Dain, J.G., Nicoletti, J., Ballard, F., 1997. Biotransformation of clozapine in humans. Drug Metab. Dispos. 25, 603–609.

Denissen, J.F., Grabowski, B.A., Johnson, M.K., Buko, A.M., Kempf, D.J., Thomas, S.B., Surber, B.W., 1997. Metabolism and disposition of the HIV-1 protease inhibitor ritonavir (ABT-538) in rats, dogs, and humans. Drug Metab. Dispos. 25, 489–501.

Dickson, M., Gagnon, J.P., 2004. Key factors in the rising cost of new drug discovery and development. Nat. Rev. Drug Discov. 3, 417–429.

Fan, P.W., Bolton, J.L., 2001. Bioactivation of tamoxifen to metabolite E quinone methide: reaction with glutathione and DNA. Drug Metab. Dispos. 29, 891–896.

Gopaul, V.S., Chowdhury, S.K., Blumenkrantz, N.B., Zhong, R., Grubb, N., Wirth, M., Alton, K.B., Patrick, J.E., McNamara, P., 2001. Profiling and Characterization of Metabolites of Ribavirin–Alanine Ester (SCH 351754) in Rats and Monkeys: Application of Stable Isotope Labeling in Structural Elucidation by LC–MS. American Society for Mass Spectrometry and Allied Topics Meeting, Chicago, IL.

Granvil, C.P., Krausz, K.W., Gelboin, H.V., Idle, J.R., Gonzalez, F.J., 2002. 4-Hydroxylation of debrisoquine by human CYP1A1 and its inhibition by quinidine and quinine. J. Pharmacol. Exp. Ther. 301, 1025–1032.

Guengerich, F.P., Parikh, A., Yun, C.-H., Kim, D., Nakamura, K., 2000. What makes P450s work? Searches for answers with known and new P450s. Drug Met. Rev. 32, 267–281.

Holm, K.J., Spencer, C.M., 2000. Bupropion: a review of its use in the management of smoking cessation. Drugs 59, 1007–1024.

Howard, H.R., Prakash, C., Seeger, T.F., 1994. Ziprasidone hydochloride. Drugs Future 19, 560–563.

Ishii, Y., Takami, A., Tsuruda, K., Kurogi, A., Yamada, H., Oguri, K., 1997. Induction of two UDP-glucuronosyltransferase isoforms sensitive to phenobarbital that are involved in morphine glucuronidation. Production of isoform-selective antipeptide antibodies toward UGT1.1r and UGT2B1. Drug Metab. Dispos. 25, 163–167.

Iyer, R.A., Mitroka, J., Malhotra, B., Bonacorsi, S., Waller, S.C., Rinehart, J.K., Roongta, V.A., Kripalani, K., 2001. Metabolism of [^{14}C]omapatrilat, a sulfhydryl-containing vasopeptidase inhibitor in humans. Drug Metab. Dispos. 29, 60–69.

Jones, B.C., Hyland, R., Ackland, M., Tyman, C.A., Smith, D.A., 1998. Interaction of terfenadine and its primary metabolites with cytochrome P450 2D6. Drug Metab. Dispos. 26, 875–882.

Jung, F., Richardson, T.H., Raucy, J.L., Johnson, E.F., 1997. Diazepam metabolism by CDNA-expressed human 2C P450S. Identification of P4502C18 and P4502C19 as low K_M diazepam *N*-demethylases. Drug Metab. Dispos. 25, 133–139.

Kobayashi, K., Yamamoto, T., Chiba, K., Tani, M., Shimada, N., Ishizaki, T., Kuroiwa, Y., 1998. Human buprenorphine *N*-dealkylation is catalyzed by cytochrome P450 3A4. Drug Metab. Dispos. 26, 818–821.

Kratzsch, C., Peters, F.T., Kraemer, T., Weber, A.A., Maur, H.H., 2003. Screening, library-assisted identification and validated quantification of fifteen neuroleptics and three of their metabolites in plasma by liquid chromatography/mass spectrometry with atmospheric pressure chemical ionization. J. Mass Spectrom. 38, 283–295.

Lagrange, F., Pehourcq, F., Baumevieille, M., Begaud, B., 1998. Determination of buprenorphine in plasma by liquid chromatography: application to heroin-dependent. J. Pharmaceut. Biomed. Anal. 16, 1295–1300.

Lawton, M.P., Cashman, J.R., Cresteil, T., Dolphin, C.T., Elfarra, A.A., Hines, R.N., Hodgson, E., Kimura, T., Ozols, J., Phillips, I.R., Philpot, R.M., Poulsen, L.L., Rettie, A.E., Shephard, E.A., Williams, D.E., Ziegler, D.M., 1994. A nomenclature for the mammalian flavin-containing monooxygenase gene family based on amino acid sequence identities. Arch. Biochem. Biopys. 308, 254–257.

Lertratanangkoon, K., Horning, M.G., 1982. Metabolism of carbamazepine. Drug Metab. Dispos. 10, 1–10.

Lin, G., McKay, G., Midha, K.K., 1996. Characterization of metabolites of clozapine *N*-oxide in the rat by micro-column high performance liquid chromatography/mass spectrometry with electrospray interface. J. Pharmaceut. Biomed. Anal. 14, 1561–1577.

Mahmood, I., Neau, S.H., Mason, W.D., 1994. An enzymatic assay for the MAO-B inhibitor selegiline in plasma. J. Pharmaceut. Biomed. Anal. 12, 895–899.

Marshall, A.G., Hendrickson, C.L., Jackson, G.S., 1998. Fourier transform ion cyclotron resonance mass spectrometry: a primer. Mass Spectrom. Rev. 17, 1–35.

Marshall, P.S., Straka, R.J., Johnson, K., 1992. Determination of dextromethorphan and its *O*-demethylated metabolite from urine therapeutic. Drug Monit. 14, 402–407.

Matsui, K., Taniguchi, S., Yoshimura, T., 1999. Correlation of the intrinsic clearance of donepezil (Aricept) between in vivo and in vitro studies in rat, dog and human. Xenobiotica 29, 1059–1072.

McClean, S., O'Kane, E.J., Smyth, W.F., 2000. Electrospray ionization-mass spectrometric characterization of selected anti-psychotic drugs and their detection and determination in human hair samples by liquid chromatography–tandem mass spectrometry. J. Chromatogr. B 740, 141–157.

Mesdjian, E., Seree, E., Charvet, B., Mirrione, A., Bourgarel-Rey, V., Desobry, A., Barra, Y., 1999. Metabolism of carbamazepine by CYP3A6: a model for in vitro drug interactions studies. Life Sci. 64, 827–835.

Miao, X.-S., Metcalfe, C.D., 2003. Determination of carbamazepine and its metabolites in aqueous samples using liquid chromatography–electrospray tandem mass spectrometry. Anal. Chem. 75, 3731–3738.

Nicholson, J.K., Lindon, J.C., Scarfe, G.B., Wilson, I.D., Abou-Shakra, F., Sage, A.B., Castro-Perez, J., 2001. High-performance liquid chromatography linked to inductively coupled plasma mass spectrometry and orthogonal acceleration time-of-flight mass spectrometry for the simultaneous detection and identification of metabolites of 2-bromo-4-trifluoromethyl-[^{13}C]-acetanilide in rat urine. Anal. Chem. 73, 1491–1494.

Patrick, J.E., Kosoglou, T., Stauber, K.L., Alton, K.B., Maxwell, S.E., Zhu, Y., Statkevich, P., Iannucci, R., Chowdhury, S., Affrime, M., Cayen, M.N., 2002. Disposition of the selective cholesterol absorption inhibitor ezetimibe in healthy male subjects. Drug Metab. Dispos. 30, 430–437.

Pols, J., Ramanathan, R., Chowdhury, S., Alton, K.B., 2004. Evaluation of Hydrogen/Deuterium Exchange and Liquid Chromatography-Mass Spectrometry for Identification of Hydroxylamine, N-Oxide, and Hydroxyl Analogs of Desloratadine. American Society for Mass Spectrometry and Allied Topics Meeting, Nashville, TN.

Prakash, C., Kamel, A., Gummerus, J., Wilner, K., 1997. Metabolism and excretion of a new antipsychotic drug, ziprasidone, in humans. Drug Metab. Dispos. 25, 863–872.

Qin, W., Li, S.F.Y., 2002. An ionic coating for determination of sildenafil and UK-103,320 in human serum by capillary zone electrophoresis-ion trap mass spectrometry. Electrophoresis 23, 4110–41166.

Ramanathan, R., Alvarez, N., Su, A.D., Chowdhury, S., Alton, K.B., Stauber, K., Patrick, J.E., 2005. Metabolism and excretion of loratadine in male and female mice, rats and monkeys. Xenobiotica 35, 155–189.

Ramanathan, R., Su, A.D., Alvarez, N., Blumenkrantz, N., Chowdhury, S.K., Alton, K.B., Patrick J.E., 2000. Liquid chromatography/mass spectrometry methods for distinguishing *N*-oxides from hydroxylated compounds. Anal. Chem. 72, 1352–1359.

Ramanathan, R., Su, A.D., Alvarez, N., Chowdhury, S., Wirth, M., Alton, K.B., Patrick, J.E., 2004. Disposition and LC–MS Characterization of Metabolites Following Oral Administration of ^{14}C-Loratadine to Healthy Male Volunteers. American Society for Mass Spectrometry and Allied Topics Meeting, Nashville, TN.

Rawlins, M.D., 2004. Cutting the cost of drug development. Nat. Rev. Drug Discov. 3, 360–364.

Scott, R.J., Palmer, J., Lewis, I.A.S., Pleasance, S., 1999. Determination of a GW cocktail of cytochrome P450 probe substrates and their metabolites in plasma and urine using automated solid phase extraction and fast gradient liquid chromatography tandem mass spectrometry. Rapid Commun. Mass Spectrom. 13, 2305–2319.

Shiiyama, S., Soejima-Ohkuma, T., Honda, S., Kumagai, Y., Cho, A.K., Yamada, H., Oguri, K., Yoshimura, H., 1997. Major role of the CYP2C isozymes in deamination of amphetamine and benzphetamine: evidence for the quinidine-specific inhibition of the reactions catalyzed by rabbit enzyme. Xenobiotica 27, 379–387.

Shin, H-.S., 1997. Metabolism of selegiline in humans. Identification, excretion, and stereochemistry of urine metabolites. Drug Metab. Dispos. 25, 657–662.

Sidelmann, U.G., Christiansen, E., Krogh, L., Cornett, C., Tjørnelund, J., Hansen, S.H., 1997. Purification and ^{1}H NMR Spectroscopic Characterization of Phase II Metabolites of Tolfenamic Acid. Drug. Metab. Dispos. 25, 725–731.

Slatter, J.G., Stalker, D.J., Feenstra, K.L., Welshman, I.R., Bruss, J.B., Sams, J.P., Johnson, M.G., Sanders, P.E., Hauer, M.J., Fagerness, P.E., Stryd, R.P., Peng, P.W., Shobe, E.M., 2001. Pharmacokinetics, metabolism, and excretion of linezolid following an oral dose of [^{14}C]linezolid to healthy human subjects. Drug Metab. Dispos. 29, 1136–1145.

Spracklin, D.K., Kharasch, E.D., 1998. Human halothane reduction in vitro by cytochrome P450 2A6 and 3A4: identification of low and high K_M isoforms. Drug Metab. Dispos. 26, 605–607.

Surapaneni, S.S., Clay, M.P., Spangle, L.A., Paschal, J.W., Lindstrom, T.D., 1997. In vitro biotransformation and identification of human cytochrome P450 isozyme-dependent metabolism of tazofelone. Drug Metab. Dispos. 25, 1383–1388.

Taylor, D.C., Cresswell, P.R., Bartlett, D.C., 1978. The metabolism and elimination of cimetidine, a histamine H2-receptor antagonist, in the rat, dog, and man. Drug Metab. Dispos. 6, 21–30.

Tong, W., Chowdhury, S.K., Chen, J.C., Zhong, R., Alton, K.B., Patrick, J.E., 2001. Fragmentation of *N*-oxides (deoxygenation) in atmospheric pressure ionization: investigation of the activation process. Rapid Comm. Mass Spect. 15, 2085–2090.

Toyo'oka, T., Kumaki, Y., Kanbori, M., Kato, M., Nakahara, Y., 2003. Determination of hypnotic benzodiazepines (alprazolam, estazolam, and midazolam) and their metabolites in rat hair and plasma by reversed-phase liquid-chromatography with electrospray ionization mass spectrometry. J. Pharm. Biomed. Anal. 30, 1773–1787.

Van Heek, M., France, C.F., Compton, D.S., Mcleod, R.L., Yumibe, N.P., Alton, K.B., Sybertz, E.J., Davis, H.R., 1997. In vivo metabolism-based discovery of a potent cholesterol absorption inhibitor, SCH 58235, in the rat and rhesus monkey through the identification of the active metabolites of SC 48461. J. Pharmacol. Exp. Ther. 283, 157–163.

Venkatakrishnan, K., Von Moltke, L.L., Greenblatt, D.J., 2001. Human drug metabolism and the cytochromes P450: application and relevance of in vitro models. J. Clin. Pharmacol. 41, 1149–1179.

Vickers, E.M., Sinclair, J.R., Zollinger, M., Heitz, F., Glänzel, U., Johanson, L., Fischer, V., 1999. Multiple cytochrome P-450s involved in the metabolism of terbinafine suggest a limited potential for drug–drug interactions. Drug Metab. Dispos. 27, 1029–1038.

Walker, D.K., Ackland, M.J., James, G.C., Muirhead, G.J., Rance, D.J., Wastall, P., Wright, P.A., 1999. Pharmacokinetics and metabolism of sildenafil in mouse, rat, rabbit, dog and man. Xenobiotica 29, 297–310.

Wang, J.-S., Backman, J.T., Taavitsainen, P., Neuvonen, P.J., Kivisto, K.T., 2000. Involvement of CYP1A2 and CYP3A4 in lidocaine *N*-deethylation and 3-hydroxylation in humans. Drug Metab. Dispos. 28, 959–965.

Whitehouse, C.M., Dreyer, R.N., Yamashita, M., Fenn, J.B., 1985. Electrospray interface for liquid chromatographs and mass spectrometers. Anal. Chem. 57, 675–679.

Wienkers, L.C., Allievi, C., Hauer, M.J., Wynalda, M.A., 1999. Cytochrome P-450-mediated metabolism of the individual enantiomers of the antidepressant agent reboxetine in human liver microsomes. Drug Metab. Dispos. 27, 1334–1340.

Wynalda, M.A., Hutzler, J.M., Koets, M.D., Podoll, T., Wienkers, L.C., 2003. In vitro metabolism of clindamycin in human liver and intestinal microsomes. Drug Metab. Dispos. 31, 878–887.

Yu, A., Dong, H., Lang, D., Haining, R.L., 2001. Characterization of dextromethorphan *O*- and *N*-demethylation catalyzed by highly purified recombinant human CYP2D6. Drug Metab. Dispos. 29, 1362–1365.

Zhong, D., Xing, J., Zhang, S., Sun, L., 2002. Study of the electrospray ionization tandem mass spectrometry of sildenafil derivatives. Rapid Commun. Mass Spectrom. 16, 1836–1843.

Ziegler, D.M., 1988. Flavin-containing monooxygenase catalytic mechanism and substrate specificities. Drug Metab. Rev. 25, 1–32.

Identification and Quantification of Drugs, Metabolites and Metabolizing
Enzymes by LC–MS
Swapan K. Chowdhury, editor.

Chapter 11

DETECTION AND CHARACTERIZATION OF HIGHLY POLAR METABOLITES BY LC–MS: PROPER SELECTION OF LC COLUMN AND USE OF STABLE ISOTOPE-LABELED DRUG TO STUDY METABOLISM OF RIBAVIRIN IN RATS

S.K. Chowdhury, V.S. Gopaul, N. Blumenkrantz, R. Zhong, K.M. Kulmatycki, and K.B. Alton

11.1. Introduction

Ribavirin (1-β-D-ribofuranosyl-1*H*-1,2,4-triazole-3-carboxamide, $RTCONH_2$), a broad spectrum antiviral agent is highly effective for the treatment of chronic hepatitis C viral infection when administered in combination with Intron A™ or PEG-Intron™ [1–3].

Since its introduction to the U.S. market, the biotransformation of $RTCONH_2$ was initially reported in two main studies [4,5]. In these publications, three metabolic pathways for $RTCONH_2$: (a) reversible phosphorylation, (b) deribosylation, and (c) amide hydrolysis were described as shown in Fig. 1. These studies were based on the isolation of metabolites from biological samples from humans and animals either by thin-layer chromatography or Sephadex® chromatography with detection by liquid scintillation spectrometry (LSS) [4] or a complex gas chromatography–mass spectrometry (GC–MS) method following derivatization (trimethylsilylation or permethylation) of each metabolite prior to analysis [5]. The highly polar nature of $RTCONH_2$ and its metabolites poses an analytical challenge to achieving on-line identification and characterization of ribavirin and its metabolites in biological matrices. For this reason, direct, on-line metabolite profiling studies of $RTCONH_2$ using current technologies such as LC–MS have only been recently reported.

Alton *et al.* presented data on the characterization of the two metabolites of ribavirin in human urine following a single 600 mg dose of ^{14}C-ribavirin [6]. Analyses were performed using selected reaction monitoring (SRM), which is, LC–MS/MS (where MS/MS stands for tandem mass spectrometry) of a selected precursor ion with a single product ion detection technique that confirmed the presence of two metabolites. The relative amounts of the drug and metabolites were independently determined from a separate radiochromatographic profile. Due to the highly specific nature of the

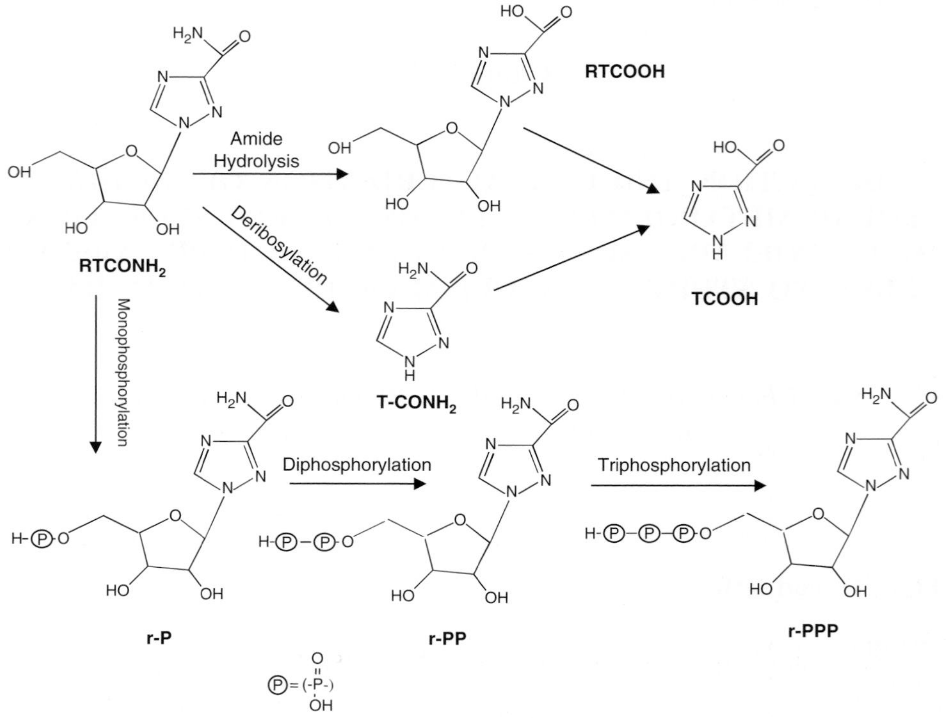

Figure 1.
Postulated metabolic pathways for $RTCONH_2$ in rats (adapted from Ref. [5])

LC–MS/MS in SRM mode, additional metabolites, if present, could not be identified using this type of analyses [6].

We present a simple, sensitive and direct reverse phase (RP) LC–MS method for the on-line identification and characterization of $RTCONH_2$ and its metabolites that does not require isolation and derivatization of individual metabolites. Ribavirin and its metabolites are present as dephosphorylated moieties in plasma and excreta. Compounds that are phosphorylated intracellularly as with the case of ribavirin and its metabolites are hydrolyzed upon leaving the tissue and/or red blood cells [7]. The method reported here differs from previously published procedures [4–6] in that it provides on-line characterization of metabolites in biological samples using full scan LC–MS and LC–MS/MS techniques in conjunction with radiometric detection, thus accounting for all radiochromatographic peaks. This approach of simultaneous detection by a radiometric detector and LC–MS allows for semi-quantitative assessment and detection of all metabolites in a single experiment.

To facilitate metabolite identification in this study, male Sprague–Dawley rats were orally dosed (60 mg/kg) with a mixture containing equal amounts of $RTCONH_2$ and $^{13}C_3$-$RTCONH_2$, and a small mass (1.29 μCi/mg of final dose) of ^{14}C-$RTCONH_2$.

Radiochromatograms from in-line flow scintillation analysis were used to determine retention times to evaluate mass spectra of drug-derived compounds. Monitoring of ribavirin and metabolites during the optimization of the extraction process was accomplished by LSS. The relative amounts of the parent drug and each metabolite present in every matrix were estimated from the corresponding radiochromatogram. The mass spectrum corresponding to each radiometric peak exhibited ion doublets with a mass-to-charge ratio (*m/z*) difference equal to the number of stable isotope-labeled carbon atoms present in the metabolite. Finally, LC–MS and MS/MS spectra were compared to those of synthetic reference standards for definitive identification of each drug-derived entity.

11.2. Experimental

11.2.1. Materials

^{14}C-$RTCONH_2$ and $^{13}C_3$-$RTCONH_2$ were synthesized by the Radiochemistry Section of Chemical Research at Schering-Plough Research Institute. 1-β-D-ribofuranosyl-1,2,4-triazole-3-carboxylic acid (RTCOOH) and 1,2,4-triazole carboxamide ($TCONH_2$) were obtained from Kobe Natural Products and Chemical Co. (Kobe, Japan).

Ammonium acetate was purchased from Sigma (St. Louis, MO) and acetonitrile and methanol from Burdick and Jackson (Muskegon, MI). Trifluoroacetic acid (TFA) was obtained from Aldrich Chemical (St. Louis, MO).

11.2.2. Preparation of Stock and Working Solutions

Stock solutions of ^{14}C-$RTCONH_2$ and RTCOOH (1 mg/ml) were prepared in water and that of $TCONH_2$ was prepared (1 mg/ml) in water:acetonitrile (4:1). Further dilution with water:methanol (2:8, v/v with 0.1% TFA) achieved the following concentrations of $RTCONH_2$ and $TCONH_2$: (a) 10 ng/μl, (b) 3.33 ng/μl, (c) 1.67 ng/μl, (d) 0.8 ng/μl, and (e) 0.4 ng/μl. Similarly, (a) 25 ng/μl, (b) 8.33 ng/μl, (c) 4.2 ng/μl, (d) 2.1 ng/μl, and (e) 1.0 ng/μl of RTCOOH were also prepared in water:methanol (2:8, v/v with 0.1% TFA). On each day of study sample analysis, 10 μl of a mixture containing all three standards (10 ng/μl of $RTCONH_2$ and $TCONH_2$, and 25 ng/μl of RTCOOH) was injected onto the high-performance liquid chromatography (HPLC) column to establish system suitability, which was assessed based on the elution time and mass spectrometric response of the reference standards.

11.2.3. Instrumentation

Table 1 provides a list of equipment used in this study.

11.2.4. Chromatographic Conditions

Mobile phase A (MPA) consisted of 10 mM ammonium acetate (pH adjusted to 5 with TFA) and Mobile phase B (MPB) consisted of acetonitrile with 0.1% TFA. Gradient elution of metabolites was achieved using linear changes in mobile phase composition as described in Table 2.

During the first 20 min, the HPLC flow (1 ml/min) was split such that 17% of the HPLC effluent was directed to the mass spectrometer and the remaining 83% was diverted to the radiometric detector. The HPLC effluent was mixed in the radiometric detector with a scintillation fluid (Flo-Scint III, Packard Instrument Co.) at a flow rate of 2.7–3.0 ml/min. This flow-split procedure permitted simultaneous detection of

Table 1.
Equipment list for the analysis of ribavirin and its metabolites

Equipment	Model and vendor
Mass spectrometer	TSQ 7000, Finnigan MAT (San Jose, CA)
HPLC pump, controller and autosampler	Waters Alliance model 2690XE (Waters Corp., Milford, MA)
Radioactivity detector	Model 525F00 (Packard Instrument Co., Meriden, CT)
Radioactivity detector cell	500 μl (Packard Instrument Co.)
Scintillation fluid	Flo-Scint III (Packard Instrument Co.)
Liquid scintillation spectrometer	Model 2750 TR/LL (Packard Instrument Co.)
HPLC column	Platinum C_{18}, 100 A, 5 μm, 250 × 4.6 mm (Alltech Associates, Inc., Deerfield, IL)
Guard column	Platinum C_{18}, 100 A, 5 μm (Alltech Associates)

Table 2.
Programmed HPLC gradient and flow of mobile phases

Time (min)	Flow (ml/min)	MPA (%)	MPB (%)	Gradient program
0.0	1.0	100	0	Initial
5.0	1.0	100	0	Start gradient
20.0	1.0	50	50	Linear
20.5	2.0	5	95	Linear
23.5	2.0	5	95	Linear
24.0	2.0	100	0	Linear
24.5	1.0	100	0	Start equilibration
45.0	1.0	100	0	End equilibration

drug-derived radiocarbon and production of associated mass spectra. After 20 min, the flow to the mass spectrometer was diverted to waste.

11.2.5. Mass Spectrometry Conditions

Tuning and *m/z* calibration of the TSQ-7000 mass spectrometer were initially performed using standard calibrants NH_2-(methionine-arginine-phenylalanine-alanine)-COOH and myoglobin. Subsequently, a compound-specific tuning was performed by direct infusion of ribavirin in a solution of methanol and water (2:8, v/v) with 0.1% TFA into the mobile phase. The optimized LC–MS and MS/MS parameters used in this study are provided in Table 3.

11.2.6. Drug Administration and Specimen Collection

Male Sprague–Dawley rats (292–334 g) were fasted overnight before dosing and up to 4 h post-dose. Selected matrices were collected following oral administration of

Table 3.
Parameters used for LC–MS and LC–MS/MS experiments

Parameter	MS	MS/MS
Ionization source	Electrospray ionization (ESI)	Electrospray ionization (ESI)
Ionization mode	Positive	Positive
Spray needle voltage (kV)	4.0	4.0
Capillary temperature (°C)	350	350
Sample flow rate (ml/min after splitting)	0.17	0.17
Sheath gas (arbitrary unit)	Nitrogen (70)	Nitrogen (70)
Auxiliary gas (arbitrary unit)	Nitrogen (30)	Nitrogen (30)
Collision gas	NA[a]	Argon
Collision cell offset voltage (V)	NA	20
Collision cell pressure	NA	1.7 mTorr (0.23 Pa)
Scan range (s)	*m/z* 50–500 or 50–625	*m/z* 50–300 for products of *m/z* 246 or 249; *m/z* 20–385 for products of *m/z* 372 or 375
Scan time (s)	1.5	1.5
Total acquisition time (min)	20	20

[a]NA=not applicable

$^{13}C_3/^{14}C$-ribavirin (~60 mg/kg and 25 μCi/rat) as follows: urine was collected from two rats prior to dosing and in block intervals of 0–8 and 8–24 h following administration of $^{13}C_3/^{14}C$-ribavirin. Urine was stored at −20°C pending analysis.

Blood was collected from six rats using chilled heparinized Vacutainer® tubes at 0.5 and 2 h post-dose by cardiac puncture under CO_2 induced anesthesia and then placed on wet ice. Within 20 min of collection, blood was centrifuged for 10 min (~2000*g*) in a refrigerated centrifuge maintained at 4°C. The plasma samples were acidified with 50 μl of 8% TFA solution per ml of plasma to prevent degradation of metabolites. Blood from a control rat was similarly harvested and plasma acidified with 50 μl of 8% per ml of plasma.

11.2.7. Sample Preparation for LC–MS Analysis

11.2.7.1. Urine

Urine samples were analyzed without extraction. Aliquots (100 μl) of the pre-dose, pooled 0–8 and 0–24 h post-dose samples were each diluted with 400 μl of water, placed in a 1.5 ml Eppendorf microcentrifuge tube and centrifuged (~16 000*g*) for 15 min at 20°C. Approximately 100 μl of each supernatant was transferred into an HPLC vial equipped with a minimum volume insert and 10 μl injected into the LC–MS system.

11.2.7.2. Plasma

Plasma samples were pooled per time point and an aliquot from each pooled sample was analyzed by LSS for the determination of radioactive content. The extraction of drug-derived material from plasma was examined using the two procedures described below.

11.2.7.2.1. Solvent Extraction (SE)

Methanol (3 ml) was added to 1.5 ml of pooled plasma, mixed, centrifuged (1090*g*) for 2 min and the supernatant transferred to a clean container. The pellet was washed with 1.5 ml of methanol:water (8:2, v/v), vortexed for 1 min and centrifuged for 15 min. Both supernatants were combined and then analyzed for radioactivity using LSS to determine extraction recovery. The combined extracts were dried under vacuum, reconstituted in 400 μl of methanol:water (8:2, v/v), and an aliquot (65–80 μl) of the solution was injected into the LC–MS system for analysis.

11.2.7.2.2. Solid Phase Extraction (SPE)

Each Oasis HLB cartridge (Waters Corp.) was conditioned with 5 ml of methanol and 5 ml of water. Plasma samples (0.5 ml) were loaded onto pre-conditioned cartridges, washed with 1 ml of water and the drug-derived material eluted with 3 ml of methanol. Following in vacuo concentration, the extract was reconstituted in 200 μl of water:methanol (2:8, v/v) and an aliquot (35–65 μl) was injected into the LC–MS system for MS and MS/MS analyses.

11.3. Results

The LC–MS method developed for the profiling and identification of metabolites of $RTCONH_2$ separated the parent drug and two of its metabolite standards ($TCONH_2$ and RTCOOH) within a 20 min run (Fig. 2). The chromatographic, mass spectral, and structural data obtained for reference standards of $RTCONH_2$ and two metabolites identified in rat plasma and urine are summarized in Table 4.

$RTCONH_2$ eluted at 6.5 min, while RTCOOH and $TCONH_2$ eluted at 2.7 and 5.0 min, respectively (Table 4 and Fig. 2). Baseline separation was achieved between each drug-related component facilitating accurate integration of chromatographic peaks necessary for the determination of peak areas and subsequent calculations of relative amounts of each compound.

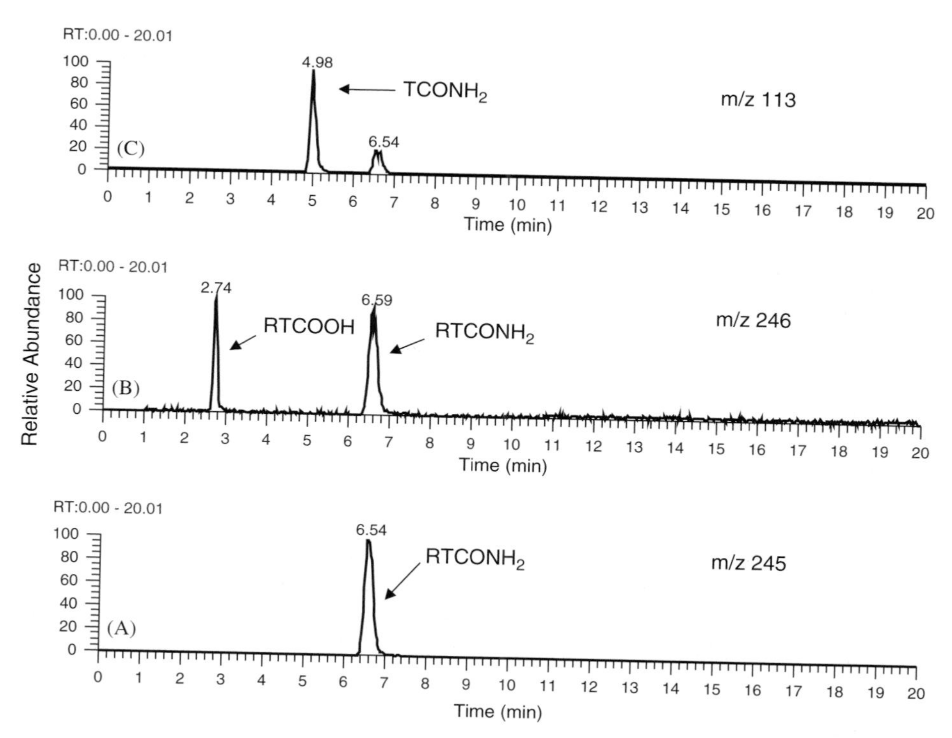

Figure 2.
Extracted ion chromatograms (MH^+) from the LC–MS analysis of a mixture of reference standards: (A) 100 ng of ^{14}C-ribavirin; (B) 250 ng of RTCOOH; and (C) 100 ng of $TCONH_2$. The peak at 6.59 min in panel B is due to the natural $^{13}C/^{2}H/^{15}N$ contribution of the $RTCONH_2$

Table 4.
Summary of chromatographic, mass spectral, and structural data obtained for $^{14}C/^{13}C_3$-$RTCONH_2$ and its metabolites

Name	Retention time[a] (min)	m/z (MH^+, $^{13}C_x$-MH^+)	Structure (* denotes the position of ^{13}C label)
$RTCONH_2$ (Ribavirin)	6.3–6.6	245, 248	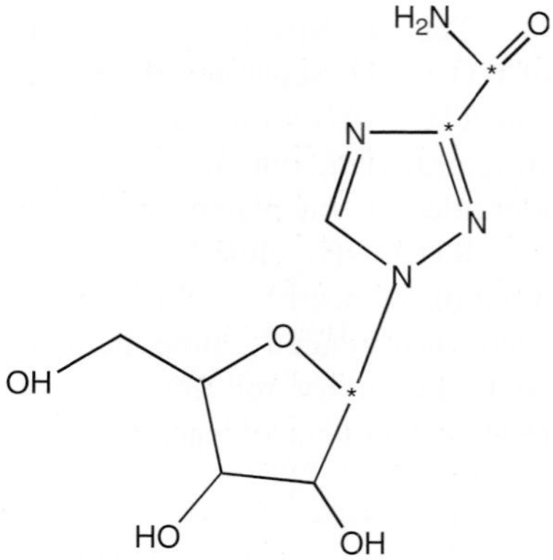
$TCONH_2$	4.9–5.0	113, 115	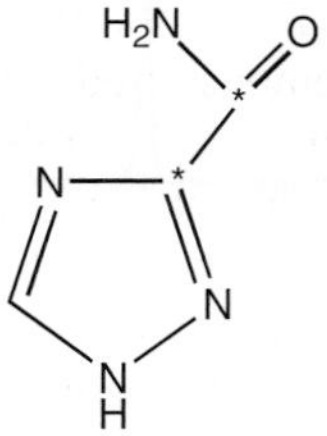
RTCOOH	2.7–2.8	246, 249	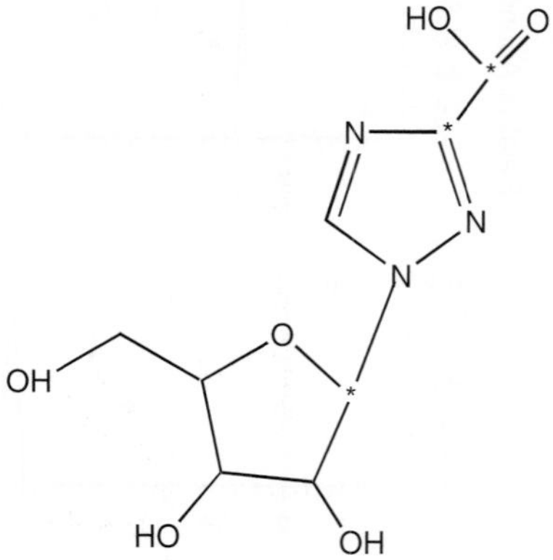

[a]Retention times are from the extracted ion chromatograms for the reference standards. In the radiochromatograms each compound eluted 0.24 min after they were detected by the mass spectrometer

The relative response of the reference standards for $RTCONH_2$ and two of its major metabolites in full-scan LC–MS spectra are shown in Fig. 3. In the absence of any biological matrix-derived interference, RTCOOH was less sensitive than $RTCONH_2$ and $TCONH_2$. Following on-column injection (4 ng each) of $RTCONH_2$ and $TCONH_2$, the full-scan LC–MS signal-to-noise ratio (S/N) was seven or better, while that for 10.5 ng of RTCOOH was three.

11.3.1. Plasma Metabolites

The recovery of drug-derived material (radioactivity) from the 2 h plasma processed by solvent extraction (88%) and that by SPE (87%) were nearly identical. The mass spectral and radiochromatographic profiles obtained from the 2 h plasma sample were similar using either extraction method and dominated by a peak with similar retention time observed for the synthetic standard of $TCONH_2$. Irrespective of the extraction procedure used, this metabolite accounted for nearly 50% of the total integrated chromatographic radioactivity (Table 5). The mass spectrum (~5.2 min) was characterized by a doublet with *m/z* values at 115 and 113 Thomson (Th), consistent with the MH^+ for the ^{13}C-enriched $TCONH_2$ (Fig. 4B).

A large number of non-drug-related ions were present in the mass spectra from a 2 h plasma extract (Fig. 4). Detection of a doublet structure (*m/z* 113/115 Th) for the MH^+ ion clearly demonstrated the presence of $TCONH_2$ in plasma. This experiment illustrates the advantage of administering stable isotope-labeled drugs for metabolite identification when diagnostic natural isotopes (such as Cl, Br, etc.) are not present in the molecule.

The radioactive peak at 6.92 min in the chromatogram revealed the presence of MH^+ ions in the mass spectrum at *m/z* 245 and 248 Th (Fig. 4A). Again, without the

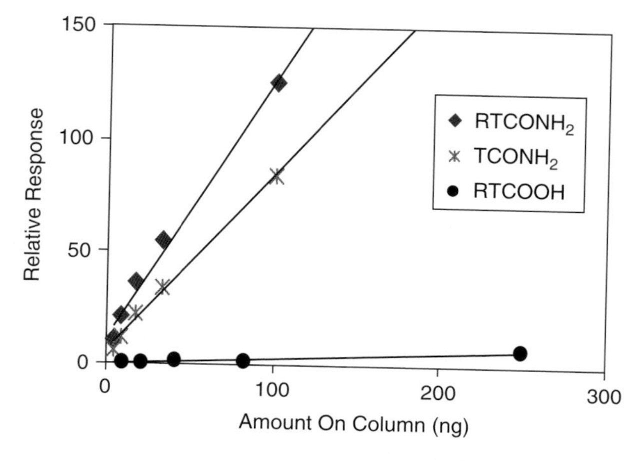

Figure 3.
Relative responses for $RTCONH_2$, $TCONH_2$, and RTCOOH. Each data point is derived from three separate analyses

Table 5.
Metabolites of $^{14}C/^{13}C_3$-RTCONH$_2$ in 2 h plasma following a single-dose administration to a male Sprague–Dawley rat

Analyte	*m/z*	Rt[a] (min)	Percent of total chromatographic radioactivity	
			Solid-phase extraction	Solvent extraction
RTCOOH	246	2.8	2.1	11.7
TCONH$_2$	113	5.1	50.8	49.1
Ribavirin (RTCONH$_2$)	245	6.8	28.1	22.3

[a]The retention times are from the radiochromatogram of 2 h pooled plasma following solid phase extraction

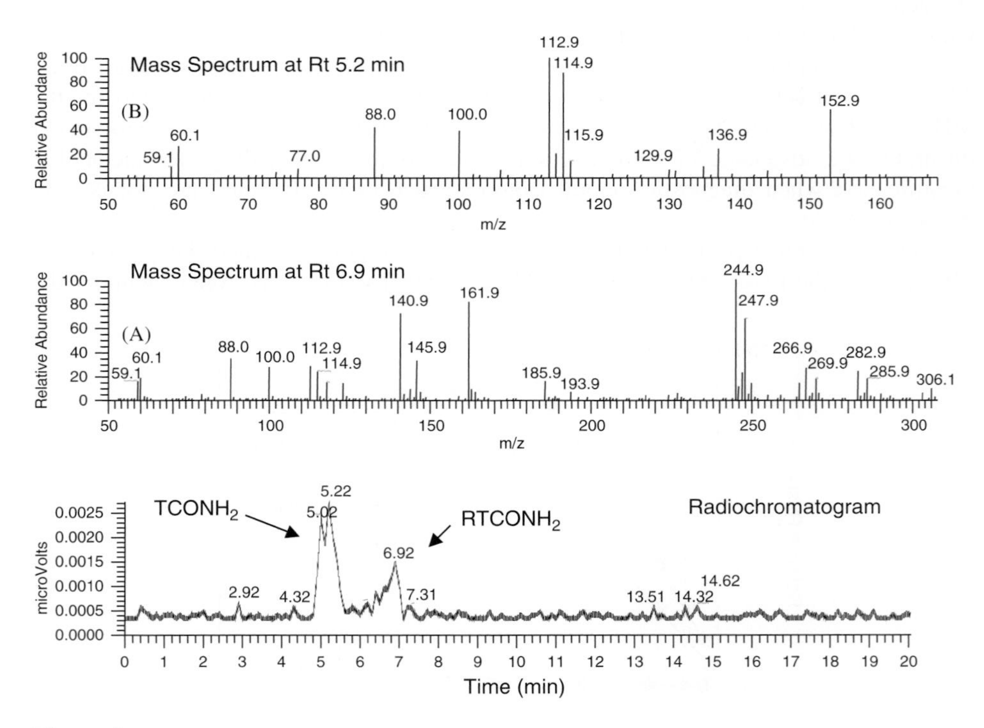

Figure 4.
The radiochromatogram of pooled 2 h plasma extract (SPE) from rats dosed with $^{14}C/^{13}C_3$-RTCONH$_2$ (bottom panel). The elution of TCONH$_2$ and RTCONH$_2$ are shown. The corresponding mass spectra of peaks labeled RTCONH$_2$ and TCONH$_2$ are shown in panels A and B, respectively

presence of the doublet at *m/z* 245 and 248 Th, an unambiguous determination for the presence of $RTCONH_2$ could not have been made. This compound represented 28.1 and 22.3% of the total circulating drug-related material following SPE and SE, respectively. Differences in the determined amounts of $RTCONH_2$ from plasma are discussed in more detail later.

Diagnostic ions corresponding to RTCOOH were not detected in the mass spectrum coincident with a low-intensity radioactive peak at 2.92 min in the radiochromatogram (Fig. 4) obtained from a 2 h plasma extract (SPE). The presence of a trace metabolite, however, was uncovered by target LC–MS/MS analysis of this sample for *m/z* 246 (^{12}C) and *m/z* 249 (^{13}C). As expected for the fragmentation of protonated RTCOOH (Fig. 5), the major product ions of both precursor ions were *m/z* 114 and 116 Th, respectively. A mechanism for the deribosylation of $^{13}C_3$ RTCOOH\RTCOOH in MS/MS yielding product ions at *m/z* 116 and 114 Th from the precursor ions at *m/z* 249 and 246 This shown in the following scheme.

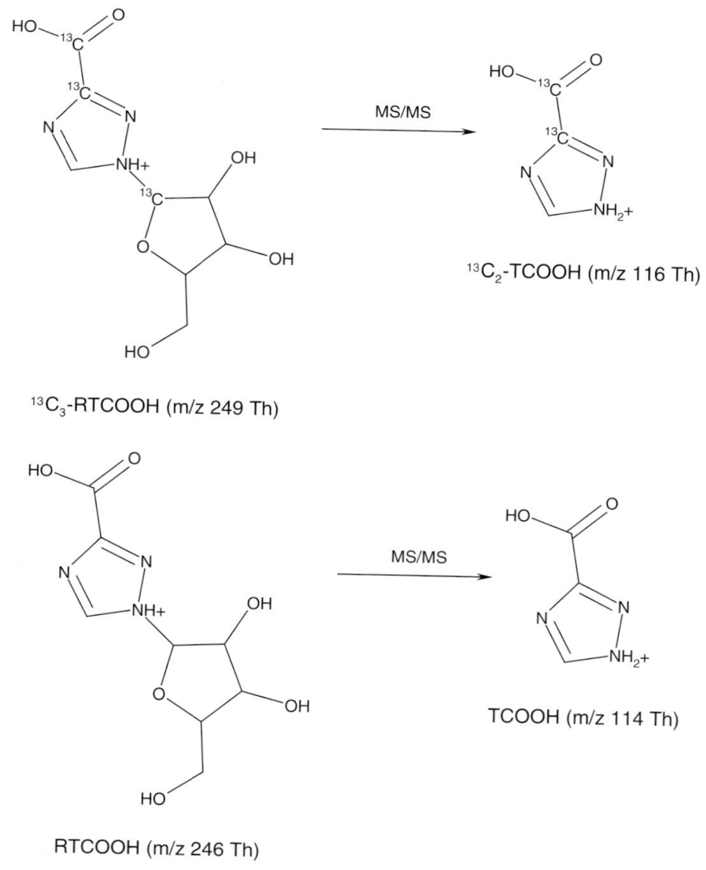

$^{13}C_3$-RTCOOH (m/z 249 Th) —MS/MS→ $^{13}C_2$-TCOOH (m/z 116 Th)

RTCOOH (m/z 246 Th) —MS/MS→ TCOOH (m/z 114 Th)

Mechanism of $^{13}C_3$-RTCOOH\RTCOOH deribosylation under MS/MS

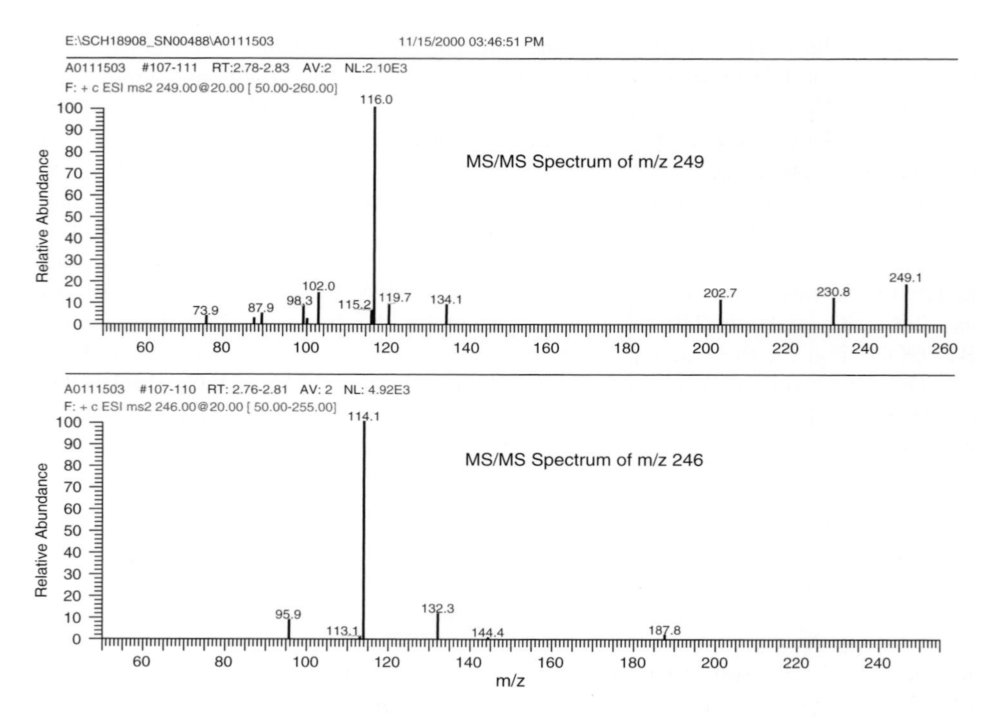

Figure 5.
LC–MS/MS spectra obtained on *m/z* 246 Th (^{12}C ions) and *m/z* 249 Th (^{13}C ions) from a 2 h plasma extract (SPE) from rats dosed with $^{14}C/^{13}C_3$- $RTCONH_2$ are shown in bottom and top panels, respectively

Table 5 shows the relative amounts as percent of total radiocarbon profile of each detected circulating metabolite following SPE and SE of rat plasma. The early eluting metabolite (RTCOOH) was detected in much smaller amounts (2.1%) in the SPE extract compared to that in the solvent extract (11.7%). In contrast, following SPE the late eluting component ($RTCONH_2$) was detected in higher amounts (28.1%) relative to that observed in the solvent extract (22.3%). Ribavirin and its metabolites are highly soluble in water such that RTCOOH was poorly retained by the solid-phase extraction cartridges. Therefore, the relative amount of each drug-derived material is skewed and resulted in apparent higher abundance of ribavirin in the SPE plasma extract. Direct extraction of plasma with methanol improved the recovery of RTCOOH.

11.3.2. Urinary Metabolites

In urine, a major peak at 5.2 min ($TCONH_2$) accounted for 67% (0–8 h) and 72% (0–24 h) of the total integrated chromatographic radioactivity (Table 6). Mass spectra for this major peak showed a prominent doublet at *m/z* 113 and 115 Th (Fig. 6C) together with corresponding ammonium (*m/z* 130/132 Th) and potassium

Table 6.
Metabolites of $^{14}C/^{13}C_3$-$RTCONH_2$ in urine following a single dose administration to male Sprague–Dawley rats[a]

Metabolite	Rt (min)	*m/z*	Percent of total chromatographic radioactivity		Percent of dose[b]	
			0–8 h	0–24 h	0–8 h	0–24 h
RTCOOH	3.1	246	22.5	19.6	3.56	8.90
$TCONH_2$	5.2	113	67.2	71.9	10.6	32.6
Ribavirin ($RTCONH_2$)	6.8	245	5.4	3.48	0.85	1.58
Total					15.0	43.2

[a]Urine samples were centrifuged prior to analysis
[b]Percent dose = % of total integrated peak area × % of dose excreted in urine for the time interval

ion adducts (*m/z* 151/153 Th). Here again, the *m/z* difference between the doublet ($\Delta = 2$) corresponds to the presence of two ^{13}C atoms in the surviving triazole ring of the metabolite. This metabolite accounted for 11 and 33% of the radiocarbon dose in pooled 0–8 and 0–24 h urine, respectively (Table 6).

The radiometric profile from both urine samples also showed a peak eluting at ~2.9 min (Fig. 6A) corresponding to RTCOOH. This metabolite accounted for 3.6 and 8.9% of the dose in pooled 0–8 and 0–24 h urine collections, respectively (Table 6). The detection of a weak doublet at *m/z* 246 and 249 Th in the mass spectrum of this peak is consistent with the presence of RTCOOH/$^{13}C_3$-RTCOOH in a ratio of 1:1 (Fig. 6B). The identity of this doublet was further confirmed by MS/MS of *m/z* 246 and 249 Th in the 0–24 h sample (Figs. 7B and C). The major product ions of these precursor ions were *m/z* 114 and 116 Th, respectively, confirming the presence of the unlabeled and ^{13}C-labeled metabolite.

For both urine samples, the radiochromatogram showed a minor peak between 6.80 and 7.00 min, similar to the retention time of $RTCONH_2$ reference standard (Figs. 6A and 7A). The mass spectrum of this peak contained a doublet at *m/z* 245 and 248 Th (Fig. 6D) consistent with the presence of an equal mixture of the ^{13}C-labeled and unlabeled compound. $RTCONH_2$ accounted for 1.6% of the administered dose excreted in the urine over 24 h.

11.4. Discussion

$TCONH_2$ and RTCOOH were the only two metabolites of $RTCONH_2$ identified in rat urine and plasma. These molecules are very polar and extremely difficult to retain

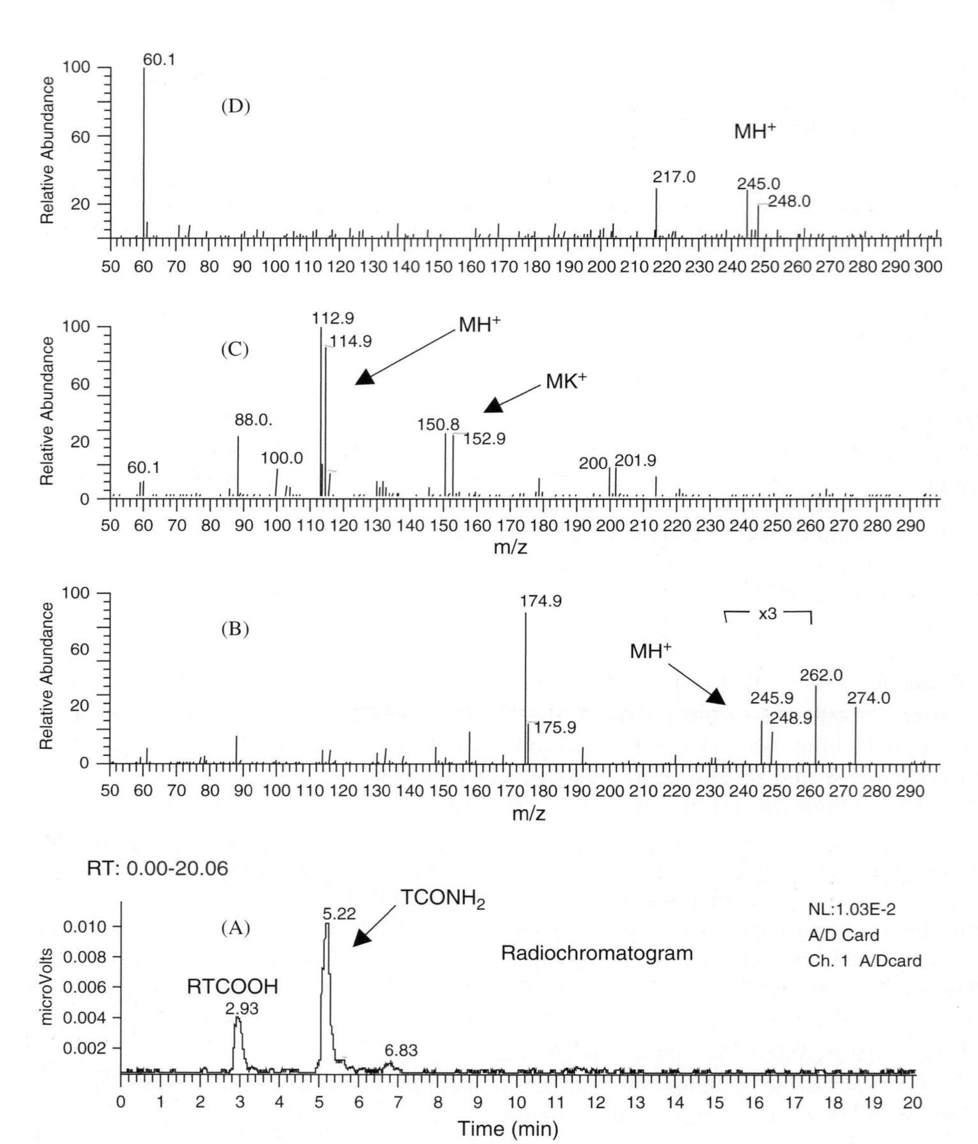

Figure 6.
Radiochromatogram (A) of pooled 0–8 h urine sample of rats following an oral administration of $^{14}C/^{13}C_3$-RTCONH$_2$ showing the radioactivity profile of RTCONH$_2$ and its metabolites; the mass spectrum (B) of RTCOOH at Rt 2.93 min, the mass spectrum (C) of TCONH$_2$ at Rt 5.22 min, and the mass spectrum of RTCONH$_2$ (D) at Rt 6.83 min

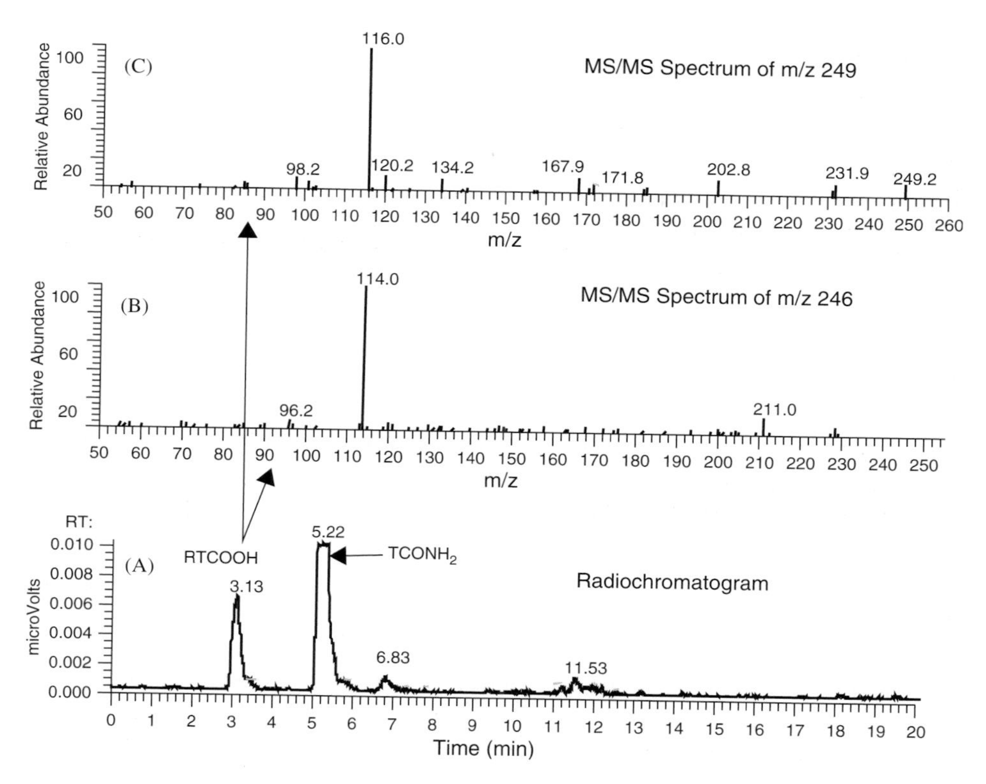

Figure 7.
Radiochromatogram (A) of pooled 0–24 h rat urine following an oral administration of $^{14}C/^{13}C_3$-RTCONH$_2$; the MS/MS spectrum (B) of *m/z* 246 Th (^{12}C-RTCOOH), and the MS/MS spectrum (C) of *m/z* 249 Th ($^{13}C_3$-RTCOOH)

by C-18 RP columns that are not modified for the analysis of highly polar molecules. A key factor, which facilitated the on-line chromatographic detection of RTCONH$_2$ and its metabolites, was the selection of an RP–HPLC column capable of retaining and separating a mixture of the compounds in a nearly aqueous mobile phase. The C_{18} Platinum column (10 A, 5 μm, 250 × 4.6 mm, Alltech Associates Inc.) was suitable for this purpose. The stationary phase, which is described to have a low carbon (6%) load bonded to a high-purity base silica end-capped with methyl groups [8]. The combination of these two features provided the retention characteristics and selectivity necessary for the separation of RTCONH$_2$ and its water-soluble metabolites.

The combined use of radiolabel, stable isotope, and unlabeled drug is arguably the most efficient experimental approach to investigate the biotransformation of xenobiotics [9]. Herein, this technique facilitated the unambiguous identification and characterization of RTCONH$_2$ and its metabolites by radiometric detection and MS. While the identification of RTCONH$_2$ and TCONH$_2$ was easily achieved from the diagnostic ion doublets (^{12}C and ^{13}C) in the LC–MS spectra that of RTCOOH necessitated analysis by the more sensitive MS/MS technique. The absence of detectable MH^+ ions for RTCOOH in the full-scan LC-mass spectrum could be attributed to co-elution of endogenous compounds, which suppressed the ionization of the metabolite. This is not uncommon during the analysis of biological samples by LC–MS in electrospray mode [10,11]. The effect is less pronounced during product ion scanning (MS/MS) because of the higher selectivity of this technique. As a result, the signal-to-noise ratio is increased and sensitivity is enhanced.

As described by Jarvis *et al.* [7] following administration of ribavirin the drug is taken up by erythrocytes. Within erythrocytes the drug exists in the form of mono-, di-, and tri-phosphorylated metabolites (Fig. 1). Studies with ribavirin and other nucleoside analogues have shown that they are substrates for the es nucleoside transporter in human erythrocytes. Since the polar phosphorylated metabolites are not substrates of es transporter, they become trapped within the cell and are released upon dephosphorylation [12]. Therefore, only dephosphorylated ribavirin and its metabolites are expected in urine and plasma samples (Fig. 1).

In rat urine, the main metabolite of $RTCONH_2$ was $TCONH_2$, which has been reported by Miller *et al.* [5] to be formed via the deribosylation of the parent drug (Fig. 1). Consistent with this report, RTCOOH (the hydrolyzed product of $RTCONH_2$) appears to be the next most prevalent metabolite identified in the urine. Furthermore, this study supports previous reports that both RTCOOH and $TCONH_2$ are circulating metabolites in rats, although the former was relatively minor [4,5].

The two metabolites together accounted for 41.5% of the dose excreted in the urine over a 0–24 h interval. Only 1.6% of the administered dose was excreted unchanged in the urine. Contrary to published reports, we did not detect TCOOH in the LC–MS spectra [5,6]. Miller *et al.* [5] observed only trace levels of this metabolite in rat serum.

We also have applied this method to examine the metabolite profiles of a prodrug of $RTCONH_2$ in various biological samples from rats and monkeys [13].

11.5. Conclusions

The direct, on-line profiling and characterization of metabolites of $^{14}C/^{13}C_3$-$RTCONH_2$ in rats can be easily achieved by LC–MS. The use of a C_{18} Platinum column (100 A, 5 μm, 250 × 4.6 mm) facilitated the retention of $RTCONH_2$ and its very polar metabolites. The C_{18} Platinum column offered unique selectivity due to interaction of its high-purity base silica with highly polar compounds. The radiolabel assisted in (a) the assessment of the recovery of drug-derived material during the extraction process, (b) provided a primary means of detection, and (c) relative quantitation of the drug and its metabolites in each biological sample. The judicious labeling of the drug with stable isotope facilitated the recognition of each metabolite of $RTCONH_2$ by the diagnostic doublet ions in the LC–MS spectra. Therefore, a combination of proper selection of an LC column and use of radiolabel and stable isotopes was key in the separation, detection, and identification of $RTCONH_2$ and its very polar metabolites.

Acknowledgment

The authors would like to thank Drs. P. McNamara, J. E. Patrick, and M. Wirth for their support of the study and helpful discussions.

References

1. Sidwell, R.W., Huffman, J.H., Khare, G.P., Allen, L.B., Witkowski, J.T., Robins, R.K., 1972. Broad spectrum antiviral activity of Virazole: 1-β-D-ribofuranosyl-1,2,4-triazole-3-carboxamide. Science 177, 705–706.
2. Physician's Desk Reference, 2005, 59th Edition, Thomson PDR, NJ, USA.
3. Poynard, T., McHutchison, J., Manns, M., Trepo, C., Lindsay, K., Goodman, Z., Ling, M.H., Albrecht, J., 2002. Impact of pegylated interferon alfa-2b and ribavirin on liver fibrosis in patients with chronic hepatitis C. Gastroenterology 22(5), 1525–1528.
4. Catlin, D.H., Smith, R.A., Samuels, A.I., 1980. ^{14}C-Ribavirin: Distribution and pharmacokinetic studies in rats, baboons and man. In: Smith, R.A., Kirkpatrick, W. (Eds.), Ribavirin: A Broad Spectrum Antiviral Agent. Academic Press, New York, pp. 83–98.
5. Miller, J., Kigwana, L., Streeter, D., 1977. The relationship between the metabolism of ribavirin and its proposed mechanism of action. Ann. NY Acad. Sci. 284, 211–229.
6. Alton, K.B., Wirth, M., Glue, P., Zambas, D., Pilon, D., Affrime, M.B., Cayen, M.N., Patrick, J.E., 1999. Disposition of ^{14}C-Ribavirin following oral administration to healthy male volunteers. Presented at the 9th North American ISSX Meeting, October 24–28, Nashville, TN, USA.
7. Jarvis, S.M., Thorn, J.A., Glue, P., 1998. Ribavirin uptake by human erythrocytes and the involvement of nitrobenzylthioinosine-sensitive (es)-nucleoside transporters. Br. J. Pharmacol. 123, 1587–1592.
8. Alltech Chromatography Catalogue 500, 2000, 377.
9. Cautreels, W., Davi, H., Berger, Y., 1987. Les isotopes stables, utilisation pour l'identification des metabolites d'un medicament. Therapie 42, 439–444.
10. Fu, I., Woolf, E.J., Matuszewski, B.K., 1998. Effect of the sample matrix on the determination of indinavir in human urine by HPLC with turbo ion spray tandem mass spectrometric detection. J. Pharm. Biomed. Anal. 18, 347–357.
11. Buhrman, D.L., Price, P.I., Rudewicz, P.J., 1996. Quantitation of SR 27417 in human plasma using electrospray liquid chromatography–tandem mass spectrometry: a study of ion suppression. J. Am. Soc. Mass Spectrom. 7, 1099–1105.
12. Glue, P., 1999. The clinical pharmacology of ribavirin. Semin. Liver Dis. 19(1), 17–24.
13. Gopaul, V.S., Chowdhury, S.K., Blumenkrantz, N.B., Zhong, R., Grubb, N., Wirth, M., Alton, K.B., Patrick, J.E., McNamara, P., 2001. Profiling and characterization of metabolites of ribavirin-alanine ester in rats and monkeys: application of stable isotope labeling in structure elucidation by LC–MS. Proceedings of the 49th ASMS Conference on Mass Spectrometry and Allied Topics.

Identification and Quantification of Drugs, Metabolites and Metabolizing
Enzymes by LC–MS
Swapan K. Chowdhury, editor.

Chapter 12

CYTOCHROME P450 (CYP) AND UDP-GLUCURONOSYLTRANSFERASE (UGT) ENZYMES: ROLE IN DRUG METABOLISM, POLYMORPHISM, AND IDENTIFICATION OF THEIR INVOLVEMENT IN DRUG METABOLISM

Anima Ghosal, Ragulan Ramanathan, Narendra S. Kishnani, Swapan K. Chowdhury, and Kevin B. Alton

12.1. Introduction

The biotransformation of drugs is broadly divided into two categories, termed Phase I and Phase II metabolism. Phase I reactions include oxidation, reduction, hydrolysis, and hydration. Metabolic oxidations usually occur through the action of cytochrome P450 (CYP) oxidative enzymes. Although there are at least 50 different P450 isoforms, drug metabolism in humans most likely involve CYP1A2, CYP3A4, CYP2C9, CYP2C19, and CYP2D6. P450 enzymes catalyze aromatic hydroxylation, aliphatic hydroxylation, N-, O-, and S-dealkylation; N-hydroxylation, N-oxidation, sulfoxidation, deamination, and dehalogenation. Although there are many non-P450 enzymes that are responsible for the metabolism of many xenobiotics, the focus of this chapter is limited to CYP enzymes for Phase I metabolism and UDP-glucuronosyltransferase (UGT) enzymes for Phase II metabolism.

Understanding the involvement of P450 enzymes in drug metabolism is critical to assessing the potential for drug interaction with concomitant drugs, food, and endogenous substances. In recent years, several drugs have been withdrawn from the U.S. market due to drug–drug interactions. These include terfenadine (Seldane®) in February 1998, mibefradil (Posicor®) in June 1998, astemizole (Hismanal®) in July 1999, cisapride (Propulsid®) in January 2000, troglitazone (Rezulin®) in 2000, and cerivastatin (Baycol®) in 2001. Although such drug–drug interactions are primarily related to inhibition of enzymatic pathways, the contribution of genetic polymorphism and/or absence of related enzymes cannot be ruled out.

With the exception of therapeutic agents designed as pro-drugs, the majority of drugs are directly biotransformed to less active or inactive metabolites. However, there are many examples in which the major metabolite retains activity similar to the parent. Fexofenadine (Allegra), the active metabolite of terfenadine, has equal potency at the histamine receptor and is presently marketed for the treatment for allergic rhinitis symptoms. In addition to the involvement of P450 enzymes, several parameters like (1) K_m, dose and local concentration of drugs in hepatocytes, (2) duration of co-treatment and, (3) CYP levels in the liver or intestine are also expected to be critical parameters in predicting clinical relevance of drug interactions. Hepatic P450 enzymes have been widely studied due to their significance in the metabolism of most clinically used drugs (Ingelman-Sundberg *et al.*, 1999). Extrahepatic CYPs may be of minor importance in drug metabolism, however they are an important consideration when assessing toxicity in extrahepatic organs (Raunio *et al.*, 1995b). Hence, thorough characterization of the enzymology provides not only the basis for prediction of potential toxicities and drug–drug interactions but also provides data critical to the design of appropriate clinical interaction studies.

Phase I enzymes primarily modify lipophilic molecules by creating polar functionalities to increase hydrophilicity, which thereby facilitates clearance from the body. Such additional functionalities, e.g. a hydroxyl group, may be readily amenable to Phase II conjugation reactions. Although such reactions are frequently observed, they are not required steps for Phase II metabolism. Phase II reactions "conjugate" or add a water-soluble entity such as acetate, sulfate or glucuronate onto a drug at newly created or preexisting sites, forming a more polar and water-soluble metabolite that can be more easily excreted in urine and/or bile.

Glucuronidation is the major conjugation pathway probably due to the relatively high natural abundance of the reaction co-factor, UDP-glucuronic acid. This process occurs with alcohols, phenols, hydroxylamines, carboxylic acids, amines, sulfonamides, and thiols (Gibson and Skett, 1994). Glucuronides are often excreted in bile and thus released into the gut where they can be broken down to the parent compound by β-glucuronidase and possibly reabsorbed (enterohepatic recirculation).

The role of these enzymes in drug metabolism will be reviewed within the context of their polymorphism. Finally, analytical technologies used today in determining their involvement in drug metabolism will be presented.

12.2. Cytochrome P450 (CYP) and Subfamilies

Cytochrome P450s constitute a superfamily of membrane-bound enzymes, mostly localized to the endoplasmic reticulum. They are responsible for the oxidative, peroxidative, and reductive metabolism of a diverse group of compounds, including endobiotics, such as steroids, bile acids, fatty acids, prostaglandins, leukotrienes, and xenobiotics, most of the therapeutic drugs and environmental pollutants (Nelson *et al.*, 1996). Historically, P450 enzymes gave a unique 450 nm absorption peak, for which it was given the name cytochrome P450 (Omura, 1999). CYPs require an electron transfer chain for the oxidation of RH to ROH as shown in Fig. 1 (Segall *et al.*, 1997). In the endoplasmic reticulum, the source of electrons is an enzyme designated NADPH-CYP reductase, previously called NADPH-cytochrome *c* reductase (Omura, 1999) while in mitochondria, electrons are transferred from NADPH by redoxin

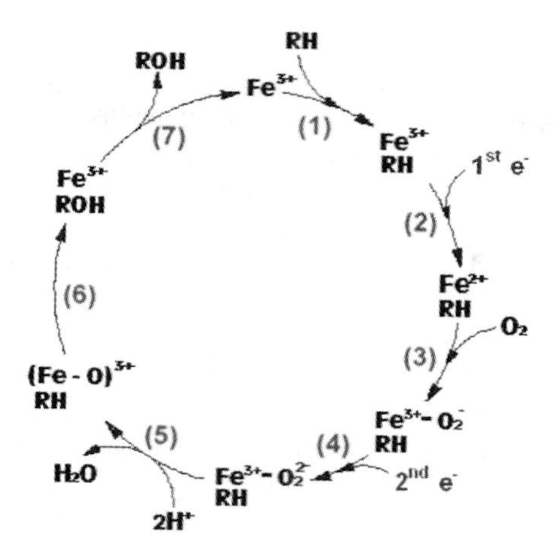

Figure 1.
Cytochrome P450 cycle. The binding of a substrate (RH) to a P450 molecule causes the transfer of an electron from NADH or NADPH (step 1). The next stage in the cycle is the reduction of the Fe^{3+} ion by an electron transferred from NADPH via an electron transfer chain (step 2). An O_2 molecule binds rapidly to the Fe^{2+} ion forming Fe^{3+}-O_2^- followed by a second reduction forming Fe^{3+}-O_2^{2-} (steps 3, 4). The O_2 reacts with two protons from the surrounding solvent forming water (step 5). The next stage in the cycle is the formation of a hydroxylated form of the substrate (step 6). In the final step (7), the product is released from the active site of the enzyme, which returns to its initial state

reductase to redoxin and then to CYP (Gonzalez, 1990). The majority of CYPs are expressed in human liver, but they are also expressed to a lesser extent in extrahepatic tissues. Only a few CYP forms involved in the metabolism of xenobiotics are found exclusively in extrahepatic tissues such as the lung, intestine, kidney, and skin.

The most current nomenclature of the CYP superfamily across species is provided in Table 1 (Nelson *et al.*, 1996). This system has evolved over many years and only after recent agreements has this nomenclature allowed consistent and unambiguous discussion related to metabolism. The nomenclature is based on divergent evolution of the superfamily, new sequences, gene mapping, and amino acid similarities. Mammalian sequences within the same subfamily are always >55% identical.

12.2.1. CYP1 Family

CYP1A1, CYP1A2, and CYP1B1 are members of the CYP1 family (Nelson *et al.*, 1996). They are induced by polycyclic aromatic hydrocarbons (PAH), TCDD (2,3,7,8-tetrachlorodibenzo-*p*-dioxin; Schmidt and Bradfield, 1996), and smoking (Anttila *et al.*, 1991; Willey *et al.*, 1997; Zevin and Benowitz, 1999). These enzymes

Table 1.
Nomenclature of cytochrome P450 (CYP) superfamily

CYPs	Human	Rat	Rabbit	Mouse	Monkey	Dog
CYP1A	CYP1A1	CYP1A1	CYP1A1	CYP1A1	CYP1A1	CYP1A1
	CYP1A2	CYP1A2	CYP1A2	CYP1A2	CYP1A2	CYP1A2
CYP1B	CYP1B1	CYP1B1	NA	CYP1B1	NA	NA
CYP2A	CYP2A6	CYP2A1	CYP2A10	CYP2A4	CYP2A	CYP2A
	CYP2A13	CYP2A2	CYP2A11	CYP2A5		
		CYP2A3				
CYP2B	CYP2B6	CYP2B1	CYP2B4	CYP2B9	CYP2B17	CYP2B11
		CYP2B2	CYP2B4P	CYP2B10		
		CYP2B3	CYP2B5	CYP2B19		
CYP2C	CYP2C8	CYP2C6				
	CYP2C9	CYP2C7	CYP2C1	CYP2C29	CYP2C20	CYP2C21
	CYP2C18	CYP2C11	CYP2C2		CYP2C37	CYP2C41
	CYP2C19	CYP2C12	CYP2C3			
		CYP2C13	CYP2C4			
			CYP2C5			
			CYP2C14			
			CYP2C15			
			CYP2C16			
			CYP2C30			
CYP2D	CYP2D6	CYP2D1	CYP2D23	CYP2D9	CYP2D17	CYP2D15
	CYP2D7P1	CYP2D2	CYP2D24	CYP2D10		
	CYP2D7P2	CYP2D3		CYP2D11		
	CYP2D8P1	CYP2D4		CYP2D12		
	CYP2D8P2	CYP2D5		CYP2D13		
		CYP2D18				
CYP2E	CYP2E1	CYP2E1	CYP2E1	CYP2E1	CYP2E1	CYP2E1
			CYP2E2			
CYP2F	CYP2F1	CYP2F	NA	CYP2F2	NA	NA
CYP2J	CYP2J2	CYP2J3	CYP2J1	CYP2J5	NA	NA
		CYP2J4		CYP2J6		
				CYP2J9		
CYP3A	CYP3A4	CYP3A1/23	CYP3A6	CYP3A11	CYP3A8	CYP3A12
	CYP3A5	CYP3A2		CYP3A13		CYP3A26
	CYP3A7	CYP3A9		CYP3A16		
	CYP3A43	CYP3A18		CYP3A25		
CYP4A	CYP4A9	CYP4A1	CYP4A4	CYP4A10	NA	NA
	CYP4A11	CYP4A2	CYP4A5	CYP4A12		
	CYP4A22	CYP4A3	CYP4A6	CYP4A14		
		CYP4A8	CYP4A7			
CYP4B	CYP4B1	CYP4B1	CYP4B1	CYP4B1	NA	NA

CYP4F	CYP4F2 CYP4F3 CYP4F8 CYP4F11 CYP4F12	CYP4F1 CYP4F4 CYP4F5 CYP4F6	NA	CYP4F13 CYP4F14 CYP4F15 CYP4F16 CYP4F18	CYP4F	NA

NA = Not available
Source: Nelson *et al.* (1996) and Shou *et al.*, *DMD* **31**, 1161–1169 (2003).

have received particular attention because they all are active in the metabolism of PAHs to intermediates that can bind to DNA and, if the damage of the DNA goes unrepaired, may produce mutations involved in neoplasmic transformation (Shimada *et al.*, 1996). Thus, they have been implicated in the formation of chemically induced cancers (Nebert *et al.*, 1996).

CYP1A1 expression in human liver is very low (Edwards *et al.*, 1998), but it is a major extrahepatic CYP enzyme (Raunio *et al.*, 1995b). The constitutive expression of CYP1A1 is also very low in extrahepatic tissues, but it is inducible by AHR (aryl hydrocarbon receptor) ligands in almost every tissue studied, including lung, lymphocytes, mammary gland, and placenta (Raunio *et al.*, 1995b).

The expression of CYP1A2 is liver-specific (Raunio *et al.*, 1995b) and constitutes about 13% of the total hepatic CYP content (Shimada *et al.*, 1994; Imaoka *et al.*, 1996). It activates PAHs, nitrosamines, aflatoxin B_1, and especially aryl amines into forms that can bind to DNA and produce mutations (Aoyama *et al.*, 1990; Shimada *et al.*, 1996). CYP1A2 is induced in vivo by cigarette smoke, charbroiled meat, cruciferous vegetables containing indole-3-carbinol, phenytoin, rifampicin, and omeprazole (Landi *et al.*, 1999). A commonly recognized substrate used as a chemical probe is caffeine (theobromine and theophylline are related substrates).

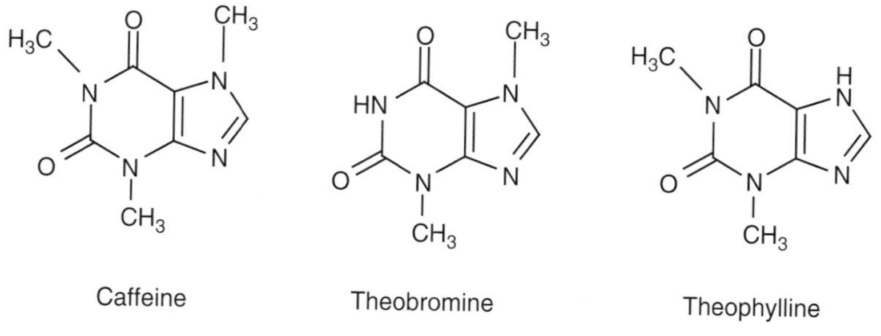

Caffeine Theobromine Theophylline

Similar to CYP1A1, CYP1B1 is an extrahepatic P450 expressed in almost every tissue, including kidney, prostate, mammary gland, and ovary (Sutter *et al.*, 1994; Shimada *et al.*, 1996; Tang *et al.*, 1999). Chang *et al.* (2003) reported that although CYP1B1 mRNA is expressed in human liver and the levels are increased in smokers, CYP1B1 protein remains undetectable.

12.2.2. CYP2 Family

The human CYP2 family includes the subfamilies CYP2A, CYP2B, CYP2C, CYP2D, CYP2E, CYP2F, and CYP2J (Nelson *et al.*, 1996).

The CYP2A6 protein has been detected in liver and constitutes about 4% of total CYP content (Shimada *et al.*, 1994; Imaoka *et al.*, 1996). There has been a growing interest in CYP2A6, due to its major role in the metabolism of nicotine in vitro (Nakajima *et al.*, 1996; Yamazaki *et al.*, 1999) and in vivo (Kitagawa *et al.*, 1999). It is induced in vivo by phenobarbital and other antiepileptic drugs (Sotaniemi *et al.*, 1995). Relatively high levels of CYP2A13 mRNA have been detected in human lung and adult and fetal nasal mucosa (Koskela *et al.*, 1999).

CYP2B6 is a minor CYP form in human liver, accounting for only 1–2% of total hepatic CYP (Mimura *et al.*, 1993; Shimada *et al.*, 1994; Imaoka *et al.*, 1996). Human CYP2B6 is also induced by phenobarbital and rifampicin in primary hepatocytes (Rodríguez-Antona *et al.*, 2000; Pascussi *et al.*, 2000).

The human CYP2C subfamily comprises CYP2C8, CYP2C9, CYP2C18, and CYP2C19 (Nelson *et al.*, 1996). CYP2C accounts for about 20% of the human total liver CYP content (Shimada *et al.*, 1994; Imaoka *et al.*, 1996). CYP2C9 is the main CYP2C in human liver, followed by CYP2C8 and CYP2C19 (Edwards *et al.*, 1998). Well-documented pharmaceutical substrates for CYP2C include diazepam, omeprazole, mephenytoin, tolbutamide, and warfarin (Guengerich, 1995) as well as many nonsteroidal antiinflammatory drugs (Pelkonen *et al.*, 1998). Selective substrates include taxol for CYP2C8, tolbutamide for CYP2C9, and mephenytoin for CYP2C19 (Pelkonen *et al.*, 1998).

CYP2D6 constitutes about 2% of total hepatic CYP (Shimada *et al.*, 1994; Imaoka *et al.*, 1996); the protein is also expressed in duodenum and brain (Pelkonen and Raunio, 1997). CYP2D6 metabolizes many cardiovascular and neurologic drugs in use today (e.g. metoprolol, diltiazem, sparteine, debrisoquine, and fluoxetine). Several commonly used medications also inhibit CYP2D6. These include quinidine (Branch *et al.*, 2000) as well as haloperidol and some other antipsychotics (Shin *et al.*, 1999, 2001). The well-described pharmacokinetic interaction between selective serotonin reuptake inhibitor (SSRI) antidepressants and tricyclic antidepressants appears to be due to the fact that SSRIs like fluoxetine and paroxetine are both potent inhibitors of CYP2D6 (Bergstrom *et al.*, 1992) and render patients metabolically equivalent to individuals who do not have the enzyme. This results in higher plasma levels of tricyclic antidepressants and increases the potential for side effects. In contrast, patients co-prescribed fluoxetine or paroxetine with codeine may experience no analgesic benefit, since codeine requires CYP2D6 for metabolism to morphine. In this respect, codeine is actually a pro-drug that is converted to morphine. In addition to causing nausea and other adverse effects, codeine in its unmetabolized form is much less active as an analgesic. Thus, the study of CYP2D6 has led to understanding the failure of codeine to relieve pain in some patients as well as the etiology of its side effects. A commonly used clinical probe substrate for CYP2D6 is bufuralol while fluoxetine and quinidine are used as inhibitors.

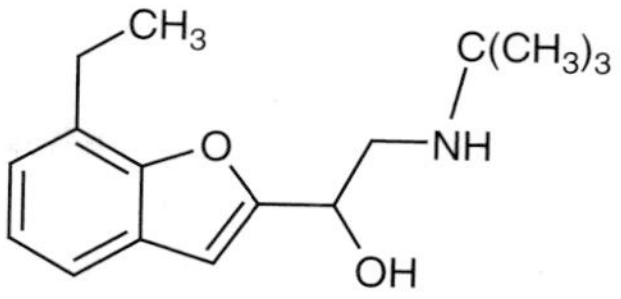

Bufuralol

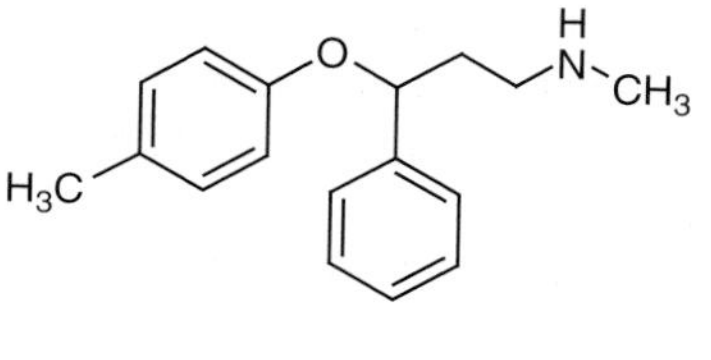

Fluoxetine

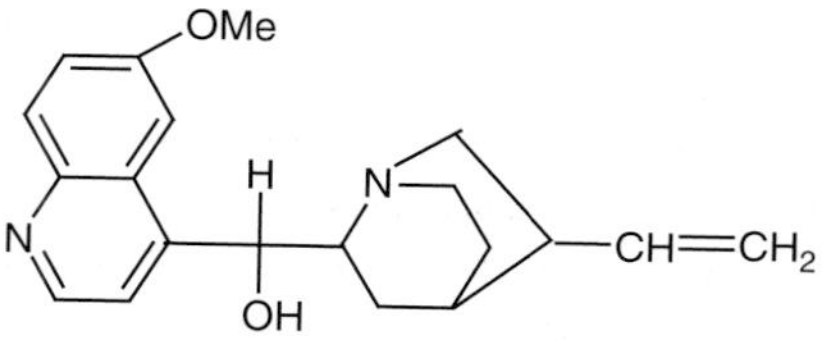

Quinidine

CYP2E1 is the only isoform in the CYP2E subfamily known to date (Nelson *et al.*, 1996). The CYP2E1 enzyme has been studied extensively due to its role in the metabolism of ethanol and also as an activator of chemical carcinogens (Lieber, 1997). Over 70 substrates were demonstrated to be activated by CYP2E1. They all are small hydrophobic compounds (Ronis *et al.*, 1996) including pharmaceuticals, such as paracetamol, chlorzoxazone, enflurane, and halothane (Guengerich, 1995). About 7% of the liver CYP content consists of CYP2E1 (Shimada *et al.*, 1994; Imaoka *et al.*, 1996). It is also expressed in lung and brain (Raunio *et al.*, 1995b). Ethanol intake increases the human CYP2E1 expression and content in liver in vivo (Perrot *et al.*, 1989). In addition to activating procarcinogens, CYP2E1 also produces free radicals causing tissue injury. These radicals are formed both in the presence and in the absence of substrates (Lieber, 1997). A commonly used chemical probe substrate of CYP2E1 is chlorzoxazone.

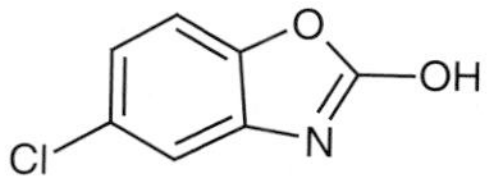

Chlorzoxazone

CYP2J2 is expressed mainly in heart, kidney (Wu *et al.*, 1996), lung (Zeldin *et al.*, 1996), pancreas (Zeldin *et al.*, 1997b), and gastrointestinal tract (Zeldin *et al.*, 1997a). CYP2J2 is involved in the metabolism of arachidonic acid into epoxyeicosatrienoic acids (EETs), which have important functional roles in cardiac physiology (Wu *et al.*, 1996; Zeldin *et al.*, 1997a; Capdevila *et al.*, 2000). CYP2J2 is also involved in the intestinal first-pass metabolism of astemizole in humans (Matsumoto *et al.*, 2002).

12.2.3. CYP3 Family

As the largest member of the CYP superfamily, human CYP3A includes four isoforms: CYP3A4, CYP3A5, CYP3A7 (Nelson *et al.*, 1996), and the recently identified

CYP3A43 (Domanski *et al.*, 2001). CYP3A enzymes have overlapping catalytic specificities; however, their tissue expression patterns differ. CYP3A4 is mainly expressed in liver, CYP3A5 in extrahepatic tissues, and CYP3A7 in fetal liver (Thummel and Wilkinson, 1998). CYP3A4 and CYP3A7 are regulated by the pregnane X receptor (PXR), whereas CYP3A5 is controlled by a glucocorticoid receptor.

CYP3A4 is the most prominent drug-metabolizing enzyme in human liver. About 30–40% of the total hepatic CYP content consists of CYP3A4 (Shimada *et al.*, 1994; Imaoka *et al.*, 1996). It is also highly expressed in the intestinal tract, especially the small intestine (Kolars *et al.*, 1992). Estimates suggest that nearly 50% of all drugs that are metabolized by CYPs are metabolized by CYP3A4 (Bertz and Granneman, 1997). Well-known substrates for this enzyme include drugs such as quinidine, nifedipine, diltiazem, lidocaine, lovastatin, erythromycin, cyclosporin, triazolam, and midazolam, and endogenous substances, including testosterone, progesterone, and androstenedione (Pelkonen *et al.*, 1998; Guengerich, 1999). Midazolam and erythromycin have been used as in vivo probes for drug interaction with CYP3A4 (Thummel and Wilkinson, 1998). CYP3A4 also activates procarcinogens, including aflatoxin B_1 (Aoyama *et al.*, 1990) and 6-aminochrysene (Yamazaki *et al.*, 1995). CYP3A4 is induced in human hepatocytes by rifampicin, dexamethasone, and phenobarbital (Schuetz *et al.*, 1993) among others. CYP3A4 is induced in vivo by rifampicin and barbiturates in liver (Perrot *et al.*, 1989) and by rifampicin in small intestine (Kolars *et al.*, 1992). Although CYP3A is not polymorphic in its distribution, its activity varies over 50-fold in the general population (Williams *et al.*, 1993). Drugs that are substrates of CYP3A can be extensively metabolized in the GI tract, and in fact, the GI tract is responsible for a large part of the metabolism that was formerly attributed totally to the liver. Inhibition of GI tract CYP3A also results in higher plasma levels of substrate drugs. Notable CYP3A inhibitors include ketoconazole, itraconazole, fluconazole, cimetidine, clarithromycin, erythromycin, troleandomycin, grapefruit juice, etc. Grapefruit juice contains a bioflavonoid that inhibits CYP3A and blocks the metabolism of many drugs. This was first described for felodipine (Bailey *et al.*, 1991) but has now been observed with several other drugs (Kane and Lipsky, 2000). This interaction can lead to reduced clearance and higher blood levels when the drugs are taken simultaneously with grapefruit juice. With regular consumption, grapefruit juice also reduces the expression of CYP3A in the GI tract (Lown *et al.*, 1997). Inhibitors of CYP3A will make patients phenotypically resemble poor metabolizers (PMs). In contrast, several commonly used drugs have been characterized as inducers of CYP3A (carbamazepine, rifampin, rifabutin, ritonavir, St. John's wort). Use of these drugs could potentially attenuate therapeutic efficacy of a CYP3A substrate.

CYP3A5 is expressed polymorphically in human liver (Wrighton *et al.*, 1989), but consistently in lung (Kivistö *et al.*, 1996), colon (Gervot *et al.*, 1996), kidney (Schuetz *et al.*, 1992), oesophagus (Lechevrel *et al.*, 1999), and anterior pituitary gland (Murray *et al.*, 1995), demonstrating CYP3A5 to be a more extrahepatic form of CYP3A. About 20–25% of human livers have substantial levels of CYP3A5 protein (Wrighton *et al.*, 1989). In comparison to CYP3A4, CYP3A5 shows roughly the same substrate preference pattern, but the turnover rates are usually lower (Wrighton *et al.*, 1990). However, CYP3A5 is unable to metabolize some CYP3A4 substrates, including erythromycin and quinidine (Wrighton *et al.*, 1990). CYP3A7 is mainly expressed in human fetal liver, where it is the major CYP form (Kitada and Kamataki 1994). Commonly used CYP3A4 chemical probe substrates (testosterone, midazolam, and nifedipine) are provided below.

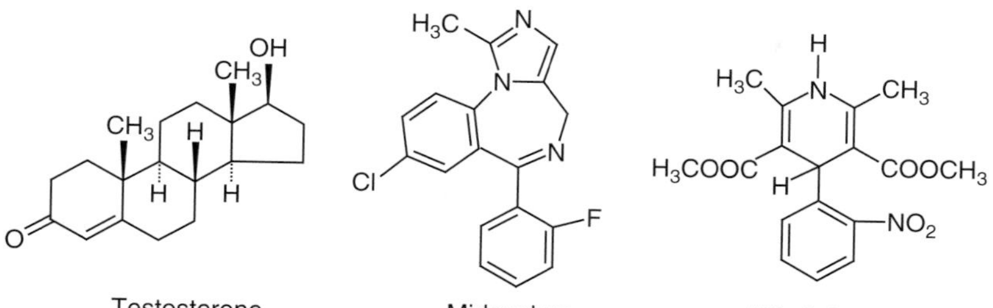

Testosterone Midazolam Nifedipine

12.2.4. CYP4 Family

The CYP4A gene subfamily was originally discovered as fatty acid ω-hydroxylase enzymes, which are induced by peroxisome proliferators (Heng *et al.*, 1997). Genes from the CYP4B and CYP4F family have been cloned and are able to hydroxylate fatty acids, thus, all three mammalian CYP4 gene families metabolize fatty acids. Metabolism of fatty acids and eicosanoids by the CYP4A family is believed to be very important in lipid homoeostasis and signalling. The metabolism of arachidonic acid to 20-hydroxyeicosatetraenoic acid by CYP4A proteins is believed to regulate blood pressure in the kidney. Moreover, the rapid induction of CYP4A protein is known to be crucial for peroxisome proliferation in the liver (Heng *et al.*, 1997). CYP4B1 is the only CYP4 family member with activity towards xenobiotics. CYP4B1 mRNA was found to be expressed in human lung, but not in liver (Nhamburo *et al.*, 1989).

The CYP4F family of CYP enzymes consists of proteins that catalyze the ω-hydroxylation of fatty acids, eicosanoids, and steroids (LeBrun *et al.*, 2002). The first member of this family to be discovered was CYP4F1, a protein that is constitutively expressed at relatively high levels in rat hepatoma. This protein catalyzes the ω-hydroxylation of leukotriene B_4, lipoxin A, prostaglandin A_1, and several hydroxyeicosatetraenoic acids (HETEs), but not of lauric, palmitic, or arachidonic acids. The human proteins CYP4F2 and CYP4F3 also catalyze the ω-hydroxylation of leukotriene B_4, but in contrast to CYP4F1, they also support the ω-hydroxylation of arachidonic acid (LeBrun *et al.*, 2002).

12.3. Role of CYPs in Drug Metabolism

The estimated level of the five major P450 isoforms in pooled human liver is shown in Fig. 2 (McGinnity *et al.*, 2000). CYP3A is the most abundant P450 enzyme in human liver and is highly expressed in the intestinal tract. This enzyme contributes substantially to the metabolic clearance of ~50% of currently marketed drugs that

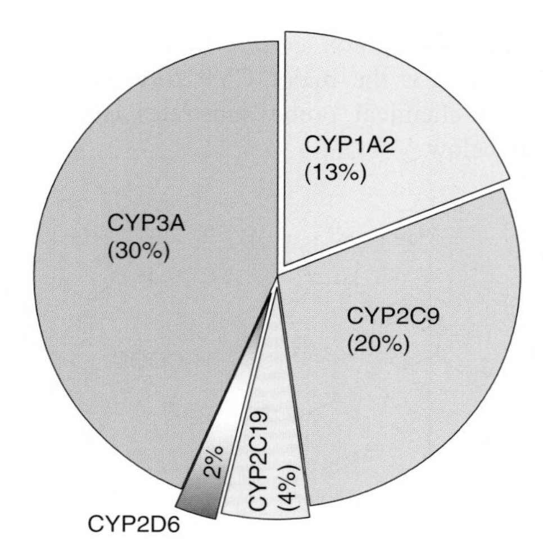

Figure 2.
Estimated level of the five major P450 isoforms in pooled human liver (from McGinnity *et al.*, 2000)

undergo oxidative metabolism (Fig. 3). Hence, drug–drug interactions involving inhibitors of CYP3A-mediated metabolism can be of great clinical consequence. Due to the broad substrate specificity of CYP3A, routine discovery and development of compounds that are completely devoid of CYP3A metabolism is not practical. Thus, it is important that scientifically valid approaches to the discovery and development of compounds metabolized by CYP3A be realized. The clinical relevance of CYP3A metabolism is dependent on a multitude of factors that include the degree of intestinal and hepatic CYP3A-mediated first-pass extraction, the therapeutic index of the compound and the adverse events associated with inhibition of CYP3A metabolism. Thus, a better understanding of the disposition of a CYP3A-metabolized compound relative to the projected or observed therapeutic index (or safety margin) can provide supportive evidence to continue development of a CYP3A substrate.

12.4. Polymorphism

Phenotypically, the polymorphism of CYP2D6 was first recognized and detected in the metabolism of debrisoquine where altered expression of CYP enzyme was noted (Gonzalez *et al.*, 1988). Since then, the profound effects of polymorphism on the metabolism of several commonly used pharmaceuticals, including several tricyclic antidepressants, haloperidol, metoprolol, propranolol, codeine, and dextromethorphan (Pelkonen *et al.*, 1998) have been reported. Among the polymorphic P450 enzymes CYP2D6, CYP2C19, CYP2C9, and CYP1A1 are most prominent and hence most intensely investigated phenotypically. In recent years though, such polymorphisms have been further characterized at the genetic level by a variety of genotyping techniques. The significant value of genotyping has been recognized beyond academia and industry consortiums. Regulatory agencies have supported the generation of such data in clinical

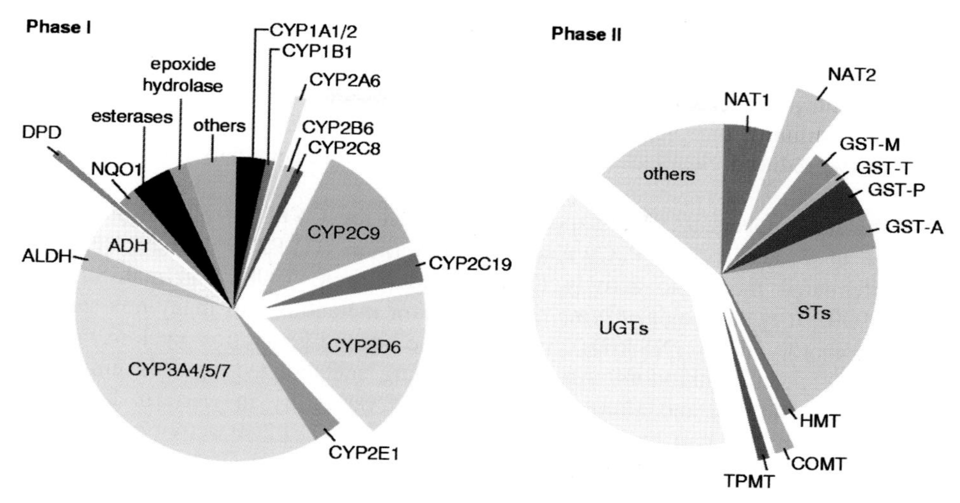

Figure 3.
The percentage of Phase I and Phase II metabolism of drugs that each enzyme contributes is estimated by the relative size of each section of the corresponding chart. ADH, alcohol dehydrogenase; ALDH, aldehyde dehydrogenase; CYP, cytochrome P450; DPD, dihydropyrimidine dehydrogenase; NQO1, NADPH:quinone oxidoreductase or DT diaphorase; COMT, catechol *O*-methyltransferase; GST, glutathione S-transferase; HMT, histamine methyltransferase; NAT, N-acetyltransferase; STs, sulfotransferases; TPMT, thiopurine methyltransferase; UGTs, uridine 5′-triphosphate glucuronosyltransferases (Reprinted with permission from Evans W.E. and Relling M.V. (1999). Copyright (1999) AAAS.)

studies as demonstrated by release of a draft guidance by the FDA in 2003 (Savage, 2003). As one of the most highly polymorphic enzymes, at a molecular level CYP2D6 has at least 30 different defective alleles, six of which contribute to 95–99% of the PM phenotype (Ingelman-Sundberg *et al.*, 1999). For example, codeine would provide little or no efficacy in individuals with decreases or no CYP2D6 activity because the metabolism of codeine to its active metabolite, morphine would be greatly minimized or absent (Caraco *et al.*, 1996). Such individuals, nonetheless, would experience adverse effects of codeine, particularly if the dose is increased in a futile attempt to obtain relief from pain. Interestingly, duplications of the CYP2D6 gene up to 13 gene copies have been reported (Johansson *et al.*, 1993), giving rise to an ultra-rapid metabolizer phenotype. Ultra-rapid metabolizers show increased metabolism and decreased drug effects of CYP2D6 substrates, such as tricyclic antidepressants (Dalen *et al.*, 1998). The PM phenotype has been reported in about 6% of the Caucasian population with about 4% having multiple CYP2D6 genes (Ingelman-Sundberg *et al.*, 1999). Thirty percent of Ethiopians studied had multiple copies of the 2D6 gene (up to13 gene copies) with increased enzyme activity resulting in ultrarapid metabolism of CYP2D6 substrates (Aklillu *et al.*, 1996). It has been speculated that since CYP2D6 is not inducible, the duplications are a way to adapt to environmental chemical pressures, most likely to alkaloids in the diet (Ingelman-Sundberg *et al.*, 1999).

The phenotype of PMs of CYP2C19 substrates is detected in 2–4% of Caucasians and in about 20% of Asians (Ingelman-Sundberg *et al.*, 1999). CYP2C19 is absent in a large percentage of Asians (~20–30%). This enzyme metabolizes many anticonvulsants, diazepam (Valium), omeprazole (Prilosec) and several of the tricyclic antidepressants. Asians have reduced clearance of diazepam compared to Caucasians (Ghoneim *et al.*, 1981).

CYP2C9 appears to be less polymorphic than the CYP2C19 and is missing in only 1% of Caucasians. CYP2C9 also has two variants (CYP2C9*2 and CYP2C9*3) with attenuated metabolic activity (Miners and Birkett, 1998; Ingelman-Sundberg *et al.*, 1999). It is the major enzyme responsible for metabolism of many non-steroidal antiinflammatory drugs (NSAIDs), including the second-generation cyclooxygenase-2 (COX-2)-specific inhibitors. An important drug metabolized by this enzyme is warfarin (Coumadin), and almost all inter-patient variability in warfarin levels and anticoagulant effects can be explained on the basis of CYP2C9 activity and not by the differences in protein binding as was originally postulated (Rolan, 1994).

Because of the significance of CYP1A1 in the activation of procarcinogens, there have been active efforts to investigate any link of the polymorphisms of the *CYP1A1* gene with the individual susceptibility to chemically induced cancers, especially lung cancer (Raunio *et al.*, 1995a). Currently, six variant alleles of the *CYP1A2* gene are known, and two of these correlate with increased and decreased induction by smoking (Sachse *et al.*, 1999; Nakajima *et al.*, 1999). These variants could partially explain the observed high level of inter-individual variation in the enzymatic activity measured in vivo with caffeine as a probe (Kalow and Tang, 1991; Butler *et al.*, 1992). CYP1A2 is the primary CYP catalyzing the metabolism of several drugs, including clozapine, theophylline, and tacrine (Landi *et al.*, 1999).

The genetic polymorphisms of CYP2A6 have been associated with inter-individual differences in smoking behavior (Pianezza *et al.*, 1998), although this report has methodological shortcomings (Oscarson *et al.*, 1998). Coumarin has been used as a probe drug to assess the activity of CYP2A6 in vivo (Cholerton *et al.*, 1992; Rautio *et al.*, 1992; Iscan *et al.*, 1994).

If the frequency of the CYP polymorphism is very low (<0.01%), and the therapeutic index of the compound is wide, then polymorphism can be considered clinically insignificant and no dose adjustment is necessary (Rodrigues and Rushmore, 2002). At the opposite end of the spectrum, when the therapeutic index is narrow and the incidence of PM subjects is high (e.g., >10%), the polymorphism will be more clinically significant and dose adjustment or therapeutic drug monitoring (TDM) is warranted in PM subjects. In reality, most drugs lie between these two extremes and, in the absence of widespread genotyping, most physicians have employed a more classical "dose titration" approach.

12.5. Identification of P450 Involvement in Drug Metabolism (Phenotyping)

There are several well-established approaches that are used to identify the specific CYP enzyme(s) responsible for the metabolism of drugs and xenobiotics which include (1) the use of chemical inhibitors and specific antibodies against human CYPs, (2) recombinant human P450s, and (3) correlation between metabolic rates and individual CYP contents (Lu *et al.*, 2003). Early identification of the major CYP isoforms involved in the metabolism of a drug candidate is useful for several purposes, including understanding ligand–enzyme structure–activity relationships, expanding the database for substrates of the polymorphic isoforms, assessing the potential intersubject variability, predicting the drug–drug interactions and, ultimately, guiding the direction of clinical trials (McGinnity *et al.*, 2000).

CYP reaction phenotyping involves an initial assessment of Michaelis-Menten kinetics using human liver microsomes. Initial reaction rates are measured over a wide range of substrate concentrations ([S]) employing incubation conditions that are linear with respect to microsomal protein concentration and time of incubation. The data are typically analyzed by linear and non-linear transformations (Rodrigues, 1999). In the majority of cases, a low apparent K_m component is most informative because this would indicate which CYP could play a major role at clinically relevant drug concentrations. Polymorphic CYPs often account for the majority of the low K_m (<20 μM) component.

12.5.1. Typical Incubation with Human Liver Microsomes

In vitro incubations of substrates are typically performed with human liver microsomes (0.1–1 nmol P450/ml) in the presence of an NADPH-regenerating system (0.5 mM $NADP^+$, 5 mM glucose-6-phosphate and 1.5 U/ml glucose-6-phosphate dehydrogenase). However, as described in the previous section, these typical conditions may change if the metabolism of the drug is not linear. In our laboratories, all incubation mixtures contain 3 mM magnesium chloride and 50–100 mM potassium phosphate buffer, pH 7.4 in a total volume of 0.5–1 ml. Prior to the addition of drug, the incubation mixture is preincubated for 2–3 min at 37°C. Reactions are initiated by addition of drug, allowed to proceed for a desired time at 37°C (based on linearity), and then terminated either by cooling and immediately subjecting to solid-phase extraction (SPE) or by addition of organic solvent followed by centrifugation. Boiled human liver microsomes and incubations without NADPH generally serve as negative controls. Samples from SPE experiment are evaporated to dryness, reconstituted and analyzed by liquid chromatography (LC) with a flow scintillation analyzer (FSA), LC–MS or LC–MS/MS. LC–MS coupled in-line with an FSA provides simultaneous detection of radioactivity with structural characterization of drug-derived material by MS. The extracts from solvent extraction procedure are often analyzed directly or following reduction of solvent volume (Ghosal *et al.*, 2003).

12.5.2. Screening Enzymes Responsible for the Metabolism of NCEs Using Recombinant Human P450 Enzymes

The data obtained with human liver microsomes are generally confirmed using recombinant CYP (rCYP, SUPERSOMES) enzymes. Use of rCYPs has become widespread as a result of the increased availability of rCYPs in high purity provided by commercial vendors. In addition, because the various CYPs in human liver microsomes have been immunoquantitated using antibodies, it is now possible to normalize data obtained with rCYPs and to fully integrate in vitro CYP reaction phenotyping data (Rodrigues, 1999). When the metabolism of a drug is catalyzed only by a single recombinant enzyme, interpretation of the result is straightforward. If the drug is metabolized by more than one recombinant enzyme, measurements of enzymatic activity alone do not provide sufficient information to estimate the relative contributions of each P450 isoform to total metabolism of the drug. Further studies with either antibodies or highly selective chemical inhibitors should be conducted in human hepatic microsomes to address the relative contribution. It should be noted that sometimes a recombinant P450, which is shown to be active in metabolism in the absence of other P450 enzymes, may play little or no role in microsomal metabolism in the presence of other P450 isoforms due to the competitive nature of the P450 enzymes. The other approach is to determine the K_m and V_{max} values of the drug with each active recombinant P450 enzyme so that the intrinsic clearance (V_{max}/K_m) values for each reaction can be calculated. Based on the intrinsic clearance and the relative abundance of each P450 isoform in human liver microsomes, the relative importance of different P450 enzymes in the metabolism of the compound of interest in human liver microsomes can be evaluated (Crespi and Miller, 1999; Rodrigues, 1999; Venkatakrishnan *et al.*, 2001).

In our laboratories, screening of recombinant human P450 enzymes (CYP1A1, CYP1A2, CYP2A6, CYP1B1, CYP2B6, CYP2C8, CYP2C9, CYP2C18, CYP2C19, CYP2D6, CYP2E1, CYP3A4, CYP3A5, CYP4A11, CYP4F2, CYP4F3A, and CYP4F3B) are generally performed with a constant amount of CYP (under steady-state condition) and new chemical entities (NCEs) are incubated for 5–120 min (based on linearity). All incubations contain 3 mM magnesium chloride and an NADPH-regenerating system in 5 ml of 50–100 mM potassium phosphate buffer, pH 7.4. Incubations with CYP2C9 and CYP2A6 are performed in Tris buffer according to the supplier's recommendation. Reactions are initiated by addition of the drug, terminated after the desired incubation time at 37°C, subjected to SPE, dried, reconstituted and analyzed by LC/FSA, LC–MS, LC–MS/FSA or LC–MS/MS. Reactions can also be terminated by addition of organic solvents, which also allows extraction of drug-derived material into organic solvents compatible with the analytical profiling method. Following centrifugation, the supernatant can be analyzed directly as described earlier. Insect microsomes without cDNA of human P450 enzymes are used as control.

12.5.3. Correlation Analysis

Typically, enzyme kinetic studies are followed by correlation analysis using a bank of human liver microsomes with known CYP content and activities at one or more substrate concentrations. In this instance, the rates of metabolism of the drug in question are correlated using regression analysis with the levels of CYP isoform-selective monooxygenase activities (Rodrigues, 1999). Complementary studies are then carried out with inhibitory antibodies and chemical inhibitors that are selective for different forms of CYP (Rodrigues, 1999; Lin and Rodrigues, 2001).

Although results from correlation studies are usually consistent with P450 identification from inhibition studies and screening of recombinant enzymes, an inconsistency has been reported for some P450 enzymes (Karanam *et al.*, 1994; Heyn *et al.*, 1996; Stevens *et al.*, 1997). Heyn *et al.* (1996) observed high correlations between *S*-mephenytoin N-demethylation and CPY2B6 ($r = 0.91$), CYP2A6 ($r = 0.88$), and CYP3A4 ($r = 0.74$). However, other studies established that CYP2B6 is the major enzyme responsible for *S*-mephenytoin N-demethylation whereas CYP2A6 and CYP3A4 play little or no role in this reaction. Broad overlap of substrate specificity among P450 enzymes and their relative abundance may significantly contribute to false-positive and false-negative results.

12.5.4. Inhibition by Antibodies

Potent, specific and inhibitory antibodies (either polyclonal or monoclonal) against various human P450 isoforms represent one of the most valuable tools for P450 identification (Gelboin *et al.*, 1999; Mei *et al.*, 1999). If the potency and specificity of a particular antibody can be verified with human liver microsomes and recombinant P450 preparations (including the closely related isoforms in the same subfamily, e.g., CYP2C8, CYP2C9, CYP2C18, and CYP2C19), then the antibody can be chosen as the only approach necessary for a P450-phenotyping study.

12.5.5. Inhibition by Chemicals

In some instances, two or more CYPs may contribute to the metabolism of a single compound and may possess similar K_m values. Therefore, the contribution of each CYP will be governed by the V_{max} and abundance of each enzyme in tissue. If one of the CYPs is polymorphic, then it is possible that drug interactions with potent CYP3A4 inhibitors (e.g., ketoconazole) will be more pronounced in PM (vs. extensive metabolizer or EM) subjects. Often chemical inhibitors are better than antibodies when considering: (1) the cost and supply of commercially available antibodies (both monoclonal and polyclonal) will probably not be greatly improved in the near future; (2) their easy availability through commercial sources or custom synthesis; and (3) finally, safe and highly selective P450 inhibitors can be used in clinical settings to assess the role of a specific P450 isoform in drug therapy. Antibodies are not suitable for the use in clinical or other in vivo studies. Similar to that described for antibody inhibition studies, the potency of a chemical inhibitor and the relative contribution of P450 enzyme involved can be assessed from a complete inhibition curve (Newton *et al.*, 1995). In addition, IC50 values can be estimated from the inhibition curve for different P450 enzymes and can be used to assess the selectivity of an inhibitor.

12.5.6. Sample Analysis

Typically, in the late discovery and early development stages of a new drug's life cycle, radiolabeled drug substrate is available and analyses of chemical inhibition and reaction phenotyping samples involve LC/FSA or LC/FSA-MS. Usually these methods are lower throughput compared to the LC–MS/MS method for the quantitation of parent drug only used in the discovery phase to assess the disappearance rate of the parent drug (non-radioactive). In contrast, LC/FSA or LC/FSA-MS profile methods allow direct measurement of the appearance rate for major metabolites, in addition to the disappearance rate for parent compound. In some pharmaceutical companies all chemical inhibition and reaction phenotyping samples are assayed using LC/FSA-MS. In others, selected samples are analyzed by LC/FSA-MS, while the majority of samples are analyzed by LC/FSA only. In many cases either an ion trap or a triple quadrupole mass spectrometer equipped with either atmospheric pressure chemical ionization (APCI) or electrospray ionization (ESI) source is commonly used (Ghosal *et al.*, 2003).

12.5.7. Characterization of Enzymology for the Metabolism of Loratadine: A Representative Example

In our laboratories, during the screening of loratadine (LOR) metabolites following incubation with human P450 SUPERSOMES, all incubates were subjected to LC-FSA analyses. Selected control samples and those which showed conversion were analyzed further by LC–MS/FSA and LC–MS^n/FSA. Structural characterization was performed using either Finnigan LCQ or TSQ Quantum mass spectrometers (ThermoElectron, San Jose, CA) nominally operated under the conditions in the following table:

Parameter	Setting
Ionization source	Electrospray ionization (ESI)
Ionization mode	Positive
Spray needle voltage	4.0–6.0 kV
Capillary temperature	200–300°C
Sample flow rate	0.18–0.25 ml/min after splitting
Sheath gas	Nitrogen (40–80)
Auxiliary gas	Nitrogen (5–20)

The HPLC column effluent was split so that 18–25% of the flow was analyzed by the mass spectrometer and 75–82% diverted to the FSA. In addition to the mass spectrometer, the components of the LC/FSA-MS system typically include an LC pump, column oven, controller, autosampler (Waters, Milford, MA) and a FSA (Perkin-Elmer, Boston, MA).

HPLC column temperature was maintained at 40°C for all LC–MS and LC–MS/MS experiments. The mobile phase, which consisted of 1 mM ammonium acetate adjusted to pH 6.0 with acetic acid (A) and acetonitrile (B), was maintained at a constant flow rate (1 ml/min).

Gradient elution of metabolites was achieved in the LC–MS/FSA system using programmed changes (linear) in mobile phase composition as summarized in the following table:

Time (min)	A (%)	B (%)	Status
0.0	90	10	Start gradient #1
25.0	72	28	Start gradient #2
40.0	10	90	End gradient #2
50.0	10	90	End data acquisition
51.0	90	10	Start equilibration
65.0	90	10	End equilibration

Following LC/FSA-MS analysis, the area of each detected radioactive peak was expressed as a percent of the total peak area integrated in the radiochromatogram. The mass spectrum corresponding to each peak in the radiochromatogram provided the characterization of metabolites.

The biotransformation pathway for LOR to desloratadine (DL) is shown below.

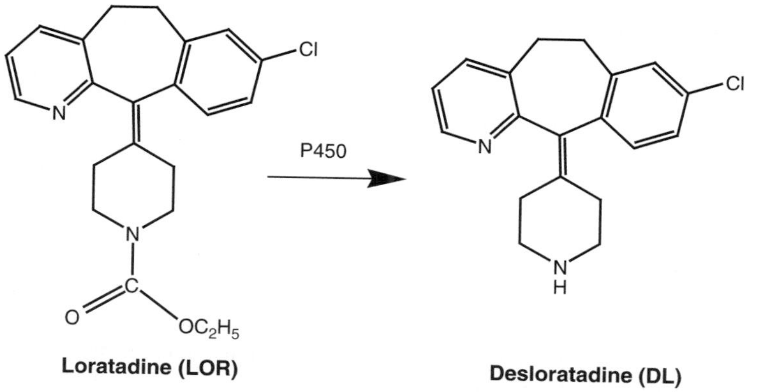

Loratadine (LOR) **Desloratadine (DL)**

The metabolism of LOR was previously reported (Yumibe *et al.*, 1996) to be mediated by CYP3A4 and CYP2D6. More recently, Ghosal *et al.* (2003) investigated the metabolism of LOR using 14 cDNA-expressed human CYPs. In these experiments, LC–MS with in-line FSA was used for definitive characterization of metabolites. Selected examples of chromatograms following incubation of LOR with CYP3A4, CYP2D6, and CYP2C19 are shown in Fig. 4.

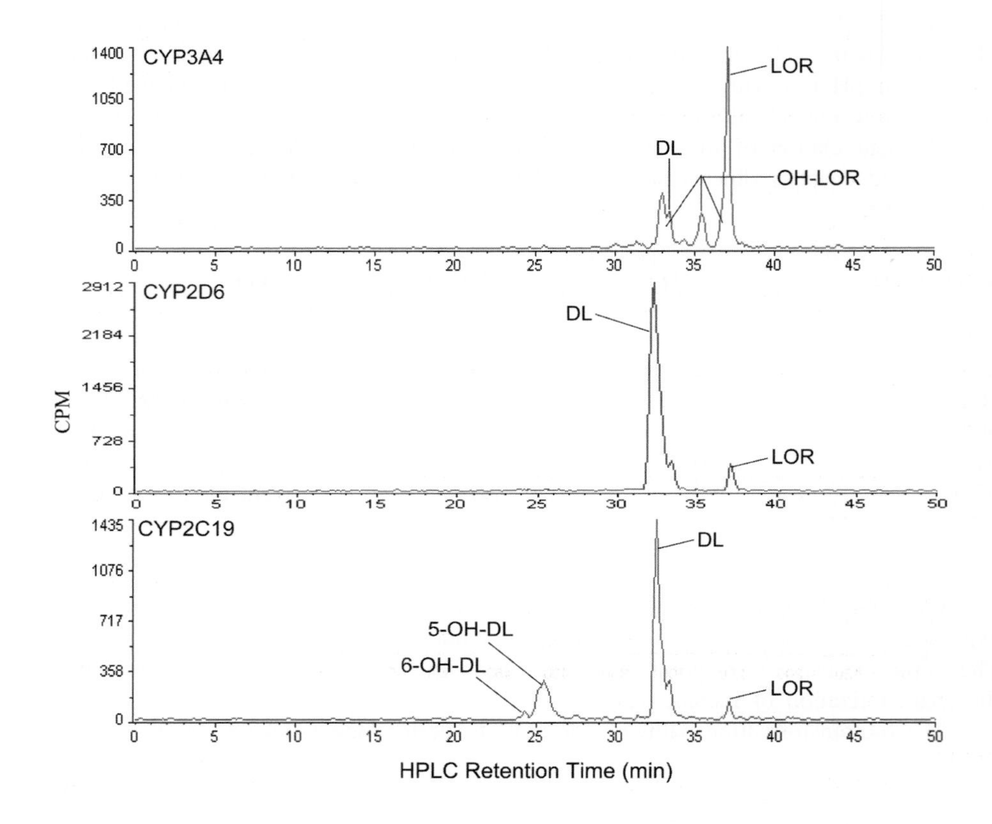

Figure 4.
Radiocarbon profiles following 120 min incubation of 26 μM ^{14}C-Loratadine (LOR) with P450 SUPERSOMES® (CYP3A4, CYP2D6, and CYP2C19) in the presence of NADPH-regenerating system

Although in some peaks multiple metabolites co-eluted, the presence of each metabolite was confirmed by LC–MS. A selection of mass spectra from these metabolites is shown in Fig. 5. The mass-to-charge ratio (*m/z*) for each compound presents its molecular weight (MW) and the primary characterization of the metabolite. For example, the 2nd panel (from bottom) shows that the *m/z* for a metabolite (DL) was detected at 311 Th. Therefore, the MW of this compound is 310, which is 72 Da lower than that of LOR and is consistent with the descarboethoxylation of LOR to DL. Following determination of MW, MS/MS experiments were performed to identify the site and nature of metabolism in DL and LOR. LC–MS spectra and chromatograms and LC elution times of metabolites from these experiments were further compared with those of reference standards (when available) to obtain definitive characterization of metabolites.

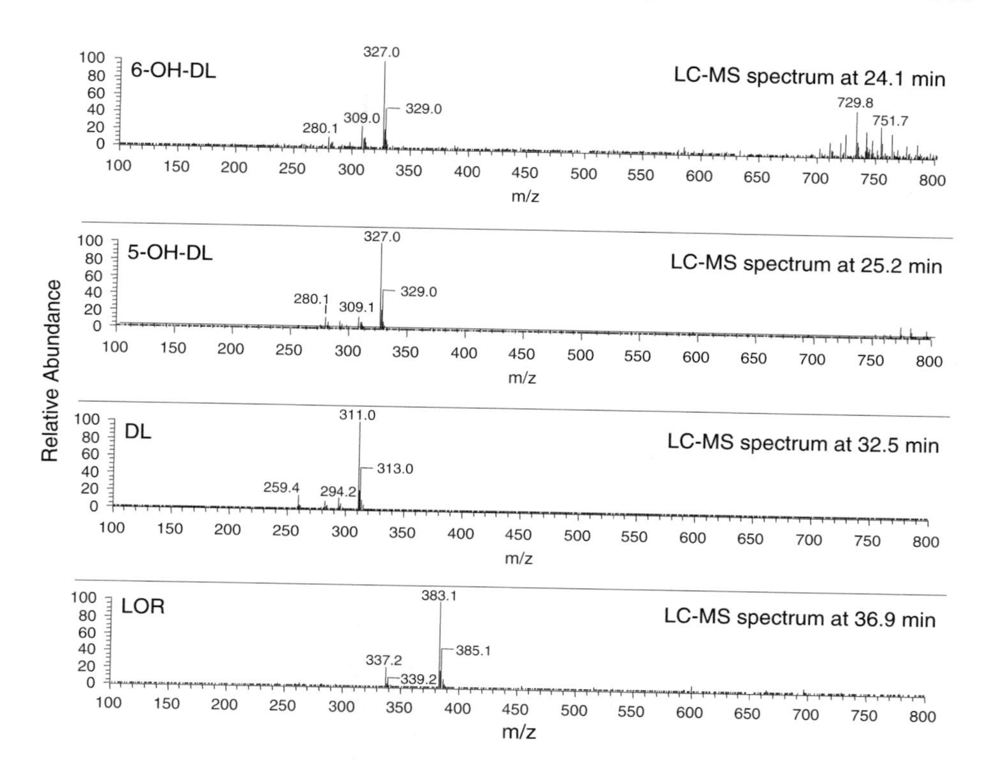

Figure 5.
LC–MS spectra of drug-derived material (Fig. 4) following incubation of LOR with CYP2C19

Data from the screening of ^{14}C-LOR with 14 human P450s are summarized in Fig. 6. The results demonstrate that LOR was catalyzed to DL mainly by CYP1A1 (extra-hepatic P-450), CYP2C19, CYP2D6, and CYP3A4 and to a lesser extent by CYP1A2, CYP2B6, CYP2C8, CYP2C9, and CYP3A5. CYP1A1, CYP2C19, and CYP2D6, however, converted LOR to DL and further to 5- and 6-hydroxy desloratadine while CYP3A4 only yielded DL.

The formation rate of DL was measured in each of the 10 human liver microsomal samples provided in HepatoScreen™ Test Kit and correlated with the biochemical activities data provided in the kit. Since the biochemical activity data were generated by specific CYP enzymes, high correlation would suggest that similar enzymes were involved in the formation of metabolites from LOR.

The highest correlation between the HepatoScreen™ Test Kit assay data ($n = 10$) and the formation of DL was noted for testosterone 6β-hydroxylation ($r = 0.67$) catalyzed by CYP3A4/5 (Fig. 7) and dextromethorphan *O*-demethylation catalyzed by CYP2D6 ($r = 0.72$, Fig. 7). The results of correlation analysis between the enzyme activities and metabolite formation confirmed the contribution of CYP3A4

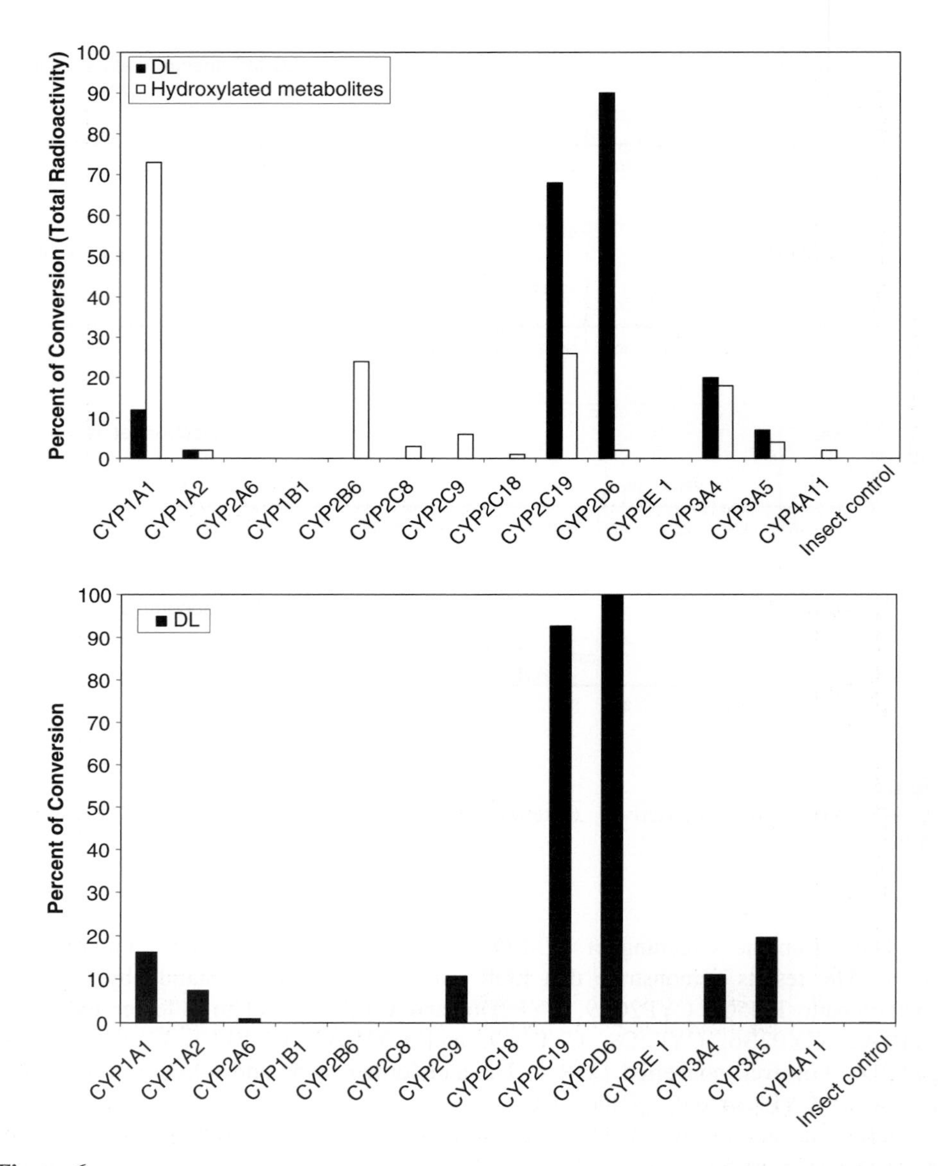

Figure 6.
Screening of loratadine (26 μM, top panel and 8 nM bottom panel) metabolites following incubation with human P450 SUPERSOMES®

and CYP2D6 in the metabolism of LOR in human liver microsomes. There was no significant correlation between formation of DL and other P450 enzyme activities.

This example illustrates experiments that were performed for the identification of CYP enzymes involved in the metabolism of LOR to DL and selected DL

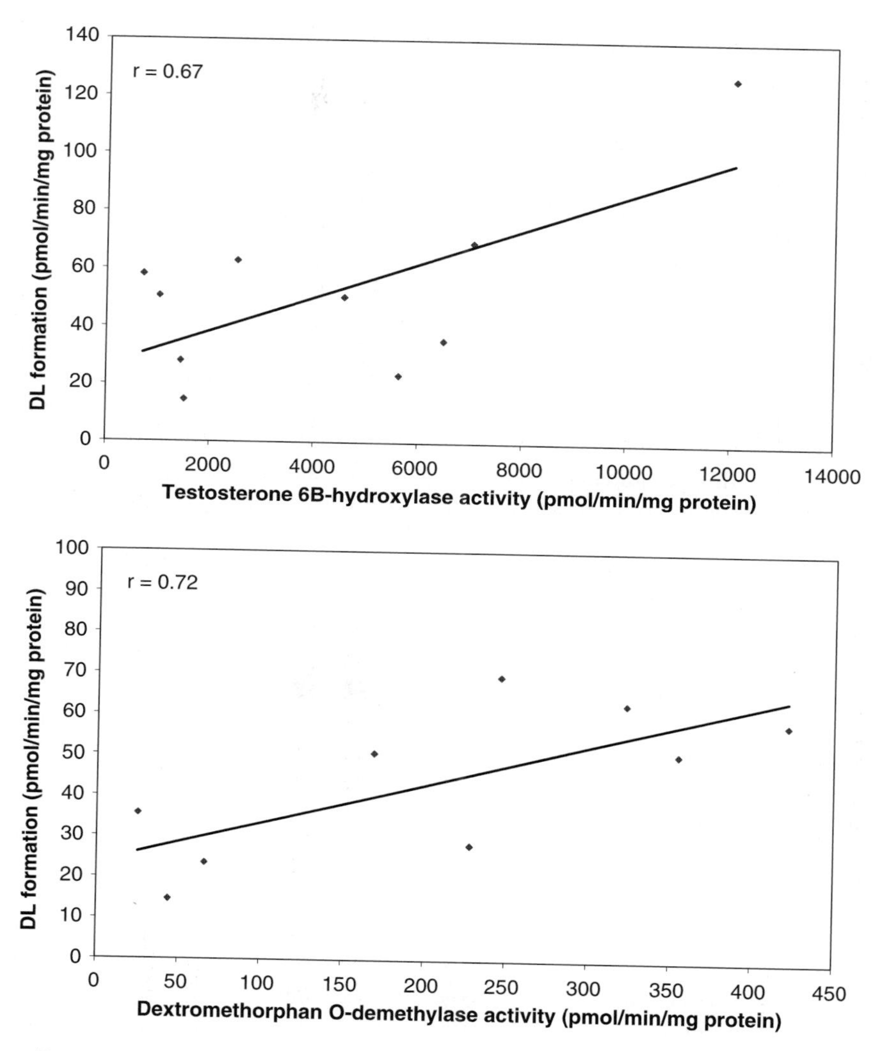

Figure 7.
Correlation analysis between desloratadine formation and testosterone 6β-hydroxylase activity (top panel) and dextromethorphan *O*-demethylase activity (bottom panel) in a panel of 10 human liver microsomes

metabolites. These experiments are by no means limited to phenotyping of LOR but can be applied to phenotyping of other compounds (Ghosal *et al*., 2003). It should be noted that although in vitro experiments generated DL, 5- and 6-OH DL from LOR, the primary in vivo human metabolite 3-OH DL (Glu) was not formed in any in vitro microsomal system. This example also highlights the limitation of in vitro systems to predict in vivo human metabolism.

12.6. High-Throughput Screening for Inhibition of CYP Enzymes

Generally in the discovery stage of drug development, radiolabeled forms of NCEs are not readily available and screening methods involve HPLC-UV, fluorescent enzyme assays or more commonly LC–MS/MS. HPLC-UV can be used at high relative substrate concentrations but its use is impractical with potent compounds exhibiting ED_{50} in the nM range. In vitro experiments are often carried out at very low concentrations (nM–μM) such that highly sensitive analytical techniques are necessary to detect and identify drug-derived material. Similarly, fluorescent enzyme assays are time consuming since only a single enzyme can be evaluated at a time. High-throughput 96-well plate-based fast gradient LC–MS/MS assays, which are capable of simultaneously evaluating the activities of all major human metabolizing enzymes have become the norm. This accelerated view of CYP enzymes responsible for the metabolism of NCEs and their inhibitory profiles is critical to the elimination of metabolically challenged new drug candidates (Dierks *et al.*, 2001; and Testino and Patonay, 2003). Each assay is usually complete in <5 min and involves detector calibration by mass spectrometric monitoring of a particular precursor and a specific product ion for the parent drug. LC–MS/MS based high-throughput methods have arisen by necessity in the pharmaceutical industry in order to accommodate the vast number of compounds entering the discovery. Although several groups have developed innovative ways to increase throughput and assess the inhibition of CYP (Ayrton *et al.*, 1998; Yin *et al.*, 2000; Chu *et al.*, 2000; Bu *et al.*, 2001; Dierks *et al.*, 2001; Weaver *et al.*, 2003), the cocktail substrate method reported by Dierks *et al.* has shown some promise in increasing the throughput in human CYP P450 screening. As shown in the following table and Fig. 8, the in vitro cocktail method involves incubation of selective substrates with human liver microsomes and monitoring a known metabolite of the substrate using LC–MS/MS methods.

Enzyme	Probe substrate	Metabolite	Precursor → product ion pair
CYP3A4	Midazolam	1′-Hydroxymidazolam	342 → 324 or 297
CYP2D6	Bufuralol	1′-Hydroxybufuralol	278 → 186
CYP2C9	Diclofenac	4′-Hydroxydiclofenac	312 → 231
CYP2C9	Tolbutamide	4-Hydroxytolbutamide	287 → 89
CYP1A2	Ethoxyresorufin	Resorufin	214 → 214
CYP1A2	Phenacetin	Paracetamol	152 → 110
CYP2C19	*S*-Mephenytoin	4′-Hydroxymephenytoin	235 → 150
CYP2C19	Omeprazole	5-Hydroxyomeprazole	362 → 214
CYP2A6	Coumarin	7-Hydroxycoumarin	163 → 107
CYP2C8	Paclitaxel	6α-Hydroxypaclitaxel	870 → 286

Recently, Walsky and Obach (2004) described validated (using good laboratory practices) LC–MS assays to measure the effect of new chemical entities on CYP activities (using standard probe substrates) useful for the assessment of CYP inhibition potential of drugs.

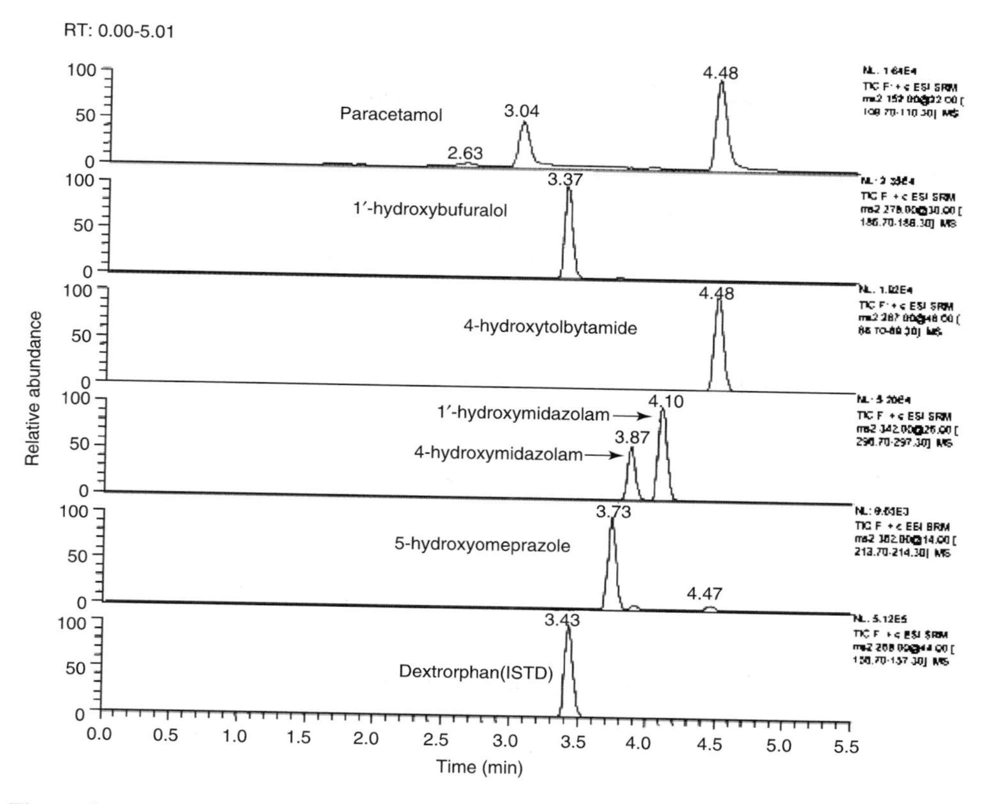

Figure 8.
Extracted ion chromatograms from the analysis of human liver microsomes incubated with a cocktail containing paracetamol (CYP1A2), 4-hydroxytolbutamide (CYP2C9), 5-hydroxyomeprazole (CYP2C19), 1′-hydroxybufuralol (CYP2D6) and 1′-hydroxymidazolam (CYP3A4) (reprinted with permission from Testino and Patonay, 2003). Dextrorphan was used as an internal standard to correct for any experimental errors and the precursor/product ion pair monitored was 258/157. The total HPLC run time of this cocktail method was 5.5 min and the sensitivity of the MS instrument allowed direct injection of the supernatant of the incubate following quenching of the reaction

12.7. UDP-glucuronosyltransferase (UGT)

Phase II conjugation introduces hydrophilic groups/molecules such as glucuronic acid, sulfate, or amino acids onto a target molecule. These reactions are catalyzed by a group of enzymes called transferases. Most transferases are located in the cytosol, except those that facilitate glucuronidation, which are microsomal enzymes. A family of enzymes called uridine diphosphate glucuronosyltransferases (UGTs), catalyze the most prevalent Phase II reaction, glucuronidation. Glucuronic acid, a six-membered cyclic sugar contains four hydroxyl groups and one carboxylic acid functionality. This molecule is extremely hydrophilic and improves the water solubility when they are covalently bound to a drug molecule. The structure of glucuronic acid is shown below:

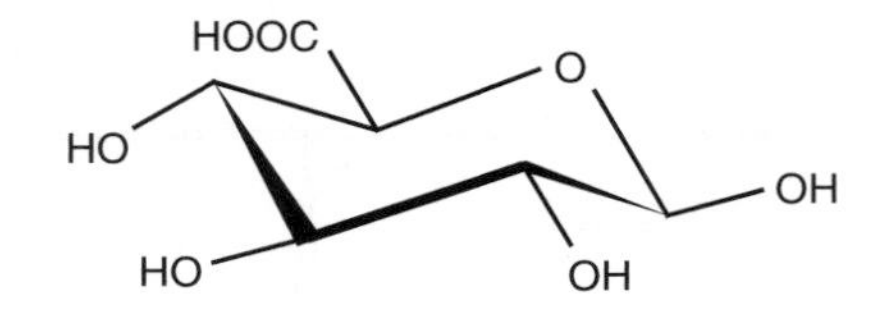

UDP-glucuronosyltransferase enzymes comprise a superfamily of key proteins that catalyze the glucuronidation reaction on a wide range of structurally diverse endogenous and exogenous chemicals. This biochemical process is also involved in the protection against environmental toxicants, carcinogens, dietary toxins and participates in the homeostasis of numerous endogenous molecules, including bilirubin, steroid hormones, and biliary acids. Significant progress has been made in the field of glucuronidation, especially with regard to the identification of human UGTs, study of their tissue distribution and substrate specificities. Substrates and tissue localization of human UGT isoforms are provided in Tables 2 and 3. More recently, the degree of allelic diversity has also been revealed for several human UGT genes. UGT isoforms have been separated into two families (UGT1A and UGT2B) based on similarities

Table 2.
Substrates and tissue localization of human UGT1A isoforms

Isoforms	Tissue localization	Major substrates (endogenous)	Major substrates (xenobiotic)	Polymorphic
UGT1A1	Liver	Bilirubin	Quercetin	Yes
	Intestine	Estriol	Naringenin	
	Colon	*β*-Estradiol	1-Naphthol	
	Gallbladder	2-Hydroxyestrone		
		2-Hydroxyestradiol	Apigenin	Yes
UGT1A3	Liver	Estrone	Naringenin	
	Intestine	2-Hydroxyestrone		
	Colon	2-Hydroxyestradiol		
	Gallbladder	4-Hydroxy-all-*trans*-retinoic acid		
UGT1A4	Liver	Androsterone	Benzidine	Not known
	Intestine	Epiandrosterone	Amitryptiline	
	Colon		Imipramine	
	Gallbladder		Trifluoperazine	
			Clozapine	
			Posaconazole	
UGT1A6	Liver	Not known	Acetaminophen	Yes
	Stomach		Methylsalicylate	
	Intestine		4-Methylumbelliferone	

	Colon Gallbladder		1-Naphthol	
UGT1A7	Stomach Intestine	Not known	HFC	Yes
UGT1A8	Intestine Colon	Estrone 2-Hydroxyestrone 4-Hydroxyestrone Dihydrotestosterone	4-Methylumbelliferone Apigenin Naringenin Emodin Propafol Eugenol	Yes
UGT1A9	Liver Colon Kidney	Thyroxine	Acetaminophen Emodin Quercetin 4-Methylumbelliferone Carvacrol	Yes
UGT1A10	Stomach Intestine Colon Gallbladder	2-Hydroxyestrone 4-Hydroxyestrone Dihydrotestosterone	Mycophenolic acid Benzo(a)pyrine Quinoline 2-Acetylaminofluorene	Not known

Source: Radominska-Pandya *et al.* (1999), Fisher *et al.* (2001) and Ghosal *et al.* (2004a).

between their primary amino acid sequences. Acetaminophen (shown below), morphine, and estradiol are glucuronidated by UGT1A6, UGT2B7, and UGT1A1, respectively (Fisher *et al.*, 2000b).

Substrates specificity and tissue localization of these UGTs are summarized in Table 2 (UGT1A) and Table 3 (UGT2B).

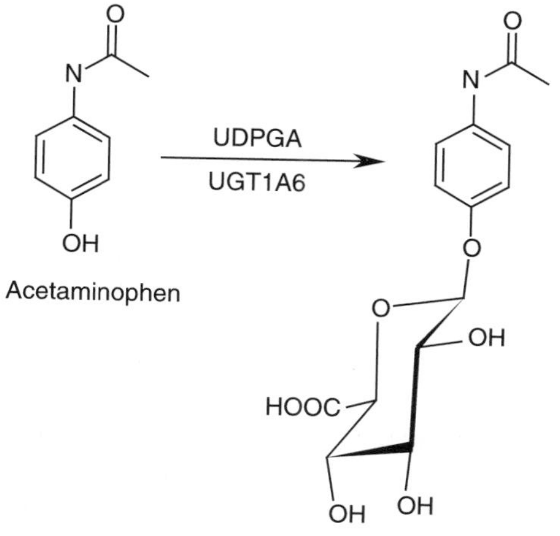

UDPGA
UGT1A6
Acetaminophen

Structures of selected substrates and inhibitors of UGTs are presented below:

Substrates

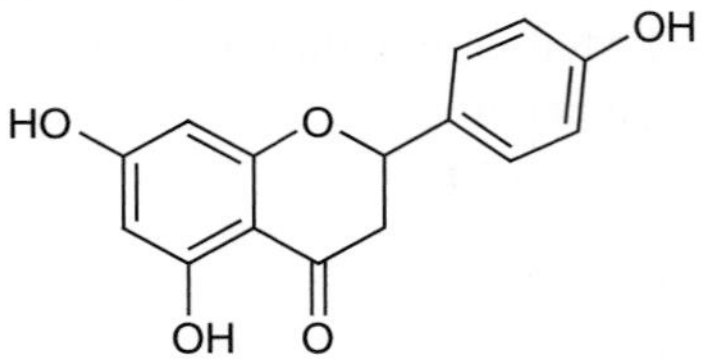

Naringenin

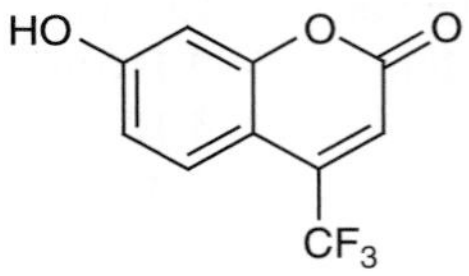

7-Hydroxy-4-(trifluoromethyl)-coumarin (HFC)

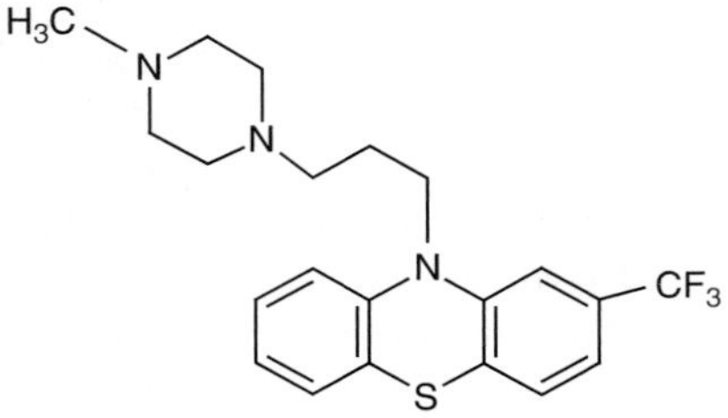

Trifluoperazine

Inhibitors

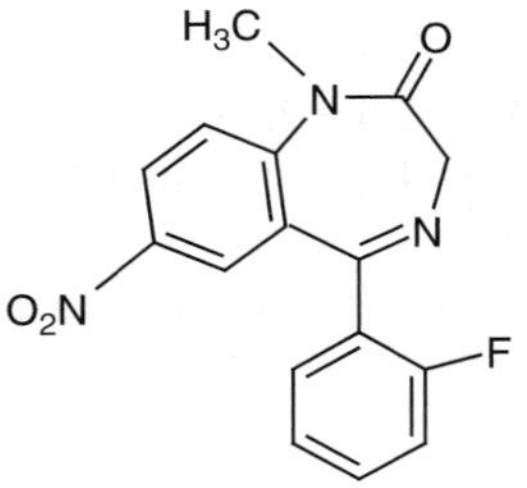

Flunitrazepam

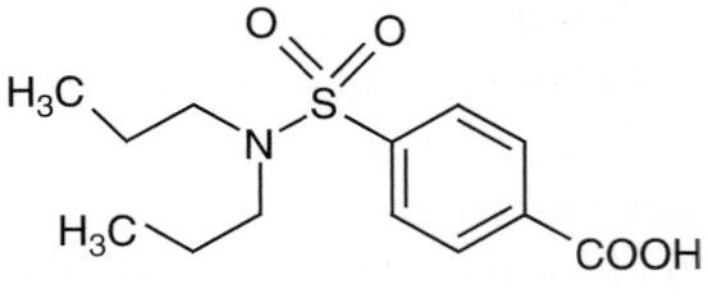

Probenecid

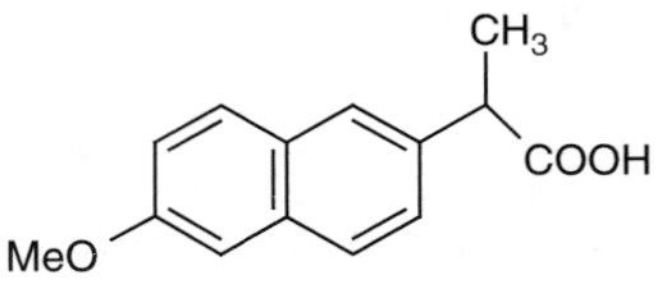

Naproxen

12.8. Identification of UGT Enzyme(s) Involved in the Metabolism

Similar to CYPs, there are several well-established approaches used to identify the specific UGT enzyme(s) responsible for the metabolism of NCE. These include (1) incubation with recombinant human UGT enzymes, (2) use of chemical inhibitors

Table 3.
Substrates and tissue localization of human UGT2B isoforms

Isoforms	Tissue localization	Major endogenous substrates	Major xenobiotic substrates	Polymorphic
UGT2B4	Liver Intestine	Androsterone	Not known	Yes
UGT2B7	Liver Intestine Kidney	Androsterone Epitestosterone 4-Hydroxyestradiol Estriol 2-Hydroxyestrione Linoleic acid	Morphine Naloxone Codeine Buprenorphine Nalbuphine Naltrexone	Yes
UGT2B10	Liver	Not known	Not known	Not known
UGT2B11	Liver Kidney Prostate Skin Adrenal Lung Mammary gland	Not known	Not known	Not known
UGT2B15	Liver Intestine Prostate Skin Breast	Dihydrotestosterone	Eugenol 4-Methylumbelliferone Naringenin SCH 23390 Esculetin 8-Hydroxyquinoline	Yes
UGT2B17	Prostate	Testosterone Dihydrotestosterone Androsterone	Eugenol	Not known

Source: Radominska-Pandya *et al.* (1999), Fisher *et al.* (2001) and Ghosal *et al.* (2004a).

against human UGTs, and (3) correlation between metabolic rates and individual UGT activity (Ghosal *et al.*, 2004a–c).

UGT phenotyping involves an initial assessment of Michaelis–Menten kinetics using native human liver microsomes. Initial reaction rates are measured over a wide range of substrate concentrations ([S]) employing incubation conditions that are linear with respect to microsomal protein concentration and time of incubation. Data are analyzed by linear and non-linear transformations (Ghosal *et al.*, 2004a). In the majority of cases, the low apparent K_m component is most informative because this indicates which UGTs play a major role at clinically relevant drug concentrations.

Generally, low K_m and/or high V_{max}/K_m (intrinsic clearance) is predictive of preferred biotransformation pathway. An example of UGT phenotyping is provided in the following section using posaconazole, a triazole antifungal agent.

12.8.1. Incubation with Pooled Human Liver Microsomes

Posaconazole concentrations of 0.5–50 μM, microsomal protein concentrations of 0.05–2 mg/ml, and incubation time of 15–240 min were used to optimize the assay condition. Microsomes are pretreated with 50 μg alamethicin/mg microsomal protein on ice for 15 min to diminish the latency of UGT activity (Fisher *et al.*, 2000a). All microsomal incubations contained microsomes treated with alamethicin, 10 mM magnesium chloride, 5 mM saccharolactone (an inhibitor of *β*-glucuronidase) and substrate (^{14}C-posaconazole) in 0.5 ml of 0.1 M Tris buffer, pH 7.4 (Pless *et al.*, 1999). Prior to the addition of UDPGA, the incubation mixtures were prewarmed for 3 min at 37°C. Reactions were initiated by the addition of 2 mM UDPGA, allowed to proceed for various time periods at 37°C, and terminated with ice-cold methanol. The incubation mixtures were mixed, centrifuged at 4°C for 10 min, and the supernatants analyzed by HPLC coupled with a radiometric detector. Incubations without UDPGA and boiled human liver microsomes served as negative control. For LC/MS analyses, the reactions were either terminated as described above or terminated by cooling in ice water, followed by solid-phase extraction. Optimal time and optimal protein concentration were 120 min and 1 mg/ml, respectively.

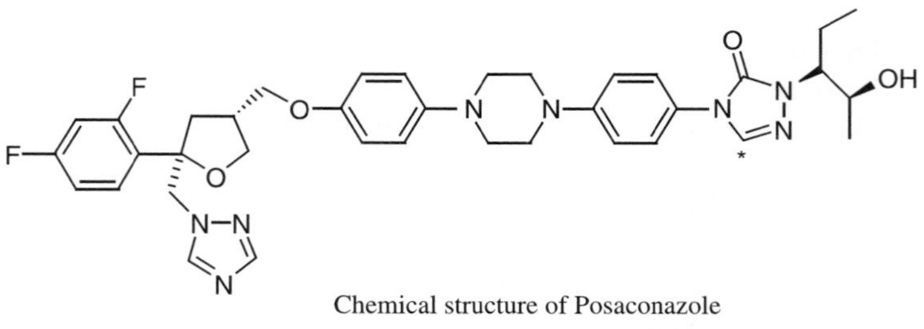

Chemical structure of Posaconazole

12.8.2. Screening of Recombinant Human UGT Enzymes

In vitro screening of 10 human UGT SUPERSOMES® (UGT1A1, UGT1A3, UGT1A4, UGT1A6, UGT1A7, UGT1A8, UGT1A9, UGT1A10, UGT2B7, and UGT2B15) was performed using a constant amount of microsomal protein (1 mg/ml) and two concentrations of substrate (^{14}C-posaconazole). All incubations were carried out as described earlier. Insect cell microsomes without cDNA for UGT served as the negative control. These samples were also analyzed by LC-FSA/MS and LC–MS to verify the formation of glucuronide conjugates. Optimal time and optimal protein concentration were determined as described in P450 section. Kinetic parameters (K_m and V_{max}) for UGTs were determined using various substrate concentrations and optimal UGT protein concentration of 1 mg/ml and proceeded for optimal time as described earlier for microsomes.

12.8.3. Correlation Study

Human liver microsomal preparations from 10 individual donors (included data for UGT-specific enzyme activities) were obtained from BD-Gentest. The ability of human liver microsomes from each donor to glucuronidate substrate is correlated with the UGT-specific enzyme activities for each sample. The assays were performed as described previously (Ghosal *et al.*, 2004a).

12.8.4. Inhibition Study with Chemical Inhibitors of UGTs

Glucuronidation of posaconazole was evaluated using known chemical inhibitors of UGTs (Ghosal *et al.*, 2004a). Human liver microsomes (1 mg/ml) or human UGT SUPERSOMES® (1 mg/ml) were preincubated with various concentrations of bilirubin for 15 min at room temperature followed by the addition of buffer, alamethicin, saccharolactone, cofactor, and substrate. The final concentration of the organic solvent in the incubation system was 1%, and all control incubations contained the same volume of appropriate vehicle. All incubations were carried out as described earlier.

In vitro incubations with 10 different recombinant human UGT SUPERSOMES™ showed that only UGT1A4 exhibited catalytic activity for the formation of posaconazole-glucuronide (Fig. 9). The major in vitro metabolite of posaconazole formed by

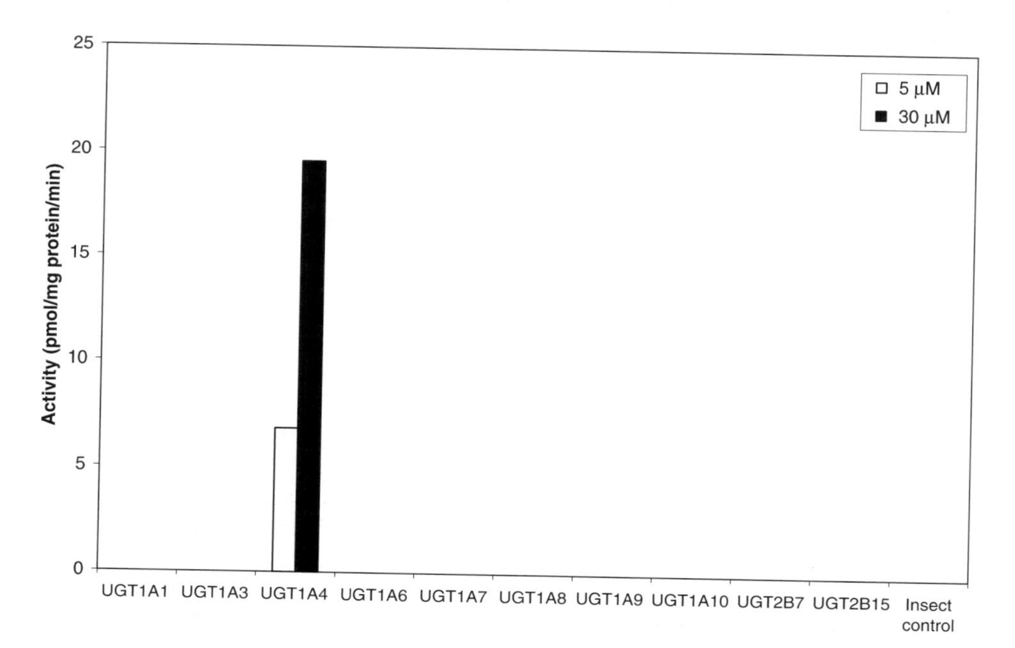

Figure 9.
Screening of UGT SUPERSOMES™ for the formation of a glucuronide conjugate of posaconazole at 5 and 30 μM (Reprinted with permission from Ghosal *et al.* (2004a). Copyright (2004) ASPET.)

human liver microsomes supplemented with UDPGA was found to be a glucuronide of posaconazole by LC–MS (*m/z* 877 Th, Fig. 10). Since the results of inhibition studies with recombinant UGT1A4 SUPERSOMES™ show that bilirubin was an inhibitor of posaconazole-glucuronide formation from UGT1A4 (Ghosal *et al.*, 2004a), it was used as a UGT1A4-inhibitor in this study. At 100 μM concentration, bilirubin inhibited formation of the glucuronide from human liver microsomes and UGT1A4 SUPERSOMES™ by 79.6 and 63.5%, respectively, with an IC50 of 11.1 μM (Fig. 11). In addition, there was a highly significant correlation between the formation of posaconazole-glucuronide and trifluoperazine glucuronidation which is known to be mediated by UGT1A4 (Fig. 12). These results confirm the involvement of UGT1A4 in the formation of posaconazole-glucuronide.

Although the preceding discussion involved the characterization of enzymology of posaconazole glucuronidation, similar procedures can be applicable to other substrates as well.

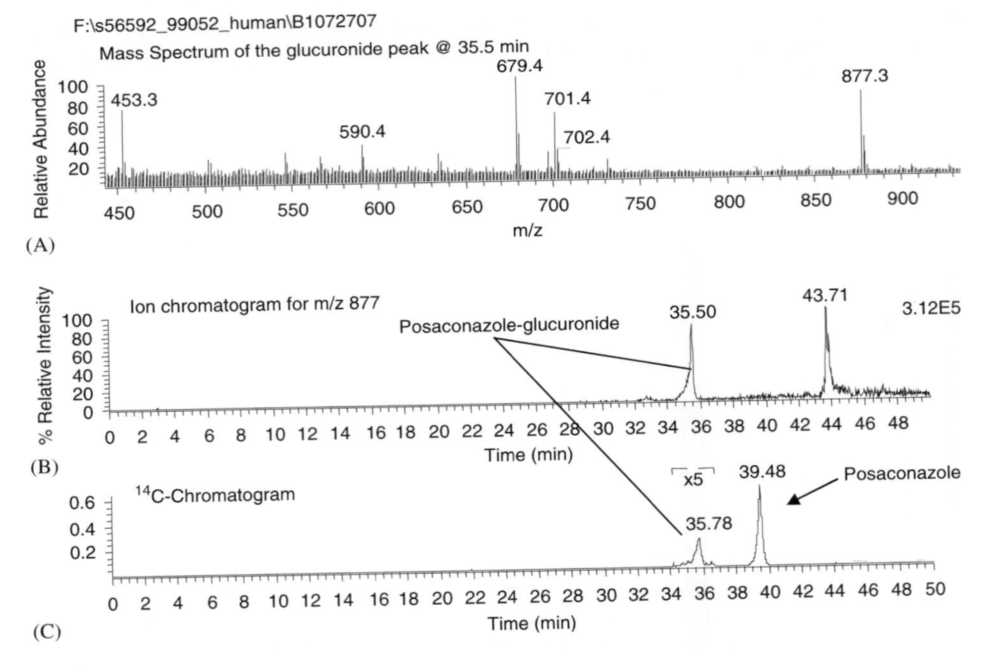

Figure 10.
LC–MS spectrum (panel A) of the peak at R_t 35.50 min. Panel B is the selected ion chromatogram at *m/z* 877 of an extract from the incubation of 30 μM ^{14}C-posaconazole with recombinant UGT1A4 SUPERSOMES® supplemented with UDPGA. Panel C is the ^{14}C-chromatogram for the simultaneous LC–MS and radiometric analysis

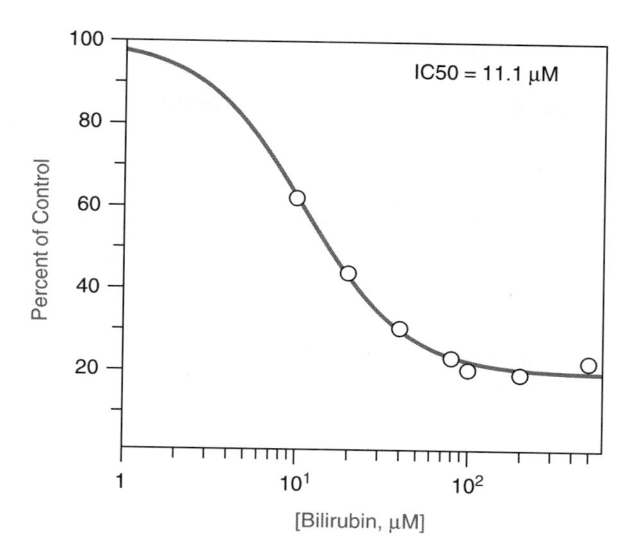

Figure 11.
Inhibition curve of bilirubin (IC50) for the formation of glucuronide from posaconazole incubated with pooled human liver microsomes (Reprinted with permission from Ghosal *et al.* (2004a). Copyright (2004) ASPET.)

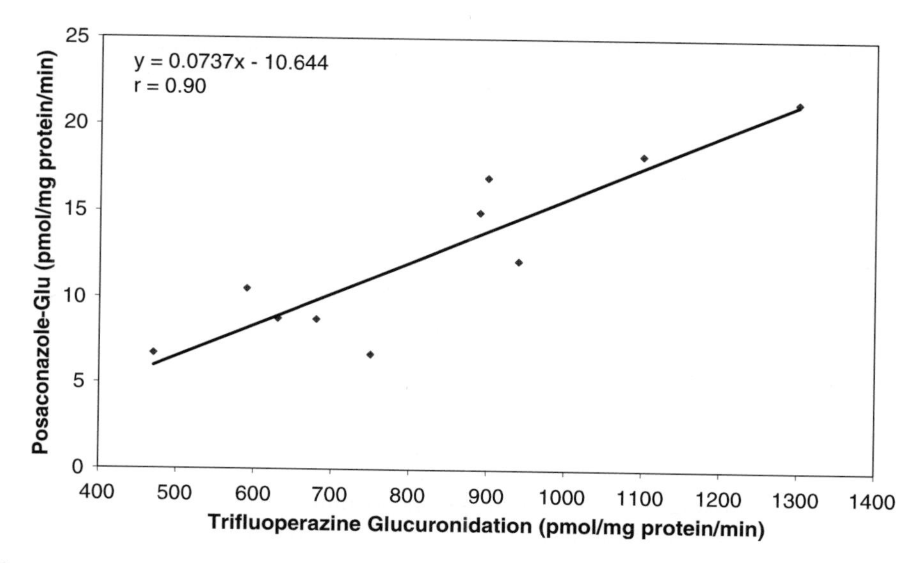

Figure 12.
Correlation between formation of glucuronide from 30 μM posaconazole incubated with ten individual human liver microsomes and trifluoperazine glucuronidation (UGT1A4 activity) (Reprinted with permission from Ghosal *et al.* (2004a). Copyright (2004) ASPET.)

12.9. UGT and Polymorphism

Genetic variations and single nucleotide polymorphisms (SNPs) within the UGT genes are remarkably common, and lead to genetic polymorphisms. The pharmacogenomic implication of polymorphic UGTs and their role in modifying cancer, have significant impact on individual risk to drug-induced toxicities. The multiplicity of transferases, some exhibiting overlapping substrate specificity, may provide functional compensation for genetic deficit in some cases. Genetic variation may cause different phenotypes by affecting expression levels or activities of individual UGTs. This inter-individual variation in UGTs has resulted in functional deficit affecting endogenous metabolism, and leads to jaundice and other diseases. Disruption of the normal metabolic physiology, by the reduction of bile acid excretion or steroid glucuronidation, may lead to cholestasis and organ dysfunction. A deficiency in the glucuronidation of drugs and xenobiotics has an important pharmacological impact which may lead to drug-induced adverse reactions, and even cancer. Additional novel polymorphisms in this gene family may yet to be revealed and could have a profound effect on the development of new drugs and therapies.

Genetic polymorphism has been described for six of the 16 functional human UGT genes characterized to date. They are UGT1A1, UGT1A6, UGT1A7, UGT2B4, UGT2B7, and UGT2B15. Since glucuronidation is an essential pathway for the elimination of a myriad of xenobiotics and endogenous compounds, genetic polymorphism of UGT is potentially of toxicological and physiological importance. However, functional significance has only been convincingly demonstrated for genetic polymorphism of UGT1A1, the major bilirubin-glucuronidating enzyme. Apart from impaired bilirubin glucuronidation, the mutations responsible for Gilbert syndrome also affect the elimination of a limited number of xenobiotics (Mackenzie *et al.*, 2000). The frequencies of individual UGT1A1 polymorphisms show extensive variability across ethnic groups (Mackenzie *et al.*, 2000). It has been proposed on the basis of altered catalytic activity of UGT1A6, UGT1A7 and UGT2B15 mutants that genetic polymorphism of these forms may be of toxicological significance, but this is yet to be proven (Miners *et al.*, 2002). However, unless the UGT in question is responsible for the exclusive metabolism of a particular drug or chemical (e.g. UGT1A1 and bilirubin) or is the predominant or only UGT present in the cell, it is unlikely that these polymorphisms will be of major clinical significance (Mackenzie *et al.*, 2000).

12.10. Future Trends

Drug metabolizing enzymes such as monoamine oxidase, flavin-containing monooxygenase, epoxide hydrolase, glutathione *S*-transferase, sulfotransferase, and *N*-acetyltransferase are most likely involved when a drug is metabolized by a non-P450 enzyme. Unlike P450, the tools necessary for in vitro phenotyping are not available for most of these enzymes. Further development of analytical procedures for in vitro reaction phenotyping for non-P450 enzymes will be of interest for predicting drug–drug interactions where the enzymes are involved in drug metabolism.

Although in vitro studies usually provide valuable information regarding biotransformation pathways of a new drug, sometimes there are discrepancies between in vitro results and in vivo findings. For example, some drugs are metabolically stable in the human liver microsomal incubation system with few or no metabolites that can be detected. Moreover, qualitatively and quantitatively different metabolites can be detected in vivo in humans. This situation can create considerable difficulty in the design and/or interpretation of in vivo drug–drug interaction studies. The potential discrepancy between in vitro and in vivo metabolism remains a challenge for future studies.

Adverse drug interactions represent a major human health issue in industrialized countries, where chronic multitherapeutic treatments are increasingly prescribed. Thus, the knowledge of the metabolic pathways and the nature of the enzyme(s) involved in the metabolism of a drug are essential to predict and avoid potential interactions between co-administered drugs. These interactions are often difficult to predict because they involve several distinct parallel and competing processes, including absorption, distribution, metabolism, and excretion. The recent withdrawal of cerivastatin (Baycol) exemplifies the difficulty in predicting clinical adverse events and toxicities. Only recently emerging in vitro and preclinical data provide possible enzymology-related explanations of toxicities associated with combination of cerivastatin with gemfibrozil (Wang *et al.*, 2002; Prueksaritanont *et al.*, 2002). These studies suggest inhibition of both Phase I, albeit only CYP2C8, and Phase II metabolism of cerivastatin by gemfibrozil as a likely reason for clinical toxicities. This was contrary to the perceived safety of cerivastatin based on the observation that two major P450s (CYP3A4 and CYP2C8) were responsible of its metabolism (Bischoll and Heller, 1998). Hence, a consolidated consideration of all possible interactions (including transporters) in the metabolic pathways of each co-administered medication may provide a better insight with improved predictivity of possible adverse reactions. Also, recently there has been a great deal of interest in the role of polymorphic P450 in the ethnic diversity and adverse drug reactions. In future, specific dosage recommendations based on genotypes will gain a lot of interest to guide the clinician. Such development will lead to a patient-tailored drug therapy, which hopefully would be more efficient and will result in fewer adverse drug reactions.

Acknowledgment

The author (Anima Ghosal) acknowledges Moumita Ghosal of Rutgers University (Piscataway, NJ) for her assistance in the preparation of the chapter.

References

Aklillu, E., Persson, I., Bertilsson, L., Johansson, I., Rodrigues, F., Ingelman-Sundberg, M., 1996. Frequent distribution of ultrarapid metabolizers of debrisoquine in an ethiopian population carrying duplicated and multiduplicated functional CYP2D6 alleles. J. Pharmacol. Exp. Ther. 278(1), 441–446.

Anttila, S., Hietanen, E., Vainio, H., Camus, A-M., Gelboin, H.V., Park, S.S., Heikkilä, L., Karjalainen, A., Bartsch, H., 1991. Smoking and peripheral type of cancer are related to high levels of pulmonary cytochrome P450IA in lung cancer patients. Int. J. Cancer 47, 681–685.

Aoyama, T., Yamano, S., Guzelian, P.S., Gelboin, H.V., Gonzalez, F.J., 1990. Five of 12 forms of vaccinia virus-expressed human hepatic cytochrome P450 metabolically activate aflatoxin B1. Proc. Natl. Acad. Sci. USA 87, 4790–4793.

Ayrton, J., Plumb, R., Leavens, W.J., Mallett, D., Dickins, M., Dear, G.J., 1998. Application of a generic fast gradient liquid chromatography tandem mass spectrometry method for the analysis of cytochrome P450 probe substrates. Rapid Commun. Mass Spectr. 12, 217–224.

Bailey, D.G., Spence, J.D., Munoz, C., Arnold, J.M., 1991. Interaction of citrus juices with felodipine and nifedipine. Lancet 337(8736), 268–269.

Bergstrom, R.F., Peyton, A.L., Lemberger, L., 1992. Quantification and mechanism of the fluoxetine and tricyclic antidepressant interaction. Clin. Pharmacol. Ther. 51(3), 239–248.

Bertz, R.J., Granneman, G.R., 1997. Use of in vitro and in vivo data to estimate the likelihood of metabolic pharmacokinetic interactions. Clin. Pharmacokinet. 32, 210–258.

Bischoll, H., Heller, A., 1998. Preclinical and clinical pharmacology of cerivastatin. Am. J. Cardiol. 82, 18J–25J.

Branch, R.A., Adedoyin, A., Frye, R.F., Wilson, J.W., Romkes, M., 2000. In vivo modulation of CYP enzymes by quinidine and rifampin. Clin. Pharmacol. Ther. 68(4), 401–411.

Bu, H.Z., Magis, L., Knuth, K., Teitelbaum, P., 2001. High-throughput cytochrome P450 (CYP) inhibition screening via a cassette probe-dosing strategy. VI. simultaneous evaluation of inhibition potential of drugs on human hepatic isozymes CYP2A6, 3A4, 2C9, 2D6 and 2E1. Rapid Commun. Mass Sp. 15, 741–748.

Butler, M.A., Lang, N.P., Young, J.F., Caporaso, N.E., Vineis, P., Hayes, R.B., Teitel, C.H., Massengill, J.P., Lawson, M.F., Kadlubar, F.F., 1992. Determination of CYP1A2 and acetyltransferase 2 phenotypes in human populations by analysis of caffeine urinary metabolites. Pharmacogenetics 2, 116–127.

Capdevila, J.H., Falck, J.R., Harris, R.C., 2000. Cytochrome P450 and arachidonic acid bioactivation: molecular and functional properties of the arachidonate monooxygenase. J. Lipid Res. 41, 163–181.

Caraco, Y., Sheller, J., Wood, A.J., 1996. Pharmacogenetic determination of the effects of codeine and prediction of drug interactions. J. Pharmacol. Exp. Ther. 278(3), 1165–1174.

Chang, T.K., Chen, J., Pillay, V., Ho, J.Y., Bandiera, S.M., 2003. Real-time polymerase chain reaction analysis of CYP1B1 gene expression in human liver. Toxicol. Sci. 71(1), 11–19.

Cholerton, S., Idle, M.E., Vas, A., Gonzalez, F.J., Idle, J.R., 1992. Comparison of a novel thin-layer chromatographic-fluorescence detection method with a spectrofluorometric method for the determination of 7-hydroxycoumarin in human urine. J. Chromatogr. 575, 325–330.

Chu, I., Favreau, L., Soares, T., Lin, C.C., Nomeir, A.A., 2000. Validation of higher-throughput high-performance liquid chromatography/atmospheric pressure chemical ionization tandem mass spectrometry assays to conduct cytochrome P-450s CYP2D6 and CYP3A4 enzyme inhibition studies in human liver microsomes. Rapid Commun. Mass Spectr. 14, 207–214.

Crespi, C.L., Miller, V.P., 1999. The use of heterologously expressed drug-metabolizing enzymes-state of the art and prospects for the future. Pharmacol. Ther. 84, 121–131.

Dalen, P., Dahl, M.-L., Ruiz, M.L., Nordin, J., Bertilsson, L., 1998. 10-hydroxylation of nortriptyline in white persons with 0, 1, 2, 3, and 13 functional CYP2D6 genes. Clin. Pharmacol. Ther. 63, 444–452.

Dierks, E.A., Stams, K.R., Lim, H.K., Cornelius, G., Zhang, H., Ball, S.E., 2001. A method for the simultaneous valuation of the activities of seven major human drug-metabolizing cytochrome P450s using an in vitro cocktail of probe substrates and fast gradient liquid chromatography tandem mass spectrometry. Drug Metab. Dispos. 29, 23–29.

Domanski, T.L., Finta, C., Halpert, J.R., Zaphiropoulos, P.G., 2001. cDNA cloning and initial characterization of CYP3A43, a novel human cytochrome P450. Mol. Pharmocol. 59, 386–392.

Edwards, R.J., Adams, D.A., Watts, P.S., Davies, D.S., Boobis, A.R., 1998. Development of a comprehensive panel of antibodies against the major xenobiotic metabolising forms of cytochrome P450 in humans. Biochem. Pharmacol. 56, 377–387.

Evans, W.E., Relling, M.V., 1999. Pharmacogenomics: translating functional genomics into rational therapeutics. Science 286, 487–491.

Fisher, M.B., Campanale, K., Ackermann, B.L., Vandenbranden, M., Wrighton, S.A., 2000a. In vitro glucuronidation using human liver microsomes and the pore-forming peptide alamethicin. Drug Metab. Dispos. 28, 560–566.

Fisher, M.B., Vandenbranden, M., Findlay, K., Burchell, B., Thummel, K.E., Hall, S.D., Wrighton, S.A., 2000b. Tissue distribution and interindividual variations in human UDP-glucuronosyltransferase activity: relationship between UGT1A1 promoter genotype and variability in a liver bank. Pharmacogenetics 10, 727–739.

Fisher, M.B., Paine, M.F., Strelevitz, T.J. Wrighton, S.A., 2001. The role of hepatic and extrahepatic UDP-glucuroninosyltransferases in human drug metabolism. Drug Metab. Rev. 33, 273–297.

Gelboin, H.V., Krausz, K.W., Gonzalez, F.J., Yang, T.J. 1999. Inhibitory monoclonal antibodies to human cytochrome P450 enzymes: a new avenue for drug discovery. Trends Pharmacol. Sci. 20, 432–438

Gervot, L., Carriere, V., Costet, P., Cugnenc, P.H., Berger, A., Beaune, P., de Waziers, I., 1996. CYP3A5 is the major cytochrome P450 3A expressed in human colon and colonic cell lines. Environ. Toxicol. Phar. 2, 381–388.

Ghoneim, M.M., Korttila, K., Chiang, C.K., Jacobs, L., Schoenwald, R.D., Mewaldt, S.P., *et al.* 1981. Diazepam effects and kinetics in Caucasians and Orientals. Clin. Pharmacol. Ther. 29(6), 749–756.

Ghosal, A., Hapangama, N., Yuan, Y., Achanfuo-Yeboah, J., Iannucci, R., Chowdhury, S., Alton, K., Patrick, J.E., Zbaida, S., 2004a. Identification of human UDP-glucuronosyltransferase enzyme(s) responsible for the glucuronidation of posaconazole (Noxafil). Drug Metab. Dispos. 32(2), 267–271.

Ghosal, A., Hapangama, N., Yuan, Y., Achanfuo-Yeboah, J., Iannucci, R., Chowdhury, S., Alton, K., Patrick, J.E., Zbaida, S., 2004b. Identification of human UDP-glucuronosyltransferase enzyme(s) responsible for the glucuronidation of ezetimibe (Zetia). Drug Metab. Dispos. 32(3), 314–320.

Ghosal, A., Yuan, Y., Hapangama, N., Su, A. D. (Iris), Alvarez, N., Chowdhury, S.K., Alton, K.B., Patrick, J.E., Zbaida, S., 2004c. Identification of human UDP-glucuronosyltransferase enzyme(s) responsible for the glucuronidation of 3-hydroxy-desloratadine Biopharm. Drug Dispos. 25(6), 243–252.

Ghosal, A., Gupta, S., Chowdhury, S., Yuan, Y., Lu, X., Horne, D., Su, A-D., Alvarez, N., Ramanathan, R., Hanna, I., Alton, K.B., Patrick, J.E., Zbaida, S., 2003. Identification of human cytochrome P450 enzymes that metabolize loratadine (LOR). Drug Metab. Rev. 35(2), 186.

Gonzalez, F.J. 1990. Molecular genetics of the P-450 superfamily. Pharmacol. Ther. 45, 1–38.

Gonzalez, F.J., Skoda, R.C., Kimura, S., Umeno, M., Zanger, U.M., Nebert, D.W., Gelboin, H.V., Hardwick, J.P. Meyer, U.A. 1988. Characterization of the common genetic defect in humans deficient in debrisoquine metabolism. Nature 331, 442–446.

Gibson, G., Skett, P., 1994. Introduction to Drug Metabolism. Blackie Academic & Professional, Glasgow, UK.

Guengerich, F.P., 1995. Cytochromes P450 of human liver. Classification and activity profiles of the major enzymes. In: Pacifici, G.M., Fracchia, G.N. (Eds.), Advances in Drug Metabolism in Man. European Commission. Office for the Official Publications of the European Communities, Luxenbourg, pp. 179–231.

Guengerich, F.P., 1999. Cytochrome P-450 3A4: regulation and role in drug metabolism. Annu. Rev. Pharmacol. Toxicol. 39, 1–17.

Heng, Y.M., Kuo, C-W.S., Jones, P.S., Savory, R., Schulz, R.M., Tomlinson, S.R., Gray, T.J.B., Bell, D.R. 1997. A novel murine P-450 gene, Cyp4a14, is part of a cluster of Cyp4a and Cyp4b, but not CYP4F, genes in mouse and humans. Biochem. J. 325, 741–749.

Heyn, H., White, R.B., Stevens, J.C., 1996. Catalytic role of cytochrome P4502B6 in the N-demethylation of *S*-mephenytoin. Drug Metab. Dispos. 24, 948–954.

Imaoka, S., Yamada, T., Hiroi, T., Hayashi, K., Sakaki, T., Yabusaki, Y., Funae, Y., 1996. Multiple forms of human P450 expressed in *Saccharomyces cerevisiae*, systematic characterization and comparison with those of the rat. Biochem. Pharmacol. 51, 1041–1050.

Ingelman-Sundberg, M., Oscarson, M., McLellan, R.A., 1999. Trends in Pharmacol. Sci. 20, 342–349.

Iscan, M., Rostami, H., Guray, T., Pelkonen, O., Rautio, A., 1994. A study on the interindividual variability of coumarin 7-hydroxylation in a Turkish population. Eur. J. Clin. Pharmacol. 47, 315–318.

Johansson, I., Lundqvist, E., Bertilsson, L., Dahl, M-L., Sjöqvist, F., Ingelman-Sundberg, M., 1993. Inherited amplification of an active gene in the cytochrome

P450 CYP2D locus as a cause of ultrarapid metabolism of debrisoquine. Proc. Natl. Acad. Sci. USA 90, 11825–11829.

Kalow, W., Tang, B-K., 1991. Use of caffeine metabolite ratios to explore CYP1A2 and xanthine oxidase activities. Clin. Pharmacol. Ther. 50, 508–519.

Kane, G.C., Lipsky, J.J., 2000. Drug-grapefruit juice interactions. Mayo Clin. Proc. 75(9), 933–942.

Karanam, B.V., Vincent, S.H., Newton, D.J., Wang, R.W., Chiu, S.H.L., 1994. FK506 metabolism in human liver microsomes: investigation of the involvement of cytochrome P450 isozymes other than CYP3A4. Drug Metab. Dispos. 22, 811–814.

Kitada, M., Kamataki, T., 1994. Cytochrome P450 in human fetal liver: significance and fetal specific expression. Drug Metab. Rev. 26, 305–323.

Kitagawa, K., Kunugita, N., Katoh, T., Yang, M., Kawamoto, T., 1999. The significance of the homozygous CYP2A6 deletion on nicotine metabolism: a new genotyping method of CYP2A6 using a single PCR-RFLP. Biochem. Biophys. Res. Commun. 262, 146–151.

Kivistö, K.T., Griese, E-U., Fritz, P., Linder, A., Hakkola, J., Raunio, H., Beaune, P., Kroemer, H.K., 1996. Expression of cytochrome P4503A enzymes in human lung: a combined RT-PCR and immunohistochemical analysis of normal tissue and lung tumors. Naunyn-Schmiedeberg Arch. Pharmacol. 353, 207–212.

Kolars, J.C., Schmiedlin-Ren, P., Schuetz, J.D., Fang, C., Watkins, P.B., 1992. Identification of rifampicin-inducible P450IIA4 (CYP3A4) in human small bowel enterocytes. J. Clin. Invest. 90, 1871–1878.

Koskela, S., Hakkola, J., Hukkanen, J., Pelkonen, O., Sorri, M., Saranen, A., Anttila, S., Fernandez-Salguero, P., Gonzalez, F.J., Raunio, H., 1999. Expression of CYP2A genes in human liver and extrahepatic tissues. Biochem. Pharmacol. 57, 1407–1413.

Landi, M.T., Sinha, R., Lang, N.P., Kadlubar, F.F., 1999. Human cytochrome P4501A2. In: Vineis, P., Malats, N., Lang, M., d"Errico, A., Caporaso, N., Cuzick, J., Boffetta, P., (Eds.), Metabolic Polymorphisms and Susceptibility to Cancer. IARC Scientific Publications No. 148, Lyon, pp. 173–195.

LeBrun, L.A., Xu, F., Kroetz, D.L., Ortiz de Montellano, P.R., 2002. Covalent attachment of the heme prosthetic group in the CYP4F cytochrome P450 family. Biochemistry 41, 5931–5937.

Lechevrel, M., Casson, A.G., Wolf, C.R., Hardie, L.J., Flinterman, M.B., Montesano, R., Wild, C.P., 1999. Characterization of cytochrome P450 expression in human oesophageal mucosa. Carcinogenesis 20, 243–248.

Lieber, C.S., 1997. Cytochrome P-4502E1: its physiological and pathological role. Physiol. Rev. 77, 517–544.

Lin, J.H., Rodrigues, A.D., 2001. In vitro models for early studies of drug metabolism. In: Testa, B., van de Waterbeemd, H., Folkers, G., Guy, R.(Eds.), Pharmacokinetic Optimization: Biological, Physicochemical, and Computational Strategies. Wiley-VCH, pp. 217–243.

Lown, K.S., Bailey, D.G., Fontana, R.J., Janardan, S.K., Adair, C.H., Fortlage, L.A., *et al.* 1997. Grapefruit juice increases felodipine oral availability in humans by decreasing intestinal CYP3A protein expression. J. Clin. Invest. 99(10), 2545–2553.

Lu, A.Y.H., Wang, R.W., Lin, J.H., 2003. Cytochrome P450 in vitro reaction phenotyping: a re-evaluation of approaches used for P450 isoform identification. Drug Metab. Dispos. 31(4), 345–350.

Mackenzie, P.I., Miners, J.O., McKinnon, R.A., 2000. Polymorphisms in UDP glucuronosyltransferase genes: functional consequences and clinical relevance. Clin. Chem. Lab. Med. 38(9), 889–892.

Matsumoto, S., Hirama, T., Matsubara, T., Nagata, K., Yamazoe, Y., 2002. Involvement of CYP2J2 on the intestinal first-pass metabolism of antihistamine drug, Astemizole. Drug Metab. Dispos. 30(11), 1240–1245.

McGinnity, D.F., Parker, A.J., Soars, M., Riley, R.J., 2000. Automated definition of the enzymology of drug oxidation by the major human drug metabolizing cytochrome P450s. Drug Metab. Dispos. 28, 1327–1334.

Mei, Q., Tang, C., Assang, C., Lin, Y., Slaughter, D., Rodrigues, D., Baillie, T.A., Rushmore, T.H. Shou, M., 1999. Role of a potent inhibitory monoclonal antibody to cytochrome P450 in assessment of human drug metabolism. J. Pharmacol. Exp. Ther. 291, 749–759.

Mimura, M., Baba, T., Yamazaki, H., Ohmori, S., Inui, Y., Gonzalez, F.J., Guengerich, F.P., Shimada, T., 1993. Characterization of cytochrome P-450 2B6 in human liver microsomes. Drug Metab. Dispos. 21, 1048–1056.

Miners, J.O., Birkett, D.J., 1998. Cytochrome P4502C9: an enzyme of major importance in human drug metabolism. Br. J. Clin. Pharmacol. 45, 525–538.

Miners, J.O., McKinnon, R.A., Mackenzie, P.I., 2002. Genetic polymorphisms of UDP-glucuronosyltransferases and their functional significance. Toxicology 27, 181182, 453–456.

Murray, G.I., Pritchard, S., Melvin, W.T. Burke, M.D. 1995. Cytochrome P450 CYP3A5 in the human anterior pituitary gland. FEBS Lett. 364, 79–82.

Nakajima, M., Yamamoto, T., Nunoya, K-I., Yokoi, T., Nagashima, K., Inoue, K., Funae, Y., Shimada, N., Kuroiwa, Y., 1996. Role of human cytochrome P4502A6 in C-oxidation of nicotine. Drug Metab. Dispos. 24, 1212–1217.

Nakajima. M., Yokoi, T., Mizutani. M., Kinoshita, M., Funayama, M., Kamataki, T., 1999. Genetic polymorphism in the 5”-flanking region of human CYP1A2 gene: effect on the CYP1A2 inducibility in humans. J. Biochem. (Tokyo) 125, 803–808.

Nebert, D.W., McKinnon, R.A., Puga, A., 1996. Human drug-metabolizing enzyme polymorphisms: effects on risk of toxicity and cancer. DNA Cell Biol. 15, 273–280.

Nelson, D.R., Koymans, L., Kamataki, T., Stegeman, J.J., Feyereisen, R., Waxman, D.J., Waterman, M.R., Gotoh, O., Coon, M.J., Estabrook, R.W., Gunsalus, I.C., Nebert, D.W., 1996. P450 superfamily: update on new sequences, gene mapping, accession numbers and nomenclature. Pharmacogenetics 6, 1–42.

Newton, D.J., Wang, R.W., Lu, A.Y.H., 1995. Cytochrome P450 inhibitors: evaluation of specificities in the *in vitro* metabolism of therapeutic agents by human liver microsomes. Drug Metab. Dispos. 23, 154–158.

Nhamburo, P.T., Gonzalez, F.J., McBride, O.W., Gelboin, H.V., Kimura, S., 1989. Identification of a new P450 expressed in human lung: complete cDNA sequence, cDNA-directed expression, and chromosome mapping. Biochemistry 28, 8060–8066.

Omura, T., 1999. Forty years of cytochrome P450. Biochem. Biophys. Res. Commun. 266, 690–698.

Oscarson, M., Gullstén, H., Rautio, A., Bernal, M.L., Sinues, B., Dahl, M-L., Stengård, J.H., Pelkonen, O., Raunio, H., Ingelman-Sundberg, M., 1998. Genotyping of human cytochrome P450 2A6 (CYP2A6), a nicotine C-oxidase. FEBS Lett. 438, 201–205.

Pascussi, J-M., Gerbal-Chaloin, S., Pichard-Garcia, L., Daujat, M., Fabre, J-M., Maurel, P., Vilarem, M-J., 2000. Interleukin-6 negatively regulates the expression of pregnane X receptor and constitutively activated receptor in primary human hepatocytes. Biochem. Biophys. Res. Commun. 274, 707–713.

Pelkonen, O., Mäenpää, J., Taavitsainen, P., Rautio, A., Raunio, H., 1998. Inhibition and induction of human cytochrome P450 (CYP) enzymes. Xenobiotica 28, 1203–1253.

Pelkonen, O., Raunio, H., 1997. Metabolic activation of toxins: tissue-specific expression and metabolism in target organs. Environ. Health Perspect 105, 767–774.

Perrot, N., Nalpas, B., Yang, C.S., Beaune, P.H., 1989. Modulation of cytochrome P450 isozymes in human liver, by ethanol and drug intake. Eur. J. Clin. Invest. 19, 549–555.

Pianezza, M.L., Sellers, E.M., Tyndale, R.F., 1998. Nicotine metabolism defect reduces smoking. Nature 393, 750.

Pless, D., Gouze, J.N., Senay, C., Herber, R., Leroy, P., Barberousse, V., Fournel-Gigleux, S., Magdalou, J., 1999. Characterization of the UDP-glucuronosyltransferases involved in the glucuronidation of an antithrombotic thioxyloside in rat and humans. Drug Metab. Dispos. 27, 588–595.

Prueksaritanont, T., Zhao, J.J., Ma, B., Roadcap, B.A., Tang, C., Qiu, Y., Liu, L., Lin, J.H., Pearson, P.G., Baillie, T.A. 2002. Mechanistic studies on metabolic interactions between gemfibrozil and statins. J. Pharm. Exp. Ther. 301, 1042–1051.

Radominska-Pandya, A., Czernik, P.J., Little, J.M., 1999. Structural and functional studies of UDP-glucuronosyltransferases. Drug Metab. Rev. 31(4), 817–899.

Raunio, H., Husgafvel-Pursiainen, K., Anttila, S., Hietanen, E., Hirvonen, A., Pelkonen, O., 1995a. Diagnosis of polymorphisms in carcinogen-activating and inactivating enzymes and cancer susceptibility-review. Gene 159, 113–121.

Raunio, H., Pasanen, M., Mäenpää, J., Hakkola, J., Pelkonen, O., 1995b. Expression of extrahepatic cytochrome P450 in humans. In: Pacifici, G.M., Fracchia, G.N. (Eds.), Advances in Drug Metabolism in Man. European Commission, Office for Official Publications of the European Communities, Luxembourg, pp.234–287.

Rautio, A., Kraul, H., Kojo, A., Salmela, E., Pelkonen, O., 1992. Interindividual variability of coumarin 7-hydroxylation in healthy individuals. Pharmacogenetics 2, 227–233.

Rodrigues, A.D., 1999. Integrated cytochrome P450 reaction phenotyping: attempting to bridge the gap between cDNA-expressed cytochrome P450 and native human liver microsomes. Biochem. Pharmacol. 57, 465–480.

Rodrigues, A.D., Rushmore, T.H., 2002. Cytochrome P450 pharmacogenetics in drug development: in vitro studies and clinical consequences. Curr. Drug Metab. 3(3), 289–309.

Rodríguez-Antona, C., Jover, R., Gómez-Lechón, M.J., Castell, J.V., 2000. Quantitative RT–PCR measurement of human cytochrome P-450s: application to drug induction studies. Arch. Biochem. Biophys. 376, 109–116.

Rolan, P.E., 1994. Plasma protein binding displacement interactions – why are they still regarded as clinically important? Br. J. Clin. Pharmacol. 37, 125–128.

Ronis, M.J.J., Lindros, K.O., Ingelman-Sundberg, M., 1996. The CYP2E subfamily. In: Ioannides, C., Parke, D.V., (Eds.), Cytochromes P450: Metabolic and Toxicological Aspects. CRC Press, Boca Raton, FL, pp. 211–239.

Sachse, C., Brockmoller, J., Bauer, S., Roots, I., 1999. Functional significance of a C—>A polymorphism in intron 1 of the cytochrome P450 CYP1A2 gene tested with caffeine. Br. J. Clin. Pharmacol. 47, 445–449.

Savage, D.R., US Food and Drug Administration, 2003. FDA guidance on pharmacogenomics data submission. Nat. Rev. Drug Discov. 2(12), 937–938.

Schmidt, J.V., Bradfield, C.A., 1996. Ah receptor signalling pathways. Annu. Rev. Cell Dev. Biol. 12, 55–89.

Schuetz, J.D., Beach, D.L., Guzelian, P.S., 1994. Selective expression of cytochrome P450 CYP3A mRNAs in embryonic and adult human liver. Pharmacogenetics 4, 11–20.

Schuetz, E.G., Schuetz, J.D., Grogan, W.M., Naray-Fejes-Toth, A., Fejes-Toth, G., Raucy, J., Guzelian, P., Gionela, K., Watlington, C.O., 1992. Expression of cytochrome P450 3A in amphibian, rat, and human kidney. Arch. Biochem. Biophys. 294, 206–214.

Schuetz, E.G., Schuetz, J.D., Strom, S.C., Thompson, M.T., Fisher, R.A., Molowa, D.T., Li, D., Guzelian, P.S., 1993. Regulation of human liver cytochromes P-450 in family 3A in primary and continuous culture of human hepatocytes. Hepatology 18, 1254–1262.

Segall, M.D., Payne, M.C., Ellis, S.W., Tucker, G.T., Boyes, R.N., 1997. Ab initio molecular modeling in the study of drug metabolism. Eur. J. Drug Metab. Pharmacokinet. 22, 283–289.

Shimada, T., Hayes, C.L., Yamazaki, H., Amin, S., Hecht, S.S., Guengerich, F.P., Sutter, T.R., 1996. Activation of chemically diverse procarcinogens by human cytochrome P-450 1B1. Cancer Res. 56, 2979–2984.

Shimada, T., Yamazaki, H., Mimura, M., Inui, Y., Guengerich, F.P., 1994. Interindividual variations in human liver cytochrome P-450 enzymes involved in the oxidation of drugs, carcinogens and toxic chemicals: studies with liver microsomes of 30 Japanese and 30 Caucasians. J. Pharmacol. Exp. Ther. 270, 414–423.

Shin, J.G., Kane, K., Flockhart, D.A., 2001. Potent inhibition of CYP2D6 by haloperidol metabolites: stereoselective inhibition by reduced haloperidol. Br. J. Clin. Pharmacol. 51(1), 45–52.

Shin, J.G., Soukhova, N., Flockhart, D.A., 1999. Effect of antipsychotic drugs on human liver cytochrome P-450 (CYP) isoforms in vitro: preferential inhibition of CYP2D6. Drug Metab. Dispos. 27(9), 1078–1084.

Sotaniemi, E.A., Rautio, A., Bäckström, M., Arvela, P., Pelkonen, O., 1995. CYP3A4 and CYP2A6 activities marked by the metabolism of lignocaine and coumarin in patients with liver and kidney diseases and epileptic patients. Br. J. Clin. Pharmacol. 39, 71–76.

Stevens, J.C., White, R.B., Hsu, S.H., Martinet, M., 1997. Human liver CYP2B6-catalyzed hydroxylation of RP73401. J. Pharmacol. Exp. Ther. 282, 1389–1395.

Sutter, T.R., Tang, Y.M., Hayes, C.L., Wo, Y-YP., Jabs, E.W., Li, X. Yin, H., Cody, C.W., Greenlee, W.F., 1994. Complete cDNA sequence of a human dioxin-inducible mRNA identifies a new gene subfamily of a cytochrome that maps to chromosome 2. J. Biol. Chem. 269, 13092–13099.

Tang, Y.M., Chen, G-F., Thompson, P.A., Lin, D-X., Lang, N.P., Kadlubar, F.F., 1999. Development of an antipeptide antibody that binds to the C-terminal region of human CYP1B1. Drug Metab. Dispos. 27, 274–280.

Testino, S.A., Patonay, G., 2003. High-throughput inhibition screening of major human cytochrome P450 enzymes using an in vitro cocktail and liquid chromatography–tandem mass spectrometry. J. Pharmaceut. Biomed. Analysis 30, 1459–1467.

Thummel, K.E., Wilkinson, G.R., 1998. In vitro and in vivo drug interactions involving human CYP3A. Annu. Rev. Pharmacol. Toxicol. 38, 389–430.

Venkatakrishnan, K., von Moltke, L.L., Greenblatt, D.J. 2001. Application of the relative activity factor approach in scaling from heterologously expressed cytochrome P450 to human liver microsomes: studies on amitriptyline as a model substrate. J. Pharmacol. Exp. Ther. 297, 326–337.

Walsky, R.L., Obach, R.S., 2004. Validated assays for cytochrome p450 activities. Drug Metab. Dispos. 32, 647–660.

Wang, J-S., Neuvonen, M., Wen, X., Backman, J.T., Neuvonen, P.J., 2002. Gemfibrozil inhibits CYP2C8-mediated cerivastatin metabolism in human liver microsomes. Drug Metab. Dispos. 30, 1352–1356.

Weaver, R., Graham, K.S., Beattie, I.G., Riley, R.J., 2003. Cytochrome P450 inhibition using recombinant proteins and mass spectrometry/multiple reaction monitoring technology in a cassette incubation. Drug Metab. Dispos. 31, 955–966.

Willey, J.C., Coy, E.L., Frampton, M.W., Torres, A., Apostolakos, M.J., Hoehn, G., Schuermann, W.H., Thilly, W.G., Olson, D.E., Hammersley, J.R., Crespi, C.L., Utell, M.J., 1997. Quantitative RT-PCR measurement of cytochromes p450 1A1, 1B1, and 2B7, microsomal epoxide hydrolase, and NADPH oxidoreductase expression in lung cells of smokers and nonsmokers. Am. J. Respir. Cell Mol. Biol. 17, 114–124.

Williams, L., Davis, J.A., Lowenthal, D.T., 1993. The influence of food on the absorption and metabolism of drugs. Med. Clin. N. Am. 77(4), 815–829.

Wrighton, S.A., Ring, B.J., Watkins, P.B., VandenBranden, M., 1989. Identification of a polymorphically expressed member of the human cytochrome P-450III family. Mol. Pharmacol. 86, 97–105.

Wrighton, S.A., Brian, W.R., Sari, M.A., Iwasaki, M., Guengerich, F.P., Raucy, J.L., Molowa, D.T., VandenBranden, M., 1990. Studies on the expression and metabolic capabilities of human liver cytochrome P450IIIA5 (HLp3). Mol. Pharmacol. 38, 207–213.

Wu, S., Moomaw, C.R., Tomer, K.B., Falck, J.R., Zeldin, D.C., 1996. Molecular cloning and expression of CYP2J2, a human cytochrome P450 arachidonic acid epoxygenase highly expressed in heart. J. Biol. Chem. 271, 3460–3468.

Yamazaki, H., Inoue, K., Hashimoto, M., Shimada, T., 1999. Roles of CYP2A6 and CYP2B6 in nicotine C-oxidation by human liver microsomes. Arch. Toxicol. 73, 65–70.

Yamazaki, H., Inui, Y., Wrighton, S.A., Guengerich, F.P., Shimada, T., 1995. Procarcinogen activation by cytochrome P450 3A4 and 3A5 expressed in *Escherichia coli* and by human liver microsomes. Carcinogenesis 16, 2167–2170.

Yin, H., Racha, J., Li, S-Y., Olejnik, N., Satoh, H., Moore, D., 2000. Automated high throughput human CYP isoform activity assay using SPE-LC/MS method: application in CYP inhibition evaluation. Xenobiotica 30, 141–154.

Yumibe, N., Huie, K., Chen, K-J., Snow, M., Clement, R.P., Cayen, M.N., 1996. Identification of human liver cytochrome P450 enzymes that metabolize the non-sedating antihistamine loratadine. Biochem. Pharmacol. 51, 165–172.

Zeldin, D.C., Foley, J., Ma, J., Boyle, J.E., Pascual, J.M., Moomaw, C.R., Tomer, K.B., Steenbergen, C., Wu, S., 1996. CYP2J subfamily P450s in the lung: expression, localization, and potential functional significance. Mol. Pharmacol. 50, 1111–1117.

Zeldin, D.C., Foley, J., Boyle, J.E., Moomaw, C.R., Tomer, K.B., Parker, C., Steenbergen, C., Wu, S., 1997a. Predominant expression of an arachidonate epoxynase in islets of Langerhans cells in human and rat pancreas. Endocrinology 138, 1338–1346.

Zeldin, D.C., Foley, J., Goldsworthy, S.M., Cook, M.E., Boyle, J.E., Ma, J., Moomaw, C.R., Tomer, K.B., Steenbergen, C., Wu, S., 1997b. CYP2J subfamily cytochrome P450s in the gastrointestinal tract: expression, localization, and potential functional significance. Mol. Pharmacol. 51, 931–943.

Zevin, S. Benowitz, N.L., 1999. Drug interactions with tobacco smoking. An update. Clin. Pharmacokinet. 36, 425–438.

Subject Index